T0327512

**Graph Database and Graph Computing
for Power System Analysis**

# Graph Database and Graph Computing for Power System Analysis

*Renchang Dai*
*Puget Sound Energy*
*Bellevue, WA, USA*

*Guangyi Liu*
*Envision Digital*
*San Jose, CA, USA*

**IEEE Press Series on Power and Energy Systems**
Ganesh Kumar Venayagamoorthy, Series Editor

For general information on our other products and services or for technical support, please contact our Customer Care Department within the United States at (800) 762-2974, outside the United States at (317) 572-3993 or fax (317) 572-4002.

Wiley also publishes its books in a variety of electronic formats. Some content that appears in print may not be available in electronic formats. For more information about Wiley products, visit our web site at www.wiley.com.

***Library of Congress Cataloging-in-Publication Data:***

Names: Dai, Renchang, author. | Liu, Guangyi (Scientist), author.
Title: Graph database and graph computing for power system analysis / Renchang Dai, Guangyi Liu.
Description: Hoboken, New Jersey : Wiley-IEEE Press, [2024] | Includes index.
Identifiers: LCCN 2023023393 (print) | LCCN 2023023394 (ebook) | ISBN 9781119903864 (cloth) | ISBN 9781119903871 (adobe pdf) | ISBN 9781119903888 (epub)
Subjects: LCSH: Graph databases. | Electric power systems.
Classification: LCC QA76.9.D32 D33 2024 (print) | LCC QA76.9.D32 (ebook) | DDC 005.75/8–dc23/eng/20230602
LC record available at https://lccn.loc.gov/2023023393
LC ebook record available at https://lccn.loc.gov/2023023394

Cover Design: Wiley
Cover Image: © Alejandro Mendoza R/Shutterstock

Set in 9.5/12.5pt STIXTwoText by Straive, Pondicherry, India

# Contents

# About the Authors

**Renchang Dai, PhD,** is a consulting engineer and project manager at Puget Sound Energy. He received his PhD degree in electrical engineering from Tsinghua University, China, in 2001.

Dr. Dai has worked on a variety of power system problems, including power system planning, operations, and control. He was the principal engineer and group manager for Global Energy Interconnection Research Institute North America, where he led a team of engineers in researching and developing graph database and graph computing technologies for power system planning and operations.

Dr. Dai was a team leader for GE Energy. In GE Energy, he designed, developed, and implemented Energy Management System. He was also a founding member of the GE Energy Consulting Smart Grid Center of Excellence, where he consulted on smart grid deployment and renewable energy grid integration projects. In 2005, when he was a lead scientist in GE Global Research, he was awarded the GE Global Technical Award for his contributions to the development of wind turbine generator fault ride through technology.

Dr. Dai is a senior member of the IEEE. He has worked intensively on graph-based power system analysis and has published over 100 papers in international journals and conferences.

**Guangyi Liu, PhD,** is a chief scientist at Envision Digital. He is leading a team of engineers to develop power system application software that is based on graph database and graph computing technologies. He received his PhD degree in electrical engineering from the China Electric Power Research Institute, China, in 1990.

Dr. Liu has worked on a variety of power system research fields, including Energy Management System (EMS), Distribution Management System (DMS), Electricity Market, Active Distribution Network, and Big Data. He was the principal engineer and chief technology officer for Global Energy Interconnection Research Institute North America, where he led a team of engineers in researching and developing graph database and graph computing technologies for power system calculation, analysis, and optimization.

Dr. Liu is a senior member of the IEEE and a fellow of the Chinese Society of Electrical Engineering. He has worked intensively on power system analysis and optimization, and he has published over 200 papers in international journals and conferences.

# Preface

We started to work on power system analysis decades ago. Improving power system analysis computation efficiency is an ongoing task and a challenge for online applications and offline analyses in the power industry. The tireless and remarkable efforts have been endeavored by researchers and engineers trying to achieve real-time steady-state, dynamic, and optimization applications. Keeping this ambition in mind, the idea of graph computing was inspired at an occasional conservation with Professor Shoucheng Zhang from Stanford University in 2015 when he introduced a start-up company and their work on graph databases to us. The perfect match of graph nature and power network structure sparked the long exploration and journey of researching and developing graph computing theory, algorithms, methods, approaches, and applications for years. This book is a comprehensive summary and knowledge sharing of our research and engineering work on graph computing for power system analysis.

This book is divided into two parts. Part I devotes the first seven chapters to highlighting the theoretical methods and approaches. Part II is composed of Chapters 8–17 on practical implementations and applications. Part I serves prerequisites of graph computing with basics and advances of graph databases, graph parallel computing, and knowledge of solving algebraic equations, optimization problems, differential equations, and their combinations. Part II provides a comprehensive illustration of graph-based power system modeling, analysis approaches, and implementations with detailed graph query scripts. The implemented applications presented in Part II cover power system topology analysis, state estimation, power flow calculation, contingency analysis, security-constrained economic dispatch, security-constrained unit commitment, automatic generation control, small-signal stability, transient stability, and deep reinforcement learning.

Currently, the practice of power system modeling focuses on using relational databases. In relational databases, data are organized and managed in tables. The relationships between tables are connected by separated tables or by using a join operation to search common attributes in different tables to find the relationships. In this mechanism, it is challenging to maintain and manipulate a large dataset in a relational database.

Contrary to the relational database, a graph database uses graph structures for semantic queries with nodes and edges to store data. The essential differences between graph databases and relational databases are that the edge directly defines the data relationship and graph databases are designed for parallel computing. Graph computing models power systems as a graph which is consistent with the fact that power system physically is a graph – buses are connected by branches as a graph. The graph data structure tells the topology of the power network and the relations of power system components naturally. The graph computing mechanism by using queries on nodes and graph partitions, promotes parallel computing for power system applications.

Graph databases are new to the power industry. Graph computing is novel to researchers and engineers. The main objective of this book is to provide a roadmap and guidance to readers to learn the alternative and innovative approaches to modeling and solving power system problems from scratch. For this purpose, the processes of defining vertex, edge, and graph schema, creating a loading job, and developing detailed graph queries are demonstrated. The scripts for each power system analysis are provided and explained in detail to facilitate readers to gradually and comprehensively master graph computing. To make this book a reference to graduate students, illustrative problems are presented and hands-on experiences in graph computing design and programming are provided by detailed scripts.

The graph computing research activities are still progressing. We believe that graph computing has great potential in various power system analyses. We hope this book can invite and inspire researchers and engineers to study and research graph databases and graph computing and apply the graph computing theory and approaches to develop power system applications.

*Renchang Dai*
*Guangyi Liu*

# Acknowledgments

This book is a result of the fascinating journey of our study and research work for decades. We firstly acknowledge our research advisors, Professor Ming Ding from the Hefei University of Technology, Professor Boming Zhang from Tsinghua University, and Professor Erkeng Yu from the China Electric Power Research Institute, for ushering us into the world of power system analysis and leading us into the research area of power system real-time applications. The knowledge and experience they shared with us along with their advice, influence our research to this day.

We also acknowledge our research team for their contributions and support on this challenging and rewarding work for many years. A great thanks goes to Dr. Ting Chen, Mr. Hong Fan, Dr. Chen Yuan, Dr. Jingjin Wu, Dr. Yiting Zhao, Dr. Jun Tan, Dr. Jiangpeng Dai, Dr. Yongli Zhu, Dr. Longfei Wei, Dr. Xiang Zhang, Dr. Peng Wei, Dr. Yachen Tang, Mr. Kewen Liu, Mr. Wendong Zhu, Mrs. Tingting Liu, Mrs. Bowen Kan, Mr. Haiyun Han, Mr. Letian Teng, Dr. Kai Xie, Mr. Zhiwei Wang, Dr. Xi Chen, Mrs. Ziyan Yao, Dr. Wei Feng, Dr. Yijing Liu, Mrs. Jing Hong, Mr. Huaming Zhang, Dr. Saeed D. Manshadi, Dr. Mariana Kamel, Dr. Yawei Wang, Mr. Yanan Lyu, and many other colleagues and interns, for their contributions on the research and development of the material presented in this book and wonderful ideas about the graph computing related research topics.

We are grateful and thankful to Professor Fran Li from the University of Tennessee, Knoxville, Professor Jianhui Wang from Southern Methodist University, Professor Hsiao-Dong Chiang from Cornell University, and Professor Yinyu Ye from Stanford University for partnering with us on the research and development work and fantastic discussion on graph computing.

Last but not least, we would like to thank our families. We understand writing this manuscript is not an easy task from day one. The real journey is even harder than expected, with unexpected detours. We express our heartfelt thanks to them for their support and understanding over the past several years.

*Renchang Dai*
*Guangyi Liu*

**Part I**

**Theory and Approaches**

# 1

# Introduction

The electrical power system has been revolutionizing over the decades into a highly interconnected, large, and complex renewable system. Populations and economic growth globally demand higher electricity. Transactions crossing large areas are encouraged to make more economic and environmental sense and result in large power flowing over a wide area. High-voltage transmission technologies boosted voltage levels to 1000 kV Ultra-High-Voltage Alternating-Current (UHVAC) and $\pm$ 800 kV Ultra-High-Voltage Direct Current (UHVDC) to transmit power over thousands of miles [1]. Advanced power electronic devices enable flexible alternating current transmission system (FACTS), for instance, static var compensator and voltage source converter-based STATCOM, being adapted to control power flow agilely and accurately in electric power grids [2, 3].

The accelerating decarbonization of energy systems promoted and promised worldwide requires a rising penetration level of renewable energy, distributed energy, and energy storage making the power system ever large and more complex. To make the large, complex, and dynamic power system secure and cost-effective, real-time monitoring, operating, and control are crucial. Accurate and fast calculation, combined with intelligent decisions on power systems is more vital than ever, shifting from the analytical Energy Management System (EMS) to the intelligent EMS [4].

While the power system has been evolving to be bigger, the power system is getting to be more intelligent. In the power industry, the smart and intelligent grid is developed based on the following technologies: big data, deep learning, and high-performance computing [5, 6]. Big data and deep learning usually involve intensive computing efforts, thus high-performance computing is the key to making smart decisions on time. Intelligent real-time analysis based on multi-source big data analysis, deep learning techniques, and high-performance computing, as the trend, makes the power system adaptive and predictive possible.

Compared with the traditional grid, the modern power system is operated under more uncertainty because of the intermittent renewable energy and power market transactions as described above. Intelligent real-time analysis and calculation need to adopt these uncertainties more quickly to anticipate extreme events to make better decisions timely. The natural reflection of human beings in complex operating environments requires more accurate and intuitive data models and powerful calculation methods. For example, to provide powerful technical support for intelligent real-time analysis and calculation, introducing the latest numerical calculation technology into the power system power flow, state estimation, contingency analysis, security-constrained automatic generation control, security-constrained unit commitment, and faster-than-real-time transient simulation require novel data structure and calculation approach. The current data

*Graph Database and Graph Computing for Power System Analysis*, First Edition. Renchang Dai and Guangyi Liu.
© 2024 The Institute of Electrical and Electronics Engineers, Inc. Published 2024 by John Wiley & Sons, Inc.

processing and calculation approaches of the existing power system applications face the following three challenges:

1) Data Management and Analysis: It is a requirement and challenge to develop data acquisition, processing, and storage technologies that can simultaneously meet the needs of grid online analysis and offline planning for converged multi-source big datasets.
2) Mathematical Methods and Computation: It is necessary and challenging to develop new mathematical tools and algorithms to achieve faster-than-real-time grid simulation.
3) Models and Simulations: The parallel dynamic simulation framework of power systems is not currently developed enough to support real-time, wide-area protection, and control.

The trends of the electric power system are challenging the existing EMS and Market Management System (MMS) in their computationally intensive applications, such as power flow, state estimation, contingency analysis, multi-time point network analysis, security-constrained automatic generation control, security-constrained unit commitment, security-constrained economic dispatch, and faster-than-real-time transient simulation. Next-generation EMS/MMS is required to be evolving to accommodate larger scale, higher complex, more constrained, and uncertain power systems with a faster than real-time manner or even look-ahead capability with future situational awareness [7].

To meet the above challenges, the requirements and goals of parallel analytical technologies and tools are urgently needed to support new data management tools and rapid computational analysis methods.

Due to their efficient data management and rapid computational analysis capabilities, graph database and graph computing technologies are gaining more and more attention. They are promising technologies to effectively solve the problem of big data rapid analysis and processing. In the field of e-commerce, graph computing technology based on a graph database plays a key role in real-time trading and real-time analysis such as anti-money laundering, bad transaction detection, intelligent navigation, and other fields. Many Internet companies have also developed their graph computing technologies and products, such as Pregel, a graph computing system developed and designed by Google. The Trinity project of Microsoft Research is about graph database and graph computing projects. Google's Pregel products have become one of the industry examples of successful graph database applications.

In the field of analysis of power grids, using graph data management and computation technology is new. This book summarizes the recent research and development achievements on this topic. Graph database architecture to power systems is first introduced, which is needed to support fine-grained parallel computing to improve power system computation efficiency. The traditional relational database is replaced by a graph database to model power systems and implement applications. Using the graph database, the program to solve large-scale algebraic equations, high-dimensional differential equations, and optimization problems is reconfigured and redesigned to accommodate graph parallel computing in this book. By using graph database and graph computing, the computational model can be integrated with the grid model, data storage, and numerical calculation, while making full use of big data technologies such as memory computing, distributed parallel computing, and decomposition aggregation. The technology has the significant advantages of large scale, high speed, and high efficiency of computing data, and provides a technical solution with great potential for data management and analysis and calculation of a giant power grid.

Without fast and accurate calculations, a timely response to real-time events is impossible and the system is running at risk. Analysis of the North American blackout of 14 August 2003 shows that delayed and missed responses are the main reasons for the wide-area blackout [8].

Sequence-of-events records show that the system experienced two and a half minutes of disturbances from the initial event to the system collapse. In the initial critical nine seconds, the hundreds of generators that tripped offline were not fully captured by the EMS system, since the EMS update cycle (typically in minutes) is much longer than the critical events interval. Clearly, the opportunity to take timely measures to prevent the blackout was missed and the possibility of a more timely response would have been enhanced by an EMS with a faster cycle time.

The time has come when it is critical to improve the computation efficiency of EMS applications to accommodate modern power systems that are of ever-increasing size and operational complexity. The goal today is to provide a cycle time equivalent to the Supervisory Control and Data Acquisition (SCADA) cycle time or faster to provide look-ahead capability with future situational awareness using forecasted and scheduled information of load forecasting, unit commitment, and outage schedule. The analytical processing time needs to be reduced from tens of seconds to subseconds [7]. To meet this requirement, technologies for the next-generation of high-performance EMS are being studied [9–11]. However, the computation capability to complete the core EMS applications, such as state estimation, power flow, and contingency analysis at a SCADA sampling rate has not yet been achieved.

To achieve EMS computation cycle times that are faster than the SCADA sampling rate, a novel database architecture along with fast computational methods are presented in this book. Among the various computational tools available to improve computation efficiency, parallel computing is a promising technology, providing abundant storage along with multiple processing paths [12, 13]. In [12], parallel state estimation using a preconditioned conjugate gradient algorithm and an orthogonal decomposition-based algorithm is proposed. The proposed algorithm can solve state estimation problems faster using parallel computing, but it is infeasible to deal with a large condition number of a gain matrix. Alves and Monticelli in [13] proposed an approach to solving contingency analysis by parallel computer and distributed network. To utilize the linearity of the power system component, current balance equations subjected to Kirchhoff's current law are used to model power flow problems in [14]. The result shows a reduction in computational time by over 20% when using the current balance equations.

The state-of-the-art communication technologies and computation technologies in information systems are brilliantly showing power system engineers a technical solution to measure, monitor, and analyze electric power systems widely and quickly. However, when the exponentially growing data is acquired at the control center, the database and computation engine consumes a longer time to process which deteriorates the computation efficiency. To achieve analysis with high computation efficiency, novel system architecture, and fast computational algorithms are needed to assist operators to ensure a reliable, resilient, secure, and efficient electric power grid promptly. Among the various computational algorithms, parallel computing is a promising technology to improve computation efficiency taking advantage of modern computation technology, abundant storage space, and parallel capability of database and GPU. Multiple-core CPUs and GPUs are available nowadays as affordable hardware configurations to facilitate parallel computing. However, the state-of-art EMS/MMS does not effectively harness the multi-threaded parallelization capability in their applications [12, 13] for the reason of the traditional relational database and computation algorithms applied by EMS/MMS were not designed for parallel computing.

To accommodate parallel computing, both the database and calculation approaches for the EMS/MMS applications need to be redesigned to fit into a parallel database management system and parallel computing. Previous works investigated the feasibility of adopting graph computing on topology processing, state estimation, power flow analysis, "$N - 1$" contingency analysis, and security-constrained economic dispatch [15–21]. Realizing that parallel processing of the power system applications needed by real-time operation and long-term planning can be enhanced by taking advantage of the embedded graph characteristics of a power system, this book has married a graph-based database with graph computing to achieve high computational effective power system analysis to accommodate the evolving power systems and power market.

To accommodate parallel computing, database and mathematical model for power system calculation need to be redesigned to fit into parallel database management, parallel analysis, and fast visualization.

In this book, the critical power system applications are revisited. The computational approaches involved in these applications are introduced in detail. These approaches are abstracted to be mathematical problems in solving large-scale algebraic equations, high-dimensional differential equations, and mixed integer linear optimization problems. Graph data structure and graph parallel computing are introduced to model the power system in graph and solve the problems in parallel.

## 1.1 Power System Analysis

### 1.1.1 Power Flow Calculation

Power flow calculation is a well-known application in power system analysis. The intention of power flow calculation is to obtain bus voltage magnitude and angle information. Once the voltage information is known, active power and reactive power flow on each branch can be analytically determined. In mathematics, the power flow calculation model is a set of high-dimensional nonlinear algebraic equations.

There are several different methods to solve nonlinear equations. The well-known Newton–Raphson method linearizes equations using a Taylor series with the linear term only. Industry-grade EMS also uses the Fast-decoupled power flow method to approximate active and reactive flow equations by decoupling voltage magnitude and angle calculations. Although decoupled power flow method takes a few more iterations than Newton–Raphson method to converge, each iteration takes much less time. For reactance-dominated transmission networks, decoupled power flow method outperforms the Newton–Raphson method on computation efficiency. The cost is the approximation on Jacobian matrices by decoupled power flow method deteriorates power flow convergence. Usually, in industry-grade EMS, decoupled power flow method is conducted first, then Newton–Raphson method second if decoupled power flow method diverges. This strategy practically provides supporting evidence of its effectiveness for contingency analysis for a large-scale system with thousands of contingencies.

### 1.1.2 State Estimation

The power system state estimation (SE) is based on real-time telemetry from SCADA. The network topology connection of the power system is determined in real-time, along with the real-time operating state of the power grid which forms the basis of the online analysis software. It serves to monitor the state of the grid and enables EMSs to perform various important control and planning tasks such as establishing near real-time network models for the grid, optimizing power flows, and bad data detection and analysis.

There are at least three major aspects of the future power grid that will directly impact SE research. First, more advanced measurement technologies like phasor measurement units have offered hope for near real-time monitoring of the power grid.

Second, new regulations and market pricing competition may require utility companies to share more information and monitor the grid over large geographical areas. This calls for distributed control, and hence, distributed SE to facilitate interconnection-wide coordinated monitoring.

Lastly, to facilitate smart grid features such as demand response and two-way power flow, utility companies will need to have more timely and accurate models for their distribution systems. This calls for SE at the distribution level, which places more stringent requirements on SE algorithms. So far, utility companies have done little in implementing SE in distribution systems, even though SE has been deployed extensively in transmission systems for decades. However, as the electric power grid becomes smarter, more distribution automation will be needed and SE at the distribution level will become more important. The control mechanism in the distribution system will most likely be distributed and active in nature, and so will be the corresponding SE functions. This necessitates the development of new distributed SE algorithms that avail themselves of the substantially increased number of real-time measurements.

### 1.1.3 Contingency Analysis

It is a challenge and a goal to operate a large-scale, complex, and dynamic power grid with safety and cost-effectiveness. Contingency analysis is one of the applications to secure power systems operating with no violation. Contingency analysis uses base case power flow driven from SE to assess the security of power systems under the contingency of a single equipment outage and their combinations. Contingency analysis is usually running periodically every one to two minutes.

The contingency analysis is usually based on an online power grid analysis to figure out the weak point and security risks of the power grid and issue an alarm when the system is running at risk. It facilitates dispatching operators to deal with potential operation issues in time to prevent cascading events and blackouts.

Contingency analysis is time-consuming as it involves a large number of computations of AC load flow. To reduce the computational time, an automatic contingency screening approach is being adopted which identifies and ranks only those outages which cause the limit violation on power flow in the lines or voltages on the buses. Practically, only selected contingencies will lead to severe conditions in the power system. Therefore, the process of identifying these severe contingencies is referred to as contingency selection and this can be done by calculating severity indices for each contingency. This is important to target the vulnerable point in a large-scale power system network with a minimum time requirement.

The potential of artificial neural networks for nonlinear adaptive filtering and control, their ability to predict solutions from past trends, their enormous data processing capability, and their ability to provide fast responses in mapping data make them a promising tool for their application to power systems.

Looking forward, real-time and intelligent technologies need to be developed for contingency analysis to promote a look ahead and predictive security awareness. The development of big data and high-performance computing technology is the key to making this goal possible.

### 1.1.4 Security-Constrained Automatic Generation Control

The automatic generation control (AGC) is used to balance active power and regulate tie-line power flow while minimizing the power generation cost. In the present state of the art, the AGC base point is determined by the economic dispatch (ED) and AGC regulates the area control error to be zero and controls the tie-line power flow to the desired command.

ED optimizes generation under network security constraints. There are two commonly used methods for active ED in power systems: (i) offline ED and (ii) online ED.

Offline ED calculates unit commitment and dispatches unit active power output for the next day or the next few days in a time interval of hours based on the generation capacity, grid network constraints, as well as the forecasted load.

Since offline ED is based on load forecasting, the generation dispatch may not accurately meet the actual load. The operating conditions of the power system are changing, and the active power output of the generators may deviate from the scheduled power generation set point. Therefore, online ED adjusts the generation output set by offline ED continuously to satisfy the power system's actual operating point in a short time interval (5–15 minutes).

In practice, for a middle-scale power system, ED calculation with network security constraints takes minutes. Due to the short cycle of AGC control conflicting with the extensive computation efforts of the security-constrained optimization, network security constraints are not modeled in AGC in real-time. When power system operation changes significantly in the ED cycle, AGC commands cannot guarantee that the network security constraints are satisfied.

The present state of the art assumes the power system operation point does not change significantly enough between two ED executions to push the AGC base point determined by ED to violate network security constraints. This assumption is not always true in power system operations.

In the case of intermittent renewable energy penetrated power systems and fast-response power electronics-based generation integrated transmission and distribution power network, the system power flow has a high probability of shifting away from the base point which is optimized in the cycle of ED. AGC does not optimize power flow within network security constraints. With the present AGC command, the power flow may result in violations. Thus, the AGC without network security constraints presents risks in power system operation.

When SE estimates violated power flow, the present state of the art heuristically changes generation limits of generators for AGC regulation to alleviate the risk of overflow under AGC command. This approach has drawbacks in that it is heuristic and the SE execution cycle in minute-scale cannot fit into AGC execution in seconds.

Taking advantage of the high performance of graph computing, the network security-constrained AGC is potentially achievable.

### 1.1.5 Security-Constrained ED

Optimization theory is an important tool in decision science and the analysis of power system operation. In security-constrained ED, objectives are set as a quantitative measure of the performance of the system under study, which could be the power system generation cost, renewable energy, or a combination of quantities. The main objective is to find values of the variables that optimize the objective. In addition, the variables are restricted or constrained to meet physical laws and security requirements. The process of identifying objectives, variables, and constraints for a given problem is known as modeling. Once the model has been formulated, an optimization algorithm can be used to find its solution.

Unconstrained optimization approaches are the basis of constrained optimization algorithms. Particularly, most of the constrained optimization problems in power system operation can be converted into unconstrained optimization problems. The major unconstrained optimization approaches that are used in power system operation are the gradient method, line search, Lagrange multiplier method, Newton–Raphson optimization, trust-region optimization, quasi-Newton method, double dogleg optimization, conjugate gradient optimization, and so on.

A general formulation of constrained optimization approaches can be modeled as:

$$\min_{x \in \Omega} f(x) \tag{1.1}$$

where $\Omega = \{x \mid c_i(x) = 0, \ i \in \mathcal{E}; c_j(x) \geq 0, \ j \in I\}$.

Linear programs have a linear objective function and linear constraints, which may include both equalities and inequalities. The feasible set is a polytope, a convex, connected set with flat, polygonal faces. The contours of the objective function are planar.

Power system operation problems are nonlinear. Thus, nonlinear programming (NLP)-based techniques can handle power system operation problems such as the optimal power flow problem and security-constrained ED with nonlinear objective and constraint functions.

The linear programming models discussed above have been continuous, in the sense that decision variables are allowed to be fractional. However, fractional solutions are not realistic in power system unit commitment. This problem is called the integer-programming problem. It is said to be a mixed integer program when some, but not all, variables are restricted to be integers.

## 1.1.6 Electromechanical Transient Simulation

Transient stability analysis assesses the state of the power system after a severe disturbance using transient simulations. The mathematical model for describing power system dynamic behavior is a nonlinear dynamic system that includes high-dimensional nonlinear differential equations and larger scale nonlinear algebraic equations (DAEs). When the time domain method is applied to power system transient simulation, differential-algebraic equations are solved by using time-consuming numerical integration methods. To improve the computation efficiency, the choice of a proper step size and parallel computing is essential.

If the step size is too large, the result will become inaccurate or even completely wrong when the large step size is not within the range of numerical stability. If the step size is too small, the transient simulation will take longer than necessary to keep the accuracy. The adaptive time step is a solution that uses the smallest possible time step to obtain an accurate result, thereby increasing the calculation speed while ensuring the calculation accuracy.

In each time step, we need to solve high order high-dimensional differential-algebraic equations. To further improve the computation efficiency, a parallel computing algorithm has been investigated in power system transient simulation. In this book, a graph-based parallel computing method is demonstrated with the adaptive time-step numerical integration method.

Using the sequential method, in each iteration, the network equations are solved to update the network bus voltage including the generator terminal voltage. Differentiate equations use generator terminal bus voltage as a boundary condition to solve the dynamic states of the generator, exciter, governor, Power System Stabilizer (PSS), and current injections from generators to networks. The updated current injections are applied to solve network equations in the next iteration until the converged solution is achieved.

Graph computing demonstrated outperformance on power system steady-state applications where the technology will be used to solve algebraic equations in the sequential method-based transient simulation. In the sequential method, since the differential equations for each generation system are independent once the terminal voltages are solved by algebraic equations, the differential equation sets can be solved by graph parallel computing naturally which will be addressed in this book in detail.

### 1.1.7 Photovoltaic Power Generation Forecast

In recent years, the rapid exhaustion of fossil fuel sources, environmental pollution concerns, and the aging of developed power plants are considered crucial global concerns. As a consequence, renewable energy resources including wind and solar have been rapidly integrated into the existing power grids. The reliability of power systems depends on the capability of handling expected and unexpected changes and disturbances in production and consumption while maintaining quality and continuity of service. The variability and stochastic behavior of photovoltaic (PV) power are caused by including voltage fluctuations, as well as local power quality and stability issues [22]. Hence, accurate photovoltaic power generation forecasting is required for the effective operation of power grids [23].

The studies in solar irradiance and photovoltaic power forecasting are mainly categorized into three major classes:

1) The persistence models sever as a baseline that assumes the irradiance values at future time steps are equal to the same values at the forecasting time [22].
2) Physical models employ physical processes to estimate future solar radiation values using astronomical relationships [24], meteorological parameters, and numerical weather predictions (NWPs) [25].
3) Statistical and artificial intelligence techniques estimate or regress solar irradiance and photovoltaic power generation [26–32].

To remove the strong smoothness assumption, increase the generalization capability, and improve the computation efficiency, in this book, the problem of spatio-temporal probabilistic solar radiation forecasting is presented as a graph distribution learning problem. In the approach, a set of solar measurement sites in a wide area is modeled as an undirected graph, where each node represents a site and each edge reflects the correlation between historical solar data of its corresponding nodes/sites to model the solar radiation spatio-temporal characteristics.

## 1.2 Mathematical Model

In general, power system analysis could be transformed to solve a linear system $Ax = b$, differential equations $\frac{dx}{dt} = f(x, t)$, and/or optimization problems $\min_{x \in \Omega} f(x)$. As a fundamental function, graph parallel computing approaches to solve the three typical mathematical problems involved in power system analysis, optimization, and simulations are key components in this book.

### 1.2.1 Direct Methods of Solving Large-Scale Linear Equations

Direct methods are widely used to solve linear equations by a finite sequence of operations. In the absence of rounding errors, direct methods would deliver an exact solution. Besides their high efficiency to solve moderate-size linear systems, direct methods are also popular to solve large sparse linear systems, like power flow, SE, and other power system problems. Sparse direct methods are a tightly coupled combination of techniques from numerical linear algebra, graph theory, graph algorithms, permutations, and other topics in discrete mathematics [33]. And such problem has been extensively studied.

This book focuses on direct methods for sparse linear systems, such as lower–upper (LU), Cholesky, and other factorization, and the implementation by graph parallel computing. It first introduces basic concepts, such as the definition and data structure of the sparse matrix, used in direct methods. As lots of graph concepts and algorithms are used in direct methods, the relationship between matrix and graph is included in the book of basic concepts.

### 1.2.2 Iterative Methods of Solving Large-Scale Linear Equations

The iterative method includes a series of techniques that use successive approximations to obtain more accurate solutions to a linear system at each step. Iterative methods can be expressed as $x^k = B \cdot x^{(k-1)} + c$, where $B$ and $c$ are constant. Direct method is widely used in solving power system problems. Iterative methods are applied in power system analysis as well including the Jacobi method, Gauss–Seidel method, successive over-relaxation method, symmetric successive over-relaxation method, conjugate gradient method, generalized minimal residual method, and bi-conjugate gradient.

### 1.2.3 High-Dimensional Differential Equations

The ordinary differential equations (ODEs) with an initial value that appeared in transient simulations are of the form

$$\frac{dx}{dt} = f(x, t), \quad x(t_0) = x_0. \tag{1.2}$$

Here the solution $x = x(t)$ needs to be solved for any time $t > t_0$. The variable $x$, the right-hand-side function $f(x, t)$, and the initial value $x_0$ can be either a scalar value or a vector. In addition, $f(x, t)$ can be either linear or nonlinear.

There are generally two classes of numerical methods for solving ODEs: (i) one-step methods and (ii) linear multistep methods.

One-step methods make use of the previously computed $x_n$ to produce a value of $x_{n+1}$. They do make use of one or more evaluations of the function at intermediate points between $t_n$ and $t_{n+1}$ in order to improve accuracy. However, these function evaluations are then discarded and are not reused in making future steps. In contrast, multistep methods make use of the previously found values $x_n, x_{n-1}, \ldots$ in order to produce a value of $x_{n+1}$.

The simplest class of one-step methods is the Euler's methods and their relatives – forward Euler, backward Euler, and Trapezoidal rule. One disadvantage of backward Euler and the Trapezoidal rule is that they require solving implicit equations at each time $t_{n+1}$. In addition to achieving higher orders of accuracy, methods in the Runge–Kutta family self-adapt the time step sizes. These methods always include two methods, one with higher order which is used for the error estimation. The other lower-order method is for the approximated solution. These two methods are designed to have the same intermediate steps, which ensure that the extra computation effort for error estimation is negligible.

### 1.2.4 Mixed Integer-Programming Problems

Consider an optimization problem modeled as a Mixed-Integer Linear Program (MILP) has the following structure:

$$\min c^T x \tag{1.3}$$

$$s.t. \ Ax + By + Cv \le a \tag{1.4}$$

$$A'x + B'y + C'v \le a' \tag{1.5}$$

$$x \in \{0,1\}^n, y \in \{0,1\}^m, v \in R^p \tag{1.6}$$

This is an optimization problem that has both continuous ($v$) and binary ($x$ and $y$) sets of variables, and only some of the binary variables ($x$) have nonzero objective function coefficients. The constraint set can be divided into two subsets. The first set of constraints (1.4) models aspects of the problem that can be represented efficiently in the MILP framework (e.g. assignment constraints) and has a significant impact on the Linear Program (LP) relaxation. The second set of constraints (1.5), on the other hand, is assumed not to significantly affect the LP relaxation and is usually large in number because of the limited expressive power of MILP methods.

## 1.3 Graph Computing

When performing power system computing, from topology process, admittance matrix formation, matrix factorization, forward and backward substitution, optimal search, and numerical integration, to state visualization, a large number of database operations are called repeatedly on data reading, writing, searching, and concurrent accessing. A relational database uses join-intensive queries for the whole database for many database operations inviting more computation time for large datasets. On the contrary, the graph database outperformed a relational database in these database operations [34]. The database operation mimeograph dataset is proportional to the number of subgraphs other than the entire graph leveraging the graph database's nodal parallel and hierarchical parallel capabilities.

In the area of power system analysis, optimization, and simulation, using graph data management and computation technology, the computational model can be smoothly integrated with data storage and numerical calculation with in-memory computing, distributed parallel computing, and decomposition aggregation. The technology has great potential for large-scale power grid analysis and calculation.

The traditional Relational Database Management System (RDBMS) uses tables to store data. The RDBMS stores the structured records and their attributes in equal-length tables, and maintains the database using Structured Query Language. Ideally, structured tabular data of arbitrary complexity can be represented by relational databases. However, RDBMS has no flexibility to define unstructured datasets. Relational databases are not very accommodating of data interconnections in a dataset that is graph-based, rather than attribute-based. To represent the relational interconnections, the linking attributes of records in the database are stored in different tables for creating the relations, and the relationship between the different records is established by querying the same attribute of the corresponding records in different tables. Therefore, adding or deleting records in the RDBMS requires updating all tables with the associated shared attributes. When compared with the performance of a Graph Database Management System (GDBMS), an RDBMS takes much longer time to support attribute searching, optimal ordering, depth-first (or breadth-first) search, which limits the efficiency of topology analysis, parallel computing, and results visualization for the traditional power system applications.

In graph computing, the relationship between nodes and edges is self-defined by the graph. Unlike relational databases, graph databases model the power system using graph-oriented data structures for semantic queries with a set of nodes and edges [35, 36]. Unstructured attributes are then stored in data structures defined by the graph's nodes and edges. For power system applications, the GDBMS is more in line with the requirements of power system computing for complex data modeling, querying, sorting, and traversal.

Commonly used data structures for power system calculations include arrays, linked lists, trees, and graphs. For example, an array is often used for matrix operations; a linked list is used to represent the path set of generators and loads when taking advantage of fast forward and fast backward substitution (sparse vector methods) while solving matrix equations [37, 38]; an elimination tree is used for identifying parallel processes when parallelizing factorization [39, 40]; and a graph is used for topological processing [41].

The graph database is concise when modeling different data structures, from simple array data to the more complex structures which store graphs and trees. Hence, the graph database is very accommodating of data interconnections within a dataset, which fits the characteristics of power systems since the power system is naturally modeled as a graph consisting of nodes and branches. Nodes are physically connected through branches as edges. The unconstructed parameters of the bus, generator, load, and branch, such as active and reactive power of generator and load, resistance and reactance of lines and transformers, and so on are stored in the node or edge. The graph structure itself naturally represents the topology of the electric power grid.

The remarkable performance of the GDBMS results from its built-in parallel computing capability. The GDBMS allows us to take advantage of two types of parallelism: nodal parallelism and hierarchical parallelism.

In graph database operation parlance, nodal parallel computing refers to the computation of quantities associated with each node, where each computation is independent; in contrast, hierarchical parallelism partitions nodes into different levels according to their computing dependency and then performs the computation in parallel on nodes at the same level.

For example, nodal parallelism can be used when searching for a specific attribute value since no dependencies exist: GDBMS compares the desired value against the attribute value of all nodes and edges in parallel. In contrast, when dependencies exist and are characterized by precedence relationship depth level, the GDBMS processes all nodes using a depth-first search, exploiting hierarchical parallelism. By leveraging the graph database's nodal parallelism and hierarchical parallelism capabilities, the database operation time on graph datasets is proportional to the size of subgraphs rather than the size of the entire graph.

The parallel solution of power system application using the graph-based approach illustrated in this book will demonstrate that it can be used not only for SE, power flow, and contingency analysis but also for other applications, such as dynamic security assessment and transient stability simulation.

### 1.3.1 Graph Modeling Basics

A graph can be simply expressed as

$$G = (V, E) \tag{1.7}$$

where $V$ denotes the set of nodes and $E$ denotes the set of edges. When we design the structure of graph data in a power system, attributes of nodes and edges are used to store static/dynamic variables, such as voltage ratings, impedance (admittance), power capacity rating, real-time voltage, and real-time power. For each $v_i$, the description of its state is represented by a set of independent attributes $p_{vi} \in P$, and for the edge $e_{ij}$ connecting with node $v_i$ and node $v_j$, the description of its state is also represented by a set of independent attributes $p_{eij} \in P$. According to the above definition, a graph can be constituted by the $V$, $E$, and $P$ sets. In numerical calculation, an element $a_{ij} \neq 0$ ($i \neq j$) in admittance matrix $A$ is equivalent to nodes $v_i$ and $v_j$ are connected, on the contrary, $a_{ij} = 0$ equivalent to the edge $e_{ij}$ does not exist, which means nodes $v_i$ and $v_j$ are not connected.

### 1.3.2 Graph Parallel Computing

Graph parallel computing exploits both nodal and hierarchical parallelism. In power system applications, the tasks of matrix formation, right-hand-side correction vector calculation, and branch power flow calculation can be nodally parallelized. The tasks of matrix factorization and forward/backward substitution can be hierarchically parallelized.

When the power system is modeled using a graph structure, calculations such as admittance matrix formation can be performed using nodal parallel computing once the sparsity-based symbolic part of the data structure has been completed.

In the graph structure, each numbered vertex represents a bus, and the edge between two vertices is a branch. In the graph structure, the counterparts of the connections between nodes are nonzero off-diagonal elements in the coefficient matrix $A$. Zero (absent) elements in the matrix schematic representation indicate that no direct connections between the nodes exist in the graph.

To form a row of the admittance matrix for any vertex, the off-diagonal nonzero elements of each row correspond only to the adjacent nodes. The graph-based approach searches for the neighboring vertex (or vertices) and the edges between them to form the admittance matrix numerical entries for all vertices simultaneously.

Other nodal parallelizable calculations in the power system analysis applications include the right-hand-side vector calculation, the convergence check, bad data detection, the branch power calculation, and the voltage and power flow violation check. These calculations, performed on node attributes, are independent from each other and can therefore be performed simultaneously.

In graph hierarchical parallel computing, computation is performed in parallel on nodes at the same depth level of the elimination tree. The nodes are partitioned into different levels in the elimination tree according to their calculation dependency. The calculation of quantities associated with the higher-indexed-level nodes depends on the calculation of quantities associated with the lower-level nodes. The calculations associated with nodes at the same level are independent and are performed in parallel.

In power system analysis, the hierarchically parallelizable tasks include matrix factorization and forward/backward substitution.

## References

1 Liu, Z. (2014). *Ultra-High Voltage AC/DC Grids*. Waltham, MA: Academic Press.
2 Jayant Baliga, B. (2011). *Advanced High Voltage Power Device Concepts*. New York: Springer.
3 Tyll, H.K. (2004). *2004 IEEE/PES Transmission and Distribution Conference and Exposition: Latin America* (IEEE Cat. No. 04EX956), Sao Paulo, Brazil, pp. 976–980. https://doi.org/10.1109/TDC.2004.1432515.
4 Erkeng, Y., Liu, G., and Zhou, J. (1998). *Energy Management System*. Science Press.
5 Najafabadi, M.M., Villanustre, F., Khoshgoftaar, T.M. et al. (2015). Deep learning applications and challenges in big data analytics. *Journal of Big Data* 2: 1.
6 Chen, Y., Huang, Z., Jin, S., and Li, A. (2022). Computing for power system operation and planning: then, now, and the future. *iEnergy* 1(3): 315–324.

**7** Mansoor, A. and Gellings, C. (2021). *Needed: A Grid Operating System to Facilitate Grid Transformation*. Palo Alto, CA: Electric Power Research Institute. 94304-1338.

**8** North American Electric Reliability Council (2004). Technical analysis of the August 14, 2003, blackout: what happened, why, and what did we learn? *Report to the NERC Board of Trustees by the NERC Steering Group*, 13 July 2004.

**9** Myrda, P.T. and Grijalva, S. (2012). The need for next generation grid energy management system. *2012 CIGRE US National Committee Grid of the Future Symposium*, Kansas City, MO (July 2021).

**10** Wix, S.D. and Plunkett, P.V. (2009). *Conference: Proposed for presentation at the Clean Technology Conference & Expo 2009*, Houston, TX (3–7 May 2009).

**11** The U.S. Department of Energy (2015). An assessment of energy technologies and research opportunities. In: *Chapter 3. Enabling Modernization of the Electric Power System*. Washington, DC. https://www.energy.gov/sites/prod/files/2017/03/f34/qtr-2015-chapter3.pdf.

**12** Chen, Y., Jin, S., Rice, M., and Huang, Z. (2013). Parallel state estimation assessment with practical data. *2013 IEEE Power & Energy Society General Meeting*, Vancouver, BC, pp. 1–5. https://doi.org/10.1109/PESMG.2013.6672742.

**13** Alves, A.C.B. and Monticelli, A. (1995). Parallel and distributed solutions for contingency analysis in energy management systems. *38th Midwest Symposium on Circuits and Systems. Proceedings*, vol. 1. Rio de Janeiro, Brazil, pp. 449–452. https://doi.org/10.1109/MWSCAS.1995.504473.

**14** da Costa, V.M., Martins, N., and Pereira, J.L.R. (1999). Developments in the Newton Raphson power flow formulation based on current injections. *IEEE Transactions on Power Systems* 14 (4): 1320–1326. https://doi.org/10.1109/59.801891.

**15** Shi, J., Liu, G., Dai, R. et al. (2018). Graph based power flow calculation for energy management system. *2018 IEEE Power & Energy Society General Meeting (PESGM)*, Portland, OR, pp. 1–5. https://doi.org/10.1109/PESGM.2018.8586233.

**16** Yuan, C., Zhou, Y., Zhang, G. et al. (2018). Exploration of graph computing in power system state estimation. *2018 IEEE Power & Energy Society General Meeting (PESGM)*, Portland, OR, pp. 1–5. https://doi.org/10.1109/PESGM.2018.8586535.

**17** Chen, T., Yuan, C., Liu, G., and Dai, R. (2018). Graph based platform for electricity market study, education and training. *2018 IEEE Power & Energy Society General Meeting (PESGM)*, Portland, OR, pp. 1–5.

**18** Zhao, Y., Yuan, C., Liu, G., and Grinberg, I. (2018). Graph-based preconditioning conjugate gradient algorithm for "N−1" contingency analysis. *2018 IEEE Power & Energy Society General Meeting (PESGM)*, Portland, OR, pp. 1–5. https://doi.org/10.1109/PESGM.2018.8586214.

**19** Qiu, H., Zhou, A., Hu, B. et al. (2018). TAnalyzer: "a graph database based topology analysis tool for power grid". *Proceedings of the 2nd International Conference on Computer Science and Application Engineering*, Hohhot, China, October 2018, pp. 1–5.

**20** Kan, B., Zhu, W., Liu, G. et al. (2017). Topology modeling and analysis of a power grid network using a graph database. *International Journal of Computational Intelligence Systems* 10: 1355–1363.

**21** Lv, X., Chen, S., Zheng, S. et al. (2018). Understanding the graph databases and power grid systems. *International Conference on Advanced Electronic Materials, Computers and Materials Engineering*, vol. 439, Issue 3, Singapore, September 2018.

**22** Wan, C., Zhao, J., Song, Y. et al. (2015). Photovoltaic and solar power forecasting for smart grid energy management. *CSEE Journal of Power and Energy Systems* 1 (4): 38–46.

**23** Jiang, Y., Long, H., Zhang, Z., and Song, Z. (2017). Day-ahead prediction of bihourly solar radiance with a Markov switch approach. *IEEE Transactions on Sustainable Energy* 8 (4): 1536–1547.

**24** Hottel, H.C. (1976). A simple model for estimating the transmittance of direct solar radiation through clear atmospheres. *Solar Energy* 18: 129–134.

**25** Pfenninger, S. and Staffell, I. (2016). Long-term patterns of European PV output using 30 years of validated hourly reanalysis and satellite data. *Energy* 114: 1251–1265.

**26** Bae, K., Jang, H., and Sung, D. (2016). Hourly solar irradiance prediction based on support vector machine and its error analysis. *IEEE Transactions on Power Systems* 32 (2): 935–945.

**27** Voyant, C., Notton, G., Kalogirou, S. et al. (2017). Machine learning methods for solar radiation forecasting: a review. *Renewable Energy* 105: 569–582.

**28** Lauret, P., Voyant, C., Soubdhan, T. et al. (2015). A benchmarking of machine learning techniques for solar radiation forecasting in an insular context. *Solar Energy* 112: 446–457.

**29** Wan, C., Xu, Z., Pierre, P. et al. (2014). Probabilistic forecasting of wind power generation using extreme learning machine. *IEEE Transactions on Power Systems* 29 (3): 1033–1044.

**30** Golestaneh, F., Pinson, P., and Gooi, H. (2016). Very short-term nonparametric probabilistic forecasting of renewable energy generation – with application to solar energy. *IEEE Transactions on Power Systems* 31 (5): 3850–3863.

**31** Zhang, Y. and Wang, J. (2015). GEFCom2014 probabilistic solar power forecasting based on k-nearest neighbor and kernel density estimator. *2015 IEEE Power & Energy Society General Meeting*, Denver, CO, USA, pp. 1–5. https://doi.org/10.1109/PESGM.2015.7285696.

**32** Lauret, P., David, M., and Pedro, H. (2017). Probabilistic solar forecasting using quantile regression models. *Energies* 10 (10): 1591. https://doi.org/10.3390/en10101591.

**33** Davis, T.A., Rajamanickam, S., and Sid-Lakhdar, W.M. (2016). A survey of direct methods for sparse linear systems. *Acta Numerica* 25: 383–566.

**34** Vicknair, C., Macias, M., Zhao, Z. et al. (2010). A comparison of a graph database and a relational database. *Proceedings of the 48th Annual Southeast Regional Conference* (April 2010), Oxford, MS.

**35** Malewicz, G., Austern, M.H., Bik, A.J. et al. (2010). Pregel: a system for large-scale graph processing. *Proceedings of the 2010 ACM SIGMOD International Conference on Management of Data*, Indianapolis, June 2010, pp. 13–146.

**36** Lu, Y., Cheng, J., Yan, D., and Wu, H. (2014). Large-scale distributed graph computing systems: an experimental evaluation. *Proceedings of the VLDB Endowment* 8 (3): 281–292.

**37** Tinney, W.F., Brandwajn, V., and Chan, S.M. (1985). Sparse vector methods. *IEEE Transactions on Power Apparatus and Systems* PAS-104 (2): 295–301.

**38** Zhang, B. and Chen, S. (1998). *Advanced Power Grid Analysis*. Science Press.

**39** Schreiber, R. (1982). A new implementation of sparse Gaussian elimination. *ACM Transactions on Mathematical Software* 8 (3): 256–276.

**40** Kumar, P.S., Kumar, M.K., and Basu, A. (1992). A parallel algorithm for elimination tree computation and symbolic factorization. *Parallel Computing* 18 (8): 849–856.

**41** Pradeep, Y., Seshuraju, P., Khaparde, S.A., and Joshi, R.K. (2011). CIM-based connectivity model for bus-branch topology extraction and exchange. *IEEE Transactions on Smart Grid* 2 (2): 244–253.

# 2

# Graph Database

A database is an integrated system for collecting, recording, and maintaining information and data. A database management system provides a set of convenient and efficient tools to define, store, and retrieve the information contained in the database.

Databases play an important role in scientific and non-scientific applications. In power system analysis, equipment such as generators, transmission lines, transformers, loads, shunt capacitors/reactors, circuit breakers, nodes, buses, and their properties are typically organized in structured tables in a database. In power system applications, relational databases are widely used to store and manage power system data and models. To adapt to the real-time application of the power system, in-memory databases and hierarchical databases are developed to maintain the real-time topology and dynamic operating conditions of the power system. In-memory databases and hierarchical databases are essentially derived relational databases for storing and managing structured data and information.

In contrast, a graph database is a database that models a system and stores information using a graph structure with vertices, edges, and attributes of vertices and edges. Edges describe the relationship between vertices. The data structure in a graph database is very different from a traditional relational database. Relationships are built directly into the graph database as given properties. In a relational database, the relationship between the vertices is implicit and indirectly established through analysis and calculation.

## 2.1  Database Management Systems History

The first commercial database management systems appeared in the late 1960s when personal computers were affordable to store and maintain large amounts of information and data. Relational models and databases were first introduced in 1970 by Edgar F. Codd [1]. Relational database management systems are an efficient way to store and process structured data. With relational database management systems developed over generations, relational databases can process data very quickly. However, unstructured data is becoming more prevalent in the scientific and non-scientific environment, including power system analysis. Unstructured data is non-relational, and relational database management systems are simply not designed to handle unstructured data.

In the 1980s, Structured Query Language (SQL) became the standard language for databases [2]. SQL is still used today to manage relational and graph databases and to operate on the data in the databases.

*Graph Database and Graph Computing for Power System Analysis*, First Edition. Renchang Dai and Guangyi Liu.
© 2024 The Institute of Electrical and Electronics Engineers, Inc. Published 2024 by John Wiley & Sons, Inc.

A key event that affected the history of databases was the creation of the World Wide Web. Online businesses require client–server database systems, as such, in 1995, MySQL was introduced as an open-source relational database management system. Today, MySQL has grown into an extremely scalable database system capable of running on multiple platforms, making it the most popular database in the world.

Over the past decade, with the rise of big data, graph databases emerged in 2006, when Tim Bernes-Lee introduced the concept of large databases that linked data. Based on graph theory, graph databases can model and visualize systems well with graph data structures.

## 2.2 Graph Database Theory and Method

In a graph database, data are organized as graphs. Graphs contain vertices and edges, and their attributes. There are two categories of graph databases: property graphs and RDF graphs. RDF stands for Resource Definition Framework. The property graph focuses on analytics and querying, while the RDF graph emphasizes knowledge integration. In this book, property graphs are used to model and analyze power systems. And the graph is used in short to represent a property graph if it is not mentioned specifically.

### 2.2.1 Graph Database Principle and Concept

Graphs are used to model relationships between data and to perform queries and data analysis based on the constructed data relationships. The graph uses vertices to store subject information and edges to depict the relationship between the vertices. In power system analysis, edges are also used to represent both subject (for example, transmission line and transformer) and connecting relationship (buses connected by transmission line or transformer) to reduce the size of the graph. A graph can be expressed as

$$G = (V, E) \tag{2.1}$$

where $V$ denotes the set of vertices and $E$ denotes the set of edges. When we design the structure of graph data in a power system, attributes of vertices and edges are used to store static/dynamic variables, such as voltage ratings, impedance (admittance), power capacity rating, real-time voltage, and real-time power. For each $v_i$, the description of its state is represented by a set of independent attributes $p_{vi} \in P$, and for the edge $e_{ij}$ connecting with vertex $v_i$ and vertex $v_j$, the description of its state is also represented by a set of independent attributes $p_{eij} \in P$. According to the above definition, a graph can be constituted by the $V$, $E$, and $P$ sets. In numerical calculation, an element $a_{ij} \neq 0$ ($i \neq j$) in admittance matrix $A$ is equivalent to vertices $v_i$ and $v_j$ that are connected, on the contrary, $a_{ij} = 0$ equivalent to the edge $e_{ij}$ does not exist, which means vertices $v_i$ and $v_j$ are not connected.

For example, a simple 5-Bus system in Figure 2.1 is represented as a graph in Figure 2.2. In the graph, each vertex represents the physical bus. These vertices are connected by edges with topological relationships. In the power system graph model, edges also represent the physical subject, for example, transmission lines in this case. Virtual edges are also built-in in a graph database to connect two subjects, for example, two circuit breakers in topology analysis.

The graph model is further established by the attributes of each vertex and edge as the semantic context, which are defined in the graph schema.

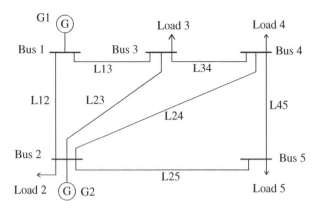

**Figure 2.1** One-line diagram of 5-bus system.

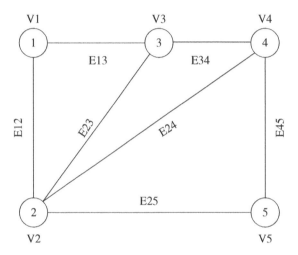

**Figure 2.2** 5-Bus system graph model.

### 2.2.1.1 Defining a Graph Schema

Before loading data into a graph database, a well-defined graph schema must be created. The graph schema defines the types of entities as either vertices or edges and defines how these entities relate to each other. Each vertex and edge type has a unique identification and a set of attributes.

The basic graph schema syntax includes `CREATE VERTEX`, `CREATE EDGE`, and `CREATE GRAPH`.

The `CREATE VERTEX` statement defines a global vertex type, with a name and an attribute list.

```
CREATE VERTEX Syntax
CREATE VERTEX vertex_type_name ( primary_id_name_type
    [, attribute_name type [DEFAULT default_value] ] )
```

`CREATE EDGE` defines a new global edge type. There are two forms of the `CREATE EDGE` statement, one for directed edges and one for undirected edges differentiated by the keywords `UNDIRECTED` and `DIRECTED`. Each edge type must specify that it connects `FROM` one vertex type `TO` another vertex type. Then attributes are listed.

```
CREATE EDGE Syntax
CREATE UNDIRECTED EDGE edge_type_name (
    FROM vertex_type_name, TO vertex_type_name
    [,attribute_name type [DEFAULT default_value] ] )
CREATE DIRECTED EDGE edge_type_name (
    FROM vertex_type_name, TO vertex_type_name
    [,attribute_name type [DEFAULT default_value] ] )
```

In power system analysis, an undirected edge represents a typical branch connecting two vertices which allows the database to bi-directionally search from one edge to another with no difference. When the system has an active device, for example, a phase-shifting transformer, it makes the admittance matrix unsymmetrical in power system analysis. In this case, two directed edges are used to model the phase-shifting transformer with opposite directions (by switching the FROM vertex and the TO vertex) and different attribute values.

CREATE GRAPH defines a graph schema, which contains the given vertex types and edge types.

```
CREATE GRAPH Syntax
CREATE GRAPH graph_name (vertex_type,edge_type)
```

For power flow calculation, the following self-described statements define a basic power system graph schema.

```
CREATE VERTEX Bus (primary_id id string, Pg double, Qg double, Vm
    double, Vr double, Qmax double, Qmin double, Pl double, Ql double,
    G double, B double)
```

```
CREATE DIRECTED EDGE Branch (FROM Bus, TO Bus, G double, B double, hB
    double, K double, BIJ double)
```

```
CREAT GRAPH powerSystemGraph (Bus, Branch)
```

Unlike the generator and load that are sitting in different tables in a relational database, in the power system graph, the generator bus, load bus, and connecting bus are not separated by vertex type. They are differentiated by the values of generator power $Pg + jQg$ and load power $Pl + jQl$. In the edge type, the branch covers the transmission line, transformer, or any other type of branch. They are differentiated by transformer turn ratio $K$ in the database or specifically defined device type attribute if needed. This flexibility simplifies to model a case that a single bus has a generator, load, neither, or both and facilitates concise connection relationships between buses and branches. The defined power system graph is scalable to expand by adding additional attributes for other power system analyses, such as state estimation, contingency analysis, security-constrained economic dispatch, automatic generation control, small-signal stability analysis, transient stability analysis, etc.

### 2.2.1.2 Creating a Loading Job
Graph schema defined the graph structure and attributes of vertices and edges as containers to load data instances. After a power system graph schema has been created, the power system data are loaded into the graph database by CREATE LOADING JOB statement.

*CREATE LOADING JOB syntax*

```
CREATE LOADING JOB job_name FOR GRAPH graph_name {
    [DEFINE statements;]
    [LOAD statements;] | [DELETE statements;]
}
```

A typical loading job example is as follows:

```
CREATE LOADING JOB job1 FOR GRAPH graph1 {
   DEFINE FILENAME file1 = "/data/v1.csv";
   DEFINE FILENAME file2 = "/data/e1.csv";
   LOAD file1 TO VERTEX v1 VALUES ($0, $1, $2);
   LOAD file2 TO EDGE e1 VALUES ($0, $1);
}
```

Taking the 5-bus system as an example, assuming the vertex and edge data are stored in two comma-separated values (CSV) files – vertexInfo.csv and edgeInfo.csv, the loading job statement defines the job to load power system data into graph powerSystemGraph defined in Section 2.2.1.1 is:

```
CREATE LOADING JOB powerFlowData FOR GRAPH powerSystemGraph {
   DEFINE FILENAME vertexInfoFile = " /vertexInfo.csv";
   DEFINE FILENAME edgeInfoFile = " /edgeInfo.csv";
   LOAD vertexInfoFile TO VERTEX Bus VALUES (id, Pg, Qg, Vm, Vr, Qmax,
   Qmin, Pl, Ql, G, B);
   LOAD edgeInfoFile TO EDGE Branch VALUES (FROM Bus, TO Bus, G, B,
   hB, K);
}
```

The data of each record in the CSV files are in the same order as the defined attributes order in the schema statements. By running the following loading job, buses and branches data stored in the vertexInfo.csv and edgeInfo.csv files are loaded into graph powerSystemGraph.

```
RUN LOADING JOB powerFlowData
```

### 2.2.1.3  Graph Query Language

Graph query language is a language for querying and analyzing large-scale graph databases. The high-level query language can easily perform various tasks in power system analysis. Unlike other programming languages such as Fortran, C/C++, etc., the graph query language consists of a few default words. They are designed to manipulate graph databases, from retrieving, sorting, filtering, inserting, deleting, and updating to traversal a set of vertices and edges. By combining multiple graph queries and external functions through graph database Application Programming Interfaces (APIs), power system analysis can be performed in parallel.

#### *2.2.1.3.1  Retrieving and Sorting Data*  The SELECT statement is used to retrieve one or a group of data from a graph database. To use the SELECT statement, what you want to select and from where you want to select it must be specified in the following syntax.

*SELECT syntax*
```
CREATE QUERY queryName() FOR GRAPH graphName{
     SELECT     sourceData
     FROM       fromClause
     ORDER BY orderByClause ASC|DESC
}
```

The SELECT statement defined in the query is to retrieve sourceData ordered by orderBy-Clause from a graph. An exemplar use case of a SELECT statement in power system analysis is listed as follows:

```
CREATE QUERY retrieveParameters() FOR GRAPH powerSystemGraph{
     SELECT   Pg, Qg
     FROM     Bus
     ORDER BY Pg ASC
}
```

The ORDER clause sorts the retrieved data in ascending by keyword ASC or descending by keyword DESC.

**2.2.1.3.2 Filtering Data**   In a SELECT statement, data is filtered by a Boolean condition in the WHERE clause. The WHERE clause is specified right after the FROM clause as follows:

```
CREATE QUERY retrieveParameters() FOR GRAPH powerSystemGraph{
     SELECT   Pg, Qg
     FROM     Bus
     WHERE    Pg>0.0
     ORDER BY Pg ASC
}
```

**2.2.1.3.3 Delete Data**   The DELETE statement deletes a given set of edges or vertices. The edges or vertices desired to be deleted are specified by the WHERE clause after the FROM clause as follows:

```
CREATE QUERY deleteEx() FOR GRAPH powerSystemGraph {
  DELETE s FROM BUS
    WHERE s.id = = "bus1"
}
```

The query above deletes vertex "bus1" from the graph powerSystemGraph.

**2.2.1.3.4 Insert Data**   The INSERT statement straightforwardly inserts an edge or vertex with given attributes. For example, the query to insert a new branch from "bus1" (fromBus) to "bus4" (toBus) is shown as follow:

```
CREATE QUERY insertEx(fromBus string, toBus string, G double,
B double, hB double, K double, BIJ double)  FOR GRAPH powerSystemGraph
{
    INSERT INTO BRUNCH (fromBus, toBus) VALUES (G, B, hB, K, BIJ)
}
```

***2.2.1.3.5 Update Data*** The UPDATE state updates the attributes of edges or vertices. A set of vertices or edges to update is described in the FROM clause, following the same rules as the FROM clause in a SELECT statement.

```
CREATE QUERY updateEx() FOR GRAPH powerSystemGraph  {
  UPDATE s FROM BUS WHERE s.id = = "UN1"
  SET s.id = "GEN1"
}
```

***2.2.1.3.6 Traversal*** A traversal statement is a special SELECT statement that can traverse a hop path from a set of vertices to their adjacent vertices under the control of the query with arrowheads on each edge set to show the direction.

A hop consists of a path from a starting set of vertices, crossing over a set of their edges, to an ending set of vertices. The traversal action is from left to right, and notates a connection with a rightward-facing arrowhead: Start:s -( Edges:e )-> Target:t

The traversal action is defined in the FROM Clause of a SELECT statement as follows:

```
    SELECT s FROM  Start:s -( Edges:e )-> Target:t
```

This statement can be interpreted as "Select all vertices s, from the vertex set source, their adjacent edges, and target vertices." The result can be interpreted as a 3-column virtual table. Each row is a 3-element tuple: (source vertex, connected edge, target vertex).

An exemplar use case of traversal function in power system analysis can be as follow:

```
    SELECT  s FROM Bus:s-(Branch:e)->Bus:t
```

This statement selects all buses in the database, their adjacent branches, and connected buses through the adjacent branches. The output is a list of a 3-element tuple in a typical format as follows:

```
{
    "Bus_ID": "Bus1",
    "attributes": {Pg, Qg, Vm, Vr, Qmax, Qmin, Pl, Q1, G, B},
    "Branch_ID": "L12",
    "attributes": {FROM Bus, TO Bus, G, B, hB, K},
    "Bus_ID": "Bus2",
    "attributes": {Pg, Qg, Vm, Vr, Qmax, Qmin, Pl, Q1, G, B}
}
```

***2.2.1.3.7 Accumulator*** The Accumulator is a unique feature of a graph query language for parallel processing. Accumulator (ACCUM) is a default word that indicates the nested statements under ACCUM accumulate information about the graph. A simple example to apply an accumulator for power system analysis parallelism is admittance matrix diagonal element calculation as follows:

```
CREATE QUERY calculateDiagonalElement() FOR GRAPH powerSystemGraph{
    SELECT s FROM Bus:s-(Branch:e)->Bus:t
      ACCUM
        s.@sumG += e.G
        s.@sumB += e.B + e.hB
      POST-ACCUM
```

```
                    s.@sumG += s.G
                    s.@sumB += s.B
}
```

The query is named as `calculateDiagonalElement()` and applied on graph `powerSys-temGraph`. The query can be called in the command terminal or called by developed functions or other queries.

As elaborated in Section 2.2.1.3.6, the second statement `SELECT s FROM Bus:s-(Branch: e)->Bus:t` selects all buses in the database, their adjacent branches, and connected buses through the adjacent branches.

Under the accumulator clause indicated by the keyword `ACCUM`, all statements are performed in parallel for all selected identities in set `s`. To form a diagonal element of the admittance matrix for any vertex, the off-diagonal nonzero elements of each row corresponding to the adjacent nodes are processed. The traversal function searches for the neighboring vertex (or vertices) and the edges between them to form the admittance matrix numerical entries for all vertices simultaneously. That is conducted by the statements `s.@sumG + = e.G` and `s.@sumB + = e.B + e.hB` which are performed for all vertices in set `s` in parallel by adding branch admittance of their adjacent branch or branches. After that, the statements under the post-accumulator clause indicated by the keyword `POST-ACCUM`, `s.@sumG + = s.G` and `s.@sumB + = s.B` add shunt capacitor or shunt reactor admittance to the diagonal elements. `s.@sumG` and `s.@sumB` are two attributes of vertex `s`.

The `ACCUM` and `POST-ACCUM` clauses support MapReduce semantics in the `SELECT` block. The Map step processes all the matched edges in parallel. The Reduce step merges the messages sent from the Map step. MapReduce parallel computing mechanism in the graph processing engine (GPE) is illustrated in Figure 2.3.

MapReduce parallel computing mechanism is a parallel programming model that splits data and operates on it in parallel before recollecting it together and aggregating it to provide centralized information in three phases.

In Phase 1, the job defined in the query is dispatched to workers. When the job request is sent to the Master, according to the available resources, the Master dispatches the job to workers by partitioning the vertices into threads. In the query calculateDiagonalElement above, all buses diagonal admittance calculations will be dispatched to each thread evenly. For instance, if the

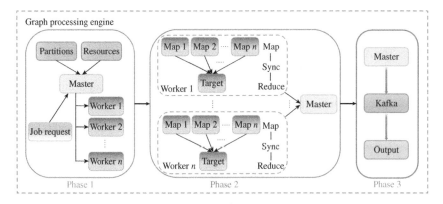

**Figure 2.3** MapReduce parallel computing mechanism.

studied power system has 200 buses and a computer has 10 available threads, each thread as a worker will be dispatched to take the job of calculating 20 buses' diagonal admittance. In Phase 2, each worker performs a MapReduce program, which is composed of a map procedure and a reduce procedure. The map procedure prepares the admittance attributes of adjacent branches simultaneously. The reduce procedure calculates the diagonal element by adding the admittance attributes. In Phase 3, the Master publishes the calculated diagonal elements to the graph as vertex attributes.

### 2.2.2  System Architecture

The graph database is a database management system that employs a graph data model to represent the study system and query large sets of highly connected data. It allows efficient execution of typical graph operations, such as single or multi-step graph traversal and shortest path between vertices. The graph database consists of the underlying graph storage engine (GSE) and GPE, as shown in Figure 2.4.

For the underlying GSE, a graph is partitioned into subgraphs for the purpose of parallel computing. A simple example, as illustrated in Section 2.2.1.3.7, is that a graph with 200 vertices/buses is partitioned into 10 subgraphs, each subgraph has 20 vertices/buses. In the GPE, which is first mentioned in Figure 2.3, jobs and tasks associated with the subgraph are scheduled and dispatched to available resources to execute in parallel. The graph functions and algorithms are implemented by high-level query language and/or interfacing with other programming languages, such as Java and C++.

### 2.2.3  Graph Computing Platform

The graph analytic platform is the core of the power system graph computing platform. The graph computing platform is the integration of a graph database, graph analytic platform, common function library, and power system analysis applications as shown in Figure 2.5. Figure 2.5 gives the ultimate goal of the graph computing platform in which the common function library supports different power system analyses from topology analysis, state estimation (SE), power flow calculation (PF), contingency analysis (CA), security-constrained automatic generation control (SCAGC), security-constrained economic dispatch (SCED), small-signal stability analysis (SSSA), transient stability analysis (TSA), and deep reinforcement learning based overload control (DRL-based overload control).

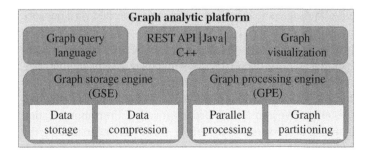

**Figure 2.4**  The graph analytic platform.

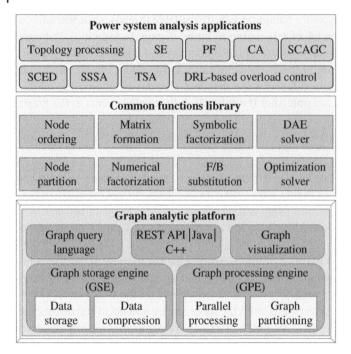

**Figure 2.5** Graph computing platform.

## 2.3 Graph Database Operations and Performance

### 2.3.1 Graph Database Management System

Graph database supports graph partition and graph storage mechanism to facilitate power system parallel computing. A single parallel machine with multi-processor or multi-thread and distributed architectures over many machines support parallel computing. Parallel architectures of a parallel machine and distributed network are typically classified into three categories: Shared-Memory Machines, Shared-Disk Machines, and Shared-Nothing Machines [3].

In the Shared-Memory Machines architecture, each processor has access to all the memory of all the processors. In the Shared-Disk Machines, every processor has its own memory, which is not accessible directly from another processor. However, the disks are accessible from any of the processors through the communication network. All processors of a Shared-Nothing Machine have their own memory and their own disk. Messages and information are communicated via a network from one processor to another.

Commercial graph database supports parallel computing sitting at a single parallel machine and distributed network and different parallel architectures. To implement power system parallel analysis with a graph database, we do not need to worry about the parallel architecture, available processors, and available machines for query execution and how the dispatched tasks are balanced among the available computation resources. We also do not need to care about the communication facilitator for passing information among processors and machines. However, understanding the high-level approaches of GPE and partition algorithms of GSE is meaningful in power system parallel analysis implementation by a graph database and graph computing.

### 2.3.1.1 Parallel Processing by MapReduce

The parallel processing in the GPE is performed by the MapReduce mechanism [4]. MapReduce is a high-level programming system that allows many important database processes to be simply performed in parallel.

MapReduce is two separate and distinct tasks. The first is the map function, which takes a set of data and transforms it into another set of data, in which individual elements are decomposed into tuples (key/value pairs). The output of the map function is taken as the input of the reduce function. And the input data tuples are combined into smaller tuples.

The outline of the map and reduce functions can be illustrated by the processing of solving distribution power flow calculation of a simple distribution power system in Figure 2.6.

The 6-bus distribution system can be modeled by a graph with vertices representing buses and edges representing branches. Additionally, we need to add virtual vertices $S$ and $T$ to indicate the start point and terminal point of the distribution system as shown in Figure 2.7. These virtual

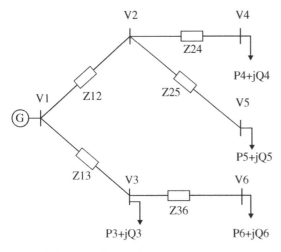

**Figure 2.6** 6-bus distribution system.

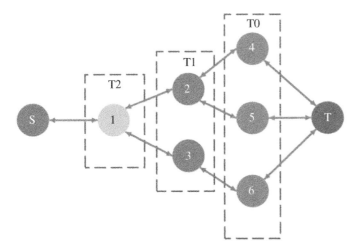

**Figure 2.7** 6-bus distribution system graph model.

vertices can also be viewed as separation points for different sections of the power distribution system. In the model, the child node-set and father node-set are defined. The child node-set is the set of nodes we are currently working on, while the father node-set is upstream nodes directly connecting with the child node-set. For instance, the father node-set of the child node-set $T_0$ is $T_1$.

In MapReduce, the map function is designed to take one key-value pair as input and produce a list of key-value pairs as output. The result of performing all the mapping processes is a collection of the intermediate result.

These key-value pairs are the outputs of the map function applied to every input pair. Each pair appears at the processor that generated it. There may be many map processes executing the same algorithm on a different part of the graph at different processors.

The reduce function is also executed by one or more processes, located at any number of processors. The input to reduce function is a single key value from the intermediate result, together with the list of all values that appear with this key in the intermediate result.

The reduce function itself combines the list of values associated with a given key. If a given map process produces more than one intermediate pair with the same key, then the reduce operation can be applied on the spot to combine the pairs, without waiting for them to be passed to the reduce process for that key.

Through a backward-forward sweep (BFS) algorithm, the power flow of the distribution system can be calculated. The parallel BFS algorithm by MapReduce mechanism is suggested in Figure 2.8.

The graph parallel computing above involves three phases. Phase 1 builds the distribution system graph database model (GDM) and allocates the resources prepared for parallel computing in Phase

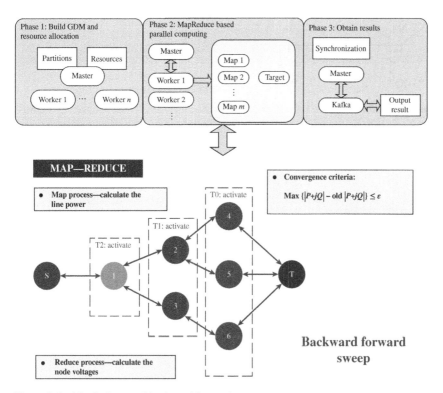

**Figure 2.8** MapReduce and backward-forward sweep.

2. Phase 2 centralizes the MapReduce process and phase 2 collects and synchronizes results. In phase 1, the whole graph, whether a large graph with millions of vertices and edges or a small graph with a few vertices and edges is optimally divided into subgraphs by graph partition to meet the needs of parallel computing and the availability of resources. Each subgraph is dispatched to an independent worker in phase 1. The workers divide their job into multiple maps in phase 2. Obviously, the map processes at the same level are parallelized, then communicate the results to the up-level by reduce process for the next-level map parallelization hierarchically. For instance, the power flow at the child nodes at the same level ($T_0$, $T_1$, ...) is calculated in parallel, then the calculated power flow is reduced (by reduce process) to the father node sets to update node voltage in parallel.

In the BFS algorithm, the map function is to calculate the line power at each edge. The reduce function is to update the node voltages, where the backward sweep is where injecting power $P, Q$ at the father node-set are calculated concurrently since only the information from the child node-set is needed for the calculation. While, in the forward sweep, voltage magnitude and angle are simply updated by (2.2) and (2.3).

$$U_i = \sqrt{\left(U_j + \Delta U\right)^2 + \delta U^2} = \sqrt{\left(U_j + \frac{PR + QX}{U_j}\right)^2 + \left(\frac{PX - QR}{U_j}\right)^2} \tag{2.2}$$

$$\theta = \arctan \frac{\delta U}{U_j + \Delta U} \tag{2.3}$$

The map and reduce functions work on a single object, so we could have as many processes and processors as the number of objects in the current node-set. In practice, it is unlikely that we have so many processors to use for a large graph. Graph partition, along with task dispatch and load balance essentially is the key to optimizing resource utilization in parallel computing.

### 2.3.1.2 Graph Partition
In practice, the graph is partitioned to subgraphs stored in shared memory, shared disk, or own memory and own disk in Shared-Nothing Machine managed by a GSE.

The graph partitioning divides vertices of a graph into groups targeting equal size and minimal cutting edges. Graph partitioning can be applied to many science and engineering problems. The number of vertices in a graph partition is roughly proportional to the computational load and the number of cutting edges represents the communication needs between the subgraphs.

To discuss graph partitioning algorithms built-in graph database management, the associated concepts are defined as follows first.

**Edge-cut**: Dividing a graph into multiple subgraphs, the edges connecting different subgraphs are called edge-cut.

The graph partitioning problem can be classified into two types: non-weighted graph partitioning and weighted graph partitioning.

**Non-weighted graph partitioning**: Given a graph $G = (V, E)$ and a positive integer $k$, $V$ is a set of vertices and $E$ is a set of edges. Graph $G$ is divided into $k$ disjoint subgraphs (each vertex set $V_i$ is referred to as a partition), such that:

① for each $V_i$, $|V_i| \approx |V|/k$, $i = 1, 2, ..., k$ ($|V|$ is the number of vertices in the vertex set $V$);
② the number of edge-cut is minimal.

As shown in Figure 2.9, the graph $G = (5, 5)$ is divided into two partitions. The edge between vertex $V_1$ and $V_4$ is cut to make the edge-cut number as 1 and the vertex number of the two

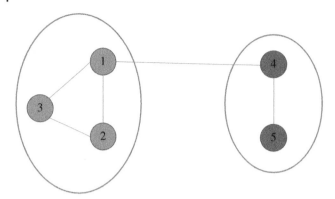

**Figure 2.9** Non-weighted graph partitioning.

subgraphs are 3 and 2, respectively. The partition is optimal with a balanced subgraph size and minimal edge cuts.

**Weighted graph partitioning**: Given a graph $G = (V, E)$ and a positive integer $k$, $V$ is a set of vertices and $E$ is a set of edges. $W(v)$ and $W(e)$ represent the weight of each vertex and edge, respectively. Graph $G$ is divided into $k$ disjoint subgraphs ($W_{G1}$, $W_{G2}$, ..., $W_{Gk}$ represent the total vertex weights of each subgraph), such that:

① for each $W_{Gi}$, $\quad |W_{Gi}| \approx \dfrac{(W_{G1} + W_{G2} + ... + W_{Gk})}{k}$, $\quad i = 1, 2, ..., k;$

② the total weight of the edge-cut is minimal.

As shown in Figure 2.10, in the weighted graph $G = (5, 5)$, the weight of each edge is illustrated on each edge, and suppose that the weight of each vertex is 1. Graph $G$ is divided into two partitions and the edge between vertex $V_1$ and $V_4$ is the edge-cut whose weight is 2. In the left partition including 3 vertices, the total vertex weight is 3; and in the right partition which contains 2 vertices, its total vertex weight is 2. The total vertex weights of these two partitions are roughly equal.

In fact, non-weighted graph partitioning is a special case of weighted graph partitioning. It can be considered as partitioning on a weighted graph whose weight of each vertex and edge is one.

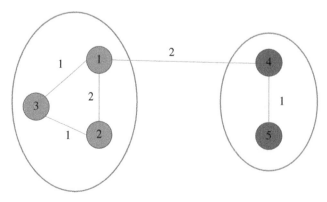

**Figure 2.10** Weighted graph partitioning.

To evaluate the result of graph partitioning, on the assumption of relatively good load balance among partitioned subgraphs, the total number or the total weight of edge-cut between different subgraphs is an important metric.

Regarding the evaluation of load balance, a common metric is a maximum load-imbalance ratio. It can be defined as the ratio of the maximum number of vertices or maximum total vertex weight among all the subgraphs to the difference between the maximum value and the average number of vertices or the average vertex weight. In the graph partitioning problem, the maximum load-imbalance ratio usually can be considered as a constraint.

In addition, there are also some other important metrics, such as algorithm complexity, execution time, memory consumption, and so on.

Since graph partitioning is an NP-hard problem, it is more practical to find a reasonable partitioning by using heuristics methods. Generally, graph partitioning methods can be divided into two categories: local methods and global methods [5].

#### 2.3.1.2.1 *Local Methods*
Basically, a local improvement method takes a partition (usually a bisection) of a graph $G$ as input and tries to decrease the cut size by local search methods. Thus, to solve the partitioning problem such an algorithm must be combined with methods that create a good initial partition. Another possibility is to generate several random initial partitions, apply the algorithm to each of them, and use the partition with the best cut size. The Kernighan–Lin algorithm [6] and Fiduccia–Mattheyses algorithm [7] are two classic local improvement methods.

1) Kernighan–Lin algorithm

In 1970, Kernighan and Lin proposed one of the earliest methods for graph partitioning, and more recent local methods are often variations of their methods. Given an initial bisection, the Kernighan–Lin method tries to find a sequence of node pair exchanges that leads to an improvement of the cut size.

Let $\{V_1, V_2\}$ be a bisection of the graph $G = (V, E)$. For each $v \in V$, we define

$$\text{int}(v) = \sum_{(v,u) \in EP(v)\,=\,P(u)} w(v,u) \tag{2.4}$$

$$\text{ext}(v) = \sum_{(v,u) \in EP(v) \neq P(u)} w(v,u) \tag{2.5}$$

where $P(v)$ is the index of the part to which node $v$ belongs, and $w(v, u)$ is the weight of edge $(v, u)$. The gain of moving a node $v$ from the part to which it currently belongs to the other part is

$$g(v) = \text{ext}(v) - \text{int}(v) \tag{2.6}$$

Thus, when $g(v) > 0$, we can decrease the cut size by $g(v)$ by moving $v$. For $v_1 \in V_1$ and $v_2 \in V_2$, let $g(v_1, v_2)$ denote the gain of exchanging $v_1, v_2$ between $V_1$ and $V_2$. That is,

$$g(v_1, v_2) = \begin{cases} g(v_1) + g(v_2) - 2w(v_1, v_2) & \text{if } (v_1, v_2) \in E \\ g(v_1) + g(v_2) & \text{if } (v_1, v_2) \notin E \end{cases} \tag{2.7}$$

One iteration of the Kernighan–Lin algorithm is as follows. The input to the iteration is a balanced bisection $\{V_1, V_2\}$. First, we unmark all nodes. Then, we repeat the following procedure $n$

times ($n = \min(|N_1|, |N_2|)$). Find an unmarked pair $v_1 \in V_1$ and $v_2 \in V_2$ for which $g(v_1, v_2)$ is maximum. Mark $v_1$ and $v_2$, and update the g-value of all the remaining unmarked nodes as if we had exchanged $v_1$ and $v_2$. After $n$ iterations, we get a series of maximal gain values $g_1, g_2, ..., g_n$. If the first $t$ node pairs have gain value $G_t = \sum_{i=1}^{t} g_i$ which is positive, we exchange the first $t$ node pairs, and begin another iteration of the Kernighan–Lin algorithm. Otherwise, the algorithm is terminated.

2) Fiduccia–Mattheyses algorithm

Fiduccia and Mattheyses improved the Kernighan–Lin method. Like the Kernighan–Lin method, the Fiduccia–Mattheyses method performs iterations during which each node moves at most once, and the best bisection observed during an iteration is used as input to the next iteration. However, instead of selecting pairs of nodes, the Fiduccia–Mattheyses method selects single nodes.

***2.3.1.2.2 Global Methods*** A global method takes a graph $G$ and an integer $k$ as inputs and generates a $k$-way partition. Most of these methods are recursive, that is, they first bisect $G$. The bisection step is then applied recursively until we have $k$ subsets of nodes. Global methods are often used in combination with local methods.

1) Geometric methods

Usually, nodes of a graph have geometric coordinates. Algorithms that use this information are called geometric partitioning algorithms. Some geometric algorithms completely ignore the edges of the graph, whereas other methods consider the edges to reduce cut size. Node and edge weights are usually assumed to be equal to 1.

Recursive coordinate bisection is the simplest example of a geometric algorithm. To obtain a bisection, first a coordinate axis is selected. Then, we find a plane, orthogonal to the selected axis, which bisects the nodes of the graph into two equal-sized subsets. This involves finding the median of a set of coordinate values.

The inertial method [8] is an elaboration of recursive coordinate bisection; instead of selecting a coordinate axis, the axis of minimum angular momentum of the set of nodes is selected. In three-dimensional space, this axis is equal to the eigenvector associated with the smallest eigenvalue of the matrix

$$I = \left[ I_{xx} I_{xy} I_{xz}; I_{yx} I_{yy} I_{yz}; I_{zx} I_{zy} I_{zz} \right] \tag{2.8}$$

where

$$I_{xx} = \sum_{v \in N} \left( (y(v) - y_c)^2 + (z(v) - z_c)^2 \right),$$

$$I_{yy} = \sum_{v \in N} \left( (x(v) - x_c)^2 + (z(v) - z_c)^2 \right),$$

$$I_{zz} = \sum_{v \in N} \left( (x(v) - x_c)^2 + (y(v) - y_c)^2 \right),$$

$$I_{xy} = I_{yx} = -\sum_{v \in N} (x(v) - x_c)(y(v) - y_c),$$

$$I_{yz} = I_{zy} = -\sum_{v \in N} (y(v) - y_c)(z(v) - z_c),$$

$$I_{xz} = I_{zx} = -\sum_{v \in N} (x(v) - x_c)(z(v) - z_c), \text{ and}$$

$$(x_c, y_c, z_c) = \frac{\sum_{v \in N}(x(v), y(v), z(v))}{|N|},$$

where, $(x_c, y_c, z_c)$ denotes the coordinates of node $v$.

Then, we continue exactly as in the recursive coordinate bisection method, that is, we find a plane, orthogonal to the axis of minimum angular momentum, that bisects the nodes of the graph into two equal-sized subsets.

Miller et al. [9] have designed an algorithm that bisects a $d$-dimensional graph. The algorithm for finding the bisecting sphere is as follows:

1) Stereographically project the nodes onto the $(d + 1)$-dimensional unit sphere. That is, node $v$ is projected to the point where the line from $v$ to the north pole of the sphere intersects the sphere.
2) Find the center point of the projected nodes. (A center point of a set of points $S$ in $d$-dimensional space is a point $c$ such that every hyperplane through $c$ divides $S$ evenly. Every set $S$ has a center point, and it can be found by linear programming.)
3) Conformally map the points on the sphere. First, rotate them around the origin so that the center point becomes a point $(0,...,0,r)$ on the $(d + 1)$-axis. Second, dilate the point by projecting the rotated points back to $d$-dimensional space, scaling the projected points by multiplying their coordinates by $\sqrt{\frac{1-r}{1+r}}$, and stereographically projecting the scaled points to the $(d + 1)$-dimensional unit sphere.
4) Choose a random hyperplane through the center of the $(d + 1)$-dimensional unit sphere.
5) The hyperplane from the previous step intersects the $(d + 1)$-dimensional unit sphere in a great circle. Transform this sphere by reversing the conformal mapping and stereographic projection. Use the obtained sphere to bisect the nodes.

A practical implementation of this algorithm is given by Gilbert et al. [10]. Their implementation includes a heuristic for computing approximate center points and a method for improving the balance of a bisection. Moreover, they do not apply the above algorithm to all nodes of the graph but to a randomly selected subset of the nodes. Also, to obtain a good partition, randomly selected hyperplanes are tried.

Bokhari et al. [11] describe sequential and parallel algorithms for parametric binary dissection, a generalization of recursive coordinate bisection that can consider the cut size. More specifically, suppose that we want to bisect the graph using a cut plane orthogonal to the x-axis. The position of the plane is chosen such that $\max(n_l + \lambda e_l, n_r + \lambda e_r)$ is minimized. Here, $n_l$ and $n_r$ are the numbers of nodes in the subset lying to the left and right of the plane, $e_l$ and $e_r$ are the numbers of the edges with exactly one end node in the left and right subset, and $\lambda$ is the parameter.

2) Coordinate-free methods
In some applications, the graphs are not embedded in space, and geometric algorithms cannot be used. Even when the graph is embedded in space, geometric methods tend to give relatively high cut sizes. In this section, algorithms that only consider the combinatorial structure of the graph are described.

The recursive graph bisection [12] method begins by finding a pseudo peripheral node in the graph, i.e. one of a pair of nodes that are approximately at the greatest graph distance from each other in the graph. (The graph distance between two nodes is the number of edges on the shortest path between the nodes.) Using a breadth-first search starting in the selected node the graph distance from this node to every other node is determined. Finally, the nodes are sorted with respect to these distances, and the sorted set is divided into two equal-sized sets. The so-called greedy method uses breadth-first search [12] (starting in a pseudo peripheral node) to find the parts one after another.

The recursive spectral bisection (RSB) [13] method uses the eigenvector corresponding to the second-lowest eigenvalue of the Laplacian matrix of the graph. (We define the Laplacian matrix $L$ of a graph as $L = D - A$, where $D$ is the diagonal matrix expressing node degree and $A$ is the adjacency matrix.) This eigenvector contains valuable information about the graph: the difference between the coordinates of the Fiedler vector provides information about the distance between the corresponding nodes. Thus, the RSB method bisects a graph by sorting its nodes with respect to their Fiedler coordinates and then dividing the sorted set into two halves. The Fiedler vector can be computed using a modified Lanczos algorithm [14]. RSB has been generalized to quadrisection and octasection and to consider node and edge weights [15]. The RSB method has been combined with the Kernighan–Lin method with satisfactory results [16].

The multilevel recursive spectral bisection (multilevel-RSB) [17] method uses a multilevel approach to speed up the computation of the Fiedler vector. This algorithm consists of three phases: coarsening, partitioning, and uncoarsening.

During the coarsening phase a sequence of graphs, $G^i = (V^i, E^i)$ is constructed from the original graph $G^0 = (V, E)$. More specifically, given a graph $G^0 = (V, E)$, an approximation $G^{i+1}$ is obtained by a maximal independent subset of $V^i$. An independent subset $V^{i+1}$ is maximal if no node can be added to the set. $N^{i+1}$ is set equal to $I^i$, and $E^{i+1}$ is constructed as follows.

With each node $v \in I^i$ is associated a domain $D_v$ that initially contains only $v$ itself. All edges in $E^i$ are unmarked. Then, if there is an unmarked edge $(u, v) \in E^i$ does as follows: If $u$ and $v$ belong to the same domain, mark $(u, v)$, and add $v$ and $(u, v)$ to that domain. If $u$ and $v$ are in different domains, say $D_x$ and $D_y$, then mark $(u, v)$, and add the edge $(x, y)$ to $E^{i+1}$. Finally, if neither $u$ nor $v$ belongs to a domain, then process the edge at a later stage.

At some point, we obtain a graph $G^m$ that is small enough for the Lanczos algorithm to compute the corresponding Fiedler vector (denoted $f^m$) in a small amount of time. To obtain $f^0$, the Fiedler vector of the initial graph, we reverse the process. Given the vector $f^{i+1}$, we obtain the vector $f^i$ by interpolation and improvement. The interpolation step is as follows. For each $v \in V^i$, if $v \in V^{i+1}$, then $f^i(v) = f^{i+1}(v)$; otherwise $f^i(v)$ is set equal to the average value of the components of $f^{i+1}$ corresponding to neighbors of $v$ in $V^i$. Next, the vector $f^i$ is improved. This is done by Rayleigh quotient iteration.

The multilevel-KL algorithm [17–21] is another example of how a multilevel approach can be used to obtain a fast algorithm. During the coarsening phase, the algorithm creates a sequence of increasingly coarser approximations of the initial graph. When a sufficiently coarse graph has been found, we enter the partitioning phase during which the coarsest graph either is bisected [19] or partitioned into $k$ parts [17, 20]. During the uncoarsening phase, this partition is propagated back through the hierarchy of the graph. A *KL*-type algorithm is invoked periodically to improve the partition. In the following, we give a more detailed description of each phase [19, 20].

During the coarsening phase, an approximation $G^{i+1}$ of a graph $G^i = (V^i, E^i)$ is obtained by first computing a maximal matching, $M^i$. Recall that a matching is a subset of $E^i$ such that no two edges in the subset share a node. A matching is maximal if no more edges can be added to the matching. Given a maximal matching $M^i$, $G^{i+1}$ is obtained by "collapsing" all matched nodes. That is, if $(u, v)$

$\in M^i$ then nodes $u$ and $v$ are replaced by a node $v'$ whose weight is the sum of the weights of $u$ and $v$. Moreover, the edges incident on $v'$ are the union of the edges incident on $v$ and $u$ minus the edge $(u, v)$. Unmatched nodes are copied over to $G^{i+1}$.

Maximal matching can be found in several ways. However, experiments indicate that so-called heavy edge matching gives the best results [19, 20]. It works as follows. Nodes are visited in random order. If a visited node $u$ is unmatched, we match $u$ with an unmatched neighbor v such that no edge between $u$ and an unmatched neighbor is heavier than the edge $(u, v)$.

During the partitioning phase, the coarsest graph $G^m$ either is bisected or portioned directly into $k$ parts [17, 19, 20]. In [17] several methods for bisecting the coarsest graph were evaluated, and the best results were obtained for a variant of the greedy algorithm. More specifically, starting in a randomly selected node they grow in part by adding fringe nodes. The fringe node whose addition to the part would result in the largest decrease in cut size is added. We have a multilevel bisection algorithm that needs to be applied recursively to obtain a $k$-way partition.

During the uncoarsening phase, the partition of $G^m$ is successively transformed into a partition of the original graph $G^0$. More specifically, for each node $v \in V^{i+1}$, let $P^{i+1}(v)$ be the index of the part to which $v$ belongs. Given $P^{i+1}$, $P^i$ is obtained as follows. First, an initial partition is constructed using projection: if $v' \in V^{i+1}$ corresponds to a matched pair $(u, v)$ of nodes in $V^i$, then $P^i(u) = P^i(v) = P^{i+1}(v')$; otherwise $P^i(v') = P^{i+1}(v')$. Next, this initial partition is improved using a variant of the Kernighan–Lin method [21]. We describe the so-called greedy refinement method. Experiments show that the greedy refinement method converges within a few iterations.

To enhance the efficiency of graph partitioning in parallel, we make each subgraph fit in a device memory and minimizing the communication requirements between subgraphs for parallel computing, the work of graph partition is explored by using deep learning, such as Rapid Neural Network Connector (RaNNC) [22] through three steps. First, the subgraphs are identified by using heuristic rules. Second, RaNNC groups the sub-optimal partitions into coarser-grained blocks. Last, the neural network searches the combinations of blocks to determine the final partitions.

### 2.3.2 Graph Database Performance

Computation efficiency is critical for power system analysis. In this section, we discuss the advantages of a graph database management system (GDBMS) and contrast its characteristics with a traditional relational database management system (RDBMS). We also provide numerical values of some performance metrics for database and core calculation operations typically found in power system applications.

The traditional RDBMS uses tables to store data. The RDBMS stores the structured records and their attributes in equal-length tables and maintains the database using SQL. To represent the relational interconnections, the linking attributes of records in the database are stored in different tables for creating the relations, and the relationship between the different records is established by querying the same attribute of the corresponding records in different tables. Therefore, adding or deleting records in the RDBMS requires updating all tables with the associated shared attributes. In addition, the performance benchmark study shows that, when compared with the performance of a GDBMS, RDBMS takes a much longer time to support attribute searching, optimal ordering, and depth-first (or breadth-first) search.

In graph computing, the relationship between nodes and edges is self-defined by the graph. Unlike relational databases, graph databases model the power system using graph-oriented data structures for semantic queries with a set of nodes and edges [23, 24]. Unstructured attributes are then stored in data structures defined by the graph's nodes and edges. For power system

**Table 2.1** Data structure modeling in RDBMS and GDBMS.

| Data structure | Array | Linked list | Tree | Graph |
| --- | --- | --- | --- | --- |
| RDBMS | One/multi-dimensional table | Value/index table | Value/multi-reference table | Adjacency/attribute table |
| GDBMS | Undirected graph | Directed graph | Directed graph | Graph |

analysis, the GDBMS is more in line with the requirements of power system computing for complex data modeling, querying, sorting, and traversal. Commonly used data structures for power system calculations include arrays, linked lists, trees, and graphs. For example, an array is often used for matrix operations; a linked list is used to represent the path set of generators and loads when taking advantage of fast forward and fast backward substitution (sparse vector methods) while solving matrix equations [25, 26]; an elimination tree is used for identifying parallel processes when parallelizing factorization [27, 28]; and a graph is used for topological processing [29]. The conceptual models of common data structures used by relational databases and graph databases are shown in Table 2.1.

There are different ways to represent a data structure by tables in a relational database. As shown in Table 2.1, relational databases are useful for storing tabular data that fit into a pre-defined schema of rows and columns. But modeling the highly connected data set into a relational database commonly results in multiple tables with complex table structures. In contrast to the relational database, the graph database is concise when modeling different data structures, from simple array data to the more complex structures which store graphs and trees. Hence, the graph database is very accommodating of data interconnections within a data set that fit the characteristics of power systems since the power system is naturally modeled as a graph consisting of nodes and branches. Nodes are physically connected through branches as edges. The unconstructed parameters of the bus, generator, load, and branch, such as active and reactive power of generator and load, resistance and reactance of lines and transformers, and so on are stored in the node or edge. The graph structure itself naturally represents the topology of the electric power grid.

To evaluate the performance of the RDBMS and the GDBMS on commonly used database operations for power system analysis, execution time metrics for a large-scale power system (MP10790 system) are taken to perform database searching, ordering, inserting, deleting, and traversal.

In power system applications, Minimum Degree (MD) Ordering (Tinney Scheme 2) [30] is used to optimize bus order for minimizing fill-ins during matrix factorization. The search operation was selected for execution time-metric evaluation because it is widely used for creating an internal list of components, controls, and/or measurements with a certain type of attribute. Searching is also widely used to find a specific study results visualization. Inserting and deleting execution time metrics are selected because they simulate adding or removing an element and removing elements for CA. Traversal is used for matrix factorization, forward/backward substitution, and topology processing.

The query return times tested on the MP10790 power systems in milliseconds for these commonly used operations by the relational database and the graph database are listed in Table 2.2 for comparison.

**Table 2.2** Performance test for RDBMS and GDBMS (ms).

| Case | Database | Search | Order | Insert | Delete | Traversal |
|------|----------|--------|-------|--------|--------|-----------|
| MP 10790 | RDBMS | 83.54 | 254.11 | 0.017 | 0.0096 | 769.17 |
| | GDBMS | 4.23 | 77.86 | 0.015 | 0.0078 | 49.64 |

**Table 2.3** Test environment.

| Hardware environment | CPU | 2 CPUs × 6 cores 2 threads at 2.10 GHz |
|---|---|---|
| | Memory | 64GB |
| Software environment | Operation system | CentOS 6.8 |
| | Relational database | Postgres |
| | Graph database | TigerGraph v0.8.1 |

In Table 2.2, the MP10790 system is a modification of the Polish 2383wp and 3012wp power systems [31, 32] which contain 10,790 nodes and 12,941 branches [33]. The tested RDBMS and GDBMS are Postgres [34] and TigerGraph [35], respectively. The test environment is shown in Table 2.3.

In the RDBMS, the MP10790 system is modeled using tables for storing various attributes of generators, loads, transmission lines, transformers, shunt capacitors/reactors, etc. In the GDBMS, the system is modeled as graphs with a set of vertices and edges. The searching operation queries all vertices and edges that match the given attribute value; the sorting operation performs the MD ordering for all nodes; the insert and delete operations randomly insert or delete a vertex or an edge; the traversal operation traverses all nodes in the system from a given node. The execution time reported in Table 2.2 is measured by averaging five independent tests. As shown in Table 2.2, the GDBMS outperforms the RDBMS in all database operations. Particularly, the execution time of the GDBMS on searching, ordering, and traversal is significantly less than that of the RDBMS.

The remarkable performance of the GDBMS results from its built-in parallel computing capability. The GDBMS allows us to take advantage of two types of parallelism: nodal parallelism and hierarchical parallelism. In graph database operation parlance, nodal parallel computing refers to the computation of quantities associated with each node, where each computation is independent; in contrast, hierarchical parallelism partitions nodes into different levels according to their computing dependency and then performs the computation in parallel on nodes at the same level. For example, nodal parallelism can be used when searching for a specific attribute value since no dependencies exist: GDBMS compares the desired value against the attribute value of all nodes and edges in parallel. In contrast, when dependencies exist and are characterized by precedence relationship depth level, the GDBMS processes all nodes using a Depth-First Search (DFS), exploiting hierarchical parallelism.

# References

1 Codd, E.F. (1970). A relational model of data for large shared data banks. *Communications of the ACM* 13 (6): 377–387.

2 ISO 9075 (1987). Information technology – database languages – SQL – part 1: framework (SQL/ framework). 1 June 1987.

3 Garcia-Molina, H., Ullman, J.D., and Widom, J. (2009). *Database Systems: The Complete Book.* Pearson Education Inc.

4 Vaidya, M. (2012). Parallel processing of cluster by MapReduce. *International Journal of Distributed and Parallel Systems* 3 (1): 167–179.

5 Fjällström, P.-O. (1998). Algorithms for graph partitioning: a survey. *Linköping Electronic Articles in Computer and Information Science* 3: 34.

6 Kernighan, B.W. and Lin, S. (1970). An efficient heuristic procedure for partitioning graphs. *Bell Labs Technical Journal* 49 (2): 291–307.

7 Fiduccia, C.M. and Mattheyses, R.M. (1982). A linear-time heuristic for improving network partitions. *19th Design Automation Conference*, Las Vegas, NV, USA, pp. 175–181. https://doi.org/10.1109/ DAC.1982.1585498.

8 Farhat, C. and Lesoinne, M. (1993). Automatic partitioning of unstructured meshes for the parallel solution of problems in computational mechanics. *International Journal for Numerical Methods in Engineering* 36 (5): 745–764.

9 Miller, G.L., Teng, S.-H., Thurston, W., and Vavasis, S.A. (1993). Automatic mesh partitioning. *Graph Theory and Sparse Matrix Computation* 56: 57–84.

10 Gilbert, J.R., Miller, G.L., and Teng, S.-H. Geometric mesh partitioning: implementation and experiments. *Proceedings of 9th International Parallel Processing Symposium*, Santa Barbara, CA, USA, pp. 418–427. https://doi.org/10.1109/IPPS.1995.395965.

11 Bokhari, S.H., Crockett, T.W., and Nicol, D.M. (1993). Parametric binary dissection. *ICASE Report No. 93–39, NASA Contractor Report 191496*, July 1993.

12 Ciarlet Jr, P. and Lamour, F. (1994). Recursive partitioning methods and greedy partitioning methods: a comparison on finite element graphs, *UCLA CAM Report*, pp. 94–99.

13 Pothen, A., Simon, H.D., and Liou, K.-P. (1990). Partitioning sparse matrices with eigenvectors of graphs. *SIAM Journal on Matrix Analysis and Applications* 11 (3): 430–452.

14 Lanczos, C. (1950). An iteration method for the solution of the eigenvalue problem of linear differential and integral operators. *Journal of Research of the National Bureau of Standards* 45: 255–282.

15 Hendrickson, B. and Leland, R. (1995). An improved spectral graph partitioning algorithm for mapping parallel computations. *SIAM Journal on Scientific Computing* 16 (2): 452–469. https://doi. org/10.1137/0916028.

16 Leland, R. and Hendrickson, B. (1994). An empirical study of static load balancing algorithms. *Proceedings of IEEE Scalable High Performance Computing Conference*, Knoxville, TN, USA, May 1994, pp. 682–685.

17 Barnard, S.T. and Simon, H.D. (1994). Fast multilevel implementation of recursive spectral bisection for partitioning unstructured problems. *Concurrency and Computation Practice and Experience* 6 (2): 101–117.

18 Hendrickson, B. and Leland, R. (1995). A multilevel algorithm for partitioning graphs. *Supercomputing '95: Proceedings of the 1995 ACM/IEEE Conference on Supercomputing (CDROM)*, San Diego, CA, USA, pp. 28–es.

19 Bui, T.N. and Jones, C. (1993). A heuristic for reducing fill-in in sparse matrix factorization. *6th Society for Industrial and Applied Mathematics (SIAM) Conference on Parallel Processing for Scientific Computing*, Norfolk, VA, USA, 21–24 March 1993, pp. 445–452.

**20** Karypis, G. and Kumar, V. (1998). A fast and high quality multilevel scheme for partitioning irregular graphs. *SIAM Journal on Scientific Computing* 20 (1): 359–392.

**21** Karypis, G. and Kumar, V. (1996). Multilevel k-way partitioning scheme for irregular graphs. *Supercomputing '96: Proceedings of the 1996 ACM/IEEE Conference on Supercomputing*, Pittsburgh, PA, USA, pp. 35–35, https://doi.org/10.1109/SUPERC.1996.183537.

**22** Tanaka, M., Taura, K., Hanawa, T., and Torisawa, K. (2021). Automatic graph partitioning for very large-scale deep learning. *The 35th IEEE International Parallel and Distributed Processing Symposium (IPDPS 2021)*, Portland, OR, USA, May 2021.

**23** Malewicz, G., Austern, M.H., Bik, A.J. et al. (2009). Pregel: a system for large-scale graph processing. *Proceedings of the 21st Annual ACM Symposium on Parallelism in Algorithms and Architectures*, Calgary, Alberta, Canada, 11–13 August 2009, pp. 135–146.

**24** Lu, Y., Cheng, J., Yan, D., and Wu, H. (2014). Large-scale distributed graph computing systems: an experimental evaluation. *Proceedings of the VLDB Endowment* 8 (3): 281–292.

**25** Tinney, W.F., Brandwajn, V., and Chan, S.M. (1985). Sparse vector methods. *IEEE Transactions on Power Apparatus and Systems* PAS-104 (2): 295–301.

**26** Zhang, B. and Chen, S. (1998). *Advanced Power Grid Analysis*. Science Press.

**27** Schreiber, R. (1982). A new implementation of sparse Gaussian elimination. *ACM Transactions on Mathematical Software* 8 (3): 256–276.

**28** Kumar, P.S., Kumar, M.K., and Basu, A. (1992). A parallel algorithm for elimination tree computation and symbolic factorization. *Parallel Computing* 18 (8): 849–856.

**29** Pradeep, Y., Seshuraju, P., Khaparde, S.A., and Joshi, R.K. (2011). CIM-based connectivity model for bus-branch topology extraction and exchange. *IEEE Transactions on Smart Grid* 2 (2): 244–253.

**30** Tinney, W.F. and Walker, J.W. (1967). Direct solutions of sparse network equations by optimally ordered triangular factorization. *Proceedings of the IEEE* 55 (11): 1801–1809.

**31** Zimmerman, R., Murillo-Sanchez, C., and Thomas, R. (2011). MATPOWER: steady-state operations, planning, and analysis tools for power systems research and education. *IEEE Transactions on Power Systems* 99: 1–8.

**32** MatPower.org. (2019). Index of docs/ref/matpower 5.0. https://matpower.org/docs/ref/matpower5.0 (accessed 24 July 2019).

**33** Li, X., Li, F., Yuan, H. et al. (2017). GPU-based fast decoupled power flow with preconditioned iterative solver and inexact Newton method. *IEEE Transactions on Power Systems* 32 (4): 2695–2703.

**34** Postgresql.org (2019). PostgreSQL: the world's most advanced open source relational database. www.postgresql.org (accessed 10 May 2022).

**35** Tigergraph.com (2019). Introducing Tigergraph: the world's fastest and most scalable graph platform. www.tigergraph.com (accessed 10 May 2022).

# 3

# Graph Parallel Computing

The power system is evolving to be of ever-increasing size and operational complexity. The time has come when it is critical to improve the computation efficiency of power system analysis to accommodate modern power systems. The analytical processing time needs to be reduced from tens of seconds to subseconds [1]. To meet this requirement, technologies for the next generation of high-performance power system analysis are being studied [2, 3]. However, the computation capability for example, in EMS, to complete the core power system applications, such as state estimation, power flow, and contingency analysis at a SCADA sampling rate has not yet been achieved.

Among the various computational tools available to improve computation efficiency, parallel computing is a promising technology, providing abundant storage along with multiple processing paths [4–6]. In [4], parallel state estimation using a preconditioned conjugate gradient algorithm and an orthogonal decomposition-based algorithm is proposed. The proposed algorithm can solve state estimate problems faster using parallel computing, but it is infeasible at the condition of a large condition number of the gain matrix. Alves and Monticelli in [5] proposed an approach to solving contingency analysis by parallel computer and distributed network. The proposed approach is tested on an 810-bus network. To utilize the linearity of power system components, current balance equations subjected to Kirchhoff's Current Law are used to model power flow problems in [6]. The result shows a reduction in computational time by over 20% when using the current balance equations. However, present power system analysis approaches do not effectively harness the available parallel-processing capability because the traditional relational database and computation methods used to solve power system problems originally were not designed for parallel computing.

To accommodate parallel computing, both the database and calculation approaches for power system analysis need to be redesigned to fit into a parallel database management system and parallel computing.

## 3.1 Graph Parallel Computing Mechanism

In Graph Processing Engine, Bulk Synchronous Parallel Computing Model (BSP) is a built-in function to support graph parallel computing. The Bulk Synchronous Parallel Computing Model was proposed by Leslie Valiant of Harvard University and Bill McColl of Oxford University [7]. It is mainly used to deal with the synchronization problem of multiple node computation in a complex network. Graph Database uses the BSP model in its parallel computing mechanism. The most basic prototype framework of the BSP is shown in Figure 3.1.

The BSP model is mainly divided into calculation processor, local computing, communication, barrier synchronization, and super-step. Its characteristic is that the calculation and network communication do not need to be sorted or synchronized in advance, and they do not send-and-receive message transmissions bilaterally through pairing [8].

MapReduce semantics are cooperating with the Bulk Synchronous Parallel model in graph parallel computation, as shown in Figure 3.2.

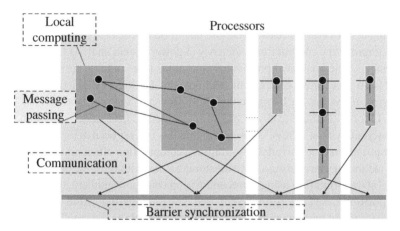

**Figure 3.1** A generic BSP model.

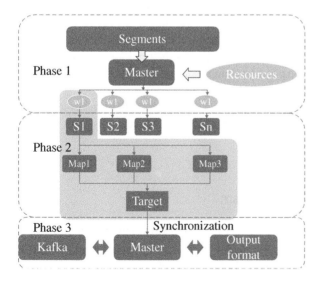

**Figure 3.2** Graph parallelism model.

The graph parallelism model is based on the graph database. For the processes, there are three phases: Map phase, BSP phase, and Reduce phase. The first and third phases are MapReduce mode. As well-known, the MapReduce model is composed of a Map procedure and a Reduce process. Map procedure, in phase one, performs filtering and sorting. Reduce procedure, in phase three, performs a result merge operation. Technically speaking, the "MapReduce System," also called "infrastructure" or "framework," orchestrates the processing by marshaling the distributed servers, running tasks in parallel, managing all communications and data transfers between the various parts of the system, and providing for redundancy and fault tolerance. This is a high-level parallel model in parallel computation.

The other parallel model is the Bulk Synchronous Parallel model which works in the servers created by phase one. The BSP model is a bridge between the software and hardware required for parallel computation. The BSP is defined as a combination of three attributes [8]:

- A number of processors
- A communication pattern between processors
- A super-step synchronizing processors

Computation consists of a sequence of super-steps. In each super-step, each processor is allocated to a task consisting of a combination of local computation steps, message transmissions, and implicit message arrivals from other processors. After each time period, a global check is made to determine whether the super-step has been completed by all the components. If it has, the machine proceeds to the next super-step. Otherwise, the next time period is allocated to the unfinished super-step.

A set of processors is commonly interpreted which may follow different threads of computation, with each processor being equipped with fast local memory and interconnected by a communication network. A BSP algorithm heavily relies on the third feature; a computation proceeds in a series of global super-steps, which consists of three components in Figure 3.3.

Concurrent local computation: every participating processor performs local computations, i.e. each process can only make use of values stored in the local memory. The computations occur asynchronously with all the others but may overlap with communication.

Communication: The processes exchange data between themselves to facilitate remote data storage capabilities.

Barrier synchronization: When a process reaches the barrier, it waits until all other processes have reached it.

Computation and communication do not require real-time sequencing. Communication is usually in the form of direct remote memory access calls rather than two-way send-and-receive messaging calls. Barrier synchronization ensures that all one-way communication ends correctly. Married with MapReduce, BSP built in the graph database management system facilitates graph nodal parallel computing and hierarchical parallel computing as fundamental functions to support power system parallel analysis. For power system parallel analysis, the focus is on reconstructing power system analysis mathematical models from the one in series to the one in parallel. The high-level challenges are decomposing the power system analysis problem into subproblems that have no computational dependence and could be performed in parallel.

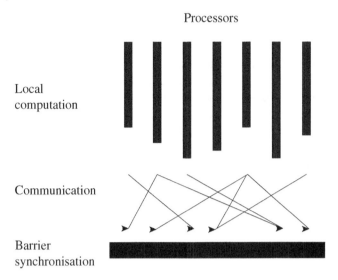

Processors

Local
computation

Communication

Barrier
synchronisation

**Figure 3.3** BSP parallel scheme.

According to computational dependence, there are two types of parallelism in graph computing: nodal parallelism and hierarchical parallelism. In graph database operation parlance, nodal parallel computing refers to the computation of quantities associated with each node, where each computation is independent; in contrast, hierarchical parallelism partitions nodes into different levels according to their computing dependency and then performs the computation in parallel on nodes at the same level. For example, nodal parallelism can be used when searching for a specific attribute value since no dependencies exist: Graph Processing Engine (GPE) compares the desired value against the attribute value of all nodes and edges in parallel. In contrast, when dependencies exist and are characterized by precedence relationship depth level, the GPE processes all nodes using a Depth First Search, exploiting hierarchical parallelism.

## 3.2   Graph Nodal Parallel Computing

Graph parallel computing exploits both nodal and hierarchical parallelism. In the power system analysis, the tasks of matrix formation, right-hand-side correction vector calculation, and branch power flow calculation can be nodally parallelized. The tasks of matrix factorization and forward/backward substitution can be hierarchically parallelized.

When the power system is modeled using a graph structure, calculations such as admittance matrix formation can be performed using nodal parallel computing once the sparsity-based symbolic part of the data structure has been completed.

The matrix (Figure 3.4a) and corresponding graph structures (Figure 3.4b) for example the 10-bus system are shown in Figure 3.4. In the graph structure, each numbered vertex represents a bus, and the edge between two vertices is a branch. As shown in the graph structure, the counterparts of the

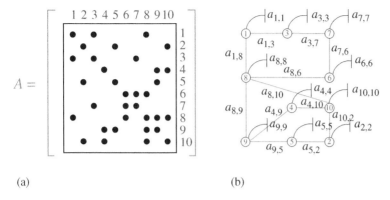

(a)                                              (b)

**Figure 3.4** Admittance matrix formation by graph. (a) admittance matrix; (b) admittance graph.

connections between nodes are nonzero off-diagonal elements in the coefficient matrix $A$. Zero (absent) elements in the matrix schematic representation in Figure 3.4a indicate that no direct connections between the nodes exist in the graph.

To form a row of the admittance matrix for any vertex, the off-diagonal nonzero elements of each row correspond only to the adjacent nodes. The graph-based approach searches for the neighboring vertex (or vertices) and the edges between them to form the admittance matrix numerical entries for all vertices simultaneously. Although the admittance matrix can obviously be formed in parallel, the traditional relational database-based power system analysis forms the admittance matrix by looping through nodes in series.

Figure 3.5 demonstrates the node-based parallel computing strategy to form the admittance matrix's numerical values using the MapReduce approach. In Phase 1, the Master dispatches the job to workers by partitioning the vertices into threads according to available computing

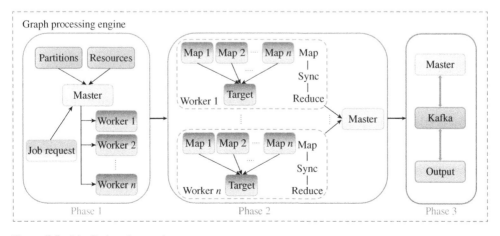

**Figure 3.5** MapReduce by graph.

resources. Ideally, vertices will be dispatched to each thread evenly. In Phase 2, each worker performs a MapReduce program, which is composed of a Map procedure and a Reduce procedure. The Map procedure calculates the off-diagonal elements of the admittance attributes of each of the node's incident edges simultaneously. The Reduce procedure calculates the diagonal element by adding the admittance attributes of the incident edges of the node of interest. In Phase 3, the Master publishes the calculated diagonal and nondiagonal elements to the graph as vertex or edge attributes.

Assume power system graph `powerSystemGraph` is built by the approach addressed in Chapter 2, the pseudo-code for admittance matrix formation using nodal parallelism is as follows:

---

**Algorithm 3.1  Form Admittance Matrix**

---

```
1: CREATE QUERY formMatrix( G(A) ) FOR
2: GRAPH powerSystemGraph
3:    SELECT NODE i AND j FROM V WHERE j∈i
4:        ACCUM
5:            aᵢᵢ+ = aᵢⱼ
```

---

where the ordered set $G(A) = G(V, E)$ is the graph model of the power system, $V$ is the vertex set, $E$ is the edge set, $E = \{(i,j) | i \in V, j \in V\}$. The algorithm complexity is $O(n)$, where $n$ is the total number of buses. The algorithm complexity is the same as the typical traditional method based on a relational database.

Other nodally parallelizable calculations in the power system analysis applications include the right-hand-side vector calculation, the convergence check, bad data detection, the branch power calculation, and the voltage and power flow violation check, etc. These calculations, performed on node attributes, are independent of each other and can therefore be performed simultaneously.

## 3.3  Graph Hierarchical Parallel Computing

In graph hierarchical parallel computing, computation is performed in parallel on nodes at the same depth level of the elimination tree. The nodes are partitioned into different levels in the elimination tree according to their calculation dependency. The calculation of quantities associated with the higher-indexed-level nodes depends on the calculation of quantities associated with the lower-indexed-level nodes. The calculations associated with nodes at the same level are independent and are performed in parallel.

In power system analysis, the hierarchically parallelizable tasks include matrix factorization and forward/backward substitution.

The core equations that need to be solved for power system state estimation, power flow, and contingency analysis, can be mathematically stated as a set of linear algebraic equations, which can be solved effectively using LU factorization (or $LDL^T$ for symmetric matrices) provided the matrices are positive definite. For hierarchical parallelism, the factorization algorithm is split into the following tasks: (i) symbolic factorization, (ii) elimination tree formation, (iii) node partitioning, (iv) numerical factorization, and (v) forward and backward substitution.

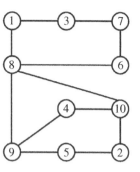

**Figure 3.6** Graph structure $G(A)$ for matrix $A$.

### 3.3.1 Symbolic Factorization

Taking the 10-bus system in Figure 3.6 as an example, matrix $A$ can be represented by the following graph, $G(A)$.

To illustrate an algorithm to determine fill-ins during the factorization, the process of matrix symbolic factorization is elaborated using the 10-bus system as an example. The general admittance matrix of the 10-bus system is given as:

$$A = \begin{bmatrix} a_{11} & & a_{13} & & & & & a_{18} & & \\ & a_{22} & & & a_{25} & & & & & a_{210} \\ a_{31} & & a_{33} & & & & a_{37} & (a_{38}) & & \\ & & & a_{44} & & & & & a_{49} & a_{410} \\ & a_{52} & & & a_{55} & & & & a_{59} & (a_{510}) \\ & & & & & a_{66} & a_{67} & a_{68} & & \\ & & a_{73} & & & a_{76} & a_{77} & (a_{78}) & & \\ a_{81} & & (a_{83}) & & & a_{86} & (a_{87}) & a_{88} & a_{89} & a_{810} \\ & & & a_{94} & a_{95} & & & a_{98} & a_{99} & (a_{910}) \\ & a_{102} & & a_{104} & (a_{105}) & & & a_{108} & (a_{109}) & a_{1010} \end{bmatrix} \tag{3.1}$$

Where, elements in the bracket will be fill-ins. In the Gaussian elimination process, when eliminating elements in column 1, in the first step, the elements in the first row are normalized.

$$a_{1j}^{(1)} = a_{1j}^{(0)} / a_{11}^{(0)} \quad (j = 1, 3, 8) \tag{3.2}$$

Then, the Gaussian elimination is performed to eliminate the nondiagonal elements in the first column by the following steps.

$$a_{3j}^{(1)} = a_{3j}^{(0)} - a_{31}^{(0)} \cdot a_{1j}^{(1)} \tag{3.3}$$

$$a_{8j}^{(1)} = a_{8j}^{(0)} - a_{81}^{(0)} \cdot a_{1j}^{(1)} \tag{3.4}$$

Element $a_{38}$ is filled in by $a_{38}^{(1)} = a_{38}^{(0)} - a_{31}^{(0)} \cdot a_{18}^{(1)}$ since $a_{38}^{(0)}$ is zero, $a_{31}^{(0)}$ and $a_{18}^{(1)}$ are nonzero making $a_{38}^{(1)}$ is a nonzero fill-in. And element $a_{83}$ is filled in by $a_{83}^{(1)} = a_{83}^{(0)} - a_{81}^{(0)} \cdot a_{13}^{(1)}$.

The elimination of the following rows (rows 2, 4, 5, 6, 7, 9, and 10) is skipped since the elements in the first column of these rows are zero.

$$a_{kj}^{(1)} = a_{kj}^{(0)} - a_{k1}^{(0)} \cdot a_{1j}^{(1)} \quad (k = 2, 4, 5, 6, 7, 9, 10) \tag{3.5}$$

where $a_{k1}^{(0)} = 0$ $(k = 2, 4, 5, 6, 7, 9, 10)$.

After the elimination of column 1, elements $a_{31}$ and $a_{81}$ in the first column are eliminated $\left(a_{31}^{(1)} = a_{81}^{(1)} = 0\right)$. And $a_{38}$ and $a_{83}$ are fill-ins. The symbolic factorization process is picturized in Figure 3.7.

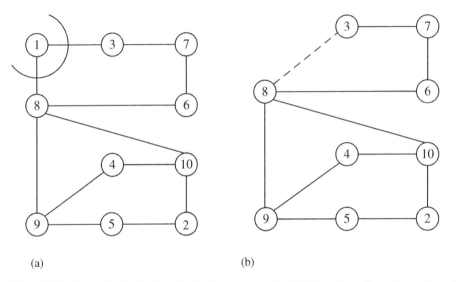

(a)                           (b)

**Figure 3.7** Column 1 elimination. (a) eliminate column 1; (b) fill-ins after eliminating column 1.

To eliminate elements in column 2, the elements in the second row are normalized first.

$$a_{2j}^{(2)} = \frac{a_{2j}^{(1)}}{a_{22}^{(1)}} \quad (j = 2, 5, 10) \tag{3.6}$$

Then, the Gaussian elimination is performed to eliminate the nondiagonal elements in the second column by the following steps.

$$a_{5j}^{(2)} = a_{5j}^{(1)} - a_{52}^{(1)} \cdot a_{2j}^{(2)} \tag{3.7}$$

$$a_{10,j}^{(2)} = a_{10,j}^{(1)} - a_{10,2}^{(1)} \cdot a_{2j}^{(2)} \tag{3.8}$$

Element $a_{5,10}$ is filled in by $a_{5,10}^{(2)} = a_{5,10}^{(1)} - a_{52}^{(1)} \cdot a_{2,10}^{(2)}$ since $a_{5,10}^{(1)}$ is zero, $a_{52}^{(1)}$ and $a_{2,10}^{(2)}$ are nonzero making $a_{5,10}^{(2)}$ is a nonzero fill-in. And element $a_{10,5}$ is filled in by $a_{10,5}^{(2)} = a_{10,5}^{(1)} - a_{10,2}^{(1)} \cdot a_{25}^{(2)}$.

The elimination of the following rows (rows 3, 4, 6, 7, 8, and 9) is skipped since the elements at the second column of these rows are zero after the first column elimination.

$$a_{kj}^{(2)} = a_{kj}^{(1)} - a_{k2}^{(1)} \cdot a_{2j}^{(2)} \quad (k = 3, 4, 6, 7, 8, 9) \tag{3.9}$$

where $a_{k2}^{(1)} = 0$ ($k = 3, 4, 6, 7, 8, 9$).

After the elimination of column 2, elements $a_{5,2}$ and $a_{10,2}$ on the second column are eliminated $\left(a_{5,2}^{(2)} = a_{10,2}^{(2)} = 0\right)$. And $a_{5,10}$ and $a_{10,5}$ are fill-ins. The symbolic factorization process is picturized in Figure 3.8.

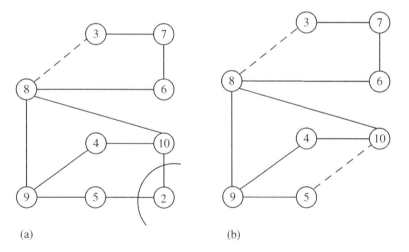

(a)  (b)

**Figure 3.8** Column 2 elimination. (a) eliminate column 2; (b) fill-ins after eliminating column 2.

To eliminate elements in column 3, the elements in the third row are normalized first.

$$a_{3j}^{(3)} = \frac{a_{3j}^{(2)}}{a_{33}^{(2)}} \quad (j = 3, 7, 8) \tag{3.10}$$

Then, the Gaussian elimination is performed to eliminate the nondiagonal elements in the third column by the following steps.

$$a_{7j}^{(3)} = a_{7j}^{(2)} - a_{73}^{(2)} \cdot a_{3j}^{(3)} \tag{3.11}$$

$$a_{8j}^{(3)} = a_{8j}^{(2)} - a_{83}^{(2)} \cdot a_{3j}^{(3)} \tag{3.12}$$

Element $a_{78}$ is filled in by $a_{78}^{(3)} = a_{78}^{(2)} - a_{73}^{(2)} \cdot a_{38}^{(3)}$ since $a_{78}^{(2)}$ is zero, $a_{73}^{(2)}$ and $a_{38}^{(3)}$ are nonzero making $a_{78}^{(3)}$ is a nonzero fill-in. And element $a_{87}$ is filled in by $a_{87}^{(3)} = a_{87}^{(2)} - a_{83}^{(2)} \cdot a_{37}^{(3)}$.

The elimination for the following rows (rows 4, 5, 6, 9, and 10) is skipped since the elements at the third column of these rows are zero after the first and second column elimination.

$$a_{kj}^{(3)} = a_{kj}^{(2)} - a_{k3}^{(2)} \cdot a_{3j}^{(3)} \quad (k = 4, 5, 6, 9, 10) \tag{3.13}$$

where $a_{k3}^{(2)} = 0$ $(k = 4, 5, 6, 9, 10)$.

After the elimination of column 3, elements $a_{73}$ and $a_{83}$ on the third column are eliminated $\left(a_{73}^{(3)} = a_{83}^{(3)} = 0\right)$. And $a_{78}$ and $a_{87}$ are fill-ins. The symbolic factorization process is picturized in Figure 3.9.

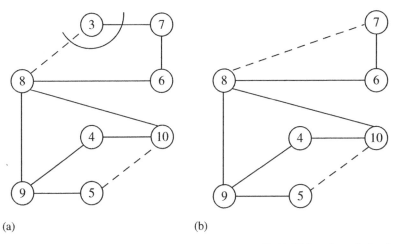

(a)  (b)

**Figure 3.9** Column 3 elimination. (a) eliminate column 3; (b) fill-ins after eliminating column 3.

The process is kept going till all columns are eliminated, and the fill-ins are determined. In the graph model, when eliminating a given node, the adjacent edges are removed, and add edges between the neighboring nodes are as fill-ins if there is no existing edge in between.

For example, when eliminating node 1, the edges $e_{13}$ and $e_{18}$ adjacent to node 1 are eliminated. Since there is no edge between the two neighbors (node 3 and node 8) of node 1, a new edge $e_{38}$ between the two neighbors is added.

Symbolic factorization is a dynamic process with dependency on previous node elimination. For example, when eliminating node 3, node 8 is a neighbor of node 3 in the updated elimination graph because of the add-in edge $e_{38}$ from the node 1 elimination. Note that node 8 was not a neighbor of node 3 in the original admittance graph. In the symbolic factorization algorithm, we must wait for node 1 elimination completed, then perform node 3 elimination.

Using the following parallel approach, the matrix symbolic factorization to determine filled graph structure $G^+(A)$ during the factorization is implemented:

---

**Algorithm 3.2  Symbolic Factorization in Graph**

---

```
1:  CREATE QUERY formFilledGraph(G(A), G+(A)) FOR
2:  GRAPH powerSystemGraph
3:      G+(A) = G(A)
4:      FOR NODE i IN V OF G(V,E)
5:         SELECT j AND k FROM V WHERE j∈i AND k∈i
6:            ACCUM
7:               DELETE E(i,j)
8:               DELETE E(i,k)
9:               INSERT INTO E+ (j, k)       //fill-in
```

---

To form the filled graph with fill-ins as added new edges, initially, the filled graph $G^+(A)$ is set as admittance graph $G(A)$ in line 3. The query loops through the vertices in the admittance graph $G(A)$ in line 4 in the numerical order in the sequence. For each vertex, neighboring vertices connecting by adjacent edges are selected (for example, $j$ and $k$. The vertex may have a single neighboring vertex or multiple neighboring vertices). The adjacent edges are removed from the admittance graph $G(A)$ but they are kept in the filled graph $G^+(A)$. The new edge between the neighboring vertices is inserted into the filled graph $G^+(A)$ if the inserted edge was not exiting in the admittance graph $G(A)$. The DELETE and INSERT statements are in the block of ACCUM indicating they are performed in parallel in the graph query.

The filled graph structure $G^+(A)$ of matrix $A$ with fill-ins after symbolic factorization is shown in Figure 3.10. The fill-in edges in the graph are fictitious/equivalent graph branches represented by fill-in elements in the $A$ matrix.

### 3.3.2  Elimination Tree

In the solution of matrix equations, matrix factorization, and forward/backward substitution are performed by column for each row. The calculation of a row is dependent only on the columns at the same level in the elimination tree. The elimination tree provides information on column dependencies in the factorization. The column dependencies are applied to the matrix factorization and forward/backward substitution using hierarchical parallel computing.

Taking the 10-bus system as an example, to continue the elimination process leftover in Section 3.3.1, the steps to eliminate columns 4–10 are shown as follows:

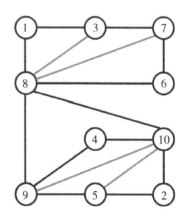

**Figure 3.10**  Filled graph structure $G^+(A)$.

*Eliminating elements in column 4*, the elements in row 4 are normalized first.

$$a_{4j}^{(4)} = \frac{a_{4j}^{(3)}}{a_{44}^{(3)}} \quad (j = 4, 9, 10) \tag{3.14}$$

Then, the Gaussian elimination is performed to eliminate the nondiagonal elements in column 4 by the following steps.

$$a_{9j}^{(4)} = a_{9j}^{(3)} - a_{94}^{(3)} \cdot a_{4j}^{(4)} \tag{3.15}$$

$$a_{10,j}^{(4)} = a_{10,j}^{(3)} - a_{10,4}^{(3)} \cdot a_{4j}^{(4)} \tag{3.16}$$

The elimination for the following rows (rows 5, 6, 7, and 8) is skipped since the elements in column 4 of these rows are zero after the previous column elimination.

$$a_{kj}^{(4)} = a_{kj}^{(3)} - a_{k4}^{(3)} \cdot a_{4j}^{(4)} \quad (k = 5, 6, 7, 8) \tag{3.17}$$

where $a_{k4}^{(3)} = 0$ ($k = 5, 6, 7, 8$).

*Eliminating elements in column 5*, the elements in row 5 are normalized first.

$$a_{5j}^{(5)} = \frac{a_{5j}^{(4)}}{a_{55}^{(4)}} \quad (j = 5, 9, 10) \tag{3.18}$$

Then, the Gaussian elimination is performed to eliminate the nondiagonal elements in column 5 by the following steps.

$$a_{9j}^{(5)} = a_{9j}^{(4)} - a_{95}^{(4)} \cdot a_{5j}^{(5)} \tag{3.19}$$

$$a_{10,j}^{(5)} = a_{10,j}^{(4)} - a_{10,5}^{(4)} \cdot a_{5j}^{(5)} \tag{3.20}$$

The elimination for the following rows (rows 6, 7, and 8) is skipped since the elements in column 5 of these rows are zero after the previous column elimination.

$$a_{kj}^{(5)} = a_{kj}^{(4)} - a_{k5}^{(4)} \cdot a_{5j}^{(5)} \quad (k = 6, 7, 8) \tag{3.21}$$

where $a_{k5}^{(4)} = 0$ ($k = 6, 7, 8$).

*Eliminating elements in column 6*, the elements in row 6 are normalized first.

$$a_{6j}^{(6)} = \frac{a_{6j}^{(5)}}{a_{66}^{(5)}} \quad (j = 6, 7, 8) \tag{3.22}$$

Then, the Gaussian elimination is performed to eliminate the nondiagonal elements in column 6 by the following steps.

$$a_{7j}^{(6)} = a_{7j}^{(5)} - a_{76}^{(5)} \cdot a_{6j}^{(6)} \tag{3.23}$$

$$a_{8j}^{(6)} = a_{8j}^{(5)} - a_{86}^{(5)} \cdot a_{6j}^{(6)} \tag{3.24}$$

The elimination for the following rows (rows 9 and 10) is skipped since the elements in column 6 of these rows are zero after the previous column elimination.

$$a_{kj}^{(6)} = a_{kj}^{(5)} - a_{k6}^{(5)} \cdot a_{6j}^{(6)} \quad (k = 9, 10) \tag{3.25}$$

where $a_{k6}^{(5)} = 0$ ($k = 9, 10$).

*Eliminating elements in column 7*, the elements in row 7 are normalized first.

$$a_{7j}^{(7)} = \frac{a_{7j}^{(6)}}{a_{77}^{(6)}} \quad (j = 7, 8) \tag{3.26}$$

Then, the Gaussian elimination is performed to eliminate the nondiagonal elements in column 7 by the following steps.

$$a_{8j}^{(7)} = a_{8j}^{(6)} - a_{87}^{(6)} \cdot a_{7j}^{(7)} \tag{3.27}$$

The elimination for the following rows (rows 9 and 10) is skipped since the elements in column 7 of these rows are zero after the previous column elimination.

$$a_{kj}^{(7)} = a_{kj}^{(6)} - a_{k7}^{(6)} \cdot a_{7j}^{(7)} \quad (k = 9, 10) \tag{3.28}$$

where $a_{k7}^{(6)} = 0$ ($k = 9, 10$).

*Eliminating elements in column 8*, the elements in row 8 are normalized first.

$$a_{8j}^{(8)} = \frac{a_{8j}^{(7)}}{a_{88}^{(7)}} \quad (j = 8, 9, 10) \tag{3.29}$$

Then, the Gaussian elimination is performed to eliminate the nondiagonal elements in column 8 by the following steps.

$$a_{9j}^{(8)} = a_{9j}^{(7)} - a_{98}^{(7)} \cdot a_{8j}^{(8)} \tag{3.30}$$

$$a_{10,j}^{(8)} = a_{10,j}^{(7)} - a_{10,8}^{(7)} \cdot a_{8j}^{(8)} \tag{3.31}$$

No row is skipped since the elements in column 8 of these rows are nonzero after the previous column elimination.

*Eliminating elements in column 9*, the elements in row 9 are normalized first.

$$a_{9j}^{(9)} = \frac{a_{9j}^{(8)}}{a_{99}^{(8)}} \quad (j = 9, 10) \tag{3.32}$$

Then, the Gaussian elimination is performed to eliminate the nondiagonal elements in column 9 by the following steps. And no row is skipped.

$$a_{10,j}^{(9)} = a_{10,j}^{(8)} - a_{10,9}^{(8)} \cdot a_{9j}^{(9)} \tag{3.33}$$

*Eliminating elements in column 10*, only normalization is performed for the last column.

$$a_{10,j}^{(10)} = \frac{a_{10,j}^{(9)}}{a_{10,10}^{(9)}} \quad (j = 10) \tag{3.34}$$

Investigating the Gaussian eliminate process for the 10-bus system, rows are skipped since the elements in the row of the working column are zero. The skipped row in the elimination is shown in Table 3.1.

Based on the elimination skipped row table, the dependent rows are the rows involved in the elimination process as shown in Table 3.2.

According to the elimination dependence information in Table 3.2, an elimination tree can be formed which provides the necessary and sufficient information on column dependencies in the Gaussian elimination.

**Table 3.1** Elimination skipped row table.

| Elimination column | Skipped rows | | | | | | | |
|---|---|---|---|---|---|---|---|---|
| 1 | 2 | | 4 | 5 | 6 | 7 | | 9 | 10 |
| 2 | | 3 | 4 | | 6 | 7 | 8 | 9 |
| 3 | | | 4 | 5 | 6 | | | 9 | 10 |
| 4 | | | | 5 | 6 | 7 | 8 | |
| 5 | | | | | 6 | 7 | 8 | |
| 6 | | | | | | | | 9 | 10 |
| 7 | | | | | | | | 9 | 10 |

**Table 3.2** Elimination dependence table.

| Elimination column number | Dependent row(s) | |
|---|---|---|
| 1 | 3 | 8 |
| 2 | 5 | 10 |
| 3 | 7 | 8 |
| 4 | 9 | 10 |
| 5 | 9 | 10 |
| 6 | 7 | 8 |
| 7 | 8 | |
| 8 | 9 | 10 |
| 9 | 10 | |
| 10 | | |

The pseudo-code to form the elimination tree of the filled graph structure, $G^+(A)$, in parallel using graph computing is shown immediately below and the formed elimination tree of the 10-bus system is shown in Figure 3.11, where *i*. parent is an attribute of vertex storing the parent of node *i*. MIN($j > i$) is a function to find the lowest-indexed node in the node set ($j \in i$). Steps 5 and 6 under ACCUM clause are performed by nodal parallelism.

---

**Algorithm 3.3  Create Elimination Tree**

---

```
1:    CREATE QUERY formEliminationTree(G⁺(A)) FOR
2:    GRAPH filledGraph
3:      SELECT i IN V FROM G⁺(A)
4:          ACCUM
5:              SELECT j FROM V WHERE j∈i
6:                  i.parent = MIN(j>i)
```

---

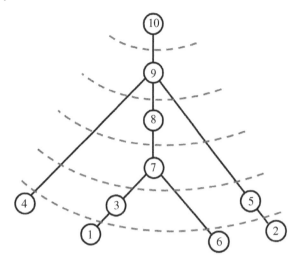

**Figure 3.11** Elimination tree *T(A)*.

### 3.3.3 Node Partition

In Figure 3.11, node layers are separated by the dotted lines and grouped in Table 3.3. Matrix factorization and forward/backward substitution are independent of the nodes at the same level/partition and dependent on the nodes at the different levels. In the factorization and forward substitution process, calculations start with the lowest-indexed level (leaf) nodes, i.e. the calculations associated with nodes 1, 2, 4, and 6 are at the same level and are performed first and in parallel. After the calculation of the lowest level is completed, nodes 3 and 5 in the upper level are processed. The calculations proceed thusly until the nodes of all layers are processed.

**Table 3.3** Node partition for hierarchical parallel.

| Hierarchical partition level | Nodes |
| --- | --- |
| 1 | 1, 2, 4, 6 |
| 2 | 3, 5 |
| 3 | 7 |
| 4 | 8 |
| 5 | 9 |
| 6 | 10 |

The pseudo-code for the node partition recursive algorithm is as follows:

---

**Algorithm 3.4   Node Hierarchical Partition**

---

```
1:   LEVEL = 1
2:   CREATE QUERY formHierarchicalPartition(T(A), LEVEL) FOR
3:   GRAPH EliminationTree
4:     WHILE T(A) != EMPTY
5:       formHierarchicalPartition(T(A), LEVEL)
6:       SELECT NODE i FROM V OF T(A) WHERE LEAVE(i) = = TRUE
7:          ACCUM
8:             i.hLevel = LEVEL
9:             DELETE i //remove node i from T(A)
10:    LEVEL ++
```

---

The query from the Hierarchical Partition is called recursively till all nodes are processed. The graph partition starts from level 1 (LEVEL = 1). The leave nodes are selected by line 6, where LEAVE($i$) is a function to check if node $i$ is a leave node. All leave nodes are deleted from the elimination tree $T(A)$ and their attribute hLevel is assigned to the current level LEVEL, where, hLevel is the node's attribute of hierarchical level.

### 3.3.4   Numerical Factorization

The matrix $A$ corresponding to the linear algebraic equations of interest can be factorized as $A = LDU$, where $L$ is the lower triangular matrix and $U$ is the upper triangular matrix.

Using the node partitioning information in the elimination tree, the pseudo-code for the numerical factorization hierarchical parallel computing approach is:

---

**Algorithm 3.5   Factorize Matrix A**

---

```
1:   CREATE QUERY FactorizeMatrixA(G(A), T(A))
2:     SELECT LEVEL l IN T(A)
3:       SELECT NODE i FROM V IN G(A) WHERE i.hLevel = =l
4:          ACCUM
5:             SELECT j FROM V IN G(A) WHERE j∈i
```

$$6: \qquad e_{ij} \cdot a^{(n)} = \frac{e_{ij} \cdot a^{(n-1)}}{v_i \cdot a^{(n-1)}}$$

$$7: \qquad e_{kj} \cdot a^{(n)} = e_{kj} \cdot a^{(n-1)} - e_{ki} \cdot a^{(n-1)} \cdot e_{ij} \cdot a^{(n)}$$

```
8:        POST-ACCUM
9:           SELECT NODE i FROM G(A)
10:             ACCUM
11:                l_ii = v_i·a
11:                SELECT NODE j FROM G(A)
12:                WHERE j∈i AND j<i
13:                   l_ij = e_ij·a
14:                   u_ij = e_ij·a
```

---

where $v_i \cdot a$ is the attribute of the vertex $v_i$ in the admittance graph $G(A)$ representing the diagonal element $a_{ii}$ and $e_{ij} \cdot a$ is the attribute of the edge $e_{ij}$ in the admittance graph $G(A)$ representing the nondiagonal element $a_{ij}$ ($i \neq j$) of admittance matrix $A$. Step 3 selects all nodes within the same level, and steps 5–7 in the block of ACCUM perform the factorization for the nodes selected in step 3 in parallel.

After the matrix is formed by the query factorize matrix $A$, in the POST-ACCUM clause, the elements are copied to matrix $L$ and $U$ under the parallel ACCUM clause.

### 3.3.5 Forward and Backward Substitution

Using the factorized matrix and node partitioning obtained from the elimination tree, the following forward/backward substitution process for solving the equation $Ax = b$ can be performed using hierarchical parallelism.

$$Ax = LDUx = b \tag{3.35}$$

$$Lz = b \tag{3.36}$$

$$Dy = z \tag{3.37}$$

$$Ux = y \tag{3.38}$$

The pseudo-code of the hierarchical parallel approach for the forward substitution is shown as follows.

---

**Algorithm 3.6   Forward Substitution**

---

```
1: CREATE QUERY ForwardSubstitution(L, G(A), T(A))
2:    SELECT LEVEL l IN T(A)
3:      SELECT NODE i FROM G(A) WHERE i.hLevel = = l
4:        ACCUM
5:          SELECT j FROM V WHERE j∈i
6:            z_i = b_i - l_ij · z_i
```

---

Similarly, the backward substitution hierarchical parallel computing method is as follows:

---

**Algorithm 3.7   Backward Substitution**

---

```
1: CREATE QUERY BackwardSubstitution (U, G(A), T(A))
2:    SELECT LEVEL l IN T(A)
3:      SELECT NODE i FROM G(A) WHERE i.hLevel = = l
4:        ACCUM
5:          SELECT j FROM V WHERE j∈i
6:            x_i = y_i - u_ij · x_i
```

---

During forward substitution, attribute $z_i$ is updated with edge attribute $l_{ij}$. Based on the node partitioning obtained from the elimination tree, the update of $z_i$ only depends on the nodes from lower levels. Nodes within the same partition level can be processed simultaneously. The forward substitution is therefore hierarchically parallelized. Similarly, using the upper triangular $U$ matrix and the node partitions, the backward substitution updates of $x_i$ for the nodes within the same level are hierarchically parallelized.

## References

**1** Electric Power Research Institute (2011). *Needed: A Grid Operating System to Facilitate Grid Transformation*. San Francisco, CA.

**2** Myrda, P.T. and Grijalva, S. (2012). The need for next generation grid energy management system. *2012 CIGRE US National Committee Grid of the Future Symposium*, Kansas City, Missouri (28–30 October 2012).

**3** U.S. Department of Energy (2015). An assessment of energy technologies and research opportunities. In: *Chapter 3. Enabling Modernization of the Electric Power System*. Washington, DC: https://www.energy.gov/sites/prod/files/2017/03/f34/qtr-2015-chapter3.pdf.

**4** Chen, Y., Jin, S., Rice, M., and Huang, Z. (2013). Parallel state estimation assessment with practical data. *2013 IEEE Power & Energy Society General Meeting*, Vancouver, BC, pp. 1–5. https://doi.org/10.1109/PESMG.2013.6672742.

**5** Alves, A.C.B. and Monticelli, A. (1995). Parallel and distributed solutions for contingency analysis in energy management systems. *38th Midwest Symposium on Circuits and Systems. Proceedings*, Rio de Janeiro, Brazil, vol. 1, pp. 449–452. https://doi.org/10.1109/MWSCAS.1995.504473.

**6** da Costa, V.M., Martins, N., and Pereira, J.L.R. (1999). Developments in the Newton Raphson power flow formulation based on current injections. *IEEE Transactions on Power Systems* 14 (4): 1320–1326.

**7** Leslie, G. (1990). Valiant, A bridging model for parallel computation. *Communications of the ACM* 33 (8).

**8** Valiant, L.G. (2011). A bridging model for multi-core computing. *Journal of Computer and System Sciences* 77 (1): 154–166.

# 4

# Large-Scale Algebraic Equations

Many scientific and engineering problems could be transformed to solve large-scale algebraic equations $f(x) = 0$. In a practical system, nonlinear equations cannot be explicitly solved by analytical solutions in most cases. In a numerical method, the solution of nonlinear equations is approached iteratively. The iterative methods start from an initial point and gradually approach the solution by simplified nonlinear form $x^{k+1} = g(x^k)$ or linearized form $f'(x^k)\Delta x = -f(x^k)$, $x^{k+1} = x^k + \Delta x$ at each iteration until the solutions are converged. The linearized update equations $f'(x^k)\Delta x = -f(x^k)$ can be rewritten into the standard linear equations $Ax = b$.

The linear equation $Ax = b$ is solvable if matrix $A$ is not singular. Direct methods are widely used to solve such systems modeled by linear equations. However, for large-scale linear equations, indirect methods are also developed to solve linear equations in which the iterative method uses successive approximations to obtain more accurate solutions to a linear system in each iteration until the solutions are converged. When the system does not change remarkably from one operating condition to another which happens in power systems when we solve state estimation, power flow, contingency analysis, transient simulation problems, indirect methods starting from the previous solution usually are converged quickly.

## 4.1 Iterative Methods of Solving Nonlinear Equations

An iterative method to solve nonlinear equations starts with an initial approximation $x^0$ to the solution $x_*$. The iterations generate a sequence of interim solutions $\{x^k, k = 1, 2, ..., n\}$ that converges to the solution $x_*$. There are two main iterative methods for solving nonlinear algebraic equations, (i) the Gauss–Seidel method, and (ii) the Newton–Raphson method.

### 4.1.1 Gauss–Seidel Method

The Gauss–Seidel method involves a process that converts the nonlinear equation $f(x) = 0$ into an equivalent form $x = g(x)$. If $x^0$ constitutes the starting point for the method, it will be seen that the solution $x_*$ for this equation, $x_* = g(x_*)$, can be reached by the numerical sequence: $x^{k+1} = g(x^k)$, $k = 1, 2, 3, ...$

*Graph Database and Graph Computing for Power System Analysis*, First Edition. Renchang Dai and Guangyi Liu.
© 2024 The Institute of Electrical and Electronics Engineers, Inc. Published 2024 by John Wiley & Sons, Inc.

Any nonlinear equations with $n$ variables can be expressed as (4.1).

$$f_1(x_1, x_2, ..., x_n) = 0$$
$$f_2(x_1, x_2, ..., x_n) = 0$$
$$\vdots$$
$$f_n(x_1, x_2, ..., x_n) = 0$$

(4.1)

In the Gauss–Seidel method, (4.1) is converted to the numerical sequence (4.2) to reach the solution.

$$x_1^{k+1} = g_1\left(x_1^k, x_2^k, ..., x_n^k\right)$$
$$x_2^{k+1} = g_2\left(x_1^k, x_2^k, ..., x_n^k\right)$$
$$\vdots$$
$$x_n^{k+1} = g_n\left(x_1^k, x_2^k, ..., x_n^k\right)$$

(4.2)

The values of the variables at the $k$th iteration are substituted into the right side of the iteration Eq. (4.2) to get the updated values of the variables until the following convergence conditions are satisfied for all variables:

$$\left| x_i^{k+1} - x_i^k \right| < \varepsilon$$

(4.3)

The calculations of the update equations are independent to update each variable. They can be straightforwardly calculated in parallel to achieve high computation efficiency. The iteration convergence can be improved by substituting the latest values of the updated variables to the equations afterward:

$$x_1^{k+1} = g_1\left(x_1^k, x_2^k, ..., x_n^k\right)$$
$$x_2^{k+1} = g_2\left(x_1^{k+1}, x_2^k, ..., x_n^k\right)$$
$$\vdots$$
$$x_n^{k+1} = g_n\left(x_1^{k+1}, x_2^{k+1}, ..., x_{n-1}^{k+1}, x_n^k\right)$$

(4.4)

The update Eq. (4.4) improve the convergence but lost the property of parallelism. Each update equation in (4.4) must be calculated in series.

### 4.1.2 PageRank Algorithm

PageRank algorithm is a derivative of the Gauss–Seidel method, which originally is an algorithm used by Google Search to rank websites in their search engine results. PageRank algorithm is a way of measuring the importance of website pages based on mutual hyperlinks between web pages.

A mutual hyperlink is considered one of the key elements for page ranking. The basic idea of the PageRank algorithm is based on the following assumptions: if a node is pointed by many other nodes, the node is defined as a key node; or if a node is pointed by another key node, the node is also considered as a key node.

PageRank estimates the importance of a web page by calculating the number and quality of links pointing to the page. The basic assumption is that more important pages are likely to receive more links from other pages.

PageRank is a link analysis algorithm that assigns a numerical weight to each element in a set of hyperlinked documents to calculate its importance. The algorithm can be applied to the use case of determining the value of an entity by mutual reference, in which its characteristics are consistent with the Gauss–Seidel algorithm.

The PageRank of a page denoted by PR($\cdot$) is defined recursively, that is, the PageRank value of a page is determined by the PageRank value of the page that links it, and at the same time, its PageRank value affects the PageRank values of other pages. This is essentially consistent with the update equations of the Gauss–Seidel method.

### 4.1.2.1 PageRank Algorithm Mechanism

Like any other iterative method, the PageRank algorithm also starts from an initial solution and requires multiple iterations until the iteration is converged to the final solution. The PageRank convergence condition is when the ranking of the PageRank values does not change.

In the PageRank algorithm process, the PageRank of each page PR($\cdot$) is set to the same value for all pages as an initialization, i.e. $\frac{1}{n}$ if the system has $n$ pages. The PageRank value is transferred equally from a given page to the target pages of its outbound links in each iteration. The PageRank determined by an outbound link is equal to the given page's own PageRank value divided by the number of its outbound links $L(\cdot)$. For example, in general, the PageRank value for any page $p_i$ at $k$th iteration can be expressed as:

$$PR(p_i; k+1) = \sum_{p_j \in B_{p_i}} \frac{PR\left(p_j; k\right)}{L\left(p_j\right)} \tag{4.5}$$

$B_{p_i}$ is the set of all pages linking to page $p_i$.
$L(p_j)$ is the number of outbound pages of page $p_j$.

The basic processing of the PageRank algorithm is as follows.

1) To assign an initial PageRank value to each given node.
2) To update the PageRank value for each node using Eq. (4.5).
3) To determine convergence. If it converges, output the result; otherwise, go back to step (2).

The PageRank of a page is calculated from the PageRank of other pages. Repetitive calculations can achieve the PageRank values of all the pages. If each page is given a random PageRank value (nonzero), then after iterative calculations, the PageRank value of these pages will converge to an unchanged number.

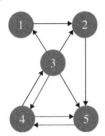

**Figure 4.1** Example of the basic principle of PageRank.

Assume a simple system contains five pages as shown in Figure 4.1.

PageRank is set to the same value (0.2) for all five pages as an initialization. The matrix of the outbound numbers is as follows:

$$
\begin{bmatrix}
0 & 0 & 4 & 0 & 0 \\
1 & 0 & 4 & 0 & 0 \\
0 & 0 & 0 & 2 & 0 \\
0 & 0 & 4 & 0 & 1 \\
0 & 1 & 4 & 2 & 0
\end{bmatrix}
\tag{4.6}
$$

Using the Eq. (4.5), in the first iteration, the $PR(1)$ is updated as follow:

$$
PR(1) = \sum \frac{1}{L(\cdot)} \cdot PR(0) = M \cdot PR(0)
\tag{4.7}
$$

where $M$ is the update matrix, $\mathcal{M}_{ij}$ is defined as follow:

$$
\mathcal{M}_{ij} =
\begin{cases}
\dfrac{1}{L(p_j)}, & i \text{ and } j \text{ are connected} \\
0, & \text{Otherwise}
\end{cases}
\tag{4.8}
$$

$$
PR(1) =
\begin{bmatrix}
0 & 0 & \tfrac{1}{4} & 0 & 0 \\
1 & 0 & \tfrac{1}{4} & 0 & 0 \\
0 & 0 & 0 & \tfrac{1}{2} & 0 \\
0 & 0 & \tfrac{1}{4} & 0 & 1 \\
0 & 1 & \tfrac{1}{4} & \tfrac{1}{2} & 0
\end{bmatrix}
\cdot
\begin{bmatrix}
\tfrac{1}{5} \\
\tfrac{1}{5} \\
\tfrac{1}{5} \\
\tfrac{1}{5} \\
\tfrac{1}{5}
\end{bmatrix}
=
\begin{bmatrix}
\tfrac{1}{20} & ⑤ \\
\tfrac{5}{20} & ② \\
\tfrac{2}{20} & ④ \\
\tfrac{5}{20} & ② \\
\tfrac{7}{20} & ①
\end{bmatrix}
\tag{4.9}
$$

The number in a circle at the most right is the rank of the page according to the calculated PageRank values.

In the second iteration, the $PR(2)$ is updated as follows:

$$
PR(2) =
\begin{bmatrix}
0 & 0 & \tfrac{1}{4} & 0 & 0 \\
1 & 0 & \tfrac{1}{4} & 0 & 0 \\
0 & 0 & 0 & \tfrac{1}{2} & 0 \\
0 & 0 & \tfrac{1}{4} & 0 & 1 \\
0 & 1 & \tfrac{1}{4} & \tfrac{1}{2} & 0
\end{bmatrix}
\cdot
\begin{bmatrix}
\tfrac{1}{20} \\
\tfrac{5}{20} \\
\tfrac{2}{20} \\
\tfrac{5}{20} \\
\tfrac{7}{20}
\end{bmatrix}
=
\begin{bmatrix}
\tfrac{1}{40} & ⑤ \\
\tfrac{3}{40} & ④ \\
\tfrac{5}{40} & ③ \\
\tfrac{15}{40} & ② \\
\tfrac{16}{40} & ①
\end{bmatrix}
\tag{4.10}
$$

In the third iteration, the *PR*(3) is updated as follow:

$$
PR(3) = \begin{bmatrix} 0 & 0 & \frac{1}{4} & 0 & 0 \\ 1 & 0 & \frac{1}{4} & 0 & 0 \\ 0 & 0 & 0 & \frac{1}{2} & 0 \\ 0 & 0 & \frac{1}{4} & 0 & 1 \\ 0 & 1 & \frac{1}{4} & \frac{1}{2} & 0 \end{bmatrix} \cdot \begin{bmatrix} \frac{1}{40} \\ \frac{3}{40} \\ \frac{5}{40} \\ \frac{15}{40} \\ \frac{16}{40} \end{bmatrix} = \begin{bmatrix} \frac{5}{160} & ⑤ \\ \frac{9}{160} & ④ \\ \frac{30}{160} & ③ \\ \frac{69}{160} & ① \\ \frac{47}{160} & ② \end{bmatrix}
\tag{4.11}
$$

In the fourth iteration, the *PR*(4) is updated as follows:

$$
PR(4) = \begin{bmatrix} 0 & 0 & \frac{1}{4} & 0 & 0 \\ 1 & 0 & \frac{1}{4} & 0 & 0 \\ 0 & 0 & 0 & \frac{1}{2} & 0 \\ 0 & 0 & \frac{1}{4} & 0 & 1 \\ 0 & 1 & \frac{1}{4} & \frac{1}{2} & 0 \end{bmatrix} \cdot \begin{bmatrix} \frac{5}{160} \\ \frac{9}{160} \\ \frac{30}{160} \\ \frac{69}{160} \\ \frac{47}{160} \end{bmatrix} = \begin{bmatrix} \frac{30}{640} & ⑤ \\ \frac{50}{640} & ④ \\ \frac{138}{640} & ③ \\ \frac{218}{640} & ① \\ \frac{204}{640} & ② \end{bmatrix}
\tag{4.12}
$$

The fourth iteration is converged at the ranking of pages in (5, 4, 3, 1, 2) which does not change the ranking from the third iteration. It is easy to prove the summation of PageRank values in a system keeps as one.

The PageRank theory holds that randomly clicking on links will eventually stop the surfer. To mimic the random surfer behavior, in which the probability that a surfer will continue is a damping factor *d*. PageRank also uses a model of a random surfer who gets bored after several clicks and switches to a random page.

A page's PageRank value reflects the chance that a surfer will randomly visit the page. A page with no outbound links to other pages is defined as a sink. When calculating the PageRank value, the sink is assumed to link to all other pages in the collection. Therefore, their PageRank values are conferred equally to all other pages.

Taking account of the damping factor, the PageRank value for any page $p_i$ is calculated as follows [1]

$$
PR(p_i) = \frac{1-d}{N} + d \cdot \left( \sum_{p_j \in B_{p_i}} \frac{PR(p_j)}{L(p_j)} \right)
\tag{4.13}
$$

*N* represents the total number of all pages.

For the complete set, PageRank can be represented as follows.

$$
PR = \begin{bmatrix} (1-d)/N \\ (1-d)/N \\ \vdots \\ (1-d)/N \end{bmatrix} + d \cdot M \cdot PR
\tag{4.14}
$$

wherein, if the page $p_i$ and page $p_j$ are not connected, the value of the $l(p_i, p_j)$ is zero; if there is a connection, $l(p_i, p_j)$ will be standardized to satisfy any $j$:

$$
\sum_{i=1}^{N} l\left(p_i, p_j\right) = 1
\tag{4.15}
$$

Equation (4.14) can be solved by the iterative method or algebraic method.

#### 4.1.2.2 Iterative Method

The iterative calculation is as follows [2, 3] which essentially is a derivative of the Jacobi method. The Jacobi method will be discussed in Section 4.3.1.1 in detail.

When $t = 0$, the original probability distribution is initialized as follows

$$
PR(p_i; 0) = \frac{1}{N}
\tag{4.16}
$$

In each subsequent iteration, the PageRank of page $p_i$ is calculated by the following equation,

$$
PR(p_i; t+1) = \frac{1-d}{N} + d \cdot \left( \sum_{p_j \in B_{p_i}} \frac{PR\left(p_j; t\right)}{L\left(p_j; t\right)} \right)
\tag{4.17}
$$

or in the matrix format as follows

$$
PR(t+1) = d \cdot M \cdot PR(t) + \begin{bmatrix} (1-d)/N \\ (1-d)/N \\ \vdots \\ (1-d)/N \end{bmatrix}
\tag{4.18}
$$

wherein, $PR_i(t) = PR(p_i; t)$.

The convergence condition of iterative calculation is $|PR(t+1) - PR(t)| < \epsilon$.

### 4.1.2.3  Algebraic Method

Equation (4.14) can be rewritten as,

$$(I - d \cdot \mathcal{M}) \cdot PR = \begin{bmatrix} (1-d)/N \\ (1-d)/N \\ \vdots \\ (1-d)/N \end{bmatrix} \tag{4.19}$$

Then PageRank value **PR** can be solved analytically as:

$$PR = (I - d \cdot \mathcal{M})^{-1} \cdot \begin{bmatrix} (1-d)/N \\ (1-d)/N \\ \vdots \\ (1-d)/N \end{bmatrix} \tag{4.20}$$

**Example 4.1**   Taking the 5-page system shown in Figure 4.1 as an example, we set the damping factor *d* as 0.8. Using the iterative method by the Eq. (4.18), in the first iteration, the *PR*(1) is updated as follows:

$$PR(1) = \begin{bmatrix} 0.2/5 \\ 0.2/5 \\ 0.2/5 \\ 0.2/5 \\ 0.2/5 \end{bmatrix} + 0.8 \begin{bmatrix} 0 & 0 & 1/4 & 0 & 0 \\ 1 & 0 & 1/4 & 0 & 0 \\ 0 & 0 & 0 & 1/2 & 0 \\ 0 & 0 & 1/4 & 0 & 1 \\ 0 & 1 & 1/4 & 1/2 & 0 \end{bmatrix} \cdot \begin{bmatrix} 1/5 \\ 1/5 \\ 1/5 \\ 1/5 \\ 1/5 \end{bmatrix} = \begin{bmatrix} 0.08 \\ 0.24 \\ 0.12 \\ 0.24 \\ 0.32 \end{bmatrix} \begin{matrix} ⑤ \\ ② \\ ④ \\ ② \\ ① \end{matrix} \tag{4.21}$$

In the second iteration, the *PR*(2) is updated as follows:

$$PR(2) = \begin{bmatrix} 0.2/5 \\ 0.2/5 \\ 0.2/5 \\ 0.2/5 \\ 0.2/5 \end{bmatrix} + 0.8 \begin{bmatrix} 0 & 0 & 1/4 & 0 & 0 \\ 1 & 0 & 1/4 & 0 & 0 \\ 0 & 0 & 0 & 1/2 & 0 \\ 0 & 0 & 1/4 & 0 & 1 \\ 0 & 1 & 1/4 & 1/2 & 0 \end{bmatrix} \cdot \begin{bmatrix} 0.08 \\ 0.24 \\ 0.12 \\ 0.24 \\ 0.32 \end{bmatrix} = \begin{bmatrix} 0.064 \\ 0.128 \\ 0.136 \\ 0.320 \\ 0.352 \end{bmatrix} \begin{matrix} ⑤ \\ ④ \\ ③ \\ ② \\ ① \end{matrix} \tag{4.22}$$

In the third iteration, the $PR(3)$ is updated as follows:

$$PR(3) = \begin{bmatrix} 0.2/5 \\ 0.2/5 \\ 0.2/5 \\ 0.2/5 \\ 0.2/5 \end{bmatrix} + 0.8 \begin{bmatrix} 0 & 0 & 1/4 & 0 & 0 \\ 1 & 0 & 1/4 & 0 & 0 \\ 0 & 0 & 0 & 1/2 & 0 \\ 0 & 0 & 1/4 & 0 & 1 \\ 0 & 1 & 1/4 & 1/2 & 0 \end{bmatrix} \cdot \begin{bmatrix} 0.064 \\ 0.128 \\ 0.136 \\ 0.320 \\ 0.352 \end{bmatrix} = \begin{bmatrix} 0.0672 \\ 0.1184 \\ 0.1680 \\ 0.3488 \\ 0.2976 \end{bmatrix} \begin{matrix} ⑤ \\ ④ \\ ③ \\ ① \\ ② \end{matrix} \tag{4.23}$$

In the fourth iteration, the $PR(4)$ is updated as follows:

$$PR(4) = \begin{bmatrix} 0.2/5 \\ 0.2/5 \\ 0.2/5 \\ 0.2/5 \\ 0.2/5 \end{bmatrix} + 0.8 \begin{bmatrix} 0 & 0 & 1/4 & 0 & 0 \\ 1 & 0 & 1/4 & 0 & 0 \\ 0 & 0 & 0 & 1/2 & 0 \\ 0 & 0 & 1/4 & 0 & 1 \\ 0 & 1 & 1/4 & 1/2 & 0 \end{bmatrix} \cdot \begin{bmatrix} 0.0672 \\ 0.1184 \\ 0.1680 \\ 0.3488 \\ 0.2976 \end{bmatrix} = \begin{bmatrix} 0.07360 \\ 0.12736 \\ 0.17952 \\ 0.31168 \\ 0.30784 \end{bmatrix} \begin{matrix} ⑤ \\ ④ \\ ③ \\ ① \\ ② \end{matrix} \tag{4.24}$$

To continue the iteration, in the tenth iteration, the $PR(10)$ is updated as follow:

$$PR(10) = \begin{bmatrix} 0.2/5 \\ 0.2/5 \\ 0.2/5 \\ 0.2/5 \\ 0.2/5 \end{bmatrix} + 0.8 \begin{bmatrix} 0 & 0 & 1/4 & 0 & 0 \\ 1 & 0 & 1/4 & 0 & 0 \\ 0 & 0 & 0 & 1/2 & 0 \\ 0 & 0 & 1/4 & 0 & 1 \\ 0 & 1 & 1/4 & 1/2 & 0 \end{bmatrix} \cdot \begin{bmatrix} 0.07372 \\ 0.13227 \\ 0.16741 \\ 0.31970 \\ 0.30690 \end{bmatrix} = \begin{bmatrix} 0.07348 \\ 0.13246 \\ 0.16788 \\ 0.31900 \\ 0.30718 \end{bmatrix} \begin{matrix} ⑤ \\ ④ \\ ③ \\ ① \\ ② \end{matrix} \tag{4.25}$$

Using the algebraic method by the Eq. (4.20), the $PR$ is calculated as follow:

$$PR = \begin{bmatrix} 1 & 0 & -0.2 & 0 & 0 \\ -0.8 & 1 & -0.2 & 0 & 0 \\ 0 & 0 & 1 & -0.4 & 0 \\ 0 & 0 & 0.2 & 1 & -0.8 \\ 0 & -0.8 & -0.2 & -0.4 & 1 \end{bmatrix}^{-1} \begin{bmatrix} 0.2/5 \\ 0.2/5 \\ 0.2/5 \\ 0.2/5 \\ 0.2/5 \end{bmatrix} \tag{4.26}$$

$$PR = \begin{bmatrix} 0.07354 \\ 0.13237 \\ 0.16770 \\ 0.31925 \\ 0.30714 \end{bmatrix} \begin{matrix} ⑤ \\ ④ \\ ③ \\ ① \\ ② \end{matrix} \tag{4.27}$$

The solution calculated by the iterative method is an excellent approximation to the solution computed by the algebraic method. The rankings by the two methods are consistent.

#### 4.1.2.4   Convergence Analysis

Convergence property is a major topic for iterative methods. In general, the converged solutions are in two categories: local converged solutions and global converged solutions. The local converged solution asserts the existence of a solution $x_*$ and a neighborhood $\mathcal{D}$ of $x_*$ where $\mathcal{D}$ : $\|x_* - x_0\| < \delta, \delta > 0$, such that from any starting point $x_0$ in the neighborhood $\mathcal{D}$, the iterations are guaranteed to converge to $x_*$. The global converged solution suggests any starting point $x_0 \in \mathbb{R}^n$, the iteration process will converge to the solution $x_*$.

In an iterative method, for the interim solution sequence $\{x^k, k = 1, 2, ..., n\}$ that converges to the solution $x_*$, the error is defined as $e^k = x^k - x_*$. If there is an exponent $p$ and constant $C > 0$, so that

$$\lim_{k \to \infty} \frac{\left|e^{k+1}\right|}{\left|e^k\right|^p} = C \tag{4.28}$$

We conclude the solution sequence is converged in the order of $p$.

PageRank algorithm is a derivative of the Gauss–Seidel method on linear systems. The global convergence of the PageRank algorithm is stringently guaranteed if $(\boldsymbol{I} - \boldsymbol{d} \cdot \mathcal{M})$ is diagonally column dominant.

We set $A = (\boldsymbol{I} - \boldsymbol{d} \cdot \mathcal{M}), b = [(1-d)/N \quad (1-d)/N \quad ... (1-d)/N]^T$, Eq. (4.20) can be rewritten as follow:

$$A \cdot PR = b \tag{4.29}$$

where we write $A = D - L$, with $D = diag(a_{11}, a_{22}, ..., a_{nn})$, then the iteration to calculate $PR$ is rewritten as follows:

$$D \cdot PR(k+1) = L \cdot PR(k) + b \tag{4.30}$$

$$PR(k+1) = D^{-1}L \cdot PR(k) + D^{-1}b \tag{4.31}$$

The solution $PR(*)$ satisfies the following equations:

$$PR(*) = D^{-1}L \cdot PR(*) + D^{-1}b \tag{4.32}$$

Subtracting Eq. (4.31) by (4.32) gives us

$$PR(k+1) - PR(*) = D^{-1}L \cdot PR(k) + D^{-1}L \cdot PR(*) \tag{4.33}$$

Thus,

$$\lim_{k \to \infty} \frac{|e^{k+1}|}{|e^k|^1} = \lim_{k \to \infty} \frac{|PR(k+1) - PR(*)|}{|PR(k) - PR(*)|^1} = D^{-1}L = C \tag{4.34}$$

For diagonally column dominant matrices, $D^{-1}L = C > 0$. In the PageRank algorithm, the damping factor $d$ is selected to make $(I - d \cdot \mathcal{M})$ a diagonally column dominant matrix which guarantees the iteration is converged in the first order.

**Example 4.2** To make a nondiagonally column dominant matrix $(I - d \cdot \mathcal{M})$, we set the damping factor $\boldsymbol{d}$ as two. Using the iterative method by Eq. (4.18), in the first iteration, the $PR(1)$ is updated as follows:

$$PR(1) = \begin{bmatrix} -\frac{1}{5} \\ -\frac{1}{5} \\ -\frac{1}{5} \\ -\frac{1}{5} \\ -\frac{1}{5} \end{bmatrix} + 2 \begin{bmatrix} 0 & 0 & \frac{1}{4} & 0 & 0 \\ 1 & 0 & \frac{1}{4} & 0 & 0 \\ 0 & 0 & 0 & \frac{1}{2} & 0 \\ 0 & 0 & \frac{1}{4} & 0 & 1 \\ 0 & 1 & \frac{1}{4} & \frac{1}{2} & 0 \end{bmatrix} \cdot \begin{bmatrix} \frac{1}{5} \\ \frac{1}{5} \\ \frac{1}{5} \\ \frac{1}{5} \\ \frac{1}{5} \end{bmatrix} = \begin{bmatrix} -0.1 \\ 0.3 \\ 0.0 \\ 0.3 \\ 0.5 \end{bmatrix} \tag{4.35}$$

In the second iteration, the $PR(2)$ is updated as follows:

$$PR(2) = \begin{bmatrix} -\frac{1}{5} \\ -\frac{1}{5} \\ -\frac{1}{5} \\ -\frac{1}{5} \\ -\frac{1}{5} \end{bmatrix} + 2 \begin{bmatrix} 0 & 0 & \frac{1}{4} & 0 & 0 \\ 1 & 0 & \frac{1}{4} & 0 & 0 \\ 0 & 0 & 0 & \frac{1}{2} & 0 \\ 0 & 0 & \frac{1}{4} & 0 & 1 \\ 0 & 1 & \frac{1}{4} & \frac{1}{2} & 0 \end{bmatrix} \cdot \begin{bmatrix} -0.1 \\ 0.3 \\ 0.0 \\ 0.3 \\ 0.5 \end{bmatrix} = \begin{bmatrix} -0.2 \\ -0.4 \\ 0.1 \\ 0.8 \\ 0.7 \end{bmatrix} \tag{4.36}$$

In the third iteration, the $PR(3)$ is updated as follows:

$$PR(3) = \begin{bmatrix} -\frac{1}{5} \\ -\frac{1}{5} \\ -\frac{1}{5} \\ -\frac{1}{5} \\ -\frac{1}{5} \end{bmatrix} + 2 \begin{bmatrix} 0 & 0 & \frac{1}{4} & 0 & 0 \\ 1 & 0 & \frac{1}{4} & 0 & 0 \\ 0 & 0 & 0 & \frac{1}{2} & 0 \\ 0 & 0 & \frac{1}{4} & 0 & 1 \\ 0 & 1 & \frac{1}{4} & \frac{1}{2} & 0 \end{bmatrix} \cdot \begin{bmatrix} -0.2 \\ -0.4 \\ 0.1 \\ 0.8 \\ 0.7 \end{bmatrix} = \begin{bmatrix} -0.15 \\ -0.55 \\ 0.60 \\ 1.25 \\ -0.15 \end{bmatrix} \qquad (4.37)$$

In the fourth iteration, the $PR(4)$ is updated as follows:

$$PR(4) = \begin{bmatrix} -\frac{1}{5} \\ -\frac{1}{5} \\ -\frac{1}{5} \\ -\frac{1}{5} \\ -\frac{1}{5} \end{bmatrix} + 2 \begin{bmatrix} 0 & 0 & \frac{1}{4} & 0 & 0 \\ 1 & 0 & \frac{1}{4} & 0 & 0 \\ 0 & 0 & 0 & \frac{1}{2} & 0 \\ 0 & 0 & \frac{1}{4} & 0 & 1 \\ 0 & 1 & \frac{1}{4} & \frac{1}{2} & 0 \end{bmatrix} \cdot \begin{bmatrix} -0.15 \\ -0.55 \\ 0.60 \\ 1.25 \\ -0.15 \end{bmatrix} = \begin{bmatrix} 0.1 \\ -0.2 \\ 1.05 \\ -0.2 \\ 0.25 \end{bmatrix} \qquad (4.38)$$

In the fifth iteration, the $PR(5)$ is updated as follows:

$$PR(5) = \begin{bmatrix} -\frac{1}{5} \\ -\frac{1}{5} \\ -\frac{1}{5} \\ -\frac{1}{5} \\ -\frac{1}{5} \end{bmatrix} + 2 \begin{bmatrix} 0 & 0 & \frac{1}{4} & 0 & 0 \\ 1 & 0 & \frac{1}{4} & 0 & 0 \\ 0 & 0 & 0 & \frac{1}{2} & 0 \\ 0 & 0 & \frac{1}{4} & 0 & 1 \\ 0 & 1 & \frac{1}{4} & \frac{1}{2} & 0 \end{bmatrix} \cdot \begin{bmatrix} 0.1 \\ -0.2 \\ 1.05 \\ -0.2 \\ 0.25 \end{bmatrix} = \begin{bmatrix} 0.325 \\ 0.525 \\ -0.400 \\ 0.825 \\ -0.275 \end{bmatrix} \qquad (4.39)$$

To continue the iteration, in the twentieth iteration, the $PR(20)$ is updated as follows:

$$PR(20) = \begin{bmatrix} -\frac{1}{5} \\ -\frac{1}{5} \\ -\frac{1}{5} \\ -\frac{1}{5} \\ -\frac{1}{5} \end{bmatrix} + 2 \begin{bmatrix} 0 & 0 & \frac{1}{4} & 0 & 0 \\ 1 & 0 & \frac{1}{4} & 0 & 0 \\ 0 & 0 & 0 & \frac{1}{2} & 0 \\ 0 & 0 & \frac{1}{4} & 0 & 1 \\ 0 & 1 & \frac{1}{4} & \frac{1}{2} & 0 \end{bmatrix} \cdot \begin{bmatrix} -10.8357 \\ 3.0424 \\ 35.9367 \\ -51.3303 \\ 24.1869 \end{bmatrix} = \begin{bmatrix} 17.7684 \\ -3.9031 \\ -51.5303 \\ 66.1422 \\ -27.4771 \end{bmatrix} \qquad (4.40)$$

The iteration diverges.

### 4.1.3 Newton–Raphson Method

The Gauss–Seidel method and the PageRank algorithm are of the one-order convergence which in general require more iterations than the Newton–Raphson method. The Newton–Raphson method is a gradient-based method to solve nonlinear equations.

A single variable nonlinear equation can be expressed as

$$f(x) = 0 \tag{4.41}$$

Assume the solution $x = x^0 + \Delta x^0$, the Eq. (4.41) can be rewritten and expanded with the Taylor series as follows:

$$f\left(x^0 + \Delta x^0\right) = 0 \tag{4.42}$$

$$f\left(x^0\right) + f'\left(x^0\right)\Delta x^0 + f''\left(x^0\right)\frac{\left(\Delta x^0\right)^2}{2!} + \dots + f^{(n)}\left(x^0\right)\frac{\left(\Delta x^0\right)^n}{n!} = 0 \tag{4.43}$$

where $f'(x^0), f''(x^0), \dots, f^{(n)}(x^0)$ are the derivatives of the function $f(x)$ at the point $x^0$.

If $x^0$ is close to the solution of Eq. (4.41), $\Delta x^0$ is small. Then its second and higher-order terms are neglected to approximate the Eq. (4.42) as a linear equation as follows:

$$f\left(x^0\right) + f'\left(x^0\right)\Delta x^0 = 0 \tag{4.44}$$

Then, we have

$$\Delta x^0 = -\frac{f\left(x^0\right)}{f'\left(x^0\right)} \tag{4.45}$$

Then, the solution is updated as

$$x^1 = x^0 + \Delta x^0 = x^0 - \frac{f\left(x^0\right)}{f'\left(x^0\right)} \tag{4.46}$$

The solution will be approached by the following iteration equation:

$$x^{k+1} = x^k + \Delta x^k = x^k - \frac{f\left(x^k\right)}{f'\left(x^k\right)} \tag{4.47}$$

The error $\Delta x^k$ is deduced until the following stop criteria is met:

$$|\Delta x^k| < \epsilon \qquad (4.48)$$

where $\epsilon$ is the tolerance.

The Newton–Raphson method can be applied to a large-scale nonlinear equation with $n$ variables.

$$f_1(x_1, x_2, ..., x_n) = 0$$

$$f_2(x_1, x_2, ..., x_n) = 0$$

$$\vdots \qquad (4.49)$$

$$f_n(x_1, x_2, ..., x_n) = 0$$

For a given initial solution $x_1^0, x_2^0, ..., x_n^0$ and a set of differences between the final solution and the initial solution $\Delta x_1^0, \Delta x_2^0, ..., \Delta x_n^0$, the Eq. (4.49) is rewritten as follows:

$$f_1\left(x_1^0 + \Delta x_1^0, x_2^0 + \Delta x_2^0, ..., x_n^0 + \Delta x_n^0\right) = 0$$

$$f_2\left(x_1^0 + \Delta x_1^0, x_2^0 + \Delta x_2^0, ..., x_n^0 + \Delta x_n^0\right) = 0$$

$$\vdots \qquad (4.50)$$

$$f_n\left(x_1^0 + \Delta x_1^0, x_2^0 + \Delta x_2^0, ..., x_n^0 + \Delta x_n^0\right) = 0$$

Like the method of the linearizing single variable nonlinear equation above, the multiple variable nonlinear equations are linearized as follows by neglecting the terms of second and higher-order derivatives:

$$f_1\left(x_1^0, x_2^0, ..., x_n^0\right) + \left.\frac{\partial f_1}{\partial x_1}\right|_{x_1^0} \Delta x_1^0 + \left.\frac{\partial f_1}{\partial x_2}\right|_{x_2^0} \Delta x_2^0 + ... + \left.\frac{\partial f_1}{\partial x_n}\right|_{x_n^0} \Delta x_n^0 = 0$$

$$f_2\left(x_1^0, x_2^0, ..., x_n^0\right) + \left.\frac{\partial f_2}{\partial x_1}\right|_{x_1^0} \Delta x_1^0 + \left.\frac{\partial f_2}{\partial x_2}\right|_{x_2^0} \Delta x_2^0 + ... + \left.\frac{\partial f_2}{\partial x_n}\right|_{x_n^0} \Delta x_n^0 = 0 \qquad (4.51)$$

$$\vdots$$

$$f_n\left(x_1^0, x_2^0, ..., x_n^0\right) + \left.\frac{\partial f_n}{\partial x_1}\right|_{x_1^0} \Delta x_1^0 + \left.\frac{\partial f_n}{\partial x_2}\right|_{x_2^0} \Delta x_2^0 + ... + \left.\frac{\partial f_n}{\partial x_n}\right|_{x_n^0} \Delta x_n^0 = 0$$

Equation (4.51) can be rewritten in matrix form as follow:

$$
\begin{bmatrix}
f_1\left(x_1^0, x_2^0, \ldots, x_n^0\right) \\
f_2\left(x_1^0, x_2^0, \ldots, x_n^0\right) \\
\vdots \\
f_n\left(x_1^0, x_2^0, \ldots, x_n^0\right)
\end{bmatrix}
= -
\begin{bmatrix}
\left.\dfrac{\partial f_1}{\partial x_1}\right|_{x_1^0} & \left.\dfrac{\partial f_1}{\partial x_2}\right|_{x_2^0} & \cdots & \left.\dfrac{\partial f_1}{\partial x_n}\right|_{x_n^0} \\[2mm]
\left.\dfrac{\partial f_2}{\partial x_1}\right|_{x_1^0} & \left.\dfrac{\partial f_2}{\partial x_2}\right|_{x_2^0} & \cdots & \left.\dfrac{\partial f_2}{\partial x_n}\right|_{x_n^0} \\[2mm]
\vdots & \vdots & \ddots & \vdots \\[2mm]
\left.\dfrac{\partial f_n}{\partial x_1}\right|_{x_1^0} & \left.\dfrac{\partial f_n}{\partial x_2}\right|_{x_2^0} & \cdots & \left.\dfrac{\partial f_n}{\partial x_n}\right|_{x_n^0}
\end{bmatrix}
\begin{bmatrix}
\Delta x_1^0 \\
\Delta x_2^0 \\
\vdots \\
\Delta x_n^0
\end{bmatrix}
\tag{4.52}
$$

Equation (4.52) is a linear equation. Solving Eq. (4.52), we get $\Delta x_1^0, \Delta x_2^0, \ldots, \Delta x_n^0$. The solution will be approached by the following iteration equations:

$$
\begin{bmatrix}
f_1\left(x_1^k, x_2^k, \ldots, x_n^k\right) \\
f_2\left(x_1^k, x_2^k, \ldots, x_n^k\right) \\
\vdots \\
f_n\left(x_1^k, x_2^k, \ldots, x_n^k\right)
\end{bmatrix}
= -
\begin{bmatrix}
\left.\dfrac{\partial f_1}{\partial x_1}\right|_{x_1^k} & \left.\dfrac{\partial f_1}{\partial x_2}\right|_{x_2^k} & \cdots & \left.\dfrac{\partial f_1}{\partial x_n}\right|_{x_n^k} \\[2mm]
\left.\dfrac{\partial f_2}{\partial x_1}\right|_{x_1^k} & \left.\dfrac{\partial f_2}{\partial x_2}\right|_{x_2^k} & \cdots & \left.\dfrac{\partial f_2}{\partial x_n}\right|_{x_n^k} \\[2mm]
\vdots & \vdots & \ddots & \vdots \\[2mm]
\left.\dfrac{\partial f_n}{\partial x_1}\right|_{x_1^k} & \left.\dfrac{\partial f_n}{\partial x_2}\right|_{x_2^k} & \cdots & \left.\dfrac{\partial f_n}{\partial x_n}\right|_{x_n^k}
\end{bmatrix}
\begin{bmatrix}
\Delta x_1^k \\
\Delta x_2^k \\
\vdots \\
\Delta x_n^k
\end{bmatrix}
\tag{4.53}
$$

$$
x_i^{k+1} = x_i^k + \Delta x_i^k \quad i = 1, 2, \ldots, n \tag{4.54}
$$

Equations (4.53) and (4.54) can be expressed as follows:

$$
\mathcal{F}\left(X^k\right) = -J^k \Delta X^k \tag{4.55}
$$

$$
X^{k+1} = X^k + \Delta X^k \tag{4.56}
$$

Where $J^k \in \mathbb{R}^{n \times n}$ is called the Jacobian matrix.

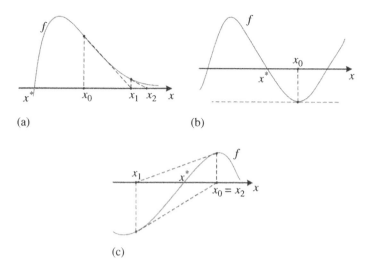

**Figure 4.2** Failure of Newton–Raphson method. (a) runaway; (b) flat spot; (c) cycling.

The Newton–Raphson method is widely used to solve power system problems. However, because of the nonconvex feasibility region of problems, failure of the Newton–Raphson method may be triggered from an inappropriate initial point or when the system is in ill condition. Figure 4.2 shows three possible scenarios of Newton–Raphson failure.

As an important topic, the failure of the Newton–Raphson method and solving ill-conditioned power system problems will be discussed in Chapter 10.

## 4.2 Direct Methods of Solving Linear Equations

### 4.2.1 Introduction

The Newton–Raphson method has been widely used in many power system applications. Each iteration of the Newton–Raphson involves a step to solve large-scale simultaneous linear algebraic equations, which is the step that requires the most computational time in the Newton–Raphson method. Direct methods are widely used to solve such linear equations.

Direct methods solve a linear system in finite steps. It has no round-off errors and provides an exact solution precisely. In addition to their high efficiency in solving medium-scale size linear systems, direct methods are also popular for solving large sparse linear systems. Sparse direct methods are a tightly coupled combination of techniques from numerical linear algebra, graph theory, graph algorithms, permutations, and other topics in discrete mathematics [4].

This chapter focuses on direct methods for sparse linear systems, such as LU, Cholesky, and other factorization. It first introduces basic concepts for direct methods, such as the definition of a sparse matrix and its data structure. Since many concepts and algorithms about graphs are used in direct methods, the interrelationship between matrices and graphs is included in the basic concepts section. Then, the theory and algorithms of direct methods for solving large sparse matrix problems are outlined. To understand the direct methods further, this chapter also introduces popular sequential and parallel solvers to solve sparse linear systems.

## 4.2.2 Basic Concepts

As defined by J.H. Wilkinson [5], a matrix is called sparse if most of its elements are zero. When storing and manipulating sparse matrices, it is often beneficial to use specialized algorithms and data structures to exploit the sparse structure of the matrix.

### 4.2.2.1 Data Structures of Sparse Matrix

Generally, a sparse matrix is stored with a compact data structure so that the numerically zero entries in the matrix can be avoided storing. The data structures for sparse matrices can be classified into two categories: static data structure and dynamic data structure [6].

The two most common formats for static structure are the triplet matrix and the compressed-row matrix (or the compressed-column matrix) [4]. In the triplet matrix, only the nonzero entries are stored. Each record of the triplet format includes three elements: the row number, the column number, and the value of a nonzero entry in the matrix.

Taking the simple 4-bus system shown in Figure 4.3 as an example, the admittance matrix of the system is as follows:

$$
\begin{bmatrix}
a_{11} & a_{12} & & \\
a_{21} & a_{22} & a_{23} & a_{24} \\
& a_{32} & a_{33} & \\
& a_{42} & & a_{44}
\end{bmatrix}
\tag{4.57}
$$

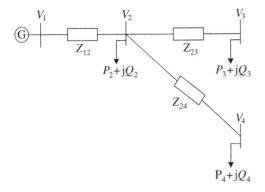

**Figure 4.3** 4-bus system.

Since the matrix is symmetrical, only the upper-triangular matrix is stored by the following triple matrix (Table 4.1).

**Table 4.1** Triple matrix.

| Value | $a_{11}$ | $a_{12}$ | $a_{22}$ | $a_{23}$ | $a_{24}$ | $a_{33}$ | $a_{44}$ |
| --- | --- | --- | --- | --- | --- | --- | --- |
| Row | 1 | 1 | 2 | 2 | 2 | 3 | 4 |
| Column | 1 | 2 | 2 | 3 | 4 | 3 | 4 |

Such a format is easy to generate but it is hard to use in most sparse direct methods, so the triplet format is often used in an interface to a package. Compared with the triplet format, the compressed-row (or the compressed-column) format is convenient to use in the internal representation for algorithm design and development. In compressed-row format (CSR), 3 one-dimensional arrays are used to describe a sparse matrix. The first array csrVal stores the values of all nonzeros. The second array csrCol stores the column indices of all nonzeros. And the third one csrRow stores the location of the first nonzeros of each row in the first two arrays (Figure 4.4).

The transposed format of CSR is a compressed sparse column (CSC), which is stored in column-major. The first array cscVal stores the values of all nonzeros. The second array cscRow stores the row indices of all nonzeros. And the third one cscCol stores the location of the first nonzeros of each column in the first two arrays (Figure 4.5).

**Figure 4.4** Sparse matrix compressed sparse row.

| csrVal | $a_{11}$ | $a_{12}$ | $a_{22}$ | $a_{23}$ | $a_{24}$ | $a_{33}$ | $a_{44}$ |
| --- | --- | --- | --- | --- | --- | --- | --- |
| csrCol | 1 | 2 | 2 | 3 | 4 | 3 | 4 |

| csrRow | 1 | 3 | 6 | 7 | 8 |
| --- | --- | --- | --- | --- | --- |

**Figure 4.5** Sparse matrix CSC.

| cscVal | $a_{11}$ | $a_{12}$ | $a_{22}$ | $a_{23}$ | $a_{33}$ | $a_{24}$ | $a_{44}$ |
| --- | --- | --- | --- | --- | --- | --- | --- |
| cscRow | 1 | 1 | 2 | 2 | 3 | 4 | 4 |

| cscCol | 1 | 2 | 4 | 6 | 8 |
| --- | --- | --- | --- | --- | --- |

The compressed sparse row and CSC formats are used in most sparse direct methods. Since the nonzeros are stored in compressed format, when an element is removed or added to the matrix, the whole compressed format must be updated. In computer science, a linked list is used to allow for the efficient insertion or removal of elements from any position in the arrays.

Linked list-based format is representative of dynamic data structure. This format stores matrix rows or columns as items connected by pointers. The linked lists can be cyclic, one-way, or two-way. It follows a dynamic behavior when embedding rows or columns embedded into a larger array.

The compressed sparse formats save storage and facilitate efficient sparse direct methods. However, the representation is not intuitive on network connectivity. In contrast to the format of arrays, a graph is concise when modeling a power network.

### 4.2.2.2 Matrices and Graphs

Graph theory is a fundamental tool for sparse matrix techniques. The duality between the canonical representation of graphs, which are abstract sets of vertices and edges. Originally, a graph is used to represent a sparse adjacency matrix in graph theory [7]. Versa Vise, an adjacency matrix can be used to represent a graph. Any matrix can be modeled as a specific graph. Different graph models, such as an undirected graph, directed graph, and bipartite graph, are modeled in different use cases [8].

#### 4.2.2.2.1 Undirected Graphs

**Definition 4.1**    A simple undirected graph is an ordered pair of sets $(V, E)$ such that $E = \{(i, j) \mid i \in V, j \in V\}$. $V$ is the vertex set and $E$ is the edge set. An example of an undirected graph is shown in Figure 4.6 to represent the power system shown in Figure 4.3.

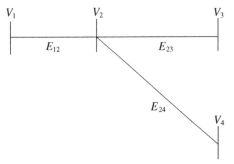

**Figure 4.6**    Undirected graph.

A symmetric matrix can be represented by an undirected graph equivalently as illustrated in Figure 4.7:

$$
\begin{bmatrix}
a_{11} & a_{12} & & \\
a_{21} & a_{22} & a_{23} & a_{24} \\
& a_{32} & a_{33} & \\
& a_{42} & & a_{44}
\end{bmatrix}
$$

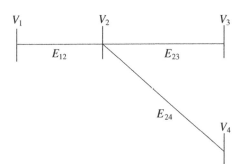

**Figure 4.7**    Relation between symmetric matrix and undirected graph.

#### 4.2.2.2.2 Directed Graphs

**Definition 4.2**  A simple directed graph is an ordered pair of sets $(V, E)$ such that $E = \{(i, j) \mid i \in V, j \in V\}$. $V$ is the vertex set and $E$ is the edge set.

When a power system has an active device, for example, a phase shifting transformer, which makes the admittance matrix unsymmetrical in power system analysis as shown in Figure 4.8 and (4.58) where $a_{23} \neq a_{32}$:

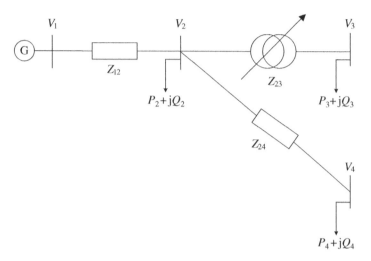

**Figure 4.8**  4-bus system with phase shifting transformer.

$$\begin{bmatrix} a_{11} & a_{12} & & \\ a_{21} & a_{22} & a_{23} & a_{24} \\ & a_{32} & a_{33} & \\ & a_{42} & & a_{44} \end{bmatrix} \tag{4.58}$$

The directed graph to represent the matrix (4.58) is shown in Figure 4.9.

As illustrated in Figure 4.9, the nonzero pattern of an asymmetric matrix can be represented by a directed graph.

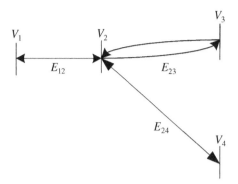

**Figure 4.9** Directed graph.

### 4.2.3 Historical Development

The earliest references to sparse matrix date back to the early 1950s in graph theory and combinatorics [9]. The first paper directly mentioned graphs with sparse elimination and was published in 1961 [10]. In the early 1970s, the first theses on sparse direct methods appeared in the world's top universities such as Harvard [11], Stanford [12], Oxford [13], and Yale [14]. Since the mid-1970s, books focusing on the topic of direct methods for sparse linear systems include those by Tewarson in 1973 [15], George and Liu in 1981 [16], Pissanetsky in 1984 [17], Duff, Erisman, and Reid in 1986 [18], Zlatev in 1991 [19], Björck in 1996 [20], and Davis in 2006 [21].

Tewarson's survey in 1970 [22] summarized many topics of direct methods for solving sparse linear systems, such as LU factorization, Householder and Givens transformations, Markowitz ordering, minimum degree ordering, and bandwidth. A survey by Reid in 1974 [23] focused on right-looking Gaussian elimination. In 1977, Duff [24] conducted an extensive survey of sparse matrix methods and their applications.

More researchers studied this area. A lot of works were created in the 1980s. In 1981, George provided a tutorial survey of sparse Cholesky factorization [25]. In 1987, Zlatev compared and contrasted the methods, according to static or dynamic data structures, for Cholesky, LU, and QR [26]. The first survey that included multi-frontal methods was published by Duff in 1989 [27].

In parallel computing, although parallel methods appeared in the work of Calahan in 1973 [28], the first survey focusing on parallel methods was written by Heath, Ng, and Peyton in 1991 [29], which focused solely on Cholesky factorization, but considers ordering, symbolic analysis, and basic factorizations as well as super-nodal and multi-frontal methods. In 1999, Duff and Van der Vorst investigated extensively on parallel algorithms [30].

Computation efficiency is an active topic in modern sparse direct solvers. Direct solvers achieve their high computation performance in several ways: (i) asymptotically efficient symbolic and graph algorithms that allow floating-point work to dominate the computation, (ii) parallelism, and (iii) operations on dense submatrices, via the super-nodal, frontal, and multi-frontal methods [31].

## 4.2.4 Direct Methods

Gaussian elimination is the basic algorithm of direct methods of solving linear systems. Using Gaussian elimination, the linear equation $Ax = b$ is reduced to an upper-triangular system $Ux = y$ by forward elimination, and then backward substitution is applied to derive a solution. Other commonly used direct methods, such as LU factorization and Cholesky factorization, essentially are derivatives of classic Gaussian elimination. Cholesky factorization applies to the Hermitian positive-definite matrix. It is essentially a variant of Gaussian elimination. In power system analysis, LU factorization is a more general algorithm to solve large sparse linear equations.

Generally, there are three steps of direct methods for solving large sparse linear systems:

1) Factorize matrix $A$ into the product of lower-triangular matrix $L$ and upper-triangular matrix $U$.
2) Forward substitution is applied to $Ly = b$ to get the value of y.
3) Backward substitution is employed in the linear equations $Ux = y$ to derive the solution.

LU factorization of the matrix mainly includes two steps: symbolic analysis and numerical factorization. During symbolic analysis, ordering algorithms are applied to reduce fill-ins. Then, based on the nonzero pattern of matrix $A$, the nonzero pattern of the factorized matrix is computed which predicts the positions of nonzeros. Finally, the numerical values of $L$ and $U$ are computed by numerical factorization.

In the following content, the methods of solving triangular systems are first introduced; then symbolic analysis is discussed covering fill-reduce ordering algorithms, elimination trees, calculation of row/column nonzero counts, and symbolic factorization.

### 4.2.4.1 Solving Triangular Systems

Solving a triangular system $Lx = b$, where $L$ is a sparse lower-triangular matrix, is the key mathematical kernel of the sparse direct methods. It is used in a sparse Cholesky factorization algorithm and a sparse LU factorization algorithm. Solving $Lx = b$ is also necessary for solving $Ax = b$ after factorizing $A$.

In power system analysis, usually, the right-hand side vector $b$ in the linear system is sparse. When the right-hand side vector $b$ is sparse, the solution $x$ is also sparse, and not all columns of $L$ are necessarily needed to take part in the computation.

When the right-hand side vector $b$ has nonzero elements, $x$ becomes nonzero determined by a sparse path set of $b$ following two rules: (i) $b_i \neq 0 \Rightarrow x_i \neq 0$, (ii) $x_j \neq 0 \wedge \exists i(l_{ij} \neq 0) \Rightarrow x_i \neq 0$.

Computing the nonzero element set $X$ of $x$ requires a depth-first search of the directed graph $G_L$, starting at nonzero node-set $B$ in $b$. A depth-first search computes $x$ in topological order, and performing the numerical solution in that order preserves the numerical dependencies. Assuming a lower-triangular matrix $L$ is in the structure of the following where "$*$" represents a nonzero element.

$$L = \begin{bmatrix} * & & & & & & & & & \\ & * & & & & & & & & \\ * & & * & & & & & & & \\ & & * & & * & & & & & \\ & & & * & & * & & & & \\ & & & & * & & & & & \\ & * & & & & & * & & & \\ & & * & & * & & & & & \\ & & & * & & * & & & * & \\ & & * & & * & & * & & & * \end{bmatrix}$$

(4.59)

The matrix $L$ can be represented by a directed graph as shown in Figure 4.10.

As shown in Figure 4.10, if $B = \{4, 6\}$, the path of node 4 is $P_4 = \{4, 9\}$ and the path of node 6 is $P_6 = \{6, 9, 10\}$, then in this example, the path set $X = P_4 \cup P_6 = \{4, 6, 9, 10\}$, which is also sparse.

### 4.2.4.2 Symbolic Factorization

Symbolic factorization finds the nonzero pattern of the Cholesky factor $L$. As shown in 3.3.1, symbolic factorization can be performed by using the elimination graph. In 1980, George and Liu [32] presented the first linear-time symbolic factorization method based on the quotient graph.

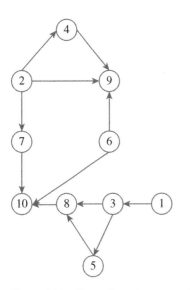

**Figure 4.10** Directed graph representing matrix $L$.

### 4.2.4.3 Fill-Reducing Ordering

The fill-minimization problem can be stated as follows. Given a matrix $A$, find a row and column permutation $P$ and $Q$, $Q = P^T$, such that the number of nonzeros in the factorization of $PAQ$ is minimized. Essentially, the permutations are reordering the row and column of matrix $A$.

The minimum fill-in problem is NP-complete [33] thus heuristics are used to solve the problem. The fill-reducing ordering algorithms can be divided into three categories of approaches: local approaches, global approaches, and hybrid approaches. For local approaches, at each step of the factorization, the pivot that is likely to minimize fill-in is selected. The Markowitz criterion for a general matrix [34] and minimum degree for symmetric matrices [35, 36] are local approaches. In global approaches, the matrix is permuted to confine the fill-in within certain parts of the permuted matrix. Cuthill–McKee [37], Reverse

Cuthill–McKeeand Nested dissection [38] are global approaches. Hybrid approaches are the combination of local and global approaches. They first permute the matrix globally to confine the fill-in, then reorder small parts of the matrix using local heuristics.

The minimum degree algorithm is a widely used heuristic. Tinney and Walker [35] developed the first minimum degree method for symmetric matrices. Huang and Wing [39] presented a variation of minimum degree that considers the number of parallel factorization steps. George and Liu [40] introduced quotient graphs to speed up the minimum degree algorithm. George and Liu [41] investigated the evolution of the minimum degree algorithm and the techniques used to improve it.

Symmetric and asymmetric ordering methods are closely related. Tewarson [42] introduced the graph $A^T A$ for ordering the columns of $A$ prior to LU factorization. To avoid forming $A^T A$, two related algorithms, COLMMD [43] and COLAMD [44, 45] operate on the pattern of $A$ instead. Nested dissection is a fill-reducing ordering that works well for matrices resulting from the discretization of two-dimensional or three-dimensional geometry problems. Kernighan and Lin [46] created a heuristic method that starts with an initial assignment of vertices to two parts and iteratively swaps vertices between the two parts of the graph. Parallel nested dissection ordering is often used and described as part of parallel Cholesky factorizations [47]. Parallelism can also be improved by hybrid methods where a graph can be embedded in a Euclidean space and ordering can be derived using a geometrical nested dissection algorithm [48]. Grigori, Boman, Donfack, and Davis proposed an approach to directly compute hypergraph-based asymmetric nested dissection [49].

## 4.3 Indirect Methods of Solving Linear Equations

The iterative method includes a series of techniques that use successive approximations to obtain more accurate solutions to a linear system at each step. There are two types of iterative methods – stationary methods and nonstationary methods. Stationary iterative methods solve a linear system using an operator that approximates the original system based on the iterative residual. Stationary methods are easier to understand, derive, analyze, and implement, but generally do not work effectively and are not guaranteed to converge.

Nonstationary methods are new. They are based on the idea of orthogonal vector sequences, except for the Chebyshev iterative method, which is based on orthogonal polynomials. In contrast to stationary methods, the computation of nonstationary methods involves information that changes in each iteration. Variables are calculated by taking the inner products of residuals or other vectors derived from each iteration.

Iterative methods can be expressed as $x^k = B \cdot x^{(k-1)} + c$, where $B$ and $c$ are constant. In this section, four stationary iterative methods: Jacobi method, Gauss–Seidel method, SOR method, SSOR method, and three nonstationary iterative methods CG method, GMRES method and BCG are presented.

### 4.3.1 Stationary Methods

#### 4.3.1.1 Jacobi Method
In numerical linear algebra, the Jacobi method is an algorithm for solving diagonally dominant linear equations. Let $A \cdot x = b$ be a square system of $n$ linear equations, where

$$A = \begin{bmatrix} a_{11} & a_{12} & \cdots & a_{1n} \\ a_{21} & a_{22} & \cdots & a_{2n} \\ \vdots & \vdots & \ddots & \vdots \\ a_{n1} & a_{n2} & \cdots & a_{nn} \end{bmatrix}, x = \begin{bmatrix} x_1 \\ x_2 \\ \vdots \\ x_n \end{bmatrix}, b = \begin{bmatrix} b_1 \\ b_2 \\ \vdots \\ b_n \end{bmatrix}$$

The $i$th equation is

$$\sum_{j=1}^{n} a_{i,j} x_j = b_i \tag{4.60}$$

Then, the variable $x_i$ is solved by assuming that the other entries $x_j$ remain fixed

$$x_i = \frac{b_i - \sum_{j \neq i} a_{i,j} x_j}{a_{i,i}} \tag{4.61}$$

In each iteration, the variable $x_i$ is solved by

$$x_i^{(k)} = \frac{b_i - \sum_{j \neq i} a_{i,j} x_j^{(k-1)}}{a_{i,i}} \tag{4.62}$$

The Jacobi method treats all equations independently. The order in which the equations are calculated is irrelevant. Because of this property, the Jacobi method is also called the simultaneous displacement method since the updates can be made simultaneously.

In matrix terms, the definition of the Jacobi method in (4.62) can be also expressed as

$$x^{(k)} = -D^{-1}(L + U)x^{(k-1)} + D^{-1}b \tag{4.63}$$

where the matrices $D$, $L$, and $U$ represent the diagonal, the lower-triangular, and the upper-triangular parts of $A$, respectively.

The pseudocode for the Jacobi method is shown in Algorithm 4.1.

---

**Algorithm 4.1  Jacobi Method**

---

```
 1: Initialize x⁽⁰⁾
 2: for k = 1, 2, ⋯
 3:        for i = 1, 2, ⋯, n
 4:               x̄ᵢ = 0
 5:               for j = 1, 2⋯, i − 1, i + 1, ⋯, n
 6:                      x̄ᵢ = x̄ᵢ + aᵢ,ⱼ xⱼ⁽ᵏ⁻¹⁾
 7:               end
 8:               x̄ᵢ = (bᵢ − x̄ᵢ)/aᵢ,ᵢ
 9:        end
10:        x⁽ᵏ⁾ = x̄
11:        Check convergence; continue if necessary
12: end
```

---

#### 4.3.1.2 Gauss–Seidel Method

According to (4.63), if each equation is processed in sequence, the calculated results can be immediately used for the next iteration. Then the Gauss–Seidel method is derived as follows:

$$x_i^{(k)} = \frac{b_i - \sum_{j < i} a_{i,j} x_j^{(k)} - \sum_{j > i} a_{i,j} x_j^{(k-1)}}{a_{i,i}} \tag{4.64}$$

There are two notable features of the Gauss–Seidel method. First, the computation of (4.64) is sequential. Unlike the Jacobi method, the update of each component $x_i^{(k)}$ not only depends upon previous iteration results $x_j^{(k-1)}$, but also depends on the results obtained from the current iteration $x_j^{(k)}$. Therefore, the update of system states cannot be performed in parallel. Second, since the calculation of $x^{(k)}$ is sequential, the Gauss–Seidel method is also called the method of successive displacements. If this ordering is changed, the components of the new iteration will change accordingly.

According to these two features, if $A$ is sparse, the dependency of each variable on previous variables is not a necessity. The zero-value elements in the matrix may weaken the dependency. By ordering the equations, we can reduce dependencies and restore the ability to update a set of components in parallel. However, reordering the equations affects the convergence of the Gauss–Seidel method.

In matrix form, the definition of the Gauss–Seidel method in (4.64) can be also expressed as follow:

$$x_i^{(k)} = (D + L)^{-1} \left( - U x^{(k-1)} + b \right) \tag{4.65}$$

where $D$, $L$, and $U$ represent the diagonal, lower-triangular, and upper-triangular parts of $A$, respectively.

The pseudocode for the Gauss–Seidel algorithm is given in Algorithm 4.2.

---

**Algorithm 4.2   Gauss–Seidel Algorithm**

---

```
1: Initialize x⁽⁰⁾
2: for k = 1, 2, ···
3:         for i = 1, 2, ···, n
4:                 σ = 0
5:                 for j = 1, 2···, i − 1
6:                         σ = σ + a_{i,j} x_j^{(k−1)}
7:                 end
8:                 for j = i + 1, ···, n
9:                         σ = σ + a_{i,j} x_j^{(k−1)}
10:                end
11:                x_i^k = (b_i − σ)/a_{i,i}
12:         end
13:         Check convergence; continue if necessary
14: end
```

---

#### 4.3.1.3  SOR Method

The SOR method is developed by using extrapolation to the Gauss–Seidel method. This extrapolation calculates a weighted average between the previous iterate and the current Gauss–Seidel iterate for each variable:

$$x_i^{(k)} = \omega \bar{x}_i^{(k)} + (1 - \omega) x_i^{(k-1)} \tag{4.66}$$

where $\bar{x}$ denotes a Gauss–Seidel iterate, and $\omega$ is the extrapolation factor.

In matrix terms, the SOR algorithm can be written as follows:

$$x^{(k)} = (D - \omega L)^{-1}(\omega U + (1 - \omega)D)x^{(k-1)} + \omega(D - \omega L)^{-1}b \tag{4.67}$$

The pseudocode for the SOR algorithm is given in Algorithm 4.3.

---

**Algorithm 4.3  SOR Method**

---

1: **Initialize** $x^{(0)}$
2: **for** $k$ = 1, 2, $\cdots$
3:     **for** $i$ = 1, 2, $\cdots$, $n$
4:         $\sigma$ = 0
5:         **for** $j$ = 1, 2$\cdots$, $i - 1$
6:             $\sigma = \sigma + a_{i,j} x_j^{(k)}$
7:         **end**
8:         **for** $j$ = $i + 1$, $\cdots$, $n$
9:             $\sigma = \sigma + a_{i,j} x_j^{(k-1)}$
10:        **end**
11:        $\sigma$ = $(b_i - \sigma)/a_{i,\ i}$
12:        $x_i^{(k)} = x_i^{(k-1)} + \omega\left(\sigma - x_i^{(k-1)}\right)$
13:     **end**
14:     Check convergence; continue if necessary
15: **end**

---

#### 4.3.1.4  SSOR Method

If the coefficient matrix $A$ is symmetric, then the SSOR method can solve the linear equations by two SOR steps – forward SOR step and backward SOR step. The forward SOR step is conducted by (4.65). The backward SOR step updates the unknown variables reversely.

In matrix form, the SSOR iteration can be expressed as follows:

$$x^{(k)} = B_1 B_2 x^{(k-1)} + \omega(2 - \omega)(D - \omega U)^{-1}D(D - \omega L)^{-1}b \tag{4.68}$$

where

$$B_1 = (D - \omega U)^{-1}(\omega L + (1 - \omega)D)$$

and

$$B_2 = (D - \omega L)^{-1}(\omega U + (1 - \omega)D)$$

The pseudocode for the SSOR algorithm is given in Algorithm 4.4.

---

**Algorithm 4.4  SSOR Method**

---

1: **Initialize** $x^{(0)}$
2: **Let** $\mathbf{x}^{\left(\frac{1}{2}\right)} = \mathbf{x}^{(0)}$
3:
4: **for** $k = 1, 2, \cdots$
5:      **for** $i = 1, 2, \cdots, n$
6:           $\sigma = 0$
7:           **for** $j = 1, 2\cdots, i - 1$
8:                $\sigma = \sigma + a_{i,j} x_j^{\left(k - \frac{1}{2}\right)}$
9:           **end**
10:          **for** $j = i + 1, \cdots, n$
11:               $\sigma = \sigma + a_{i,j} x_j^{(k-1)}$
12:          **end**
13:          $\sigma = (b\_i - \sigma) / a\_(i, i)$
14:          $x_i^{\left(k - \frac{1}{2}\right)} = x_i^{(k-1)} + \omega \left(\sigma - x_i^{(k-1)}\right)$
15:      **end**
16:      **for** $i = n, n - 1, \cdots, 1$
17:           $\sigma = 0$
18:           **for** $j = 1, 2, \cdots, i - 1$
19:                $\sigma = \sigma + a_{i,j} x_j^{\left(k - \frac{1}{2}\right)}$
20:           **end**
21:           **for** $j = i + 1, \cdots, n$
22:                $\sigma = \sigma + a_{i,j} x_j^{(k)}$
23:           **end**
24:           $x_i^{(k)} = x_i^{\left(k - \frac{1}{2}\right)} + \omega \left(\sigma - x_i^{\left(k - \frac{1}{2}\right)}\right)$
25:      **end**
26:      Check convergence; continue if necessary
27: **end**

---

### 4.3.2 Nonstationary Methods

#### 4.3.2.1 CG Method

The CG method is an efficient method for symmetric positive-definite systems. It is one of the well-known nonstationary methods. The iterates $x^{(i)}$ are updated in each iteration by adding a multiplication of $\alpha_i$ and $p^{(i)}$:

$$x^{(i)} = x^{(i-1)} + \alpha_i p^{(i)} \tag{4.69}$$

The corresponding residuals $r^{(i)} = b - Ax^{(i)}$ are updated as

$$r^{(i)} = r^{(i-1)} - \alpha q^{(i)} \tag{4.70}$$

where $q^{(i)} = Ap^{(i)}$.

The choice $\alpha = \alpha_i = r^{(i-1)^T} r^{(i-1)} / p^{(i)^T} Ap^{(i)}$ minimizes $r^{(i)^T} A^{-1} r^{(i)}$ over all possible choices for $\alpha$ in Eq. (4.69).

The search directions are updated using the residuals

$$p^{(i)} = r^{(i)} - \beta_{i-1} p^{(i-1)} \tag{4.71}$$

where the choice $\beta_i = r^{(i)^T} r^{(i)} / r^{(i-1)^T} r^{(i-1)}$ ensures that $p^{(i)}$ and $Ap^{(i)}$, or equivalently, $r^{(i)}$ and $r^{(i-1)}$ are orthogonal. In fact, this choice of $\beta_i$ makes $p^{(i)}$ and $r^{(i)}$ orthogonal to all previous $Ap^{(i)}$ and $r^{(i)}$, respectively.

The pseudocode for the Preconditioned CG Method is given in Algorithm 4.5.

---

**Algorithm 4.5   Preconditioned CG Method**

---

```
1: Initialize x^(0)
2: Compute   r^(0) = b - Ax^(0)
3: for   i = 1, 2, ···
4:           solve   Mz^(i-1) = r^(i-1)
5:           ρ_{i-1} = r^(i-1)^T z^(i-1)
6:           if   i = 1
7:               p^(1) = z^(0)
8:           else
9:               β_{i-1} = ρ_{i-1}/ρ_{i-2}
10:              p^(i) = z^(i-1) + β_{i-1}p^(i-1)
11:          endif
12:          q^(i) = Ap^(i)
13:          α_i = ρ_{i-1}/p^(i)^T q^(i)
14:          x^(i) = x^(i-1) + α_i p^(i)
15:          r^(i) = r^(i-1) - α_i q^(i)
16:          check convergence, continue if necessary
17: end
```

---

### 4.3.2.2  GMRES

The GMRES method generates a sequence of orthogonal vectors. All previously computed vectors in the orthogonal sequence must be preserved.

In the CG method, the residuals form an orthogonal basis for the spatial $span\{r^{(0)}, Ar^{(0)}, A^2r^{(0)}, ...\}$.

$$\omega^{(i)} = Av^{(i)}$$

For $k = 1, 2, \cdots, i,$

$$\omega^{(i)} = \omega^{(i)} - \left(\omega^{(i)}, v^{(k)}\right)v^{(k)} \tag{4.72}$$

$$v^{(i+1)} = \frac{\omega^{(i)}}{\|\omega^{(i)}\|} \tag{4.73}$$

The GMRES iterates are as follows:

$$x^{(i)} = x^{(0)} + y_1 v^{(1)} + \cdots + y_i v^{(i)} \tag{4.74}$$

The pseudocode for the restarted GMRES (m) algorithm with preconditioner $M$ is given in Algorithm 4.6.

---

**Algorithm 4.6  Preconditioned GMRES(m) Method**

---

1: **Initialize** $x^{(0)}$
2: **for** $j = 1, 2, \cdots$
3:     solve $r$ from $Mr = b - Ax^{(0)}$
4:     $v^{(1)} = r/\|r\|_2$
5:     $s := \|r\|_2 e_1$
6:     **for** $i = 1, 2, \cdots, m$
7:         **solve** $\omega$ from $M\omega = Av^{(i)}$
8:         **for** $k = 1, 2, \cdots, i$
9:             $h_{k,i} = (\omega, v^{(k)})$
10:             $\omega = \omega - h_{k,i}v^{(k)}$
11:         **end**
12:         $h_{i+1,i} = \|\omega\|_2$
13:         $v^{(i+1)} = \omega/h_{i+1,i}$
14:         apply $J_1, \cdots, J_{i-1}$ on $(h_{1,i}, \cdots, h_{i+1,i})$ construct $J_i$, acting on $i$th and $(i+1)$st component of $h_{\cdot,i}$, such that $(i+1)$st component of $J_i h_{\cdot,i}$ is 0
15:         $s := J_i s$

```
16:                    if s(i+1) is small enough then (UPDATE(x̃, i) and quit)
17:             end
18:             UPDATE(x̃,m)
19: end
20: In this scheme UPDATE(x̃,i) replaces the following computations:
21: Compute y as the solution of Hy = s̃, in which the upper i×i
      triangular part of H has h_{i,j}
      as its elements (in least squares sense if H is singular),
      s̃ represents the first i components of s
```

22: $\tilde{x} = x^{(0)} + y_1 v^{(1)} + y_2 v^{(2)} + \cdots + y_{(i)} v^{(i)}$, $s^{(i+1)} = \|b - A\tilde{x}\|_2$, if $\tilde{x}$ is an accurate enough approximation then quit, else $x^{(0)} = \tilde{x}$

### 4.3.2.3 BCG (bi-CG)

The CG method is inapplicable for nonsymmetric systems because the residual vectors cannot be orthogonalized by short recurrences [50, 51]. The GMRES method maintains the orthogonality of the residuals by using long recurrences. In contrast, the BCG method replaces the orthogonal sequence of residuals with two mutually orthogonal sequences.

The update relation of residuals in the CG method is augmented in the BCG method with similar relationship based on $A^T$ instead of $A$. Therefore, it updates both residual sequences as follows:

$$r^{(i)} = r^{(i-1)} - \alpha_i A p^{(i)} \tag{4.75}$$

$$\tilde{r}^{(i)} = \tilde{r}^{(i-1)} - \alpha_i A \tilde{p}^{(i)} \tag{4.76}$$

and two sequences of search directions are calculated by

$$p^{(i)} = r^{(i-1)} + \beta_{i-1} p^{(i-1)} \tag{4.77}$$

$$\tilde{p}^{(i)} = \tilde{r}^{(i-1)} + \beta_{i-1} \tilde{p}^{(i)} \tag{4.78}$$

The coefficients $\alpha_i$ and $\beta_i$ are calculated to ensure the bi-orthogonality relations in (4.81)

$$\alpha_i = \frac{\tilde{r}^{(i-1)^T} r^{(i-1)}}{\tilde{p}^{(i)^T} A p^{(i)}} \tag{4.79}$$

$$\beta_i = \frac{\tilde{r}^{(i)^T} r^{(i)}}{\tilde{r}^{(i-1)^T} r^{(i-1)}} \tag{4.80}$$

$$\tilde{r}^{(i)^T} r^{(j)} = \tilde{p}^{(i)^T} A p^{(j)} = 0, \quad \text{if } i \neq j \tag{4.81}$$

The pseudocode for the Preconditioned BCG Method with preconditioner $M$ is given in Algorithm 4.7.

---
**Algorithm 4.7    Preconditioned GMRES(m) Method**

---
```
1: Initialize x⁽⁰⁾
2: Compute   r⁽⁰⁾ = b − Ax⁽⁰⁾
3: Choose r̃⁽⁰⁾ (for example, r̃⁽⁰⁾ = r⁽⁰⁾).
4: for   i=1, 2, ···
5:        solve  Mz⁽ⁱ⁻¹⁾ = r⁽ⁱ⁻¹⁾
6:        solve  Mᵀz̃⁽ⁱ⁻¹⁾ = r̃⁽ⁱ⁻¹⁾
7:        ρᵢ₋₁ = z⁽ⁱ⁻¹⁾ᵀr̃⁽ⁱ⁻¹⁾
8:        if  ρᵢ₋₁ = 0, method fails
9:        if  i = 1
10:               p⁽ⁱ⁾ = z⁽ⁱ⁻¹⁾
11:               p̃⁽ⁱ⁾ = z̃⁽ⁱ⁻¹⁾
12:           else
13:               βᵢ₋₁ = ρᵢ₋₁/ρᵢ₋₂
14:               p⁽ⁱ⁾ = z⁽ⁱ⁻¹⁾ + βᵢ₋₁p⁽ⁱ⁻¹⁾
15:               p̃⁽ⁱ⁾ = z̃⁽ⁱ⁻¹⁾ + βᵢ₋₁p̃⁽ⁱ⁻¹⁾
16:           endif
17:           q⁽ⁱ⁾ = Ap⁽ⁱ⁾
18:           q̃⁽ⁱ⁾ = Aᵀp̃⁽ⁱ⁾
19:           αᵢ = ρᵢ₋₁/p̃⁽ⁱ⁾ᵀq⁽ⁱ⁾
20:           x⁽ⁱ⁾ = x⁽ⁱ⁻¹⁾ + αᵢp⁽ⁱ⁾
21:           r⁽ⁱ⁾ = r⁽ⁱ⁻¹⁾ − αᵢq⁽ⁱ⁾
22:           r̃⁽ⁱ⁾ = r̃⁽ⁱ⁻¹⁾ − αᵢq̃⁽ⁱ⁾
23:           check convergence; continue if necessary
24: end
```
---

# References

**1** Brin, S. and Page, L. (1998). The anatomy of a large scale hypertextual web search engine. *Computer Networks and ISDN Systems* 30 (1/7): 107–117.

**2** Franceschet, M. (2011). PageRank: standing on the shoulders of giants. *Communications of the ACM* 54 (6): 92–101.

**3** Arasu, A., Novak, J., Tomkins, A., and Tomlin, J. (2002). PageRank computation and the structure of the web: experiments and algorithms. *Proceedings of the Eleventh International World Wide Web Conference*, Honolulu, Hawaii, USA (7–11 May 2002). ACM 2002, pp. 1–5. ISBN 1-58113-449-5.

**4** Davis, T.A., Rajamanickam, S., and Sid-Lakhdar, W.M. (2016). A survey of direct methods for sparse linear systems. *Acta Numerica* 25: 383–566.

**5** Wilkinson, J.H., Reinsch, C., and Bauer, F.L. (1971). *Linear Algebra*. Springer.

**6** Tůma, M. (2005). *Direct Methods for Sparse Matrices*. Ostrava: http://www.karlin.mff.cuni.cz/~mirektuma/ps/direct.pdf.

**7** Kepner, J.V. and Gilbert, J.R. (2011). *Graph Algorithms in the Language of Linear Algebra*. Society for Industrial and Applied Mathematics.

**8** Saad, Y. (2014). Background: a brief introduction to graph theory. https://www-users.cselabs.umn.edu/classes/Spring-2019/csci8314/FILES/LecN4.pdf.

**9** Duff, I.S. (2009). Development and history of sparse direct methods. *The SIAM Conference on Applied Linear Algebra*, Monterey, California (26–29 October 2009), pp. 1–34.

**10** Parter, S.V. (1961). Multi-line; iterative methods for elliptic difference equations and fundamental frequencies. *Numerische Mathematik* 3 (1): 305–319.

**11** Rose, D.J. (1970). *Symmetric Elimination on Sparse Positive Definite Systems and the Potential Flow Network Problem*. Harvard University.

**12** George, J.A. (1971). *Computer Implementation of the Finite Element Method*. Stanford University.

**13** Duff, I.S. (1972). *Analysis of Sparse Systems*. University of Oxford.

**14** Sherman, A.H. (1975). *On the Efficient Solution of Sparse Systems of Linear and Nonlinear Equations*. Yale University.

**15** Tewarson, R.P. (1973). *Sparse Matrices*. Academic Press.

**16** George, A. and Liu, J.W.H. (1981). *Computer Solution of Large Sparse Positive Definite Systems*. Prentice-Hall: Google Books.

**17** Pissanetzky, S. (1984). *Sparse Matrix Technology*. Elsevier Science.

**18** Duff, I.S., Erisman, A.M., and Reid, J.K. (1986). *Direct Methods for Sparse Matrice*. London: Oxford University Press.

**19** Zlatev, Z. (1991). *Computational Methods for General Sparse Matrices*. Netherlands: Springer.

**20** Bjorck, A. (1996). *Society for Industrial and Applied Mathematics., Numerical Methods for Least Squares Problems*. Philadelphia, PA: Society for Industrial and Applied Mathematics.

**21** Davis, T.A. (2006). *Direct Methods for Sparse Linear Systems*. Society for Industrial and Applied Mathematics.

**22** Tewarson, R.P. (1970). Computations with sparse matrices. *SIAM Review* 12 (4): 527–544.

**23** Reid, J.K. (1974). Direct methods for sparse matrices. In: *Software Numerical Mathematics*, 29–48. Academic Press.

**24** Duff, I.S. (1977). A survey of sparse matrix research. *Proceedings of the IEEE* 65 (4): 500–535.

**25** George, A. (1981). *Direct Solution of Sparse Positive Definite Systems: Some Basic Ideas and Open Problems*, 283–306. New York: Academic Press.

**26** Zlatev, Z. (1987). A survey of the advances in the exploitation of the sparsity in the solution of large problems. *Journal of Computational and Applied Mathematics* 20: 83–105.

**27** Duff, I.S. (1989). Direct solvers. *Computer Physics Reports* 11: 1–20.

**28** Calahan, D.A. (1973). Parallel solution of sparse simultaneous linear equations. *Proceedings of the 11th Annual Allerton Conference on Circuits and System Theory,* Monticello, Illinois (3–8 October 1973), pp. 729–735.

**29** Heath, M.T., Ng, E., and Peyton, B.W. (1991). Parallel algorithms for sparse linear systems. *SIAM Review* 33 (3): 420–460.

**30** Duff, I.S. and Van der Vorst, H.A. (1999). Developments and trends in the parallel solution of linear systems. *Parallel Computing* 25 (13–14): 1931–1970.

**31** Duff, I.S. (2000). The impact of high-performance computing in the solution of linear systems: trends and problems. *Journal of Computational and Applied Mathematics* 123 (1–2): 515–530.

**32** George, A. and Liu, J.W.H. (1980). An optimal algorithm for symbolic factorization of symmetric matrices. *SIAM Journal on Computing* 9 (3): 583–593.

**33** Yannakakis, M. (1981). Computing the minimum fill-in is NP-complete. *SIAM Journal on Discrete Mathematics* 2: 77–79.

**34** Markowitz, H.M. (1957). The elimination form of the inverse and its application to linear programming. *Management Science* 3 (3): 255–269.

**35** Tinney, W.F. and Walker, J.W. (1967). Direct solutions of sparse network equations by optimally ordered triangular factorization. *Proceedings of the IEEE* 1801–1809.

**36** Rose, D.J. (1972). A graph-theoretic study of the numerical solution of sparse positive definite systems of linear equations. In: *Graph Theory and Computing*, 183–217. New York: Academic Press.

**37** Cuthill, E. and McKee, J. (1969). Reducing the bandwidth of sparse symmetric matrices. *ACM'69: Proceedings of the 1969 24th National Conference* (August 1969), pp. 157–172.

**38** George, A. (1973). Nested dissection of a regular finite element mesh. *SIAM Journal on Numerical Analysis* 10 (2): 345–363.

**39** Huang, J. and Wing, O. (1979). Optimal parallel triangulation of a sparse matrix. *IEEE Transactions on Circuits and Systems* 26 (9): 726–732.

**40** George, A. and Liu, J.W.H. (1980). A fast implementation of the minimum degree algorithm using quotient graphs. *ACM Transactions on Mathematical Software* 6 (3): 337–358.

**41** George, A. and Liu, J.W.H. (1989). The evolution of the minimum degree ordering algorithm. *SIAM Review* 31 (1): 1–19.

**42** Tewarson, R.P. (1967). Solution of a system of simultaneous linear equations with a sparse coefficient matrix by elimination methods. *BIT* 7 (3): 226–239.

**43** Gilbert, J.R. and Schreiber, R. (1992). Highly parallel sparse Cholesky factorization. *SIAM Journal on Scientific and Statistical Computing* 13 (5): 1151–1172.

**44** Davis, T.A., Gilbert, J.R., Larimore, S.I., and Ng, E.G. (2004). Algorithm 836: COLAMD, a column approximate minimum degree ordering algorithm. *ACM Transactions on Mathematical Software* 30 (3): 377–380.

**45** Davis, T.A., Gilbert, J.R., Larimore, S.I., and Ng, E.G. (2004). A column approximate minimum degree ordering algorithm. *ACM Transactions on Mathematical Software* 30 (3): 353–376.

**46** Kernighan, B.W. and Lin, S. (1970). An efficient heuristic procedure for partitioning graphs. *Bell System Technical Journal* 49 (2): 291–307.

**47** MConroy, J. (1990). Parallel nested dissection. *Parallel Computing* 16 (2–3): 139–156.

**48** Heath, M.T. and Raghavan, P. (1995). A cartesian parallel nested dissection algorithm. *SIAM Journal on Matrix Analysis and Applications* 16 (1): 235–253.

**49** Grigori, L., Boman, E.G., Donfack, S., and Davis, T.A. (2010). Hypergraph-based unsymmetric nested dissection ordering for sparse LU factorization. *SIAM Journal on Scientific Computing* 32 (6): 3426–3446.

**50** Voevidub, V. (1983). The problem of non-self-adjoint generalization of the conjugate gradient method is closed, U.S.S.R. *Computational Mathematics and Mathematical Physics* 23: 143–144.

**51** Faber, V. and Manteuffel, T. (1984). Necessary and sufficient conditions for the existence of a conjugate gradient method. *SIAM Journal on Numerical Analysis* 21: 315–339.

# 5

# High-Dimensional Differential Equations

When the time domain method is applied to power system transient simulation, the mathematical model for describing power system dynamic behavior is a nonlinear dynamic system that involves high-dimensional nonlinear differential equations and larger-scale nonlinear algebraic equations (DAEs). In general, differential equations are solved by numerical integration methods.

## 5.1 Integration Methods

The ordinary differential equations (ODEs) with an initial value that appears in power system analysis are standardized into the form as follows:

$$\frac{dx}{dt} = f(x, t), \quad x(t_0) = x_0 \tag{5.1}$$

Here we need to solve the solution $x = x(t)$ for any time $t > t_0$. The variable $x$, the right-hand-side function $f(x, t)$, and the initial value $x_0$ can be either a scalar value or a vector. In addition, $f(x, t)$ can either be linear or nonlinear.

Suppose the ODEs (5.1) are to be solved on the time interval $[t_0, T]$. The numerical methods for solving ODEs begin by dividing the interval by the time points $t_n = t_0 + nh$, $n = 0, 1, \cdots, N$, where the time step size $h = (T - t_0)/N$ and $N$ is a positive integer. The task of the numerical method for solving (5.1) is that for each $n$ we seek a numerical approximation $x_n$ to approximate $x(t_n)$. Initially, $x_0 = x(t_0)$ is given, and the values at the following time are solved recursively, i.e. assuming we have calculated $x_n$, $0 \leq n \leq N - 1$, we then solve for $x_{n+1}$.

### 5.1.1 An Overview of Integration Methods and their Accuracy

There are generally two types of numerical methods for solving ODEs: (i) One-step methods and (ii) Linear multistep methods.

The one-step methods use the previously computed $x_n$ to generate the value of $x_{n+1}$. They use one or more function evaluations at intermediate points between $t_n$ and $t_{n+1}$ in order to improve accuracy. However, these function evaluations are then discarded and are not reused in future steps.

*Graph Database and Graph Computing for Power System Analysis*, First Edition. Renchang Dai and Guangyi Liu.
© 2024 The Institute of Electrical and Electronics Engineers, Inc. Published 2024 by John Wiley & Sons, Inc.

Instead, multistep methods utilize previously calculated values $x_n$, $x_{n-1}$, ... to calculate the value of $x_{n+1}$.

Since the simplified form is used at each time point, a truncation error is introduced during the integration process to solve ODEs. Analyzing truncation error is meaningful to designing an adaptive time step method to improve computation efficiency. Truncation error is the error made by replacing the time derivatives with a discrete-time approximation. It is useful to separate the error produced at each time step by using a finite difference derivative, and the cumulative effect of the error produced at each step. The local truncation error is the truncation error produced at a single step assuming all previous steps are accurate, and it is defined as (5.2):

$$e_{n+1}(h) := x(t_{n+1}) - x_{n+1} \tag{5.2}$$

Therefore, $e_{n+1}(h)$ is the new error that arises when stepping from $t_n$ to $t_{n+1}$, due to the local truncation of the Taylor series of $x(t)$.

The global truncation error is the maximum cumulative truncation error. The global truncation error is related to the local truncation error produced at each step.

### 5.1.1.1 One-Step Methods

The simplest one-step method is the Euler method and its derivatives. Express $x(t)$ by a Tayler series expansion at $t_n$ as follows:

$$x(t) = x(t_n) + (t - t_n)x'(t_n) + \frac{(t-t_n)^2}{2}x''(\tau) \tag{5.3}$$

Where $t - t_n = h$ is the time step, and $\tau \in [t_n, t]$. Since $f_n = f(x(t_n), t_n) = x'(t_n)$, Eq. (5.3) at $t_{n+1}$ is rewritten as:

$$x(t_{n+1}) = x(t_n) + hf_n + 0(h^2) \tag{5.4}$$

Simply, the forward Euler method ignores the reminder term $0(h^2)$ to solve the ODEs (5.1) as:

$$x_{n+1} = x_n + hf_n \tag{5.5}$$

The forward Euler method is simple but rarely used in solving power system transient problems because of its lack of numerical stability. As an improved method, the backward Euler method is preferred with a Taylor series expansion at $t_{n+1}$ other than at $t_n$ in the forward Euler method:

$$x(t) = x(t_{n+1}) + (t - t_{n+1})x'(t_{n+1}) + \frac{(t-t_{n+1})^2}{2}x''(\tau) \tag{5.6}$$

Where $t_{n+1} - t = h$ is time step, and $\tau \epsilon [t, t_{n+1}]$. Since $f_{n+1} = f(x(t_{n+1}), t_{n+1}) = x'(t_{n+1})$, Eq. (5.3) at $t_n$ is rewritten as:

$$x(t_n) = x(t_{n+1}) - hf_{n+1} + 0(h^2) \tag{5.7}$$

Simply, the backward Euler method ignores the reminder term $0(h^2)$ to solve the ODEs (5.1) as:

$$x_{n+1} = x_n + hf_{n+1} \tag{5.8}$$

Both forward and backward Euler have first-order accuracy. To achieve higher-order accuracy, the trapezoidal rule is introduced by keeping the second-order term in a Taylor series expansion at $t_n$ for $x(t)$ and $x'(t)$ as follows:

$$x(t) = x(t_n) + (t - t_n)x'(t_n) + \frac{(t-t_n)^2}{2}x''(t_n) + \frac{(t-t_n)^3}{6}x'''(\tau) \tag{5.9}$$

$$x'(t) = x'(t_n) + (t - t_n)x''(t_n) + \frac{(t-t_n)^2}{2}x'''(\tau) \tag{5.10}$$

Where $t - t_n = h$ is the time step and $\tau \epsilon [t_n, t]$. Equation (5.9) multiplies 2, then subtracts (5.10), and after it multiplies $h$ gives us:

$$x(t) = x(t_n) + \frac{h}{2}\left[x'^{(t)} - x'(t_n)\right] - \frac{h^3}{12}x'''(\tau) \tag{5.11}$$

Since $f_n = f(x(t_n), t_n) = x'(t_n), f_{n+1} = f(x(t_{n+1}), t_{n+1}) = x'(t_{n+1})$, Eq. (5.11) at $t_{n+1}$ is rewritten as:

$$x(t_{n+1}) = x(t_n) + \frac{h}{2}[f_{n+1} + f_n] + 0(h^3) \tag{5.12}$$

The trapezoidal rule ignores the reminder term $0(h^3)$ to solve the ODEs (5.1) as:

$$x_{n+1} = x_n + \frac{h}{2}[f_{n+1} + f_n] \tag{5.13}$$

Figure 5.1 on the right is a graphical view of the trapezoidal rule. The area of the shaded region represents the value of $\int_{t_n}^{t_{n+1}} f(x, t)dt$. The trapezoidal rule has an approximation accuracy of order 2, while both forward and backward Euler methods have first-order accuracy.

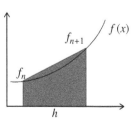

**Figure 5.1** Trapezoidal rule.

One disadvantage of the backward Euler and the trapezoidal rule is that they require solving implicit equations each time $t_{n+1}$. The Runge–Kutta methods avoid solving implicit equations by estimating $f(x(t), t)$ using known values of $x_n$. The classical Dormand-Prince45 method, a member of the Runge–Kutta family of ODE solvers takes the form as follows which has an accuracy of order five:

$$x_{n+1} = x_n + h\left(\frac{5179}{57600}F_1 + \frac{7571}{16695}F_3 + \frac{393}{640}F_4 - \frac{92097}{339200}F_5 + \frac{187}{2100}F_6 + \frac{1}{40}F_7\right) \quad (5.14)$$

where

$$F_1 = f(t_n, x_n)$$

$$F_2 = f\left(t_n + \frac{h}{5}, x_n + \frac{h}{5}F_1\right)$$

$$F_3 = f\left(t_n + \frac{3h}{10}, x_n + h\left(\frac{3}{40}F_1 + \frac{9}{40}F_2\right)\right)$$

$$F_4 = f\left(t_n + \frac{4v}{5}, x_n + h\left(\frac{44}{45}F_1 - \frac{56}{15}F_2 + \frac{32}{9}F_3\right)\right)$$

$$F_5 = f\left(t_n + \frac{8v}{9}, x_n + h\left(\frac{19372}{6561}F_1 - \frac{25360}{2187}F_2 + \frac{64448}{6561}F_3 - \frac{212}{729}F_4\right)\right)$$

$$F_6 = f\left(t_n + h, x_n + h\left(\frac{9017}{3168}F_1 - \frac{355}{33}F_2 + \frac{46732}{5247}F_3 + \frac{49}{176}F_4 - \frac{5103}{18656}F_5\right)\right)$$

$$F_7 = f\left(t_n + h, x_n + h\left(\frac{35}{384}F_1 + \frac{500}{1113}F_3 + \frac{125}{192}F_4 - \frac{2187}{6784}F_5 + \frac{11}{84}F_6\right)\right)$$

The corresponding error estimator is

$$e_{n+1} = h\left(\left(\frac{35}{384} - \frac{5179}{57600}\right)F_1 + \left(\frac{500}{1113} - \frac{7571}{16695}\right)F_3 + \left(\frac{125}{192} - \frac{393}{640}\right)F_4\right.$$

$$\left. + \left(-\frac{2187}{6784} + \frac{92097}{339200}\right)F_5 + \left(\frac{11}{84} - \frac{187}{2100}\right)F_6 - \frac{1}{40}F_7\right) \quad (5.15)$$

$x_{n+1}$ is the approximation of $x(t_{n+1})$. Since the Dormand-Prince45 method is capable of achieving higher-order accuracy and estimating the integration error, during the integration, the step size is adapted to keep the estimated error within a predefined threshold. If the local truncation error is smaller than the threshold, the step is accepted. If the local truncation error is too small, the time step is increased by a certain rule for coming steps. This results in an optimal step size, which saves computation time. The adaptive time step control method will be discussed in Section 5.2 in detail.

### 5.1.1.2 Linear Multistep Methods

Classical numerical linear multistep methods for solving ODEs are the Backward Differentiation Formulas (BDFs), the Adams–Bashforth methods, and the Adams–Moulton methods.

In contrast to Runge–Kutta methods making use of the previously computed $x_n$ to calculate the value of $x_{n+1}$, multistep methods use of the previous values $x_n, x_{n-1},...$ to produce the value of $x_{n+1}$. Linear multistep methods have better computation efficiency than Runge–Kutta methods and are more powerful to solve stiff power system transient problems. Assuming that the time step size $h$ is fixed, the simplest multistep method is the second-order Backward Differentiation Formula, which utilizes the second-order endpoint finite difference approximation as follows:

$$f(x(t_{n+1}), t_{n+1}) = \frac{dx(t_{n+1})}{dt} \approx \frac{3x(t_{n+1}) - 4x(t_n) + x(t_{n-1})}{2h} \tag{5.16}$$

Therefore, the $x(t_{n+1})$ approximation formula is

$$x_{n+1} = \frac{4}{3}x_n - \frac{1}{3}x_{n-1} + \frac{2h}{3}f_{n+1} \tag{5.17}$$

Since $f_{n+1}$ is the function of $x_{n+1}$. The second-order Backward Differentiation Formula, which is denoted as BDF2 or Gear2, is an implicit method that has an order of accuracy 2.

The second, third, and fourth-order of Adams–Bashforth methods are as follows:

$$x_{n+1} = x_n + h\left(\frac{3}{2}f_n - \frac{1}{2}f_{n-1}\right) \tag{5.18}$$

$$x_{n+1} = x_n + h\left(\frac{23}{12}f_n - \frac{16}{12}f_{n-1} + \frac{5}{12}f_{n-2}\right) \tag{5.19}$$

$$x_{n+1} = x_n + h\left(\frac{55}{24}f_n - \frac{59}{24}f_{n-1} + \frac{37}{24}f_{n-2} - \frac{9}{24}f_{n-3}\right) \tag{5.20}$$

These methods use the known values of $x_n$ and the solution computed in the previous steps. They are explicit methods.

In contrast, Adams–Moulton methods are all implicit methods. The second-order accuracy of Adams–Moulton is the trapezoidal method described above, while the third- and fourth-order of Adams–Moulton methods are defined as follows:

$$x_{n+1} = x_n + h\left(\frac{5}{12}f_{n+1} + \frac{8}{12}f_n - \frac{1}{12}f_{n-1}\right) \tag{5.21}$$

$$x_{n+1} = x_n + h\left(\frac{9}{24}f_{n+1} + \frac{19}{24}f_n - \frac{5}{24}f_{n-1} + \frac{1}{24}f_{n-2}\right) \tag{5.22}$$

### 5.1.2 Integration Methods for Power System Transient Simulations

In power system transient simulation, typically, four integration methods are commonly used to solve the power system model's differential-algebraic equations, forward Euler method, backward Euler method, trapezoidal rule, and BDFs.

To solve nonlinear system transient problems, transient simulators employ Newton's method. Newton's method is an iterative procedure that starts with an initial approximated solution and updates the solution by iterations until it converges to the solution. If the transient analysis has convergence difficulties at a particular time point, as long as the system is solvable and numerically stable, the iterations should achieve convergence by taking a smaller time step to enter the region of convergence.

### 5.1.3 Transient Analysis Accuracy

In general, a smaller time step indicates a smaller local truncation error, however, the error may accumulate as we move forward with simulation time steps, leading to a global error, thus the time step must be chosen carefully. The accuracy of a power system transient analysis is difficult to estimate. It depends on the topology, parameters, and characteristics of the power system being simulated, as well as the choice of integration method and integration time step. For first-order integration methods, such as the backward Euler method, the truncation error is proportional to the square of the time step. For second-order methods such as the trapezoidal rule and the second-order backward difference formula, the truncation error is proportional to the cube of the time step. And how the error accumulates is determined by the characteristics of the power system.

The integration methods used in power system transient simulation are either single-step or multistep methods. These methods are formulated approximately by a low-order polynomial. For example, backward Euler method is exact if the solution is a line. The trapezoidal rule and the second-order backward difference formula are exact for straight lines and quadratic equations, and so on.

Unlike steady-state analysis of power systems, the solution calculated at a current time point in a transient analysis strongly depends on the solution calculated in the previous steps. Therefore, reducing error propagation from one step to the next becomes critical. Desired integration methods with high numerical stability ensure that the effect of an error made on one step fades on succeeding steps.

### 5.1.4 Transient Analysis Stability

The numerical methods used for the transient analysis are required to be stiffly stable and accurate as well. The transient analysis stability here is related to numerical methods other than the power system transient problem itself. In the integration methods above, whatever single-step or multistep methods, the differential equations are approximated to be difference equations at each time step $h$. The method is stable when the solution is not sensitive to the presence of perturbations. As time step $h$ is approaching zero, the difference equation is currently solving the differential equation.

#### 5.1.4.1 Absolute Stability

Intuitively, the *absolute stability* for a fixed time step size $h > 0$ of a numerical integration method refers to the property that given a small perturbation $\varepsilon$ to the initial condition, the approximated solutions remain bounded at any time. In particular, for a specific numerical method solving the simple ODE $\dfrac{dx(t)}{dt} = \lambda x(t)$, with $\lambda$ a complex value, we use the absolute stability and the region of absolute stability to characterize its stability property.

The forward Euler method solves the ODE $\dfrac{dx(t)}{dt} = \lambda x(t)$ by (5.5), in which $f_n = \lambda x_n$ in solving this ODE case, thus (5.5) is rewritten as

$$x_{n+1} = x_n + h\lambda x_n = (1 + h\lambda)x_n \tag{5.23}$$

The forward Euler method is said to be absolutely stable when $|1 + h\lambda| \leq 1$, otherwise it is unstable.

For a region $\mathcal{R}_A$ in the complex plane, if for all $z = h\lambda \in \mathcal{R}_A$, the method is absolutely stable, then the method is said to have a region of absolute stability $\mathcal{R}_A$. Taking the forward Euler method as an example, the value of z lies in the absolute stability region $-2 \leq z \leq 0$, and the method is guaranteed to be stable. The stability regions of the forward Euler method, backward Euler method, Trapezoidal method, and Gear2 method are shown in Figure 5.2 [1]. In general, the larger the absolute stability region is, the weaker will be the restriction on the time step size.

Furthermore, a linear multistep method is said to be *A-stable* if its region of absolute stability includes the entire left half of the complex plane. The A-stability of a method guarantees that there is no limit to the time step size. The forward Euler method has a very limited stability region, while the backward Euler method covers the largest region in the complex plane. The backward Euler method, the trapezoidal rule, and the Gear2 method are all A-stable.

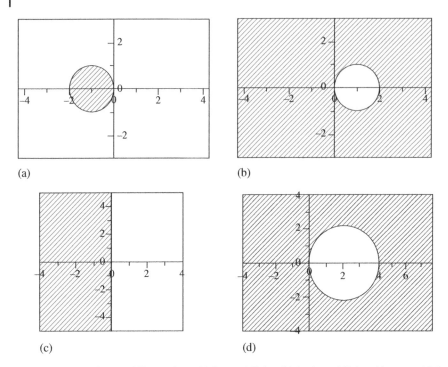

(a)                  (b)

(c)                  (d)

**Figure 5.2** Absolute stability regions. (a) forward Euler; (b) backward Euler; (c) trapezoidal rule; (d) gear2.

### 5.1.4.2 Stiff Stability

When fast transients and slow dynamics are coupled in the power system, the phenomena of stiffness will be observed in the power system transient simulation. The definition of a stiff system can be described as when a numerical method with an absolute stability region $\mathcal{R}_A$, is applied to solve an initial value problem, a time step must be exceedingly small to achieve the exact solution in a certain time interval, then the system is said to be *stiff* in that time interval.

There is no stringent mathematical definition for stiffness. Specifically, stiffness can be defined for an ODE system of the form $\dfrac{dx(t)}{dt} = Ax(t)$, with $A$ a constant matrix of size $m \times m$. Suppose the eigenvalues of $A$ are denoted by $\lambda_1, \lambda_2, \cdots, \lambda_m$ with corresponding eigenvectors $\varphi_1, \varphi_2, \cdots, \varphi_m$. Also, assume that $Re\,(\lambda_k) < 0$ for all $k = 1, 2, \cdots, m$. The general solution to the ODE system is given as follows:

$$x(t) = \sum_{k=1}^{m} c_k e^{\lambda_k t} \varphi_k \tag{5.24}$$

Here $c_k$ are arbitrary constants. $x(t)$ is the summation of $m$ terms. Each term decays at the rate of $|Re\,(\lambda_k)|$. If $\max |Re\,(\lambda_k)| \gg \min |Re\,(\lambda_k)|$, then the terms decay at significantly different rates. The exceedingly small time step is selected to ensure stability and the fastest vanished term is

**Table 5.1** Accuracy and stability of selected integration methods.

| Method | Integration formula | Order of approximation | A-stable | Stiff stable | Category |
|---|---|---|---|---|---|
| Forward Euler | $x_{n+1} = x_n + hf_n$ | 1 | No | No | One-step |
| Backward Euler | $x_{n+1} = x_n + hf_{n+1}$ | 1 | Yes | Yes | One-step |
| Trapezoidal rule | $x_{n+1} = x_n + \dfrac{h}{2}(f_{n+1} + f_n)$ | 2 | Yes | Yes | One-step |
| Gear2 | $x_{n+1} = \dfrac{4}{3}x_n - \dfrac{1}{3}x_{n-1} + \dfrac{2h}{3}f_{n+1}$ | 2 | Yes | Yes | Multistep |

captured, while the slowest term changes negligibly in one step. This motivates the definition of system stiffness by the stiff ratio as:

$$stiffness = \frac{\max \mid Re(\lambda_k) \mid}{\min \mid Re(\lambda_k) \mid} \tag{5.25}$$

It is clear that an A-stable method is also stiffly stable, but not vice versa. In particular, since all explicit methods are not A-stable, they will be limited by time step size for stiff systems. Stiffness can cause large errors in simulation when using explicit integration methods if the integration step size is too large. This problem can be solved by selecting a smaller integration step size, but it will obviously increase the computational cost. In summary, the characteristics of the integration methods mentioned above are shown in Table 5.1.

## 5.2 Time Step Control

When a time domain method is applied to power system transient simulation, the mathematical model for describing power system dynamic behavior is a nonlinear dynamic system that includes high-dimensional nonlinear differential equations and large-scale nonlinear algebraic equations (DAEs).

Typically, DAEs are solved by using time-consuming numerical integration methods. To improve the computation efficiency, the choice of a proper step size and parallel computing is essential.

If the step size is too large, the result will become inaccurate or even completely wrong when the large step size is not within the range of numerical stability. If the step size is too small, the transient simulation will take longer than necessary to keep the accuracy. The adaptive time step is a solution that uses the smallest possible time step to obtain an accurate result, thereby increasing the calculation speed while ensuring the calculation accuracy.

## 5.2.1 Adaptive Time Step

General-purpose ODEs solvers always include the ability to vary the time step size, as this provides a considerable speed advantage. This applies to power system transient simulation, especially given the widely varying time constant components employed. In general, a variable time step approach requires three components:

1) A way to estimate the local truncation error. Note that (5.2) is not suitable for error estimation, since usually, the solution $x(t)$ is unknown.
2) A strategy to decide whether to increase or decrease the time step and by how much based on the estimated local truncation error and the predefined threshold.
3) A method to implement the time step change by updating the difference equations accordingly.

In general, there are three methods to adaptively change time step: change by iteration number [2, 3], change by estimated local truncation error [4–6], or change by state variable derivative [1, 7].

Practical experience has confirmed that when the state variable changes rapidly, a smaller time step should be used; when the state variable changes slowly, a larger time step can be used. The time step can be adjusted according to the number of iterations required to solve the ODE by Newton's method. Therefore, an increase in the number of iterations indicates that a smaller time step is required, or otherwise, a larger time step is required.

The method of time step control by the number of iterations of the Newton method selects a time step according to the number of iterations required for the previous time step. However, there is no direct functional relationship between the number of iterations and the truncation error. Therefore, it is difficult to control the truncation error using this method.

In the second method, in order to improve the calculation efficiency, based on the local truncation error metric, a variable time step method is used in the simulation. By using this method, for all time steps taken in the simulation process, the resulting local truncation error can be controlled at an acceptable level. This is the advantage of time step control based on local truncation errors.

In the third method, the time step size can be adjusted by the aggregated partial derivative of state variables, so that small time step sizes can be used when the state variables are changing rapidly, while larger time step sizes can be used when state variables are undergoing slow changes.

In practice, this method specifies the range of feasible time steps and the range of state variable gradients. If the calculated gradient is less than the minimum state variable gradient, the time step should be increased; if the calculated gradient is greater than the maximum state variable gradient, the time step should be decreased. However, the actual performance of this method largely depends on the setting of the time step and state variable gradient range. In addition, the aggregated partial derivatives do not indicate truncation errors. The actual performance of this method still needs further study.

Critical factors driving an effective adaptive time step approach are the absolute stability, convergence, and accuracy of numerical integration methods. The larger the absolute stability region, the better the convergence, and the higher order of accuracy methods support larger time step size. In [8], variable step numerical simulation is adapted to a predictor–corrector method by estimated truncation error to simulate power system transient and demonstrates the effectiveness of the proposed method. However, in order to achieve better computation efficiency improvement, the high-order multistep method, the Dormand-Prince45 method is needed and taken as an example in this chapter to illustrate the adaptive time step control approach. The Dormand-Prince45 method is A-stable and has local truncation errors of $O(h^5)$.

### 5.2.1.1 Change by Iteration Number
Practical experience confirms the intuitive notion that small time steps must be taken when the solution is changing quickly, while large time steps can be taken when the solution is changing slowly. Since the difference equations are nonlinear in solving power system transient, Newton's method is applied to solve the nonlinear equations. When the nonlinear equations are updated significantly from the previous time point to the current point, it supposes to take more iterations to converge to a solution. This phenomenon motivates the approach to adjust the time step size according to the number of iterations needed in Newton's method. So, an increase in the number of iterations indicates a large time step size, and otherwise, a small step size is required.

SPICE provides a time step control method called iteration-count time step control [7], which selects a time step based on the number of iterations required to solve the system for the previous time step. However, there is no direct relationship between iteration number and local truncation error. Thus, it is hard to develop an approach to quantify the time step size by the iteration number at each step. When using iteration-count time step control, SPICE intuitively limits the time step to be no greater than $T_{max}$. $T_{max}$ is specified or a defaulted value [7].

### 5.2.1.2 Change by Estimated Truncation Error
To update the time step by estimated truncation error by (5.15) in the Dormand-Prince45 method, a simple step-error quotient is formed.

$$q = \frac{e_n}{e_{tol}} \tag{5.26}$$

where $e_n$ is the local truncation error, $e_{tol}$ is the error tolerance.

If $q > 1$, the integration must be repeated with a necessary reduction of the current step size, as the new step size by the following expression is used

$$h^n = q^{\frac{1}{k+1}} \cdot h^n \tag{5.27}$$

with $k$ denoting the order of the Dormand-Prince45 method.

If $q < 1$, then the calculated value in the current step is accepted and the new step size is updated as

$$h^{n+1} = a \cdot q^{\frac{1}{k+1}} \cdot h^n \tag{5.28}$$

Since estimating the local truncation error is computationally expensive, and each time an infeasible time step is rejected, it is necessary to recalculate and estimate the local truncation error.

Therefore, only a wide range of time steps can compensate for the estimated local truncation error calculation time consumption. In addition, practical experience shows that methods based on local truncation errors may be too conservative, resulting in too small time steps. To compensate for the conservation, an accelerator $a$ is used.

### 5.2.1.3 Change by State Variable Derivative

The partial derivative of the state variables shows how fast the state variables change. The time step size can be adjusted based on the aggregated partial derivative of all state variables so that smaller time step sizes can be used when the state variables are changing rapidly, while larger time step sizes can be used when state variables are undergoing slow changes.

Let $x_{n+1}$ and $x_n$ be the state variables at $(n+1)^{th}$ and $n^{th}$ time step respectively. The aggregated partial derivative (APD) can be estimated as:

$$APD = \frac{\|x_{n+1} - x_n\|_1}{t_{n+1} - t_n} \tag{5.29}$$

The first-order norm quantifies the aggregated change of all state variables. When it is divided by the time step size, the result gives a good estimation of APD.

In practice, we can specify a range of feasible time step sizes, e.g. $\Delta t_{min}$ and $\Delta t_{max}$, and a preferred range of state variable gradient, e.g. $APD_{min}$ and $APD_{max}$. If the calculated gradient is smaller than $APD_{min}$, then we should increase the time step size; if it is larger than $APD_{max}$, then we should decrease the time step size.

The time complexity for evaluating the time step size is $O(N)$. However, the actual performance of this approach strongly depends on the settings of the state variable gradient range. Besides, the APD does not represent local truncation error. The practical performance of this approach remains to be further investigated.

### 5.2.2 Multiple Time Step

In the transient simulation model, the time step of the mechanical system, electromechanical transient simulation, and electromagnetic transient simulation is in the order of seconds, milliseconds, and microseconds, respectively. The multiple time step method can be used to coordinate these simulations with different time scales. In order to speed up the calculation, different time step sizes are employed in the electromechanical-electromagnetism hybrid simulation. The process of hybrid simulation can be described as: while the electromagnetism transient (EMT) models are simulated in microseconds time step, extrapolated transient stability (TS) models equivalent circuit will be served as the external system of the EMT models. At each TS time step, EMT models are synchronized by an equivalent circuit interfacing with TS models.

There are two types of interaction protocols used for exchanging data between the EMT model and the TS model, i.e. the serial and the parallel type protocols, as shown in Figure 5.3. The interaction period is $\Delta T$, which is usually set to be the same as the integration time step in TS simulation.

**Figure 5.3** TS-EMT interaction protocols. Adapted from [9].

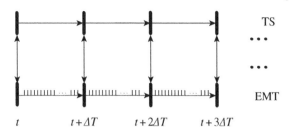

Figure 5.3 demonstrates the interaction protocol – direct data exchange in both directions before each TS simulation step. Both programs run asynchronously for each $\Delta T$ period, thus good simulation efficiency is achieved. However, The TS program uses the previous time step simulation results to update the equivalents for the following time step. This may result in large discrepancies, particularly during the period when the system experiences significant changes, as the equivalents may have not been updated in a timely fashion. To mitigate the discrepancies, an iterative process is involved to update the equivalent Thévenin voltage and current till they are converged.

The parameters used to bridge the TS and EMT models are terminal voltage and injection current at the interface between the TS model and the EMT model as shown in Figure 5.4.

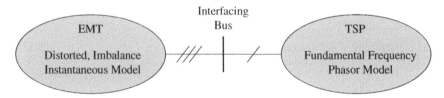

**Figure 5.4** TS-EMT interface.

The EMT simulated instantaneous values of $i_a(t)$, $i_b(t)$, $i_c(t)$ and $v_a(t)$, $v_b(t)$, $v_c(t)$ can be expanded as following forms by the Fourier series:

$$i_a(t) = A_{ia0} + \sum_{k=1}^{n} A_{iak} \cos(k\omega t + \varphi_{iak}) \tag{5.30}$$

$$i_b(t) = A_{ib0} + \sum_{k=1}^{n} A_{ibk} \cos(k\omega t + \varphi_{ibk}) \tag{5.31}$$

$$i_c(t) = A_{ic0} + \sum_{k=1}^{n} A_{ick} \cos(k\omega t + \varphi_{ick}) \tag{5.32}$$

$$v_a(t) = A_{va0} + \sum_{k=1}^{n} A_{vak} \cos(k\omega t + \varphi_{vak}) \tag{5.33}$$

$$v_b(t) = A_{vb0} + \sum_{k=1}^{n} A_{vbk} \cos(k\omega t + \varphi_{vbk}) \tag{5.34}$$

$$v_c(t) = A_{vc0} + \sum_{k=1}^{n} A_{vck} \cos(k\omega t + \varphi_{vck}) \tag{5.35}$$

Since the TS models are constructed in the fundamental components, the equivalent fundamental phasors of the current and voltage are $\dot{I}_a = I_a \angle \theta_{Ia}, \dot{I}_b = I_b \angle \theta_{Ib}, \dot{I}_c = I_c \angle \theta_{Ic}$, and $\dot{V}_a = V_a \angle \theta_{Va}$, $\dot{V}_b = V_b \angle \theta_{Vb}, \dot{V}_c = V_c \angle \theta_{Vc}$. The phase magnitude and phase angle of the equivalent current of voltage can be estimated by EMT instantaneous current and voltage samples. Take $\dot{I}_a = I_a \angle \theta_{Ia}$ as an example:

$$\min E = \sum_{k=1}^{n} \left( \sqrt{2} I_a \cos(\omega t_k + \theta_{Ia}) - i_a(t_k) \right)^2 \tag{5.36}$$

where $n$ is the number of samples before the TS simulation time point $t$. The phase magnitude and phase angle can be estimated by:

$$\frac{\partial E}{\partial I_a} = \sum_{k=1}^{n} 2 \left( \sqrt{2} I_a \cos(\omega t_k + \theta_{Ia}) - i_a(t_k) \right) \cos(\omega t_k + \theta_{Ia}) = 0 \tag{5.37}$$

$$\frac{\partial E}{\partial \theta_{Ia}} = \sum_{k=1}^{n} 2 \left( \sqrt{2} I_a \cos(\omega t_k + \theta_{Ia}) - i_a(t_k) \right) \sin(\omega t_k + \theta_{Ia}) = 0 \tag{5.38}$$

Similarly, $\dot{I}_b = I_b \angle \theta_{Ib}, \dot{I}_c = I_c \angle \theta_{Ic}$ and $\dot{V}_a = V_a \angle \theta_{Va}, \dot{V}_b = V_b \angle \theta_{Vb}, \dot{V}_c = V_c \angle \theta_{Vc}$ can be estimated. Then, the estimated three-phase current $I_{EMT(t)}^{abc}$ are transferred to a three-sequence current $I_{EMT(t)}^{120}$ which is used as the input for the three-sequence TS simulation as shown in Figure 5.5.

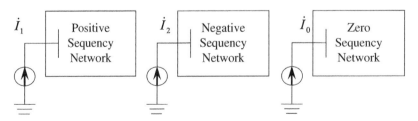

**Figure 5.5** EMT equivalent sequence current injection to TS.

A new three-sequence voltage vector of the interfacing buses is obtained by solving the TS three-sequence equations.

$$Y^{(1)}V^{(1)}_{TS(t)} = I^{(1)}_{EMT(t)} \tag{5.39}$$

$$Y^{(2)}V^{(2)}_{TS(t)} = I^{(2)}_{EMT(t)} \tag{5.40}$$

$$Y^{(0)}V^{(0)}_{TS(t)} = I^{(0)}_{EMT(t)} \tag{5.41}$$

Subsequently, the three-sequence Thévenin equivalent voltages $V^{abc}_{TS(t)}$ are derived by the network equivalent voltage $V^{120}_{TS(t)}$ and sent back to EMT as instantaneous waveforms:

$$v^{a}_{TS(t)} = V^{a}_{TS(t)} \cos\left(\omega t + \varphi^{a}_{TS(t)}\right) \tag{5.42}$$

$$v^{b}_{TS(t)} = V^{b}_{TS(t)} \cos\left(\omega t + \varphi^{b}_{TS(t)}\right) \tag{5.43}$$

$$v^{b}_{TS(t)} = V^{b}_{TS(t)} \cos\left(\omega t + \varphi^{b}_{TS(t)}\right) \tag{5.44}$$

The EMT simulator uses the three-phase instantaneous voltage $v^{abc}_{TS(t)}$ to update the instantaneous injection current $i_a(t)$, $i_b(t)$, $i_c(t)$ to TS simulator. The process is repeated till the interfacing current and voltage are converged. Then the EMT and TS differential equations are solved simultaneously and the simulator moves to the next time step $t + \Delta T$.

$$\dot{X}(t) = f(X(t)) \tag{5.45}$$

### 5.2.3 Break Points

Break points are used to ensure that transient sharp corners in between two predefined time steps are simulated faithfully. They also ensure that events that occur between two predefined time steps are not missed, such as short circuit faults, topology changes, control signals, etc. Whenever an abrupt change in slope occurs, the simulator generates break points to capture these sudden events and transient sharps.

Since the simulator has to set a time point at each breakpoint, two points that are close to each other in time require the simulator to take a very small time step to catch the two points. For the sake of stability, the simulator continues to take small steps for a while by typically limiting the time step size to no more than twice that of its predecessor.

## 5.3  Initial Operation Condition

TS analysis calculates the initial operating point to start with a transient simulation. Clearly, the ability to reliably calculate an accurate initial point is important for power system transient simulation. In TS analysis, the initial operating point is assuming a stable equilibrium point that does not change over time. Thus, the basic algorithm to calculate the initial operating point is to make the $\dfrac{dx(t)}{dt} = 0$ in the differential-algebraic equations representing the dynamic behavior of the power system.

$$\dot{X} = f(X, V) = 0 \tag{5.46}$$

$$YV = I \tag{5.47}$$

where, $X$ is the state variable vector of the system, $V$ is the voltage vector, $Y$ is the admittance matrix of the network and $I$ is the generator injection current which is the function of $X$ and $V$.

When we ignore the transient of stator windings, $p\psi_d = p\psi_q = 0$, the generator transient is represented by the following flux-linkage equations and voltage equations:

$$\psi_d = E_q'' - x_d'' i_d \tag{5.48}$$

$$\psi_q = -E_d'' - x_q'' i_q \tag{5.49}$$

$$V_d = -\omega\psi_q - R_a i_d \tag{5.50}$$

$$V_q = \omega\psi_d - R_a i_q \tag{5.51}$$

To simplify the model, assuming $\omega = 1$ and $R_a = 0$, we have:

$$V_d = E_d'' + x_q'' i_q \tag{5.52}$$

$$V_q = E_q'' - x_d'' i_d \tag{5.53}$$

Rewritten the Eqs. (5.52) and (5.53) in matrix form, we get

$$
\begin{bmatrix} V_d \\ V_q \end{bmatrix} = \begin{bmatrix} E_d'' \\ E_q'' \end{bmatrix} - \begin{bmatrix} 0 & -x_q'' \\ x_d'' & 0 \end{bmatrix} \begin{bmatrix} I_d \\ I_q \end{bmatrix}
\tag{5.54}
$$

Converting the equations in *d*–*q* coordinates to *x*–*y* coordinates, we have

$$
\begin{bmatrix} \sin\delta & -\cos\delta \\ \cos\delta & \sin\delta \end{bmatrix} \begin{bmatrix} V_x \\ V_y \end{bmatrix} = \begin{bmatrix} E_d'' \\ E_q'' \end{bmatrix} - \begin{bmatrix} 0 & -x_q'' \\ x_d'' & 0 \end{bmatrix} \begin{bmatrix} \sin\delta & -\cos\delta \\ \cos\delta & \sin\delta \end{bmatrix} \begin{bmatrix} I_x \\ I_y \end{bmatrix}
\tag{5.55}
$$

Multiplying $\begin{bmatrix} \sin\delta & \cos\delta \\ -\cos\delta & \sin\delta \end{bmatrix}$ at both sides of Eq. (5.55) in *d*–*q* coordinates to *x*–*y* coordinates, we have

$$
\begin{bmatrix} \sin\delta & \cos\delta \\ -\cos\delta & \sin\delta \end{bmatrix} \begin{bmatrix} \sin\delta & -\cos\delta \\ \cos\delta & \sin\delta \end{bmatrix} \begin{bmatrix} V_x \\ V_y \end{bmatrix} = \begin{bmatrix} \sin\delta & \cos\delta \\ -\cos\delta & \sin\delta \end{bmatrix} \begin{bmatrix} E_d'' \\ E_q'' \end{bmatrix}
$$

$$
- \begin{bmatrix} \sin\delta & \cos\delta \\ -\cos\delta & \sin\delta \end{bmatrix} \begin{bmatrix} 0 & -x_q'' \\ x_d'' & 0 \end{bmatrix} \begin{bmatrix} \sin\delta & -\cos\delta \\ \cos\delta & \sin\delta \end{bmatrix} \begin{bmatrix} I_x \\ I_y \end{bmatrix}
\tag{5.56}
$$

$$
\begin{bmatrix} E_d''\mathrm{Sin}\delta + E_q''\cos\delta - V_X \\ -E_d''\cos\delta + E_q''\sin\delta - V_y \end{bmatrix} = \begin{bmatrix} x_d''\sin\delta\cos\delta - x_q''\sin\delta\cos\delta & -x_d''\cos^2\delta - x_q''\sin^2\delta \\ x_d''\sin^2\delta + x_q''\cos^2\delta & -x_d''\mathrm{Sin}\delta\cos\delta + x_q''\sin\delta\cos\delta \end{bmatrix} \begin{bmatrix} I_x \\ I_y \end{bmatrix}
\tag{5.57}
$$

Thus,

$$
\begin{bmatrix} I_x \\ I_y \end{bmatrix} = \frac{1}{x_d''x_q''} \begin{bmatrix} 0 & \frac{1}{2}\left(x_d'' + x_q''\right) \\ -\frac{1}{2}\left(x_d'' + x_q''\right) & 0 \end{bmatrix} \begin{bmatrix} E_d''\mathrm{Sin}\delta + E_q''\cos\delta - V_X \\ -E_d''\cos\delta + E_q''\sin\delta - V_y \end{bmatrix}
$$

$$
+ \frac{x_q'' - x_d''}{x_d''x_q''} \begin{bmatrix} \sin 2\delta & -\cos 2\delta \\ -\cos 2\delta & -\sin 2\delta \end{bmatrix} \begin{bmatrix} E_d''\mathrm{Sin}\delta + E_q''\cos\delta - V_X \\ -E_d''\cos\delta + E_q''\sin\delta - V_y \end{bmatrix}
\tag{5.58}
$$

$$\dot{I}_t = I_x + jI_y = -j\frac{x_q'' + x_d''}{x_d''x_q''}\left[\left(E_q'' - jE_d''\right)e^{j\delta} - \left(V_X + jV_y\right)\right]$$
$$+ j\frac{x_q'' - x_d''}{x_d''x_q''}\left[\left(E_q'' + jE_d''\right)e^{-j\delta} - \left(V_X - jV_y\right)\right]e^{j2\delta} \tag{5.59}$$

$$\dot{I}_t'' = Y_t''\dot{E}_t'' - j\frac{x_q'' - x_d''}{x_d''x_q''}\left(\overset{*}{E_t''} - \overset{*}{V}\right)e^{j2\delta} \tag{5.60}$$

$$\dot{I}_t = \dot{I}_t'' - Y_t''\dot{V} \tag{5.61}$$

Where

$$Y_t'' = -j\frac{x_q'' + x_d''}{x_d''x_q''} \tag{5.62}$$

$$\dot{E}_t'' = \left(\dot{E}_q'' - j\dot{E}_d''\right)e^{j\delta} \tag{5.63}$$

$$\dot{V} = V_X + jV_y \tag{5.64}$$

$Y_t''$ is the generator virtual admittance. $\dot{I}_t''$ is the generator's virtual injection current. The generator equivalent circuit is then shown in Figure 5.6 as follows.

Taking account of resistance $R_a$, the virtual admittance is

$$Y_t'' = \frac{R_a - j\left(x_d'' + x_q''\right)}{2\left(R_a^2 + x_d''x_q''\right)} \tag{5.65}$$

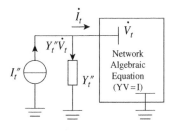

The virtual current injection is

$$\dot{I}_t'' = -j\frac{x_q'' + x_d''}{x_d''x_q''}\left(E_q'' - jE_d''\right)e^{j\delta}$$
$$+ j\frac{x_q'' - x_d''}{x_d''x_q''}\left[\left(E_q'' + jE_d''\right)e^{-j\delta} - \left(V_X - jV_y\right)\right]e^{j2\delta} \tag{5.66}$$

**Figure 5.6** Generator equivalent circuit.

To improve the network equation solving efficiency, the generator virtual admittance is added to the diagonal element of network equations in practice.

When the generator terminal voltage is solved by Eq. (5.47), its equilibrium point can be derived using the generator state vector scheme in Figure 5.7.

**Figure 5.7** Generator state vector.

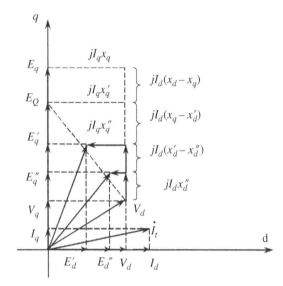

The generator's initial operating points are

$$\dot{E}_{Q(0)} = \dot{V}_{(0)} + \left(R_a + jx_q\right)\dot{I}_{t(0)} \tag{5.67}$$

$$\delta_{(0)} = arctg\left(\frac{E_{Qy(0)}}{E_{Qx(0)}}\right) \tag{5.68}$$

$$E_{fq(0)} = V_{q(0)} + R_a I_{q(0)} + x_d I_{d(0)} \tag{5.69}$$

$$E'_{q(0)} = V_{q(0)} + R_a I_{q(0)} + x'_d I_{d(0)} \tag{5.70}$$

$$E'_{d(0)} = V_{d(0)} + R_a I_{d(0)} - x'_q I_{q(0)} \tag{5.71}$$

$$E''_{q(0)} = V_{q(0)} + R_a I_{q(0)} + x''_d I_{d(0)} \tag{5.72}$$

$$E''_{d(0)} = V_{d(0)} + R_a I_{q(0)} + x''_q I_{q(0)} \tag{5.73}$$

$$P_{e(0)} = Re\left(\dot{V}\dot{I}_t\right) \tag{5.74}$$

When the resistance $R_a$ is ignored, the initial condition is simplified as

$$E''_{d(0)} = I_{q(o)} \cdot \left( x_q - x''_q \right) = \left( x_q - x''_q \right) \left( I_{tx} \cos \delta + I_{ty} \sin \delta \right) \tag{5.75}$$

$$I''_{t(0)x} = \frac{1}{x''_d} \left( E''_q \sin \delta - E''_d \cos \delta \right) \tag{5.76}$$

$$I''_{t(0)y} = \frac{1}{x''_d} \left( -E''_q \cos \delta - E''_d \sin \delta \right) \tag{5.77}$$

$$I''_{t(0)x} \cos \delta + I''_{t(0)y} \sin \delta = \frac{1}{x''_d} \left( E''_q \sin \delta \cos \delta - E''_d \cos^2 \delta \right) + \frac{1}{x''_d} \left( -E''_q \sin \delta \cos \delta - E''_d \sin^2 \delta \right) \tag{5.78}$$

$$I''_{t(0)x} \cos \delta + I''_{t(0)y} \sin \delta = -\frac{1}{x''_d} E''_d = -\frac{1}{x''_d} \left( x_q - x''_q \right) \left( I_{tx} \cos \delta + I_{ty} \sin \delta \right) \tag{5.79}$$

Since,

$$\cos \delta \left( I''_{t(0)x} x''_d + \left( x_q - x''_d \right) I_{tx} \right) = \sin \delta \left( I_{ty} \left( x_q - x''_q \right) - I''_{t(0)y} x''_d \right) \tag{5.80}$$

$$tg\delta = \frac{I''_{t(0)(x)} x''_d + \left( x_q - x''_q \right) I_{tx}}{I_{ty} \left( x''_q - x_q \right) - I''_{t(0)y} x''_d} \tag{5.81}$$

Essentially, the initial operating point problem of TS analysis is to solve voltage in the large-scale nonlinear Eq. (5.46), then initialize the generator state variables by (5.67)–(5.81) above. Solving Eq. (5.46) is challenging. Usually, the Newton method is adopted to solve the nonlinear equations iteratively. How to use the Newton method to solve the above equations in parallel is well illustrated in Chapter 4. Once the generator terminal voltages are solved by (5.46), the initialization of generator state variables can be performed for all generators in parallel.

## 5.4 Graph-Based Transient Parallel Simulation

The power system is a multi-dimensional high-order nonlinear dynamic system. The behavior of a dynamic system is described by a set of high-order nonlinear ODEs and high-dimensional algebraic equation DAEs which are rewritten as follows:

$$\frac{dX}{dt} = f(X, V, t), \quad X_0 = X(t_0) \tag{5.82}$$

$$YV = I, \quad V_0 = V(t_0) \tag{5.83}$$

$$\dot{I}''_t = -j\frac{x''_q + x''_d}{x''_d x''_q}\left(E''_q - jE''_d\right)e^{j\delta} + j\frac{x''_q - x''_d}{x''_d x''_q}\left[\left(E''_q + jE''_d\right)e^{-j\delta} - \left(V_X - jV_y\right)\right]e^{j2\delta} \tag{5.84}$$

The DAE structure for power system transient simulation is shown in Figure 5.8.

Using the sequential method, in each iteration, the network Eq. (5.83) are solved to update the network bus voltage including the generator terminal voltage. Based on the generator terminal bus voltage, differentiate Eq. (5.82) solve the dynamic states of the dynamic equipment, and current injection Eq. (5.84) update the injection current from the generator to the network. The updated current injections are applied to solve network equations in the next iteration until the converged solution is achieved. The flow chart of the sequential method for power system transient simulation is shown in Figure 5.9.

Solving DAEs is time-consuming. To improve the computation efficiency, a parallel computing algorithm has been investigated in power system transient simulation [10, 11]. And graph parallel computing is promising to achieve the goal.

By using graph computing, power flow initialization in step 1, admittance matrix formation and factorization in step 3, and bus voltage calculation can be performed in nodal and hierarchical parallel as described in Chapter 4. This chapter will focus on graph parallel computing on initializing

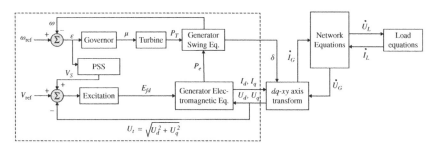

**Figure 5.8** DAE construction of power system.

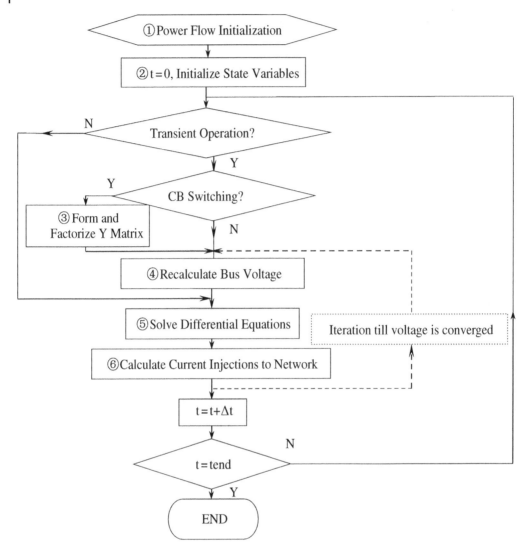

**Figure 5.9** Power system transient simulation flow chart.

state variables in step 2, solving differential equations in step 5, and calculating injection current in step 6 in the flow chart.

The essence of the power system transient simulation algorithm is to solve large-scale nonlinear algebraic equations and high-order differential equations. When the sequential method is used for calculation, the solution of high-order differential equations expressing dynamic elements and current injection calculation can be easily achieved by nodal parallelism, that is, all dynamic equipment in the system does not depend on the calculation results of other dynamic equipment and can be calculated in parallel at the same time in each time step. The pseudo-code to solve differential equations using adaptive time step by graph nodal parallelism is as follows:

---

**Algorithm 5.1   Numerical Integration Nodal Parallel Computing**

---

```
1:    T1 = select v from T0:v where v.type = unit
2:        ACCUM
3:            v.x0 := Eqs. (5.67)-(5.81)
4:    for (t = t0; t <= tend; t = t + h)
5:        T1 = select v from T0:v where v.type = unit
6:            ACCUM
7:                v.x :=  Eq. (5.82)
8:                v.i :=  Eq. (5.84)
9:                v.e :=  Eq. (5.15)
10:           POST-ACCUM
11:               v.v := Eq. (5.83)
12:               @h:= Eq. (5.27) or (5.28)
```

---

In the graph database, the admittance graph automatically defines the sparse adjacency relationship of the admittance matrix. The graph database query statement SELECT in the algorithm queries all generator nodes $v$ in the admittance graph nodes $T_0$. The third line initializes state variables for all units in parallel. Line 4 loops through time steps in the simulation time interval $[t_0,$ tend]. The fifth line of the algorithm loops through all the generators. The keyword ACCUM is used to indicate the initial conditions v.x0 calculation, state variables v.x update, truncation errors v.e estimation, and injection currents v.i calculation are in nodal parallel by graph computing. Finally, the time step $h$ as a global variable (with prefix @ in graph computing syntax) is updated by the keyword POST-ACCUM. The Eq. (5.83) is calculated under POST-ACCUM to update the bus voltage v.v as vector attribute. Note that the algebraic Eq. (5.83) are solved by nodal and hierarchical parallel as described in Chapter 4 which is not elaborated in the pseudo-code above.

## 5.5   Numerical Case Study

To verify the benefit of the variable time step, the IEEE 36-bus modified test system is used. It has 6 synchronous machines with IEEE type-1 exciters, 4 of which are synchronous compensators, 36 buses, 37 transmission lines, 10 transformers, and 21 constant impedance loads. The total load demand is 283.4 MW and 126.2 MVAr.

The IEEE 30-bus modified test system machine data, exciter data, and governor data are listed in Tables 5.2–5.4 [12].

To examine the efficiency of the variable time step, we simulate a three-phase fault at bus 28 at 2.0 seconds and clean the fault at 2.2 seconds. The simulation result is shown in Figures 5.10 and 5.11.

**Table 5.2**  IEEE 30-bus modified test system machine data.

| Type<br>Operation | GENROU<br>Sync. gen. | GENROU<br>Sync. gen. | GENROU<br>Condenser | GENROU<br>Condenser |
|---|---|---|---|---|
| Unit no. | 1 | 2 | 5, 8 | 11, 13 |
| Rated power (MVA) | 270 | 51.2 | 40 | 25 |
| Rated voltage (kV) | 18 | 13.8 | 13.8 | 13.8 |
| Rated pf | 0.85 | 0.8 | 0.0 | 0.0 |
| $H$ (s) | 4.130 | 5.078 | 1.520 | 1.200 |
| $D$ | 2.000 | 2.000 | 0.000 | 0.000 |
| $ra$ (p.u) | 0.0016 | 0.000 | 0.000 | 0.0025 |
| $xd$ (p.u) | 1.700 | 1.270 | 2.373 | 1.769 |
| $xq$ (p.u) | 1.620 | 1.240 | 1.172 | 0.855 |
| $x'd$ (p.u) | 0.256 | 0.209 | 0.343 | 0.304 |
| $x'q$ (p.u) | 0.245 | 0.850 | 1.172 | 0.5795 |
| $x''d$ (p.u) | 0.185 | 0.116 | 0.231 | 0.2035 |
| $x''q$ (p.u) | 0.185 | 0.116 | 0.231 | 0.2035 |
| $xl$ or $xp$ (p.u) | 0.155 | 0.108 | 0.132 | 0.1045 |
| $T'd0$ (s) | 4.800 | 6.600 | 11.600 | 8.000 |
| $T'q0$ (s) | 0.004 | 0.004 | 0.159 | 0.008 |
| $T''d0$ (s) | 0.004 | 0.004 | 0.058 | 0.0525 |
| $T''q0$ (s) | 0.004 | 0.004 | 0.201 | 0.0151 |
| $S(1.0)$ | 0.125 | 0.2067 | 0.295 | 0.304 |
| $S(1.2)$ | 0.450 | 0.724 | 0.776 | 0.667 |

**Table 5.3**  IEEE 30-bus modified test system exciter data.

| Type | IEEET1 | IEEET1 | IEEET1 | IEEET1 |
|---|---|---|---|---|
| Unit no. | 1 | 2 | 5, 8 | 11, 13 |
| Rated power (MVA) | 270 | 51.2 | 40 | 25 |
| Rated voltage (kV) | 18 | 13.8 | 13.8 | 13.8 |
| $Tr$ (s) | 0.000 | 0.000 | 0.000 | 0.000 |
| $Ka$ (p.u) | 30 | 400 | 400 | 400 |
| $Ta$ (s) | 0.400 | 0.050 | 0.050 | 0.050 |
| $VRmax$ (p.u) | 4.590 | 0.613 | 6.630 | 4.407 |
| $VRmin$ (p.u) | −4.590 | −0.613 | −6.630 | −4.407 |
| $Ke$ (p.u) | −0.02 | −0.0769 | −0.170 | −0.170 |
| $Te$ (s) | 0.560 | 1.370 | 0.950 | 0.950 |
| $Kf$ (p.u) | 0.050 | 0.040 | 0.040 | 0.040 |
| $Tf$ | 1.300 | 1.000 | 1.000 | 1.000 |
| $E1$ | 2.5875 | 3.0975 | 6.375 | 4.2375 |
| $SE(E1)$ | 0.7298 | 0.1117 | 0.2174 | 0.2174 |
| $E2$ | 3.450 | 4.130 | 8.500 | 5.650 |
| $SE(E2)$ | 1.3496 | 0.2248 | 0.9388 | 0.9386 |

**Table 5.4** IEEE 30-bus modified test system governor data.

| Type | BPA_GG | BPA_GG |
| --- | --- | --- |
| Unit no. | 1 | 2 |
| Rated power (MVA) | 270 | 51.2 |
| Rated voltage (kV) | 18 | 13.8 |
| *Pmax* (p.u) | 1.0018 | 1.035 |
| *R* (p.u) | 0.0185 | 0.1523 |
| *T*1 (s) | 0.100 | 0.200 |
| *T*2 (s) | 0.000 | 0.000 |
| *T*3 (s) | 0.259 | 0.300 |
| *T*4 (s) | 0.100 | 0.090 |
| *T*5 (s) | 10.000 | 0.000 |
| *F* | 0.272 | 1.000 |

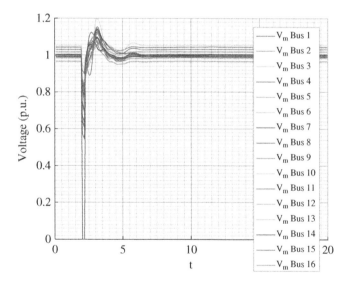

**Figure 5.10** Simulation results by fixed time step.

As shown in Figures 5.10 and 5.11, the simulation results by fixed time step and variable time step are consistent. However, when we zoom into the details as shown in Figures 5.12 and 5.13, the simulation by fixed time step missed transient during the fault. Since the time step is fixed, the fixed time step simulation cannot accurately simulate the fault state time and clear time. In the fixed time

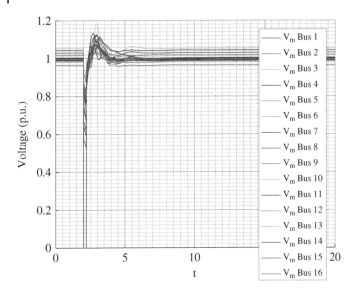

**Figure 5.11** Simulation results by variable time step.

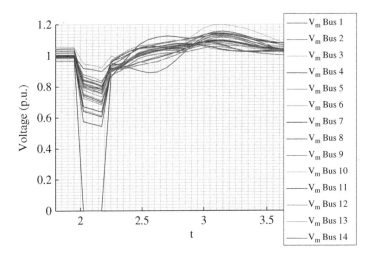

**Figure 5.12** Simulation result details by fixed time step.

step simulation, the time step is fixed at 0.015 seconds. The nearest time points of the fault starting time and the end time are 1.995 seconds which is the 133rd step ($133 \times 0.015 = 1.995$) and 2.205 seconds which is the 147th step ($147 \times 0.015 = 2.205$). In the variable time step simulation as shown in Figure 5.13, two break points at 2.0 and 2.2 seconds are placed to assure the short circuit fault event is not missing.

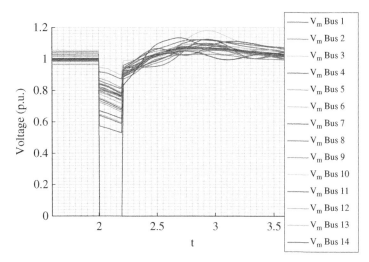

**Figure 5.13** Simulation result details by variable time step.

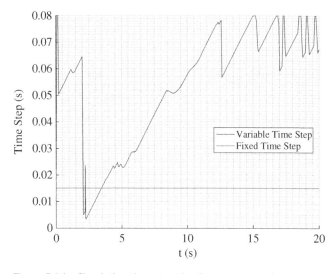

**Figure 5.14** Simulation time steps by the two approaches.

Figure 5.14 shows the simulation time steps by the two approaches. The fixed time step approach fixes the time step at 0.015 seconds and the variable time step approach starts the time step at 0.01 seconds initially and changes the time step upon the estimated truncation error adaptively and is limited by the maximum time step at 0.08 seconds.

Figures 5.15 and 5.16 show the truncation errors by the two approaches. By variable time step approach, the truncation error is controlled within $1.0 \times 10^{-4}$. However, by fixed time step, during

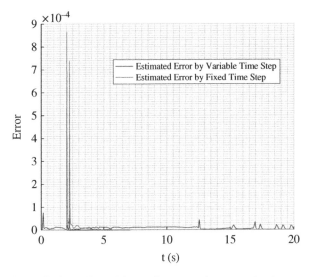

**Figure 5.15** Differential equation truncation errors by the two approaches.

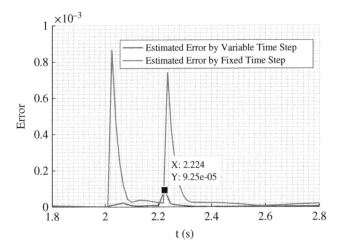

**Figure 5.16** Differential equation detailed truncation errors.

and after fault, the truncation errors jump up close to $1.0 \times 10^{-3}$. The variable time step approach has better overall simulation accuracy.

The simulator takes 1335 steps by fixed time step, but 569 steps by variable time step to simulate 20 seconds, resulting in 0.7648 seconds of calculation time by fixed time step, and 0.4107 seconds by variable time step. By taking graph parallel computing, the simulation time is further reduced from 0.4107 to 0.2566 seconds.

## 5.6 Summary

The mathematical model describing the dynamics of the power system is an ordinary differential equation (ODE) with initial values. There are generally two classes of numerical methods for solving ODEs: the single-step method and the linear multistep method. They have different local truncation errors, which are errors introduced by replacing the time derivative with a discrete-time approximation. In the integration step size control, the error generated at each time step using the finite difference derivative and the cumulative effect of the error generated at each step are considered separately.

In order to select the best integration method, one metric that matters are numerical stability. The absolute stability of a numerical integration method at a given time step means that under given initial conditions and small disturbances, the approximate solution remains bounded at all times. In addition, if the absolute stability region of the linear multistep method includes the entire left half of the complex plane, it is called A-stable. A-stability ensures that there is no limit to the time step. The region of absolute stability of the forward Euler method is very limited, while the backward Euler method covers the largest area in the complex plane. The backward Euler method, the trapezoidal method, and the Gear2 method are all A-stable.

In terms of algorithm efficiency, the linear multistep method has higher computational efficiency than the single-step method and is more suitable for solving rigid systems. The trapezoidal rule is the most commonly used integration method in power system transient simulation. The trapezoidal rule usually allows a larger time step, so the running time is shorter.

For all numerical integration methods used in power grid transient analysis, choosing an appropriate time step is very important. The choice of a time step is important to obtaining accurate results. If the time step is too large, the result will become inaccurate or even completely wrong in the absolutely stable region. If the time step is too small, the calculation will take longer without improving the accuracy. The adaptive time step shows advantages to obtain accurate results with the largest possible time step, so as to save calculation time.

In the transient simulation model, the time constants of mechanical system, electromechanical transient simulation, and electromagnetic transient simulation are seconds, milliseconds, and microseconds respectively. The multiple time step method can be used to coordinate these simulations with different time scales. In order to speed up the calculation, different time steps are used in the electromechanical-electromagnetic hybrid simulation. The process of hybrid simulation can be described as: when the electromechanical model is simulated, the electromagnetic model needs to be replaced by an appropriate equivalent circuit equivalent, and vice versa.

The breakpoint is also an issue that should be paid attention to in transient simulation. To ensure accurate simulation of events such as faults, the time step should be set at the correct breakpoint.

The transient simulation starts from the initial operating point of the system. Obviously, it is very important to reliably calculate the initial operating point for power system transient simulation. In TS analysis, the initial operating point is an equilibrium point that is assumed to be stable, and the equilibrium point does not change with time. Therefore, the basic algorithm for calculating the initial operating point is to make the differential-algebraic equation representing the dynamic behavior of the power system a zero solution. Solving this system of equations is challenging. These equations are nonlinear, and convergence is the primary consideration when solving nonlinear equations. Usually, Newton's method is used to iteratively solve nonlinear equations.

The power system transient simulation involves large-scale differential-algebraic equations calculation. To improve computation efficiency, adaptive time step, and parallel computing are essential. The graph-based power adaptive time step approach in this chapter exploited the high-order multistep method, the Dormand-Prince45 method, and the parallel computing capability using graph computing. The study results show evidence of the high computation efficiency of the method.

## References

1 Najm, F.N. (2010). *Circuit Simulation*. Hoboken, NJ/Canada: Wiley.
2 Chuan, F. (2011). High-speed extended-term time-domain simulation for online cascading analysis of power system. PhD dissertation, Iowa State University.
3 Lin, N. and Dinavahi, V. (2019). Variable time-stepping modular multilevel converter model for fast and parallel transient simulation of multi terminal DC grid. *IEEE Transactions on Industrial Electronics* 66 (9): 6661–6670.
4 Duan, T. and Dinavahi, V. (2020). Adaptive time-stepping universal line and machine models for real time and faster-than-real-time hardware emulation. *IEEE Transactions on Industrial Electronics* 67 (8): 6173–6182.
5 Christoffersen, C.E. and Alexander, J. (2004). An adaptive time step control algorithm for nonlinear time domain envelope transient. *Canadian Conference on Electrical and Computer Engineering 2004 (IEEE Cat. No.04CH37513)*, Niagara Falls, ON, Canada, vol. 2, pp. 883–886. https://doi.org/10.1109/CCECE.2004.1345256.
6 Sanchez-Gasca, J.J., D'Aquila, R., Paserba, J.J. et al. (1993). Extended-term dynamic simulation using variable time step integration. *IEEE Computer Applications in Power* 6 (4): 23–28.
7 Kundert, K.S. (1995). *The Designer's Guide to SPICE and SPECTRE*. Boson/Dordrecht/London: Kluwer Academic Publishers.
8 Cai, Y., Zhang, J., and Yu, W. (2019). A predictor-corrector method for power system variable step numerical simulation. *IEEE Transactions on Power Systems* 34 (4): 3283–3285.
9 Huang, Q. (2016). Electromagnetic transient and electromechanical transient stability hybrid simulation: design, development and its applications. PhD dissertation, Arizona State University.
10 Wu, J., Bose, A., Huang, J. et al. (1995). Parallel implementation of power system transient stability analysis. *IEEE Transactions on Power Apparatus and Systems* 10 (3): 1226–1233.
11 Aristidou, P., Fabozzi, D., and Van Cutsem, T. (2014). Dynamic simulation of large-scale power systems using a parallel schurcomplement-based decomposition method. *IEEE Transactions on Parallel and Distributed Systems* 25 (10): 2561–2570.
12 Demetriou, P., Asprou, M., Quiros-Tortos, J., and Kyriakides, E. (2017). Dynamic IEEE test systems for transient analysis. *IEEE Systems Journal* 11 (4): 2108–2117.

# 6

# Optimization Problems

## 6.1 Optimization Theory

Optimization theory is an important tool in decision science and power system operation analysis. In order to use this tool, we first need to identify optimization objectives, quantitatively measure and the performance of the system under study. These optimization metrics can be power generation costs, energy losses, or a combination of them. Furthermore, in an optimization problem, the controlled and observed variables are limited or constrained to some extent, such as power balance, generator output upper and lower limits, ramp up and ramp down limit, minimum uptime, minimum downtime, regulation reserve, contingency reserve, network security constraints, water balance constraints, dropped water and discarded water limit and water head limit. The process of identifying objectives, variables, and constraints for a given problem is called modeling. Once the model is formulated, an optimization algorithm can be used to find its solution. In this chapter, we focus on optimization methods, including linear programming and nonlinear programming, and how to solve optimization problems in parallel.

## 6.2 Linear Programming

Linear programming is widely used in solving power system optimization problems. Linear programming finds optimal solutions to problems that have a linear objective function and linear constraints, which may include both equalities and inequalities. Although few practical power system optimization problems can be perfectly modeled by a set of linear objective functions and linear constraints, linear programs provide fundamental means to solve nonlinear and mixed integer programming problems in power systems.

The standard linear programming problem is formulated as follows:

$$\min \ c_1 x_1 + c_2 x_2 + \ldots + c_n x_n \tag{6.1}$$

Subject to

$$
\begin{aligned}
a_{11} x_1 + a_{12} x_2 + \ldots + a_{1n} x_n &= b_1 \\
a_{21} x_1 + a_{22} x_2 + \ldots + a_{2n} x_n &= b_2 \\
&\vdots \\
a_{m1} x_1 + a_{m2} x_2 + \ldots + a_{mn} x_n &= b_m
\end{aligned}
\tag{6.2}
$$

*Graph Database and Graph Computing for Power System Analysis*, First Edition. Renchang Dai and Guangyi Liu.
© 2024 The Institute of Electrical and Electronics Engineers, Inc. Published 2024 by John Wiley & Sons, Inc.

and

$$x_1 \geq 0, x_2 \geq 0, ..., x_n \geq 0 \tag{6.3}$$

where $c_i$, $b_i$, and $a_{ij}$ are constants, and $x_i$ are variables to be optimized. In vector notation, the standard problem is formed as:

$$\min\ c^T x \tag{6.4}$$

Subject to

$$Ax = b \tag{6.5}$$

and

$$x \geq 0 \tag{6.6}$$

where $x \in \mathbb{R}^n$, $c \in \mathbb{R}^n$, $b \in \mathbb{R}^m$, $A \in \mathbb{R}^{m \times n}$. $x \geq 0$ ensures $x$ is a nonnegative variable.

Linear problems may have various other forms which can be converted to the standard form.

Example 1: Maximum problem $\max\ c^T x$ can be converted to be minimum problem $\min - c^T x$.

Example 2: Any inequality constraint $a_{i1}x_1 + a_{i2}x_2 + ... + a_{in}x_n \leq b_i$ can be converted to an equality constraint with a new positive slack variable $y_i$ as:

$$a_{i1}x_1 + a_{i2}x_2 + ... + a_{in}x_n + y_i = b_i, y_i \geq 0$$

Example 3: Similar to Example 2, any inequality constraint $a_{i1}x_1 + a_{i2}x_2 + ... + a_{in}x_n \geq b_i$ can be converted to an equality constraint with a new positive slack variable $y_i$ as:

$$a_{i1}x_1 + a_{i2}x_2 + ... + a_{in}x_n - y_i = b_i, y_i \geq 0$$

Example 4: $x_i$ is not restricted by $x_i \geq 0$. Thus $x_i$ is free to take on either positive, negative values, or zero. Two methods can be taken to convert the case to the linear programming standard form.

In the first method, two nonnegative variables $u_i$ and $v_i$ are introduced to represent $x_i$ as $x_i = u_i - v_i$, $u_i \geq 0$, and $v_i \geq 0$. In this method, the standard form is maintained but expressed in terms of $n + 1$ nonnegative variables which introduces redundancy since any constant added to $u_i$ and $v_i$ does not change the value of $x_i = u_i - v_i$. But this does not hinder the simplex method to solve the converted linear problem [1].

The second method eliminates $x_i$ with any one of the linear equality constraints in which the coefficient of the $x_i$ term is not zero.

$$a_{k1}x_1 + a_{k2}x_2 + ... + a_{ki}x_i + ... + a_{kn}x_n = b_k$$

where $a_{ki} \neq 0$. Then $x_i$ can be expressed by

$$x_i = \frac{b_k - (a_{k1}x_1 + a_{k2}x_2 + ... + a_{k,i-1}x_{i-1} + a_{k,i+1}x_{i+1} + ... + a_{kn}x_n)}{a_{ki}} \tag{6.7}$$

All $x_i$'s in (6.4) and (6.5) are substituted by (6.7). The standard linear program is simplified with $n - 1$ variables and $m - 1$ constraints. The value of $x_i$ is determined by other $n - 1$ variables through (6.7) after the simplified linear program is solved.

### 6.2.1 The Simplex Method

The feasible set defined by $Ax = b$, $x \geq 0$ is vertices of a convex polytope, or possibly unbounded which represent the basic feasible solutions of the linear program. For a linear program in standard form $\min_x c^T x, Ax = b, x \geq 0$, if the objective function has a minimum value, it must be on one of the vertices [2]. This self-evident theorem guarantees the objective function can be optimized or proven unbounded in a finite computation since the number of vertices is finite. However, for real power system optimization problems, the number of vertices is large. An exhaustive search on all vertices is impractical to solve real problems. If a vertex is not a minimum point of the objective function, then there is an edge the vertex connects leads to its adjacent vertex where the objective function decreases or is unbounded [3]. The searching strategy is repeated till the objective function is no longer decreasing where the linear program is solved or unbounded where the linear program has no solution. The method moves along the edges in the objective function decreasing direction, thus the number of vertices visited will be small. Motivated by this search strategy, the simplex method was developed by George Dantzig to solve linear programming problems [4–7]. The simplex method search process is depicted in Figure 6.1.

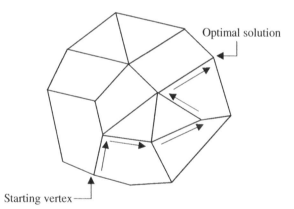

Optimal solution

Starting vertex

**Figure 6.1** Search process on polytope by simplex method.

The classic simplex method takes two steps to find the optimal, infeasible, or unbounded solution to a linear programming problem. The first step is to find a start basic solution or infeasibility of the problem if no solution is found. In the second step, starting from the basic feasible solution found in the first step, the simplex algorithm is applied to either find the optimal solution to the linear programming problem or the objective function is unbounded.

#### 6.2.1.1 Basic Feasible Solution

Consider the linear programming problem equality constraints $Ax = b$, where $x \in \mathbb{R}^n$, $b \in \mathbb{R}^m$, $A \in \mathbb{R}^{m \times n}$, $n > m$. For notational simplicity, assume that there exists matrix $A_B$ which is formed by the first $m$ columns of $A$ and matrix $A_B$ is nonsingular. If we cannot form $A_B$ by selecting any independent $m$ columns from $A$, it can be formed by adding artificial variables. Assume the matrix $A$ and vector $x$ are partitioned as $A = [A_B, A_N]$, $x = (x_B, x_N)$. $x_B$ and $x_N$ are basic variables and non-basic variables. Since $A_B$ is nonsingular, by Gauss elimination, the equality constraints $Ax = b$ can be converted to the following canonical form:

$$[A_B\,A_N]\begin{bmatrix} x_B \\ x_N \end{bmatrix} = b \tag{6.8}$$

$$A_B x_B + A_N x_N = b \tag{6.9}$$

$$x_B + A_B^{-1} A_N x_N = A_B^{-1} b \tag{6.10}$$

$$[I\,A_B^{-1}A_N]\begin{bmatrix} x_B \\ x_N \end{bmatrix} = A_B^{-1} b \tag{6.11}$$

By assigning the non-basic variables $x_N$ to be equal to 0 and the values of the basic variables $x_B$ to be equal to the right-hand sides of the constraints (6.11), $x = (x_B, x_N) = \left(A_B^{-1}b, 0\right)$ is a basic solution to the linear programming problem. However, the basic solution may be infeasible since $A_B^{-1}b$ could be negative which does not meet the requirement that $x$ is nonnegative. To mitigate the conflict, an artificial variable is added. For example, assume the component $b_k$ of vector $A_B^{-1}b$ in row $k$ is negative, and two nonnegative variables $u_k$ and $v_k$ are introduced to represent the corresponding basic variable $x_k$ as $x_k = u_k - v_k$, $u_k \geq 0$ and $v_k \geq 0$. And the basic solution for $u_k$ and $v_k$ are 0 and $-b_k$ ($-b_k > 0$) respectively to ensure the basic solution is feasible. In this method, the canonical form is maintained but expressed in terms of $n + 1$ nonnegative variables.

### 6.2.1.2 The Simplex Iteration
Without loss of generality, consider the linear programming problem in canonical form:

$$\min_x c^T x \tag{6.12}$$

Subject to

$$\begin{bmatrix} 1 & & & a_{1,m+1} & a_{1,m+2} & \cdots & a_{1,n} \\ & 1 & & a_{2,m+1} & a_{2,m+2} & \cdots & a_{2,n} \\ & & \ddots & \vdots & \vdots & \vdots & \vdots \\ & & 1 & a_{m,m+1} & a_{m,m+2} & \cdots & a_{m,n} \end{bmatrix} \begin{bmatrix} x_1 \\ x_2 \\ \vdots \\ x_m \\ x_{m+1} \\ x_{m+2} \\ \vdots \\ x_n \end{bmatrix} = \begin{bmatrix} b_1 \\ b_2 \\ \vdots \\ b_m \end{bmatrix} \tag{6.13}$$

$$x \geq 0$$

Assume the matrix $A$, vector $x$ and $c^T$ are partitioned as $A = [I, A_N]$, $x = (x_B, x_N)$, $c^T = [c_B^T, c_N^T]$. To simplify the illustration, the tableau format is adopted instead of writing the linear objective function and linear constraints explicitly.

$$\begin{bmatrix} -c^T & 0 \\ A & b \end{bmatrix} \tag{6.14}$$

The tableau form of the linear programming problem representing by basic variables and non-basic variables is:

$$\begin{bmatrix} -c_B^T & -c_N^T & 0 \\ I & A_N & b \end{bmatrix} \tag{6.15}$$

#### 6.2.1.2.1 Pivot Operation

Where the first $m$ columns of $A$ are linearly independent, the linear programming problem is in canonical form. The definition of canonical form is relaxed as that a system is in canonical form if by reordering the equations and the variables, it can be converted to the form as in (6.15). To replace the basic variable $x_p$ with the non-basic variable $x_q$, row $p$ is divided by $a_{pq}$, and then eliminate the $x_q$ term in all other equations by the following pivot equations.

$$\begin{cases} a_{ij}' = a_{ij} - \dfrac{a_{iq}}{a_{pq}} a_{pj}, \forall i \neq p \\ a_{pj}' = \dfrac{a_{pj}}{a_{pq}}, a_{pq} \neq 0 \end{cases} \tag{6.16}$$

The pivot operation can generate a new basic solution by replacing one basic variable (leaving variable) with a non-basic variable (entering variable). However, we need to determine the leaving variable and entering variable to maintain feasibility and decrease the objective function.

#### 6.2.1.2.2 Select Entering Variable

The entering variable $x_q$ is selected from $x_N$ where $q$ corresponds to the most negative number of vector $c^T$ which indicates the entering basic variable $x_q$ will decrease the objective function the fastest. Then $x_N$ is updated by removing $x_q$, while $x_B$ is updated by adding $x_q$.

#### 6.2.1.2.3 Select Leaving Variable

After selecting the entering basic variable $x_q$, the following ratio test is performed to select the leaving variable:

$$p = \min_{1 \leq i \leq m} \left( \frac{b_i}{a_{iq}} : a_{iq} > 0 \right) \tag{6.17}$$

In the test, all right-hand side components $b_i$ is divided by the corresponding coefficients in column $q$ of matrix $A$. This test is not performed with non-positive column coefficients. The basic variable $x_p$ with the smallest positive ratio is selected as leaving variable. If the ratio test is a tie, the row can be selected arbitrarily. Then $x_B$ is updated by removing $x_p$, while $x_N$ is updated by adding $x_p$.

The ratio test ensures the right-hand side will stay positive. The simplex iteration is repeated until there is no longer a positive number in the transformed $-c^T$, the first row of the tableau form.

The pseudocode for the simplex method is shown in Algorithm 6.1.

---

**Algorithm 6.1 Simplex Method**

---

1: Form standard linear program $\min_x c^T x$, $Ax = b$, $x \geq 0$

2: Build the initial simplex table $\begin{bmatrix} -c^T & 0 \\ A & b \end{bmatrix}$

3: Do While negative number in the vector $c^T$

4:    Select the entering variable $x_q$ by the most negative number of vector $c^T$

5:    Select leaving variable $x_p$ with the smallest positive ratio by ratio test (6.17)

6:    Normalize row $p$ by $a'_{pj} = \dfrac{a_{pj}}{a_{pq}}$, $a_{pq} \neq 0$

7:    Eliminate the $x_q$ term in all other rows by pivoting
$a'_{ij} = a_{ij} - \dfrac{a_{iq}}{a_{pq}} a_{pj}$, $\forall i \neq p$

---

The following simple example is taken to illustrate how to apply the simplex method to solve a linear program problem.

$$\min x_1 + x_2 = z \tag{6.18}$$

Subject to

$$2x_1 + x_2 \leq 12 \tag{6.19}$$

$$x_1 + 2x_2 \leq 9 \tag{6.20}$$

$$x_1 \geq 0, x_2 \geq 0 \tag{6.21}$$

The linear programming problem is formed as the following standard problem by adding slack variables $x_3$ and $x_4$.

$$\min x_1 + x_2 = z \tag{6.22}$$

Subject to

$$2x_1 + x_2 + x_3 = 12 \tag{6.23}$$

$$x_1 + 2x_2 + x_4 = 9 \tag{6.24}$$

$$x_1 \geq 0, x_2 \geq 0, x_3 \geq 0, x_4 \geq 0 \tag{6.25}$$

The initial tableau to represent the linear programming problem is in Table 6.1.

In this tableau, the basic variables are $x_3$ and $x_4$. And the non-basic variables are $x_1$ and $x_2$. $C_1 = C_2 = -1$ are tied, we select $x_1$ as the entering basic variable.

**Table 6.1** Initial tableau.

|       | $x_1$ | $x_2$ | $x_3$ | $x_4$ | $b$ | Ratio |
|-------|-------|-------|-------|-------|-----|-------|
| $C$   | $-1$  | $-1$  | 0     | 0     | 0   | –     |
| $x_3$ | 2     | 1     | 1     | 0     | 12  | $12/2 = 6$ |
| $x_4$ | 1     | 2     | 0     | 1     | 9   | $9/1 = 9$ |

**Table 6.2** Tableau after the first pivot.

|       | $x_1$ | $x_2$ | $x_3$ | $x_4$ | $b$ | Ratio |
|-------|-------|-------|-------|-------|-----|-------|
| $C$   | 0     | $-1/2$ | $1/2$ | 0    | 6   | –     |
| $x_3$ | 1     | $1/2$ | $1/2$ | 0    | 6   | $6/(1/2) = 12$ |
| $x_4$ | 0     | $3/2$ | $-1/2$ | 1   | 3   | $3/(3/2) = 2$ |

**Table 6.3** Tableau $x_4$ is selected as leaving variable.

|       | $x_1$ | $x_2$ | $x_3$ | $x_4$ | $b$ |
|-------|-------|-------|-------|-------|-----|
| $C$   | 0     | $-1/2$ | $1/2$ | 0    | 6   |
| $x_3$ | 1     | $1/2$ | $1/2$ | 0    | 6   |
| $x_4$ | 0     | 1     | $-1/3$ | $2/3$ | 2 |

**Table 6.4** Tableau after pivot.

|       | $x_1$ | $x_2$ | $x_3$ | $x_4$ | $b$ | Ratio |
|-------|-------|-------|-------|-------|-----|-------|
| $C$   | 0     | 0     | $1/3$ | $1/3$ | 7   | —     |
| $x_3$ | 1     | 0     | $2/3$ | $1/6$ | 5   | —     |
| $x_4$ | 0     | 1     | $-1/3$ | $2/3$ | 2  | —     |

According to the ratio test in the last column, $x_3$ is selected as the leaving variable. By the pivot operation, the next tableau is shown in Table 6.2.

$C_2 = -\dfrac{1}{2}$ is the most negative number of $C^T$. Thus $x_2$ is selected as the entering basic variable in the second search. According to the ratio test, $x_4$ is selected as the leaving variable. The new tableau is shown in Table 6.3.

By the pivot operation, the tableau is shown in Table 6.4.

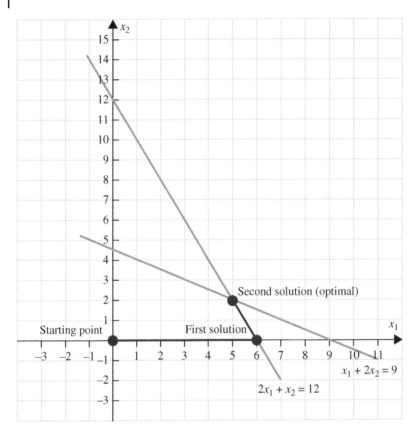

**Figure 6.2** Simplex method solving process.

Since all elements of $C^T$ are non-negative, no further ratio test is necessary. The solution $x_1 = 5$, $x_2 = 2$, $z = 7$ in Table 6.4 is optimal. The solving process is illustrated in Figure 6.2.

## 6.2.2 Interior-Point Methods

The simplex method is popular, practical, and very efficient to solve linear programming problems. However, it's not a polynomial algorithm. In the 1970s, Khachiyan and other mathematicians developed and proved that the ellipsoid method is a polynomial algorithm for linear programming [8, 9]. However, the ellipsoid method is hard to implement and shows much slower than the simplex method in almost all cases. The first practical polynomial algorithm to solve linear programming problems, known as interior-point methods, was proposed by Karmarker in 1984 [10].

By the early 1990s, the primal-dual interior point method was invented to be the most effective and practical method for solving large-scale linear programming problems [11–13]. Unlike simplex methods which search along the edges of a polytope defined by the constraints, interior point methods iterate inside the polytope toward the optimal point by which the number of iterations is reduced to reach the optimal solution. Another merit of interior point methods is they can not only solve linear programming problems but also solve certain types of conic programming problems, such as second-order conic programming and semidefinite programming.

Consider the primal linear programming problem as follow,

$$\min_x c^T x \tag{6.26}$$

Subject to

$$Ax = b, x \geq 0 \tag{6.27}$$

where $c$ and $x$ are vectors in $\mathbb{R}^n$, $b$ is a vector in $\mathbb{R}^m$, and $A \in \mathbb{R}^{m \times n}$ with full-row rank. The dual problem for (6.26) and (6.27) is

$$\max_y b^T y \tag{6.28}$$

Subject to

$$A^T y + s = c, \ s \geq 0 \tag{6.29}$$

where $y$ is a vector in $\mathbb{R}^m$ and $s$ is the slack variable vector in $\mathbb{R}^n$. Based on the logarithmic barrier method, the Lagrangian equation for the primal problem is [13–16]

$$L_p(x,y) = c^T x - \mu \sum_{j=1}^n \log(x_j) - y^T(Ax - b) \tag{6.30}$$

The Lagrangian equation for the dual problem is

$$L_d(x,y,s) = b^T y + \mu \sum_{j=1}^n \log(s_j) - x^T \tag{6.31}$$

where $\mu > 0$ is the weighting parameter. Clearly, when $\mu = 0$, the Lagrangian Eqs. (6.30) and (6.31) correspond to the original primal-dual problems (6.26) (6.27) and (6.28) (6.29).

The solutions of (6.26) (6.27) and (6.28) (6.29) are characterized by the Karush–Kuhn–Tucker (KKT) conditions by taking partial derivatives of $L_p(x, y)$ and $L_d(x, y, s)$ in respect of variables $x$, $y$, $s$, and forcing the derivatives to zero, we get:

$$Ax - b = 0, x \geq 0 \tag{6.32}$$

$$A^T y + s - c = 0, s \geq 0 \tag{6.33}$$

$$\frac{\mu}{s_j} - x_j = 0, j = 1, 2, ..., n \tag{6.34}$$

For each $\mu > 0$, the solutions satisfying the above Eqs. (6.32)–(6.34) form a trajectory in the interior of the feasible region. That's the name of the interior point method comes from. Starting with a positive value, as the parameter $\mu$ approaches 0, the optimal point is reached. The last Eq. (6.34) implies $\mu = x_j s_j$, $\forall j = 1, 2, \ldots, n$ which are relaxed complementary slackness conditions. In interior-point methods, the average distance is used to control the weighting parameter.

$$\mu = \frac{1}{n} \sum_{j=1}^{n} x_j s_j \tag{6.35}$$

To find solutions $(x^*, y^*, s^*)$ of the system, Newton's method is applied to the three equalities in KKT conditions and modifying the search directions and step lengths so that the inequalities $x \geq 0$ and $s \geq 0$ are satisfied strictly at every iteration.

By defining $X = \text{diag}(x_1, \ldots, x_n)$, $S = \text{diag}(s_1, \ldots, s_n)$, and $e = (1, \ldots, 1)^T$, the optimality conditions of primal-dual problems are restated as follows:

$$F(x, y, s) = \begin{bmatrix} Ax - b \\ A^T y + s - c \\ XSe - \mu e \end{bmatrix} = 0 \tag{6.36}$$

$$x \geq 0, s \geq 0 \tag{6.37}$$

Primal-dual methods generate iterates $(x_k, y_k, s_k)$ that satisfy Eq. (6.36) and the bounds (6.37). Like most iterative algorithms in optimization, primal-dual interior-point methods have two basic elements: a procedure for determining the searching direction and a measure of finding the step size in the search space. The procedure for determining the search direction is using Newton's method to solve the nonlinear Eqs. (6.36). Newton's method forms a linear model for $F(x, y, s)$ around the current point and obtains the search direction $(\Delta x, \Delta y, \Delta s)$ by solving the following linearized system:

$$J(x, y, s) \begin{bmatrix} \Delta x \\ \Delta y \\ \Delta s \end{bmatrix} = -F(x, y, s) \tag{6.38}$$

where $J(x, y, s)$ is the Jacobian of $F(x, y, s)$. The iteration equation to update the Newton direction is to solve the following linear equation:

$$\begin{bmatrix} A & 0 & 0 \\ 0 & A^T & I \\ S & 0 & X \end{bmatrix} \begin{bmatrix} \Delta x \\ \Delta y \\ \Delta s \end{bmatrix} = \begin{bmatrix} b - Ax_k \\ c - A^T y_k - s_k \\ \mu_k e - X_k S_k e \end{bmatrix} \tag{6.39}$$

Usually, a full step in this direction may violate the bound $(x, s) \geq 0$, so we perform a line search along the Newton direction and define the new iterate as:

$$
\begin{bmatrix} x_{k+1} \\ y_{k+1} \\ s_{k+1} \end{bmatrix} = \begin{bmatrix} x_k \\ y_k \\ s_k \end{bmatrix} + \begin{bmatrix} \alpha_p \Delta x \\ \alpha_d \Delta y \\ \alpha_d \Delta s \end{bmatrix}
\tag{6.40}
$$

To ensure $(x_{k+1}, s_{k+1}) \geq 0$, $\alpha_p$ and $\alpha_d$ are calculated as follows:

$$
\alpha_p = \min\left\{ 1, \min_{\Delta x_j < 0}\left( \frac{x_{j,k}}{-\Delta x_j} \right) \right\}
\tag{6.41}
$$

$$
\alpha_d = \min\left\{ 1, \min_{\Delta s_j < 0}\left( \frac{s_{j,k}}{-\Delta s_j} \right) \right\}
\tag{6.42}
$$

The iteration is taken until the duality gap is within a given small tolerance $\varepsilon$.

$$
\frac{c^T x_k - b^T y_k}{1 + \left| b^T y_k \right|} < \varepsilon
\tag{6.43}
$$

The pseudocode for the primal-dual interior-point method is shown in Algorithm 6.2.

---

**Algorithm 6.2    Primal-Dual Interior-Point Method**

---

1: Initialize $(x_0, \ y_0, \ s_0, \ \alpha_p, \ \alpha_d)$ with $(x_0, \ s_0) \geq 0$, $0 < (\alpha_p, \ \alpha_d) \leq 1$

2: $\mu_0 = \dfrac{1}{n}\sum_{j=1}^{n} x_{0j} s_{0j}$

3: Loop through steps 4-7 until $\dfrac{c^T x_k - b^T y_k}{1 + \left| b^T y_k \right|} < \varepsilon$

4:      Solve $\begin{bmatrix} A & 0 & 0 \\ 0 & A^T & I \\ S & 0 & X \end{bmatrix} \begin{bmatrix} \Delta x \\ \Delta y \\ \Delta s \end{bmatrix} = \begin{bmatrix} b - A x_k \\ c - A^T y_k - s_k \\ \mu_k e - X_k S_k e \end{bmatrix}$

5:      Set $\begin{bmatrix} x_{k+1} \\ y_{k+1} \\ s_{k+1} \end{bmatrix} = \begin{bmatrix} x_k \\ y_k \\ s_k \end{bmatrix} + \begin{bmatrix} \alpha_p \Delta x \\ \alpha_d \Delta y \\ \alpha_d \Delta s \end{bmatrix}$

6:      Update $\alpha_p = min\left\{ 1, \min_{\Delta x_j < 0}\left( \frac{x_{j,k}}{-\Delta x_j} \right) \right\}$   and   $\alpha_d = min\left\{ 1, \min_{\Delta s_j < 0}\left( \frac{s_{j,k}}{-\Delta s_j} \right) \right\}$

7:      Update $\mu_{k+1} = \mu_k / 10$

---

Taking the same linear programming problem in 6.2.1 as a numerical example,

$$\min c^T x = -x_1 - x_2 \tag{6.44}$$

Subject to

$$\begin{aligned} 2x_1 + x_2 &\leq 12 \\ x_1 + 2x_2 &= 9 \end{aligned} \tag{6.45}$$

and

$$(x_1, x_2) \geq 0 \tag{6.46}$$

Adding slack variables $x_3$ and $x_4$, the problem is converted into standard form

$$\min c^T x = -x_1 - x_2 \tag{6.47}$$

Subject to

$$\begin{aligned} 2x_1 + x_2 + x_3 &= 12 \\ x_1 + 2x_2 + x_4 &= 10 \end{aligned} \tag{6.48}$$

and

$$(x_1, x_2, x_3, x_4) \geq 0 \tag{6.49}$$

The matrix $A$, the vectors $c$ and $b$ of the problem in the standard form setting is:

$$A = \begin{bmatrix} 2 & 1 & 1 & 0 \\ 1 & 2 & 0 & 1 \end{bmatrix} \tag{6.50}$$

$$b = \begin{bmatrix} 12 \\ 9 \end{bmatrix} \tag{6.51}$$

$$c^T = \begin{bmatrix} -1 & -1 & 0 & 0 \end{bmatrix} \tag{6.52}$$

The initial condition is set as $x^T = (0, 0, 12, 9)$, $y^T = [-1\,-1]'$, $s^T = (1, 1, 1, 1)$. It's easy to verify the initial condition satisfies the primal problem but not the dual problem. At the initial point, the iteration equation $\begin{bmatrix} A & 0 & 0 \\ 0 & A^T & -I \\ S & 0 & X \end{bmatrix} \begin{bmatrix} \Delta x \\ \Delta y \\ \Delta s \end{bmatrix} = \begin{bmatrix} b - Ax_k \\ c - A^T y_k + s_k \\ \mu_k e - X_k S_k e \end{bmatrix}$ is formed numerically as:

$$
\begin{bmatrix}
2 & 1 & 1 & & & & & & & & \\
1 & 2 & & 1 & & & & & & & \\
 & & 2 & 1 & 1 & & & & & & \\
 & & 1 & 2 & & 1 & & & & & \\
 & & 1 & & & & 1 & & & & \\
 & & & 1 & & & & 1 & & & \\
1 & & & & & & & & & & \\
 & 1 & & & & & & & & & \\
 & & 1 & & & & & & 12 & & \\
 & & & 1 & & & & & & 9 &
\end{bmatrix}
\begin{bmatrix} \Delta x \\ \Delta y \\ \Delta s \end{bmatrix}
=
\begin{bmatrix}
0 \\ 0 \\ 1 \\ 1 \\ 0 \\ 0 \\ 2.1 \\ 2.1 \\ -9.9 \\ -6.9
\end{bmatrix}
\tag{6.53}
$$

Solving the iteration Eq. (6.53), the Newton direction is

$$
\begin{bmatrix} \Delta x \\ \Delta y \\ \Delta s \end{bmatrix}
=
\begin{bmatrix}
2.1 \\ 2.1 \\ -6.3 \\ -6.3 \\ 0.3 \\ 0.0667 \\ 0.3333 \\ 0.5667 \\ -0.3 \\ -0.0667
\end{bmatrix}
\tag{6.54}
$$

**Table 6.5** Search iteration by interior-point method.

| Iteration | x | | | | y | | s | | | |
|---|---|---|---|---|---|---|---|---|---|---|
| 0 | 0.000 | 0.000 | 12.000 | 9.000 | −1.000 | −1.000 | 1.000 | 1.000 | 1.000 | 1.000 |
| 1 | 2.100 | 2.100 | 5.700 | 2.700 | −0.700 | −0.933 | 1.333 | 1.567 | 0.700 | 0.933 |
| 2 | 2.628 | 2.284 | 4.461 | 1.805 | −0.266 | −0.469 | 0.000 | 0.203 | 0.266 | 0.469 |
| 3 | 5.191 | 1.618 | 0.000 | 0.574 | −0.313 | −0.383 | 0.008 | 0.078 | 0.313 | 0.383 |
| 4 | 5.027 | 1.940 | 0.007 | 0.094 | −0.335 | −0.333 | 0.002 | 0.000 | 0.335 | 0.333 |
| 5 | 5.000 | 2.000 | 0.000 | 0.000 | −0.333 | −0.333 | 0.000 | 0.000 | 0.333 | 0.333 |

To ensure $(x_{k+1}, s_{k+1}) \geq 0$, $\alpha_p$ and $\alpha_d$ are calculated as 1 and 1 respectively and the new point is updated in the first iteration.

$$\begin{bmatrix} x_1 \\ y_1 \\ s_1 \end{bmatrix} = \begin{bmatrix} x_0 \\ y_0 \\ s_0 \end{bmatrix} + \begin{bmatrix} \alpha_p \Delta x \\ \alpha_d \Delta y \\ \alpha_d \Delta s \end{bmatrix} = \begin{bmatrix} 2.1 \\ 2.1 \\ 5.7 \\ 2.7 \\ -0.7 \\ -0.933 \\ 1.333 \\ 1.567 \\ 0.7 \\ 0.933 \end{bmatrix} \tag{6.55}$$

After five iterations, the solution is converged at the optimal solution $x_1 = 5$, $x_2 = 2$. The iteration results are shown in Table 6.5. And the search path is shown in Figure 6.3.

## 6.3 Nonlinear Programming

The mathematical model of power system optimization problems, such as security-constrained unit commitment, security-constrained economic dispatch, reactive power optimization, etc. is nonlinear programming. The general form of a nonlinear programming problem is formed by nonlinear objective function and/or nonlinear constraints, which may include both equalities and inequalities:

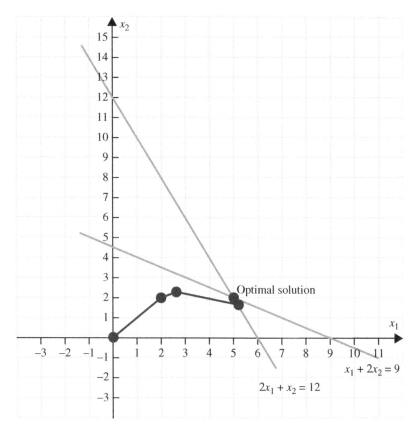

**Figure 6.3** Search path on polytope by interior-point method.

$$\min f(x) \tag{6.56}$$

Subject to

$$g_i(x) \leq 0 \ \forall i \in (1, 2, ..., m) \tag{6.57}$$

$$h_j(x) = 0 \ \forall j \in (1, 2, ..., p) \tag{6.58}$$

Nonlinear programming is well studied from unconstrained optimization approaches to constrained optimization approaches.

### 6.3.1 Unconstrained Optimization Approaches

Unconstrained optimization approaches are the basis of constrained optimization algorithms. Particularly, some of the constrained optimization problems in power system operation can be converted into unconstrained optimization problems. The major unconstrained optimization approaches that are used in power system operation are the gradient method, line search, Newton–Raphson optimization, trust-region optimization, Quasi-Newton method, double dogleg optimization, conjugate gradient optimization, and so on.

In unconstrained optimization, we minimize an objective function that depends on real variables, with no restrictions at all on the values of these variables. The mathematical formulation of an unconstrained optimization problem is $\min_x f(x)$, where $x \in R^N$ is a vector with $N$ components and $f(x)$ is an objective function.

The classical algorithm to solve the problem is to search a sequence of points $\{x_k : k \geq 0\}$, such that when $k \to \infty$, $x_k \to x^*$ where $\nabla f(x^*) = 0$. Each next point in this sequence is obtained from the previous point by moving a distance along with a feasible direction $p_k$ as follows:

$$x_{k+1} = x_k + \alpha p_k \tag{6.59}$$

where $\alpha$ is a scalar step length.

There are two fundamental strategies for deriving the distance and the direction, so as to solve the unconstrained optimization problems: line search and trust region.

### 6.3.1.1 Line Search

The line search algorithm selects a direction $p_k$ and searches along this direction in a new iterate with a lower function value from the current iterate $x_k$. The step size $\alpha$ to travel along the direction $p_k$ is approximately solved by the following one-dimensional minimization problem:

$$\min_{\alpha > 0} f(x_k + \alpha p_k) \tag{6.60}$$

By solving (6.60) exactly, we would derive the maximum benefit from the direction $p_k$, but an exact minimization may be expensive and is usually unnecessary. Instead, the line search algorithm generates a limited number of trial step lengths until it finds one that loosely approximates the minimum of (6.60). At the new point, a new search direction and step size are computed, and the process is repeated. Most line search algorithms require $p_k$ to be a descent direction, because this property guarantees that the function $f(x)$ can be reduced along this direction. The line search methods include the steepest descent method, Newton's method, Quasi-Newton methods, Newton's method with Hessian Modification, etc.

In Newton's method, the objective function is approximately expressed by the second-order Taylor series expansion at the point $x_k$:

$$f(x) = f(x_k) + f'(x_k)(x - x_k) + \frac{1}{2}f''(x_k)(x - x_k)^2 \tag{6.61}$$

In matrix format, that is

$$f(x) = f(x_k) + J(x_k)\Delta x + \frac{1}{2}\Delta x^T H(x_k)\Delta x \tag{6.62}$$

where $J(x_k) = f'(x_k)$ is the Jacobian matrix and $H(x_k) = f''(x_k)$ is the Hessian matrix. The necessary condition to achieve $\min_x f(x)$ is $\nabla f(x^*) = 0$. Applied the necessary condition to the Eq. (6.62), we get

$$\nabla f(x) = J(x_k) + H(x_k)\Delta x = 0 \tag{6.63}$$

$$\Delta x = -[H(x_k)]^{-1}J(x_k) \tag{6.64}$$

$$x_{k+1} = x_k + \Delta x \tag{6.65}$$

When the objective function $f(x)$ is quadratic, the Hessian matrix $H(x)$ is constant. The optimal solution will be obtained through one iteration. Otherwise, multiple iterations are needed to obtain the minimum of the objective function $f(x)$. $J(x)$ and $H(x)$ need to be updated in each iteration. In Newton's method, the search direction $p_k$ is

$$p_k = -[H(x_k)]^{-1}J(x_k) \tag{6.66}$$

The advantage of Newton's method is its second-order convergence but with the expensive cost of computing and memory usage of the inverse of the Hessian matrix. As illustrated in Chapter 3, using matrix factorization and substitution in graph parallel computing can expedite the process of solving linear Eq. (6.63) and save memory.

Most line search algorithms require $p_k$ to be a descent direction, one for which $p_k \nabla f_k < 0$, because this property guarantees that the function $f(x)$ can be reduced along this direction. Moreover, the search direction often has the form: $p_k = -B_k^{-1}\nabla f_k$, where $B_k$ is a symmetric and nonsingular matrix. In the steepest descent method, $B_k$ is simply the identity matrix $I$, while in Newton's method, $B_k$ is the exact Hessian $\nabla^2 f(x_k)$. In Quasi-Newton methods, $B_k$ approximates the Hessian that is updated at every iteration by means of a low-rank formula.

### 6.3.1.2 Trust Region Optimization

In the second algorithmic strategy, known as the trust region, the information gathered about $f(x)$ is used to construct a model function $m_k$ whose behavior near the current point $x_k$ is similar to that of the actual objective function $f(x)$. Because the model $m_k$ may not be a good approximation of $f(x)$ when $x$ is far from $x_k$, we restrict the search for a minimizer of $m_k$ to some regions around $x_k$. In other words, we find the candidate step $p$ by approximately solving the following sub-problem:

$$\min m_k(x_k + \Delta x) = \min f(x_k) + \Delta x^T \nabla f(x_k) + \frac{1}{2}p\Delta x^T H(x_k)\Delta x \tag{6.67}$$

Subject to

$$\|\Delta x\|_2 \leq \Delta \tag{6.68}$$

where the scalar $\Delta > 0$ is the trust-region radius to ensure $x_k + \Delta x$ lies inside the trust region. As the notation indicates, $f(x_k)$ and $\nabla f(x_k)$ are chosen to be the function and gradient values of the function at the point $x_k$, and $H(x_k)$ is the Hessian matrix $\nabla^2 f(x_k)$, so that $m_k$ and $f(x)$ are in agreement to second-order at the current iteration $x_k$.

Since the direction and the length of the step are optimized simultaneously, if the candidate solution does not produce a decrease in $f(x)$, the size of the region is reduced and (6.67) is resolved to find a new solution.

### 6.3.1.3 Quasi-Newton Method

Line search and trust region optimization calculate the Hessian matrix $H(x_k)$ at each iterate to proceed in the search direction. The Hessian matrix builds up the curvature information of the objective function $f(x)$ at the point $x_k$, but the computation of the Hessian matrix is intensive.

In recent decades, countless results on simple Newton methods have been developed due to the increasing scale of the problem size. The advantage of Quasi-Newton methods over the Hessian-based method is having low per-iteration cost since there is no need to directly calculate the Hessian matrix. In many cases, the curvature information of the objective function $f(x)$ at the point $x_k$ can even be replaced with approximation and convergence can still be guaranteed. The most notable work is the formula of Broyden, Fletcher, Goldfarb, and Shanno which by combining certain previous gradients with the current gradients to estimate and update Hessian [17–22].

$$H(x_{k+1}) = H(x_k) + \frac{QQ^T}{Q^T S} - \frac{(H(x_k))^T S^T S H(x_k)}{S^T H(x_k) S} \tag{6.69}$$

$$S = \Delta x = x_{k+1} - x_k \tag{6.70}$$

$$Q = \nabla f(x_{k+1}) - \nabla f(x_k) \tag{6.71}$$

In the approach to estimating the Hessian above, the starting point $H(x_0)$ can be exactly calculated by $\nabla^2 f(x_0)$ or set to be any symmetric positive definite matrix, for example, the simplest identity matrix $I$.

### 6.3.1.4 Double Dogleg Optimization

The double dogleg optimization method combines the steepest descent method (one of the line search methods) and the Quasi-Newton method in an explicit trust region [23, 24]. In the steepest descent method, the search direction is given by $p_{sd,k} = -\nabla f_k$ at iterate $k$. In Newton's methods, the search direction is $p_{gn,k} = -[H(x_k)]^{-1}J(x_k)$, where the Hessian $H(x_k)$ can be estimated by (6.69)–(6.71) in the Quasi-Newton method.

Along the steepest descent direction $p_{sd,k}$, the objective function is linearized as follows:

$$\min_{\alpha > 0} f(x_k + \alpha p_{sd,k}) = \min_{\alpha > 0} \frac{1}{2} \|f(x_k + \alpha p_{sd,k})\|^2 = \min_{\alpha > 0} F(x_k + \alpha p_{sd,k}) \tag{6.72}$$

$$\min_{\alpha > 0} F(x_k + \alpha p_{sd,k}) = \min_{\alpha > 0} F(x_k) + \alpha p_{sd,k}^T J^T(x_k) f(x_k) + \frac{1}{2}\alpha^2 \|J(x_k)p_{sd,k}\|^2 \tag{6.73}$$

The necessary condition to achieve the minimization in (6.73) is:

$$\frac{\partial F(x_k + \alpha p_{sd,k})}{\partial \alpha} = 0 \tag{6.74}$$

$$p_{sd,k}^T J^T(x_k) f(x_k) + \alpha \|J(x_k)p_{sd,k}\|^2 = 0 \tag{6.75}$$

$$\alpha = -\frac{p_{sd,k}^T J^T(x_k) f(x_k)}{\|J(x_k)p_{sd,k}\|^2} \tag{6.76}$$

$\alpha$ is the scaled step at the Cauchy point.

Given a trust region radius $\Delta > 0$, in the double dogleg optimization method, the final update step is selected in the following conditions:

If $\|p_{gn,k}\|_2 \leq \Delta$,

$$p_k = p_{gn,k} \tag{6.77}$$

**Figure 6.4** Dogleg step.

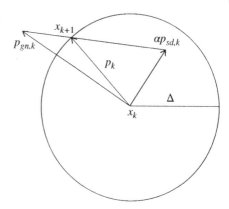

If $\|p_{gn,k}\|_2 > \Delta$ and $\|\alpha p_{sd,k}\|_2 > \Delta$,

$$p_k = \frac{\Delta p_{gn,k}}{\left\|p_{gn,k}\right\|_2} \tag{6.78}$$

If $\|p_{gn,k}\|_2 > \Delta$ and $\|\alpha p_{sd,k}\|_2 \leq \Delta$,

$$p_k = \alpha p_{sd,k} + \beta\left(p_{gn,k} - \alpha p_{sd,k}\right) \tag{6.79}$$

With $\beta$ such that the dogleg step $\|p_k\|_2 = \Delta$ as shown in Figure 6.4.

### 6.3.1.5 Conjugate Gradient Optimization

As described in 4.3.2.1, the conjugate gradient method is an efficient nonstationary method for solving symmetric positive definite systems of linear equations as follows [25, 26]:

$$Ax = b \tag{6.80}$$

where $A \in \mathbb{R}^{n \times n}$ is a symmetric positive definite matrix such that $x^T A x > 0 \; \forall x \in \mathbb{R}^n$ or equivalently all of the eigenvalues of $A$ are positive.

To solve symmetric positive definite linear Eq. (6.80) by conjugate gradient method, the iterates $x_k$ are updated in each iteration by adding a multiplication of $\alpha_k$ and $p_k$:

$$x_{k+1} = x_k + \alpha_k p_k \tag{6.81}$$

The corresponding residuals $r_k = b - Ax_k$ are updated as

$$r_{k+1} = r_k - \alpha_k A p_k \tag{6.82}$$

Where the scaled step $\alpha_k$ is given by

$$\alpha_k = \frac{r_k^T r_k}{p_k^T A p_k} \tag{6.83}$$

The search direction in the current iterate is updated using the previous search direction

$$p_{k+1} = r_{k+1} - \beta_k p_k \tag{6.84}$$

where

$$\beta_k = \frac{r_{k+1}^T r_{k+1}}{r_k^T r_k} \tag{6.85}$$

The choice of $\beta_k$ makes $r_{k+1}$ and $r_k$ orthogonal.

The conjugate gradient optimization method to minimize the following quadratic objective function (6.86) is motivated by the above conjugate gradient method.

$$\min f(x) = \min \frac{1}{2} x^T A x - x^T b \tag{6.86}$$

In the steepest descent method, the search direction is given by the negative gradient of the function $f(x)$ at the given point $x$. The scaled step is chosen to move the next iterate along the local optimal search direction until $f(x)$ is no longer descending. The steepest descent method convergence is guaranteed for a symmetric positive definite problem with a zigzag search. But the zigzag search strategy converges slowly toward the optimal solution. Thus, the conjugate gradient optimization method is developed to minimize the quadratic objective function $f(x)$ in $n$ steps by successively descending $f(x)$ along each of the conjugate directions.

Conjugate directions are defined as a set of nonzero vectors $\{p_1, p_2, ..., p_n\}$ in which any two unequal vectors are orthogonal with respect to a symmetric positive definite matrix $A$. That is

$$p_i^T A p_j = 0 \; \forall i \neq j \tag{6.87}$$

Since a set of $n$ such vectors are linearly independent and thus span the whole space $\mathbb{R}^n$, the difference between the exact solution $x^*$ and the initial guess $x^0$ can be expressed as a linear combination of the conjugate vectors, as named conjugate directions.

$$x^* - x^0 = \sum_{k=1}^{n} a_k p_k \tag{6.88}$$

Where the scaled step $\alpha_k$ is the same as the step length that minimizes the quadratic function $f(x)$ in (6.83) as rewritten as follow.

$$\alpha_k = \frac{r_k^T r_k}{p_k^T A p_k} \tag{6.89}$$

The eigenvectors of the matrix $A$ form the conjugate vectors but it invites a lot of computations to find the eigenvectors. The alternative is the conjugate vector $p_k$ which can be updated by using precious vector $p_{k-1}$ as the same as the conjugate gradient method to solve symmetric positive definite linear equations.

$$p_k = r_{k+1} - \beta_k p_{k-1} \tag{6.90}$$

where

$$\beta_k = \frac{r_{k+1}^T r_{k+1}}{r_k^T r_k} \tag{6.91}$$

The $\beta_k$ is found to impose $r_{k+1}$ and $r_k$ are orthogonal with respect to matrix $A$, i.e.

$$r_{k+1}^T A r_k = 0 \tag{6.92}$$

## 6.3.2 Constrained Optimization Approaches

A general formulation of constrained optimization approaches can be modeled as:

$$\min_{x \in \Omega} f(x) \tag{6.93}$$

where $\Omega = \{x | g_i(x) \leq 0 \, \forall i \in (1, 2, ..., m); h_j(x) = 0 \, \forall j \in (1, 2, ..., p)\}$.

### 6.3.2.1 Karush–Kuhn–Tucker Conditions

We define the Lagrangian function for the general problem (6.93):

$$L(x, \lambda) = f(x) - \sum_{i=1}^{m} \lambda_i g_i(x) - \sum_{j=1}^{p} \lambda_j h_j(x) \tag{6.94}$$

Assume that $x^*$ is an optimal solution of (6.93). The functions $f$ and $g_i$, $h_j$ in (6.93) are continuously differentiable, and the set of active constraint gradients $\{\nabla g_i(x), i \in (1, 2, ..., m)\}$ and $\{\nabla h_j(x), j \in (1, 2, ..., m)\}$ is linearly independent at $x^*$. Then there is a Lagrange multiplier vector $\lambda^*$, such that the following conditions are satisfied at $(x^*, \lambda^*)$.

$$\frac{\partial \mathcal{L}(x^*, \lambda^*)}{\partial x_i} = 0, \quad i \in (1, 2, ..., n) \tag{6.95}$$

$$g_i(x^*) \geq 0, \quad i \in (1, 2, ..., m) \tag{6.96}$$

$$h_j(x^*) = 0, \quad j \in (1, 2, ..., p) \tag{6.97}$$

$$\lambda_i^* g_i(x^*) = 0, \quad \lambda_i^* \geq 0 \,\, i \in (1, 2, ..., m) \tag{6.98}$$

The above conditions are known as the Karush–Kuhn–Tucker conditions, or KKT conditions for short. Condition (6.95) states that partial derivatives of the Lagrange function must equal to zero at the optimal point. Conditions (6.96) and (6.97) restate the nonlinear programming problem constraints. The fourth condition (6.98) is the complementary slackness condition.

### 6.3.2.2 Linear Approximations of Nonlinear Programming with Linear Constraints

Linear programming algorithms, such as the simplex method and the interior-point method have been successfully developed. Nonlinear programming problems in power system optimization can be approximated to be linear problems by linearizing nonlinear objective functions and/or nonlinear constraints.

When DC power flow is applied, security-constrained economic dispatch, optimal power flow, and reactive power optimization can be modeled as a nonlinear programming problem with nonlinear objective function constrained by linear equalities and inequalities which is generalized as:

$$\min f(x_1, x_2, ..., x_n) \tag{6.99}$$

Subject to

$$\sum_{j=1}^{n} a_{ij} x_j \leq b_i, i = 1, 2, ..., m \tag{6.100}$$

$$x_j \geq 0, j = 1, 2, ..., n \tag{6.101}$$

By feasible direction method [27, 28], assume the initial point $x_0 = (x_{01}, x_{02}, ..., x_{0n})$ is a feasible solution to the above nonlinear programming problem. The objective function $f(x)$ is linearly approximated at the point $x_0$ as:

$$f(x) = f(x_0) + \sum_{j=0}^{n} c_j (x_j - x_{0j}) \tag{6.102}$$

where $c_j$ is the partial derivative of $f(x)$ with respect to $x_j$ at the point $x_0$

$$c_j = \frac{\partial f(x)}{\partial x_j}\bigg|_{x = x_0} \tag{6.103}$$

Since $f(x_0)$, $c_j$ and $x_{0j}$ are fixed, the objective function is reduced to:

$$\min z = \sum_{j=0}^{n} c_j x_j \tag{6.104}$$

The linear approximation problem (6.104) subjecting to (6.100) and (6.101) is solved, giving an optimal solution $y = (y_1, y_2, ..., y_n)$. Note that the line segment joining $x_0$ to $y$ lies inside the feasible region of the original convex programming problem. Assume the solution of the line-segment optimization is $x_1 = (x_{11}, x_{12}, ..., x_{1n})$. The iteration is repeated by determining a new linear approximation to the objective function with $c_j$ at the new solution $x_k$ until the local optimal solution for the nonlinear programming problem defined by (6.99)–(6.101) is reached.

In solving the line-segment optimization problem, since the current feasible solution $x = (x_1, x_2, ..., x_n)$ and the direction of the line segment $d = y - x = (d_1, d_2, ..., d_n)$ are given, the next feasible solution is the solution of the following updated objective function:

$$\min_{0 \leq \theta \leq 1} f(x_1 + \theta d_1, x_2 + \theta d_2, ..., x_n + \theta d_n) \tag{6.105}$$

Since the current point $x$ and the direction $d$ are fixed by the given, this problem is a one-dimensional problem. The optimal solution of (6.105) is given by

$$\frac{\partial f(x)}{\partial \theta}\bigg|_{0 \leq \theta \leq 1} = 0 \tag{6.106}$$

Then, the new optimal solution is updated by $\tilde{x} = x + \theta d$.

### 6.3.2.3 Linear Approximations of Nonlinear Programming with Nonlinear Constraints

The linear approximations can be extended to general nonlinear programs with both the nonlinear objective function and nonlinear constraints.

Assume $x_0 = (x_{01}, x_{02}, ..., x_{0n})$ is a feasible solution to the above nonlinear programming problem.

$$\min f(x_1, x_2, ..., x_n) \tag{6.107}$$

Subject to

$$g_i(x_1, x_2, ..., x_n) \leq 0, i = 1, 2, ..., m \tag{6.108}$$

The nonlinear constraints can be linearized by Taylor's expansion as:

$$\tilde{g}_i(x) = g_i(x_0) + \sum_{j=1}^{n} a_{ij}(x_j - x_{0j}) \leq 0 \tag{6.109}$$

Where $a_{ij}$ is the partial derivative of $g_i(x)$ with respect to $x_j$ at the point $x_0$. Since $g_i(x_0)$, $a_{ij}$, and $x_{0j}$ are fixed, the constraints can be approximated as:

$$\sum_{j=1}^{n} a_{ij}x_j \leq b_{0i} \equiv -g_i(x_0) + \sum_{j=1}^{n} a_{ij}x_{0j} \tag{6.110}$$

In the same way by (6.102) and (6.103), the objective function can be linearized as the same as (6.104). Together with constraint linear approximation (6.110), the linear approximation of the nonlinear programming is:

$$\min z = \sum_{j=0}^{n} c_j x_j \tag{6.111}$$

$$\sum_{j=1}^{n} a_{ij}x_j \leq -g_i(x_0) + \sum_{j=1}^{n} a_{ij}x_{0j} = b_{0i} \tag{6.112}$$

$$x_{0j} - \delta_j \leq x_j \leq x_{0j} + \delta_j \tag{6.113}$$

Constraints (6.113) limit the step size to keep the solution in the feasible region. The partial derivatives $a_{ij}$, $b_i$, and $c_j$ are updated at each iteration and the new solution is given by the above linear approximation till the process is converged.

## 6.4 Mixed Integer Optimization Approach

The linear programming models discussed in 6.2 have been continuous, in the sense that decision variables are allowed to be fractional. Usually, this is a realistic assumption. In some power system optimization problems, however, fractional solutions are not practicable, for example in power system long-term generation capacity expansion planning, we build new generation resources to meet demand, energy, and clean energy needs. The decision variable to build or not build a resource is a 0 or 1 integer. In security-constrained unit commitment, units will be committed or de-committed to minimize the production cost meanwhile meet the security constraints. The unit on or off status has to be modeled as an integer variable, usually in 0 or 1. In more generic cases, the transformer tap ratio and phase shifter angle are not continuous. The programming problem with integer variables is called the integer-programming problem. In power system optimization and many other decision science problems, some, but not all, variables are restricted to be an integer. The optimization problem can be formulated as the following mixed integer programming problem:

$$\min f(x, y) \tag{6.114}$$

Subject to

$$g(x, y) = 0 \tag{6.115}$$

$$x \geq 0, y \in \mathbb{Z} \tag{6.116}$$

Mixed integer programs are very difficult to solve. Mixed integer programming problems are usually solved by using a linear programming-based branch-and-bound algorithm.

### 6.4.1 Branch-and-Bound Approach

Basic linear programming-based branch-and-bound begins with the original mixed integer program with all the integrality restrictions removed to form linear programming relaxation of the original mixed integer programming problem.

We have discussed the approaches to solving linear programs and nonlinear programs in 6.2 and 6.3. Assume the linear programming relaxation is solved. If the result meets all of the integrality restrictions, the solution is an optimal solution to the original mixed integer programming problem. If the result does not satisfy all of the integrality restrictions which happen in most cases, then the total set of feasible solutions is partitioned into subsets of solutions (branching) with upper or lower bounds (bounding). The subsets are solved to check the satisfaction of the integrality restrictions. The process is repeated until the optimal solution is reached.

The following example is taken to elaborate on the branch-and-bound approach.

$$\min c^T x = -x_1 - x_2 \tag{6.117}$$

Subject to

$$\begin{aligned} 2x_1 + x_2 &\leq 12.5 \\ x_1 + 2x_2 &\leq 10 \end{aligned} \tag{6.118}$$

and

$$(x_1, x_2) \geq 0, \quad (x_1, x_2) \in \mathbb{Z} \tag{6.119}$$

The optimal solution of its linear programming relaxation is $(x_1, x_2) = (5, 2.5)$, with objective $-7.5$. Variable $x_2 = 2.5$ is not satisfied the integrality restriction. The value $x_2 = 2.5$ is excluded by imposing the two restrictions $x_2 \leq 2$ and $x_2 \geq 3$.

The two new restrictions partition the original solution set into two solution subsets. The original mixed integer programming denoted $P_0$ is then branched to two mixed integer programming problems denoted by $P_1$, where $x_2 \leq 2$ is imposed, and $P_2$, where $x_2 \geq 3$ is imposed. The feasible solution set of $P_0$ must be equal to the union of the feasible solution sets of $P_1$ and $P_2$.

The variable $x_2$ is called a branching variable, and we are said to have branched on $x_2$, producing the two sub-problems $P_1$ and $P_2$.

Linear programming relaxation of $P_1$:

$$\min f = c^T x = -x_1 - x_2 \tag{6.120}$$

Subject to

$$\begin{aligned} 2x_1 + x_2 &\leq 12.5 \\ x_1 + 2x_2 &\leq 10 \\ x_2 &\leq 2 \end{aligned} \tag{6.121}$$

and

$$(x_1, x_2) \geq 0 \tag{6.122}$$

Linear programming relaxation of $P_2$:

$$\min f = c^T x = -x_1 - x_2 \tag{6.123}$$

Subject to

$$\begin{aligned} 2x_1 + x_2 &\leq 12.5 \\ x_1 + 2x_2 &\leq 10 \\ x_2 &\geq 3 \end{aligned} \tag{6.124}$$

and

$$(x_1, x_2) \geq 0 \tag{6.125}$$

The optimal solution of $P_1$ and $P_2$ linear programming relaxation is $P_1 : (x_1, x_2, f) = (5.25, 2, -7.25)$ and $P_2 : (x_1, x_2, f) = (4, 3, -7)$. The lower objective leave is kept and the binary tree of sub-problems is continuing to build and the leave with a smaller objective will be partitioned into two sub-problems. Since the linear programming relaxation optimal solution satisfying the integrality constraints is also the optimal solution of original mixed integer programming problems, the procedure terminates when integer variables take inter values in the linear programming relaxation optimal solution.

The final binary tree is shown in Figure 6.5 to illustrate the basic idea behind the branch and bound method for solving mixed integer programming problems.

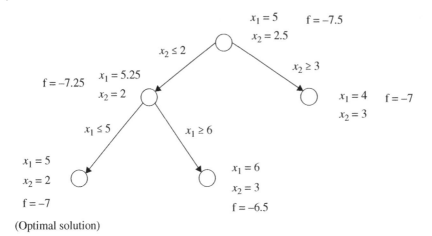

Figure 6.5 The illustration of branch and bound method.

### 6.4.2 Machine Learning for Branching

Branching is the core of mixed-integer optimization approaches. A branching rule is a decision rule to split the current optimization problem into two or more disjoint sub-problems. A good branching rule could potentially decrease the size of the branching tree, therefore reducing the running time. Popular branching rules used in mixed integer programming are:

- Most Infeasible Branching: choose the variable whose fractional part is closest to 0.5. However, this rule usually does not perform better than choosing randomly.
- Pseudo-Cost Branching: This rule keeps a history of how much improvement is made for branching on a certain variable. The improvement for variable $x_i$ could be measured by combinations of the fractional part of $x_i$ and the increase in the objective function after branching on $x_i$. The decision is made by comparing the historical performance of branching on each variable.
- Strong Branching: Strong branching evaluates which of the fractional candidate variables gives the best progress on optimization before actually branching. This approach locally solves a set of linear programming relaxations and chooses the best. However, the computational costs are high when all candidates are evaluated. In practice, strong branching could significantly reduce the number of nodes on the branching tree, while increasing the total running time since each node needs more time to process.

Among the branching rules, pseudo-cost branching is a reasonable trade-off between strong branching and the most infeasible branching. In power system optimization, such as security-constrained unit commitment [29], optimal allocation of FACTS devices [30], electrical vehicle charging station planning and optimization [31–34], and Volt-Var optimization [35], mixed integer programming is involved. Particularly for security-constrained unit commitment, decisions on branching and bounding are taken every day in day-ahead market simulation and study. The historical search paths on branching trees are great resources to learn a more favorable branching decision by machine learning.

A machine learning-based branching approach can be supervised or unsupervised. Supervised learning, for example, supports vector machine model, and computes the branching score by using the strong branching approach. The computed strong-branching score is labeled to the candidate variables at each node. The corresponding variable features and the branching score are compiled in a training set to train the machine learning model. The trained machine learning model is applied to make the branching decision in the optimization process.

Besides the supervised-learning approach, unsupervised reinforcement learning is viable to learn a branching policy by policy network, and the trained actor network makes a decision of branching. At each decision point, the right variable is chosen to branch based on the current states.

## 6.5 Optimization Problems Solution by Graph Parallel Computing

Most power system optimization problems can be formed as constrained nonlinear programming or mixed integer programming problems. Using the branch-and-bound approach, mixed integer programming problems are partitioned into two constrained nonlinear or linear programming sub-problems. A constrained nonlinear program is approximated to a constrained linear program. Ideally, optimization problems in general could be reduced to solve multiple linear programming problems in finite steps. The simplex method and interior-point method are the most practical methods to effectively solve a linear program. The benefits of improving computation efficiency by parallelizing the simplex method and the interior-point method can be leveraged to most practical optimization methods in power system optimization applications.

### 6.5.1 Simplex Method Based on Graph Parallel Computing

The canonical form of the simplex table is an asymmetrical matrix $\begin{bmatrix} -c^T & 0 \\ A & b \end{bmatrix}$. By graph theory, an asymmetrical matrix is equivalent to a directed graph.

In the simplex method, the canonic table matrix is in $\mathbb{R}^{(m+1)\times n}$, where $m$ is the number of constraints, and $n$ is the number of variables (including the slack variables). Typically, the matrix is square. Taking the same linear programming problem in 6.2.1 as a numerical example,

$$\min c^T x = -x_1 - x_2 \tag{6.126}$$

Subject to

$$\begin{aligned} 2x_1 + x_2 + x_3 &= 12 \\ x_1 + 2x_2 + x_4 &= 10 \end{aligned} \tag{6.127}$$

and

$$(x_1, x_2, x_3, x_4) \geq 0 \tag{6.128}$$

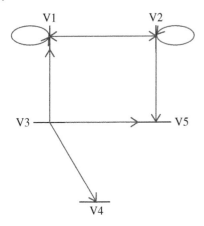

**Figure 6.6** Directed graph.

The initial simplex table is

$$\begin{bmatrix} 1 & 1 & 0 & 0 & 0 \\ 2 & 1 & 1 & 0 & 12 \\ 1 & 2 & 0 & 1 & 9 \end{bmatrix}$$  (6.129)

The directed graph to represent the matrix (6.129) is shown in Figure 6.6.

To create a graph for a simplex table, the following statements defined a simplex graph schema.

```
CREATE VERTEX V (primary_id id string)
CREATE DIRECTED EDGE E (FROM V, TO V, double a_ij)
CREAT GRAPH SimplexGraph (V, E)
CREATE LOADING JOB linearProgramData FOR GRAPH SimplexGraph {
    DEFINE FILENAME vertexInfoFile = " /vertexInfo.csv";
    DEFINE FILENAME edgeInfoFile = " /edgeInfo.csv";
    LOAD vertexInfoFile TO VERTEX V VALUES (id);
    LOAD edgeInfoFile TO EDGE E VALUES (FROM V, TO V, a_ij);
}
vertexInfo.csv is simply as
"V1"
"V2"
"V3"
"V4"
"V5"
edgeInfo.csv is
"V1", "V1", 1
"V1", "V2", 1
"V2", "V1", 2
"V2", "V2", 1
"V2", "V3", 1
"V2", "V5", 12
"V3", "V1", 1
```

```
"V3", "V2", 2
"V3", "V4", 1
"V3", "V5", 9
```

In pivot operation, since the leaving variable and entering variable are determined in run-time, the filled graph and the elimination tree cannot be created in front. When a new nonzero element `a_ij` from vertex *I* to vertex *J* is calculated during the pivot operation, the fill-in is added to the graph by an `INSERT` statement as follows:

```
INSERT INTO E (I, J) VALUES (a_ij)
```

Meanwhile, when element `a_ij` from vertex *I* to vertex *J* is eliminated during the pivot operation, the corresponding edge is deleted from the graph by a `DELETE` statement as follows:

```
DELETE s FROM E
    WHERE s.FROM = = "I" AND s.TO = = "J"
```

The entering variable $x_q$ is selected by the most negative number of vector $c^T$ which is the first row of the canonic table, equivalently the vertex "V1" and its adjacent. The entering variable selection can be parallelized by the following query:

```
CREATE QUERY selectEnteringVariable() FOR GRAPH SimplexGraph{
    SELECT s FROM V:s-(E:e)->V:t WHERE V.id="V1"
      SELECT q FROM e ORDER by e.a_ij ASC LIMIT 1
}
```

In the query, the `LIMIT` clause ensures one entering variable is returned when there are tied lowest cost coefficients.

When the lowest cost coefficient is non-negative, the optimal solution is found. Otherwise, the leaving variable is selected and the pivot operation is performed iteratively. The leaving variable $x_p$ is selected by the ratio test on the last column of the canonic table. The leaving variable selection can be parallelized by the following query.

```
CREATE QUERY selectLeavingVariable(string lastVertex) FOR GRAPH
SimplexGraph{
      SELECT b FROM E WHERE b.TO=lastVertex
        SELECT eq FROM E WHERE b.TO=q.id
          SELECT p ORDER by b.a_ij/eq.a_ij WHERE eq.a_ij>0 ASC LIMIT 1
}
```

where the last vertex identification is passed as argument `lastVertex` to the query. In the ratio test $p = \min\limits_{1 \le i \le m} \left(\dfrac{b_i}{a_{iq}}\right)$, zero element $a_{iq}$ is not tested since no corresponding edge exists in the graph which is an advantage of using a graph to keep the sparsity merit in solving a programming problem. When the entering variable and leaving variable are selected, the pivot operation on the canonic table (canonic graph) is performed as follows:

```
CREATE QUERY pivotOperation(p, q) FOR GRAPH SimplexGraph{
   SELECT e FROM E WHERE e.FROM=p.id AND e.TO=q.id
      ACCUM
         p.a_ij = p.a_ij/e.a_ij
```

```
    SELECT sr FROM V:sr-(E:er)->V:t WHERE sr.id=p.id
       SELECT sl FROM V:sl-(E:el)->V:t WHERE sl=sr
          SELECT e FROM E WHERE e.FROM =sr.id AND e.TO=sl.id
             IF e IS NULL
                   INSERT INTO E (sl, sr) VALUES (0)
                el.a_ij = el.a_ij - p.a_ij*er.a_ij
      DELETE s FROM E WHERE s.FROM = = q.id OR s.TO = = q.id
}
```

The pseudocode for the simplex method by graph computing is shown in Algorithm 6.3.

---

**Algorithm 6.3   Simplex Method in Graph**

---

```
1: CREATE QUERY simplexMethodGraph(V,E) FOR GRAPH SimplexGraph
2:    WHILE TRUE
3:       q = selectEnteringVariable()
4:       SELECT e FROM E WHERE e.FROM = "V1" and e.TO=q.id
5:       IF e.a_ij > 0 EXIT
6:       p = selectLeavingVariable()
7:       pivotOperation(p, q)
```

---

### 6.5.2   Interior-Point Method Based on Graph Parallel Computing

The interior-point method involves two major steps: updating Newton's direction and search steps.

The iteration equation to update the Newton direction is (6.39) and rewritten here for easy reading.

$$\begin{bmatrix} A & 0 & 0 \\ 0 & A^T & I \\ S & 0 & X \end{bmatrix} \begin{bmatrix} \Delta x \\ \Delta y \\ \Delta s \end{bmatrix} = \begin{bmatrix} b - Ax_k \\ c - A^T y_k - s_k \\ \mu_k e - X_k S_k e \end{bmatrix} \tag{6.130}$$

Since the matrix $\begin{bmatrix} A & 0 & 0 \\ 0 & A^T & I \\ S & 0 & X \end{bmatrix}$ is sparse, when the Gauss Elimination method is applied to solve the linear equation above, factorization, and forward and backward substitution are involved.

When $G(A)$ is formed to represent the matrix $\begin{bmatrix} A & 0 & 0 \\ 0 & A^T & I \\ S & 0 & X \end{bmatrix}$, the factorization, forward and backward substitution along with elimination tree and node partition can be performed in graph nodal and hierarchical parallel which are addressed in Chapter 3.

Since the right-hand side calculation is independent for each element, it can be updated in nodal parallel computing.

To store the right-hand side vector, graph $G(A)$ is extended to have residual as an additional attribute of each vertex. An attribute type ("Primal," "Dual," "Slack") is used to differentiate the primal variables, dual variables, and slack variables. To create graph $G(A)$, the following statements defined an interior point graph schema.

```
CREATE VERTEX V (primary_id id string, type string, a_ii double, b
   double, c double, r double, xys double, delta_xys double)
CREATE DIRECTED EDGE E (FROM V, TO V, a_ij)
CREAT GRAPH InteriorPointGraph (V, E)
CREATE LOADING JOB linearProgramData FOR GRAPH InteriorPointGraph {
   DEFINE FILENAME vertexInfoFile = " /vertexInfo.csv";
   DEFINE FILENAME edgeInfoFile = " /edgeInfo.csv";
   LOAD vertexInfoFile TO VERTEX V VALUES (id, a_ii, r);
   LOAD edgeInfoFile TO EDGE E VALUES (FROM V, TO V, a_ij);
}
```

To support the interior-point method, different from the simplex graph schema, the diagonal element of a matrix is not defined as a self-loop edge of a vertex, but as an attribute of the vertex.

In each iteration, the residuals are updated by the following query.

```
CREATE QUERY updateResiduals() FOR GRAPH interiorPointGraph {
     SELECT s FROM V WHERE s.TYPE="Primal"
       SELECT sw FROM V:s-(E:e)->V:t
         ACCUM
             s.r = s.b - e.a_ij*sw.x
     SELECT s FROM V WHERE s.TYPE="Dual"
       SELECT sw FROM V:s-(E:e)->V:t
         ACCUM
             s.r = s.c - e.a_ij*sw.y - s.s
     SELECT s FROM V WHERE s.TYPE="Slack"
       SELECT sw FROM V:s-(E:e)->V:t
         ACCUM
             s.r = @@mu - sw.x*sw.s
}
```

Where global variable @@mu is the weighting parameter.

Assume the `interiorPointGraph` is created and the vertex and edges are loaded to the graph. The pseudocode for the interior-point method by graph computing is straightforward as shown in Algorithm 6.4.

---

**Algorithm 6.4 Interior-Point Method in Graph**

---

```
1: CREATE QUERY interiorPointMethod(G(A)) FOR GRAPH interiorPointGraph
2:     Initialize (x_0, y_0, s_0, α_p, α_d) with (x_0, s_0) ≥ 0, 0 < (α_p, α_d) ≤ 1
```
3:  $\mu_0 = \frac{1}{n}\sum_{j=1}^{n}x_{0j}s_{0j}$,
4:
5:  WHILE $\dfrac{c^T x_k - b^T y_k}{1 + |b^T y_k|} < \varepsilon$

```
6:        G+(A) = formFilledGraph(G(A))
7:        T(A) = formEliminationTree(G+(A))
8:        [L(A) U(A)] = factorizeMatrixA(G+(A), T(A))
9:        updateResiduals()
10:       forwardSubstitution(L(A), G(A), T(A))
```

```
11:     backwardSubstitution (U(A), G(A), T(A))
12:     SELECT s FROM V
13:        ACCUM
14:           @a_pd = min(1, s.xys/(-s.delta_xys)) WHERE s.delta_xys
15:           s.xys = s.xys + a_pd * s.delta_xys
16:     @@mu = @@mu/10
```

# References

**1** Luenberger, D.G. and Ye, Y. (2015). *Linear and Nonlinear Programming*, 4e. Springer.

**2** Murty, K.G. (1983). *Linear Programming*. New York: Wiley.

**3** Dantzig, G.B. (1963). *Linear Programming and Extensions.* Princeton, NJ: Princeton University Press.

**4** Dantzig, G.B. (1951). Maximization of a linear function of variables subject to linear inequalities. In: *Chapter 21. Activity Analysis of Production and Allocation*, Cowles Commission Monograph 13 (ed. T.C. Koopmans). New York: Wiley.

**5** Dantzig, G.B. (1953). Computational algorithm of the revised simplex method. *RAND Report RM−1266*, The RAND Corporation, Santa Monica, CA.

**6** Dantzig, G.B. (1954). Variables with upper bounds in linear programming. *RAND Report RM−1271*, The RAND Corporation, Santa Monica, CA.

**7** Dantzig, G.B., Orden, A., and Wolfe, P. (1954). Generalized simplex method for minimizing a linear form under linear inequality restraints. *RAND Report RM−1264*, The RAND Corporation, Santa Monica, CA.

**8** Khachiyan, L.G. (1979). A polynomial algorithm for linear programming, Doklady Akad. Nauk USSR 244, 1093–1096, 1979, pp. 1093–1096. Translated in Soviet Math. Doklady 20, 1979, pp. 191–194.

**9** Bland, R.G., Goldfarb, D., and Todd, M.J. (1981). The ellipsoidal method: a survey. *Operations Research* 29: 1039–1091.

**10** Karmarkar, N. (1984). A new polynomial - time algorithm for linear programming. *Combinatorica* 4 (8): 373–395.

**11** Renegar, J. (1988). A polynomial-time algorithm, based on Newton's method, for linear programming. *Mathematical Programming* 40: 59–93.

**12** Kojima, M., Mizuno, S., and Yoshise, A. (1989). A polynomial-time algorithm for a class of linear complementarity problems. *Mathematical Programming* 44: 1–26.

**13** Monteiro, R.D.C. and Adler, I. (1989). Interior path following primal-dual algorithms: part I: linear programming. *Mathematical Programming* 44: 27–41.

**14** Megiddo, N. (1989). *Pathways to the Optimal Set in Linear Programming, in Progress in Mathematical Programming: Interior Point and Related Methods*, 131–158. New York: Springer.

**15** Gill, P.E., Murray, W., Saunders, M.A. et al. (1986). On projected Newton barrier methods for linear programming and an equivalence to Karmarkar's projective method. *Mathematical Programming* 36: 183–209.

**16** Ye, Y. (1997). *Interior Point Algorithms.* New York: Wiley.

**17** Fletcher, R. (1987). *Practical Method of Optimization.* Wiley.

**18** Broyden, C.G. (1970). The convergence of a class of double-rank minimization algorithms. *Journal of the Institute of Mathematics and its Applications* 6: 70–90.

**19** Fletcher, R. (1970). A new approach to variable metric algorithms. *Computer Journal* 13: 317–322.

**20** Goldfarb, D. (1970). A family of variable metric methods derived by variational means. *Mathematics of Computing* 24: 23–26.

**21** Shanno, D.F. (1970). Conditioning of Quasi-Newton methods for function minimization. *Mathematics of Computing* 24: 647–656.

**22** Fletcher, R. and Powell, M.J.D. (1963). A rapidly convergent descent method for minimization. *Computer Journal* 6: 163–168.

**23** Bellavia, S. and Pieraccini, S. (2015). On affine-scaling inexact dogleg methods for bound-constrained nonlinear systems. *Optimization Methods and Software* 30 (2): 276–300.

**24** Pawlowski, R.P., Simonis, J.P., Walker, H.F., and Shadid, J.N. (2008). Inexact Newton dogleg methods. *Society for Industrial and Applied Mathematics Journal on Numerical Analysis* 46 (4): 2112–2132.

**25** Hestenes, M.R. and Stiefel, E. (1952). Methods of conjugate gradients for solving linear systems. *Journal of Research of the National Bureau of Standards* 49 (6): 409–436.

**26** Reid, J.K. (1972). The use of conjugate gradients for systems of linear equations possessing 'Property A.'. *Journal of the Institute of Mathematics and its Applications* 9 (2): 325–332.

**27** Zangwill, W.I. (1969). *Nonlinear Programming: A Unified Approach*. Englewood Cliffs, NJ: Prentice-Hall.

**28** Frank, M. and Wolfe, P. (1956). An algorithm for quadratic programming. *Naval Research Logistics Quarterly* 3: 95–110.

**29** Yang, N., Dong, Z., Wu, L. et al. (2022). A comprehensive review of security-constrained unit commitment. *Journal of Modern Power Systems and Clean Energy* 10 (3): 562–576.

**30** Valle, Y.D., Perkel, J., Venayagamoorthy, G.K., and Harley, R.G. (2009). Optimal allocation of facts devices: classical versus metaheuristic approaches. *SAIEE Africa Research Journal* 100 (1): 12–23. https://doi.org/10.23919/SAIEE.2009.8531637.

**31** Tan, J., Liu, G., Dai, R., and Wang, Z. (2017). Bi-level charging station planning for integrated power distribution and transportation system. *2017 IEEE Conference on Energy Internet and Energy System Integration (EI2)*, Beijing, China, pp. 1–5. https://doi.org/10.1109/EI2.2017.8245744.

**32** Mao, D., Wang, J., Tan, J. et al. (2019). Location planning of fast charging station consideri+9+ng its impact on the power grid assets. *2019 IEEE Transportation Electrification Conference and Expo (ITEC)*, pp. 1–5.

**33** Bayani, R., Manshadi, S.D., Liu, G. et al. (2022). Autonomous charging of electric vehicle fleets to enhance renewable generation dispatchability. *CSEE Journal of Power and Energy Systems* 8 (3): 669–681.

**34** Saber, A.Y. and Venayagamoorthy, G.K. (2010). Intelligent unit commitment with vehicle-to-grid—a cost-emission optimization. *Journal of Power Sources* 195 (3): 898–911.

**35** Dharmawardena, H. and Venayagamoorthy, G.K. (2022). Distributed volt-var curve optimization using a cellular computational network representation of an electric power distribution system. *Energies* 15 (12): 4438.

# 7

# Graph-Based Machine Learning

Making machines mimic human intelligence has been the goal pursued by artificial intelligence research [1]. For decades, scientists and engineers have been exploring theories and algorithms of machine learning and artificial intelligence from different perspectives. Learning is the process of using prior knowledge to make posteriori decisions. In the energy industry, machine learning has been applied to power system load forecasting [2]; wind and photovoltaic (PV) power generation prediction [3–5]; static and dynamic security assessment [6–9]; security enhancement control [10]; unit commitment [11]; and many other power system applications [12].

Among these machine learning applications, PV power generation forecasting is challenging in light of the variability and stochastic behavior of PV power caused by solar radiation uncertainty [13]. In recent years, renewable energy resources including wind and solar have been rapidly integrated into the power grids to mitigate the issue of fossil fuel resource exhaustion and environmental pollution concerns. The reliability of power systems depends on the capability of predicting renewable power generation intermittence and uncertainty. Hence, accurate solar irradiance forecasting for PV production estimation is critical for power system planning and operation.

## 7.1 State of Art on PV Generation Forecasting

The studies in the area of solar irradiance and PV power forecasting are mainly categorized into four major classes:

1) Persistence Model: The model assumes that the irradiance values at future time steps are equal to the same values at the forecasting time. Due to such a strong smoothness assumption, the persistence scheme is only effective for intra-hour applications [14].
2) Physical Model: Physical models employ physical processes to estimate future solar radiation values using astronomical relationships [15], meteorological parameters, and numerical weather predictions [16–18].
3) Artificial Intelligence Model: The nonstationary and highly nonlinear characteristics of solar radiation time series lead to the superiority of artificial intelligence approaches over traditional

*Graph Database and Graph Computing for Power System Analysis*, First Edition. Renchang Dai and Guangyi Liu.
© 2024 The Institute of Electrical and Electronics Engineers, Inc. Published 2024 by John Wiley & Sons, Inc.

physical models [19]. Machine learning algorithms are employed as approximation functions to estimate solar irradiance and PV power [20, 21].

4) Ensemble Method: Ensemble methods aggregate multiple predictors to increase the prediction accuracy over individual prediction models. The ensemble models generally apply bootstrap sampling to obtain data subsets for training the individual base predictor [22], so as to build up probabilistic prediction models [23]. Three typical probabilistic methods including the Bayesian model [24], Markov chain model [13], and quantile regression [25] were trained as base predictors to obtain an ensemble of the predictive distribution with optimal sharpness and reliability metrics.

Learning methods highly rely on the training dataset to generalize the prediction models. However, most solar farms are built recently. Historical data are limited to build a single solar farm prediction model. In addition, typically, PV generation is forecasted at each solar farm individually and independently. The spatial correlations of different solar farms are not captured. Graph machine learning in graph-structured data is developed to conquer the challenges.

In the graph machine learning model, the solar radiation probability distribution pattern is captured by a generative graph neural network as a probabilistic prediction model first. The generative graph neural network learns patterns from a historical dataset to generate new samples under the observed data distribution. Mathematically, the generative graph neural network is trained as an optimization problem where the probability of observed data in a given dataset is estimated. In this approach, a convolutional graph auto-encoder is developed as a graph learning approximator of the probability distribution which is mathematically proved to learn continuous probability distribution functions from the nodes in an arbitrary graph. The convolutional graph auto-encoder is defined based on the first-order approximation of graph convolutions and standard function approximation. This trained deep-learning model is able to generate new samples corresponding to each node, after observing historical graph-structured data, while learning the nodal distributions. Thus, the challenge of spatiotemporal probabilistic solar radiation forecasting is converted as a graph distribution learning problem and solved by the convolutional graph auto-encoder.

## 7.2 Graph Machine Learning Model

The spatiotemporal correlation of $n$ neighboring solar sites in a region is modeled by an undirected graph $X$ with node set $V = \{v_i \mid 1 \leq i \leq n\}$ and edge set $E$ where each node represents a solar site and each edge reflects the correlation between the neighboring solar sites. The adjacency matrix $A$ is defined as:

$$A_{i,j} = \begin{cases} e^{-d_{i,j}} & cor(i,j) \geq T \\ 0 & cor(i,j) < T \end{cases} \tag{7.1}$$

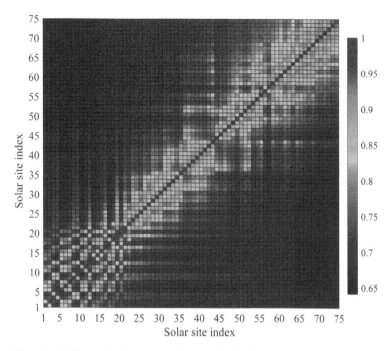

**Figure 7.1** Photovoltaic power station correlation heat map.

where $0 \leq T \leq 1$ is a given sparsity parameter; $cor(i, j)$ is the correlation of the historical time-series solar radiation of solar sets $i$ and $j$; and $d_{i,j}$ is the distance between the two sets. The more two nodes are correlated, the higher the value $cor(i, j)$ is. Figure 7.1 shows an exemplary correlation heat map of 75 PV power stations.

The adjacency matrix $A$ is equivalent to an undirected graph as shown in Figure 7.2.

In Figure 7.2, each node represents a solar site and each edge represents the correlation of corresponding nodes when the correlation is greater than the sparsity parameter $T$ ($T$ is given as 0.8 in the demonstrated graph). No edge in between two nodes indicates $A_{i,j} = 0$ or equivalently $cor(i,j) < T$.

**Figure 7.2** Graph model of the 75 nodes and 464 edges.

The graph structure is further clustered into six subgraphs using the Girvan–Newman algorithm [26]. Each subgraph consists of a subset of nodes densely connected to each other with relatively large edge weights due to their high mutual information. The historical time-series radiation data for each subgraph will be used to train the corresponding probabilistic prediction model to preserve the spatiotemporal correlation of densely connected PV sites.

After modeling the subgraphs by the definition of $A$, for each subgraph, at each time step $t$, each node $v_i$ contains a time-series attribute $T(v_i, t)$ corresponding to the historical solar radiation data used as the input to the forecasting model to predict future solar radiation value $T(v_i, t + \Delta t)$. The key is to estimate a conditional probability distribution $P(T(v_i, t + \Delta t)|\pi)$, $\pi$ is the historical solar radiation of all solar sites in the subgraph, where $\pi = \langle T(v_1, t), T(v_2, t), ..., T(v_n, t) \rangle$.

The solar radiation conditional probability distribution function $P(T(v_i, t + \Delta t)|\pi)$ can be approximated by the convolutional graph auto-encoder.

## 7.3 Convolutional Graph Auto-Encoder [27]

### 7.3.1 Auto-Encoder

Auto-encoder in general is a neural network to reproduce the input as its output preserving the desired characteristics of the input. In the solar radiation probabilistic prediction model case, the desired characteristic is the solar radiation probability distribution function. The auto-encoder will be trained to represent the solar radiation probability distribution function.

A typical auto-encoder is implemented by one or multiple hidden layer neural networks as shown in Figure 7.3.

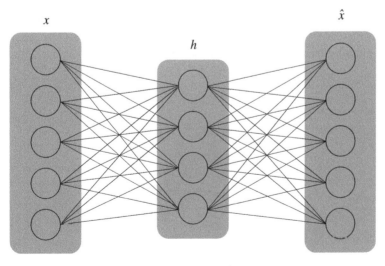

**Figure 7.3** General structure of the auto-encoder.

Auto-encoder consists of two components: an encoder $h = f(X)$ and a decoder $\hat{X} = g(X)$. To preserve the characteristics of input $X$, the network is trained by minimizing the reconstruction error:

$$l(X, \hat{X}) = l(X, g(f(x))) = \|X - g(f(x))\|_2 \tag{7.2}$$

To avoid the output reproducing the input, the number of nodes in the hidden layer is limited to compress the information of the input $X$ to a low-dimensional representation, but preserve the desired characteristics of the input $X$.

### 7.3.2 Auto-Encoder on Graphs

Graph auto-encoder uses graph neural network structure as shown in Figure 7.4.

For each node $v_i \in V$ in a graph, the corresponding attribute of node $i$ and its neighbors $\boldsymbol{a}_i = (a_{i1}, a_{i2}, ..., a_{ii}, ..., a_{in})$ is the input of the encode function:

$$\boldsymbol{z_i} = f(\boldsymbol{a_i}; \boldsymbol{\Theta_e}) \tag{7.3}$$

where $\boldsymbol{a_i}$ is the solar radiation time-series vector of site $i$ and its adjacent solar sites. $\boldsymbol{\Theta_e}$ is the encoder feedforward neural network parameters which will be trained by the historical time-series data $\boldsymbol{a_i}$. The output $\boldsymbol{z_i}$ of the encoder is served as the input of the decoder to reconstruct $\boldsymbol{a_i}$ as:

$$\hat{\boldsymbol{a}}_i = g(\boldsymbol{z_i}; \boldsymbol{\Theta_d}) \tag{7.4}$$

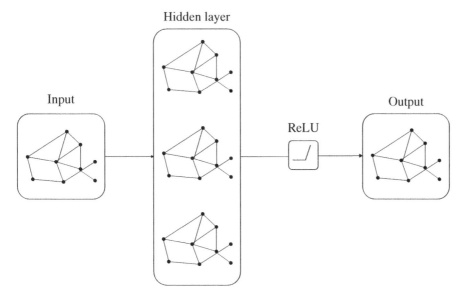

**Figure 7.4** Structure of graph auto-encoder neural network.

where $\Theta_d$ is the decoder neural network parameter which will be trained by minimizing the following reconstruction loss:

$$\mathcal{L} = \sum_{v_i \in V} \|a_i - \hat{a}_i\|_2 \tag{7.5}$$

Compared with the traditional auto-encoder, the graph auto-encoder makes full use of the spatial distribution relationship and spatial correlation of the dataset.

### 7.3.3 Probability Distribution Function Approximation

Auto-encoder serves as a generative model to approximate the probability distribution $P(X)$ of a given dataset $X = \{X_1, X_2, ..., X_1\}$. It is also a latent variable model which generates samples from latent variables. The challenge is to capture a probability distribution $P(X)$ over n-dimensional data points $X$ in a potentially high-dimensional vector space $X \subseteq \mathbb{R}^n$. To conquer the challenge, a latent variable is defined based on a model in which the hidden random vector $Z \in Z$ embodies the major characteristics of $P(X)$ (e.g. the probability distribution function of the future solar radiation, or any desired nodal probability distribution function). More specifically, $Z$ is sampled following unknown distribution $P(Z)$ over the high-dimensional space $Z$, given a variable $Z$ from the distribution $P(Z)$.

To justify that the approach can generate sample $X^*$ which is representative of the given dataset $X$, we ensure that there exists at least one configuration $\hat{Z} \in Z$ that causes the model to generate some sample $\hat{X}$ in $X$. Assuming a family of deterministic functions $f(Z; \theta)$ with parameters $\theta \in \Theta$, each latent variable–parameter pair is mapped to a sample using $f: Z \times \Theta \longrightarrow X$. We find an optimal $\theta^* \in \Theta$ such that when $Z \sim P(Z)$, the value of $X^* = f(Z; \theta = \theta^*)$ is as close as possible to $X \in X$. In other words, the probability of $f$ creating an output $X^*$ similar to the observed data $X$ is maximized; hence, the optimization is written as:

$$\theta^* = \underset{\theta \in \Theta}{\text{argmax}} \left[ P(X) = \int f(Z; \theta) P(Z) dZ \right] \tag{7.6}$$

$P(X)$ in (7.6) can be written as:

$$P(X) = \int P(X|Z; \Theta) P(Z) dZ \tag{7.7}$$

To ensure that the model is representative of each $X_i$ in the given dataset $X$, the following log-likelihood is maximized for each sample $X_i$ in dataset $X$:

$$\underset{\theta \in \Theta}{\text{argmax}} \left[ \log P(X_i) = \log \int P(X_i|Z; \Theta) P(Z) dZ \right] \tag{7.8}$$

where $\Theta$ is the parameter to be learned by the auto-encoder. However, the integral in (7.8) is intractable. To remedy this issue, a distribution function $Q(Z \mid X; \Phi)$ parametrized with $\Phi$ is defined to approximately estimate $P(X_i \mid Z; \Theta)$ by combinations of simple Gaussian distributions $Q(Z \mid X; \Phi) = N(\mu(X; \Phi), \Sigma(X; \Phi))$, considering Theorem 7.1, where the mean and covariance matrix are learned through deterministic function by an artificial neural network parameterized with $\Phi$. And $P(X \mid Z; \Theta)$ is modeled as a normal distribution $P(X \mid Z; \Theta) = N(0, I)$.

**Theorem 7.1** In any space $\Lambda$, any complicated probability density function over samples can be modeled using a set of $\dim(\Lambda)$ random variables with normal distribution, mapped through a high capacity function.

As a consequence, an approximator can be learned to map $Z$ to a hidden variable $\xi$ and further mapped to $\hat{X} \in X$ to maximize the likelihood of sample $X$. Here, $f$ is modeled by an artificial neural network as a standard function approximator which is capable of learning highly nonlinear target functions using multiple hidden layers.

The first layers of these architectures provide a nonlinear mapping from $Z \in \mathbf{Z}$ to $\xi$ which is an unknown distribution. $\xi$ is further mapped to a sample $X \in \mathbf{X}$. Note that if the model has sufficient capacity (ample number of hidden layers, as in the case of deep neural networks), then the neural network is able to solve the maximization in (7.8) to obtain optimal parameters $\theta^*$.

The expected value of $P(X \mid Z; \Theta)$ in (7.8) with respect to Z, $E_{Z \sim Q}[P(X \mid Z; \Theta)]$, can be computed using the Kullback–Leibler divergence [28]:

$$KL[Q(Z \mid X; \Phi) \mid\mid P(Z \mid X; \Theta)] = E_{Z \sim Q}[logQ(Z \mid X; \Phi) - logP(Z \mid X; \Theta)] \tag{7.9}$$

Applying the Bayesian rule [29] for $P(Z \mid X)$, (7.9) can be rewritten as:

$$KL[Q(Z \mid X; \Phi) \mid\mid P(Z \mid X; \Theta)]$$

$$= E_{Z \sim Q}\left[logQ(Z \mid X; \Phi) - log\left(\frac{P(X \mid Z; \Theta)P(Z; \Theta)}{P(X; \Theta)}\right)\right]$$

$$= E_{Z \sim Q}[logQ(Z \mid X; \Phi) - logP(X \mid Z; \Theta) - logP(Z; \Theta) + logP(X; \Theta)] \tag{7.10}$$

This equality is further written as:

$$logP(X) = KL[Q(Z \mid X) \mid\mid P(Z \mid X)] + E_{Z \sim Q}[logP(X \mid Z)] - KL[Q(Z \mid X) \mid\mid P(Z)] \tag{7.11}$$

where the parameters $\Theta$ and $\Phi$ are removed to make the equality concise. To generate $X$ (i.e. create samples $X^* \approx X$), the objective of maximizing $logP(X)$ in (7.8) is converted to maximize the right-hand side of (7.11) with respect to the parameters $\Theta$ and $\Phi$ using stochastic gradient descent (SGD) method to train $\Theta$ and $\Phi$. In the formulation of (7.11), $Q(Z\,|\,X;\,\Phi)$ is the encoder encoding $X$ into $Z$, and $P(X\,|\,Z;\,\Theta)$ is the decoder decoding $Z$ to obtain $X$. According to Theorem 7.1, to solve the optimization, Q is defined as:

$$Q(Z\,|\,X) = N(\mu(X;\Phi), \Sigma(X;\Phi)) \tag{7.12}$$

where the mean and covariance matrix are learned through deterministic function by artificial neural network parameterized with $\Phi$ by SGD method. As both $Q(Z\,|\,X;\,\Phi)$ and $P(X\,|\,Z;\,\Theta)$ are $d$-dimensional multivariate Gaussian distributions, the term $KL[Q(Z\,|\,X)\|P(Z\,|\,X)]$ in (7.11) is computed by:

$$KL\Big[Q(Z\,|\,X)\Big\|P(Z\,|\,X)\Big] = KL[N(\mu(X;\Phi), \Sigma(X;\Phi))\,||\,N(0,I)]$$

$$= \frac{1}{2}\left[log\,\frac{det(I)}{det(\Sigma)} - d + tr(\Sigma) + (0-\mu)^T(0-\mu)\right]$$

$$= \frac{1}{2}\left[-log\big(det(\Sigma)\big) - d + tr(\Sigma) + \mu^T\mu\right] \tag{7.13}$$

where $d$ is the dimension of the dataset. Therefore, to maximize (7.11), the following optimization problem is solved:

$$\theta^* = \underset{\theta \in \Theta}{argmax} E_{X\sim D}[E_{Z\sim Q}[logP(X\,|\,Z;\,\Theta)] - KL[Q(Z\,|\,X;\,\Phi)\|P(Z;\,\Theta)]] \tag{7.14}$$

Applying the re-parametrization technique, (7.14) can be written as:

$$\theta^* = \underset{\theta \in \Theta}{argmax} E_{X\sim D}\big[E_{\varepsilon\sim N(0,I)}\big[logP(X\,|\,Z = \mu(X) + \Sigma^{\frac{1}{2}}(X)\cdot\varepsilon;\,\Phi) - KL[Q(Z\,|\,X;\,\Phi)\|P(Z;\,\Theta)]\big]$$
$$\tag{7.15}$$

Based on (7.12) and (7.15), the convolutional auto-encoder training network is constructed in Figure 7.5.

Figure 7.5 shows the training structure of the generative model to generate $X^* \approx X$. The encoder neural network Q takes $X$ observed in dataset $D$ to train the outputs $\mu$ and $\Sigma$ in (7.12). The error of the encoder neural network is $KL[Q(Z\,|\,X)\|P(Z\,|\,X)]$ computed in (7.13).

The gradient of this error function is used by the SGD method to train the neural network. After computing $\mu$ and $\Sigma$ using Q, the latent variable $Z = \mu(X;\Phi) + \Sigma^{\frac{1}{2}}(X)\cdot\varepsilon$ is obtained using (7.15).

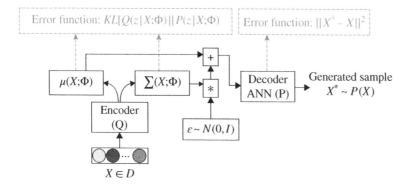

**Figure 7.5** Learning structure of convolutional auto-encoder.

Then, $Z$ is fed to the decoder neural network $P$ to obtain the generated sample $X^* \approx X$. The error function of this neural network is computed by $\|X - X^*\|^2$ to reflect the distance between the generated sample $X^*$ and the corresponding observed value $X$. When encoder $Q$ and decoder $P$ are trained by the SGD method, to generate a new sample $X^* \approx X$, one can simply feed samples meeting the distribution $Z \sim N(0, I)$ to $P$ to obtain $X^*$ in the testing structure of the convolutional encoder as shown in Figure 7.6.

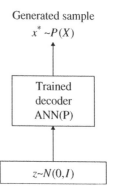

**Figure 7.6** Testing structure of convolutional auto-encoder.

### 7.3.4 Convolutional Graph Auto-Encoder

In Section 7.3.3, the probability distribution $P(X)$ of a given dataset $X$ is approximated by a convolutional auto-encoder. To learn $P^*(V^*|\pi)$, that is, probability distribution function of $V^*$ in graph $G$ by given observed historical data $\pi$, the convolutional graph auto-encoder shown in Figure 7.7 is developed on top of the convolutional auto-encoder in Figure 7.5.

Convolutional graph auto-encoder consists of three artificial neural networks: (i) graph feature extraction network, which gives us a compact representation of $\pi$ stored in $G$, denoted by $R(G)$, (ii) encoder network $Q$ that implements $\mu'$ and $\sigma'$ to capture $Q(Z \mid \pi, V^*)$, and (iii) decoder network $P$ which implements $\mu(\pi, Z)$ to produce samples $\hat{V}$ drawn from the true future global horizontal irradiance (GHI) distribution $P^*(V^*(t')|\pi)$.

Given historical solar radiation data GHI $\pi$, the objective is to train the convolutional graph auto-encoder to approximate $P^*(V^* \mid \pi)$ and generate samples $\hat{V} \approx V^*$ as shown in Figure 7.8.

Where

$$\hat{V} = \mu(\pi, Z) + \varepsilon \ \ s.t. \ \ Z \sim N(0,1), \varepsilon \sim N(0,1) \tag{7.16}$$

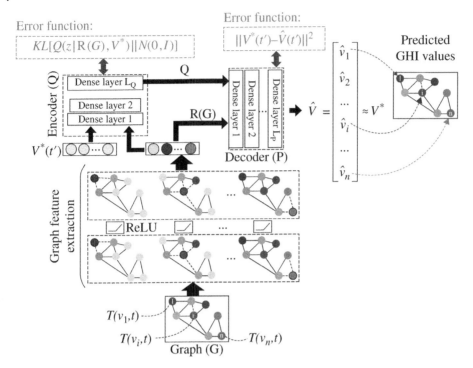

**Figure 7.7** Convolutional graph auto-encoder.

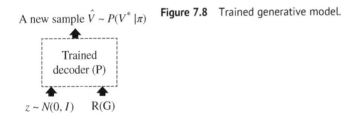

**Figure 7.8** Trained generative model.

Both Z and $\varepsilon$ are white Gaussian noises. $\mu$ is implemented by a trained artificial neural network as shown in Section 7.3.3. To compute $E_{Z \sim Q}[logP(V^*) \mid Z, \pi]$, Bayes rule [29] is applied:

$$
\begin{aligned}
E_{Z \sim Q}[logP(V^*(t')|Z,\pi)] &= E_{Z \sim Q}\left[log\,\frac{P(Z|V^*(t'),\pi)P(V^*(t')|\pi)}{P(Z|\pi)}\right] \\
&= E_{Z \sim Q}[logP(Z|V^*(t'),\pi) - logP(Z|\pi) + logP(V^*(t')|\pi)]
\end{aligned}
\tag{7.17}
$$

Assuming $Z \sim Q$ using probability distribution function $Q(Z)$, (7.17) can be written as:

$$
\begin{aligned}
& logP(V^*(t')\,|\,\pi) - E_{Z \sim Q}[logQ(Z) - logP(Z\,|\,V^*(t'),\pi)] \\
& = E_{Z \sim Q}[logP(V^*(t')\,|\,Z,\pi) + logP(Z\,|\,\pi) - logQ(Z)]
\end{aligned}
\tag{7.18}
$$

Following (7.12), we have $Q = N(\mu'(\pi, V^*(t')), \sigma'(\pi, V^*(t')))$ where $\mu'$ and $\sigma'$ are trained by the artificial neural network. Let us denote $Q$ by $Q(Z|\pi, V^*)$, (7.18) is written as:

$$
\begin{aligned}
& logP(V^*\,|\,\pi) - KL[Q(Z|\pi, V^*)\,||\,P(zZ\,|\,\pi, V^*)] = E_{z \sim Q}[logP(V^*|Z, \pi)] \\
& \quad - KL[Q(Z|\pi, V^*)\,||\,P(Z\,|\,\pi)]
\end{aligned}
\tag{7.19}
$$

To maximize $logP(V^*|\pi)$ in (7.19), the convolutional graph auto-encoder is trained by SGD to maximize $E_T = logP(V^*|z, \pi) - KL[Q(z|\pi, V^*)\,|\,|P(z\,|\,\pi)]$, which leads to maximize the likelihood of $V^*$ while training $Q$ to accurately estimate $P(Z\,|\,V^*, \pi)$.

According to Theorem 7.1, we assign $P(Z|\pi) = N(0, 1)$. The latent vector is $Z = \mu'(\pi, V^*(t')) + \alpha \otimes \sigma'(\pi, V^*(t'))$, where $\alpha \sim N(0, 1)$ and $\otimes$ is the element-wise product operation. $E_T = logP(V^* | z, \pi) - KL[Q(z|\pi, V^*)\,|\,|P(z\,|\,\pi)]$ is differentiable with respect to the whole parameters of the convolutional graph auto-encoder; hence, the whole convolutional graph auto-encoder model can be tuned by SGD method to maximize $E_T$.

### 7.3.5 Graph Feature Extraction Artificial Neural Network (R(G))

At each training step $t$, the spectral graph convolutions of $G$, which stores $\pi = \; < T(v_1,t), T(v_2,t), ..., T(v_n,t) >$ inside its nodes, is computed by $\psi_\theta * \pi = U\psi_\theta U^T \pi$. Here, $U$ is the eigenvector matrix of the normalized Laplacian $L = U\Omega U^T$ and $\theta \in \mathbb{R}^n$ is the parameter vector for the convolutional filter $\psi_\theta = diag\,(\theta)$ in the Fourier domain. Note that the Fourier transform of $\pi$ is computed by $U^T \pi$. $\psi_\theta$ is defined as a function of $L'$s eigenvalues; hence, our filter is denoted by $\psi_\theta(\Omega)$. Estimating $\psi_\theta(\Omega)$ by Chebyshev polynomials [30, 31] $P_j$, we have $\psi_\omega \approx \sum_{j=0}^{J} \omega_j P_j \left(\frac{2}{\gamma_{max}}\Omega - I\right)$, where $\gamma_{max}$ is the maximum eigenvalue of $L$, and $\omega_j$ is the $j$th Chebyshev coefficient. Therefore, the spectral graph convolution function on $G$ is:

$$
\psi_\omega * \pi \approx \sum_{j=0}^{J} \omega_j P_j \left(\frac{2}{\gamma_{max}}\Omega - I\right)\pi
\tag{7.20}
$$

The convolution in (7.20) is further simplified by $\delta = \omega_0 = -\omega_1$ which decreases the size of the parameters while $\gamma_{max} = 2$ for $J = 1$; as a result, (7.20) can be computed by:

$$\psi_\omega * \pi \approx \omega_0 P_0 (L - I)\pi + \omega_1 P_1 (L - I)\pi = \delta \left( I + D^{-\frac{1}{2}} A D^{-\frac{1}{2}} \right) \pi \tag{7.21}$$

Based on the convolution (7.21), a graph feature extraction neural network with $L_G$ hidden layers is defined to extract spatiotemporal features from solar radiation observations at all nodes/sites of $G$. Here, the output of each layer $1 \leq k \leq L_G$ is:

$$O^k = ReLU \left( M O^{k-1} W^k \right) \text{ s.t. } M = \widetilde{D}^{-\frac{1}{2}} (A + I) \widetilde{D}^{-\frac{1}{2}} \tag{7.22}$$

where $\widetilde{D}_{ii} = \sum_j (A + I)_{ij}$. The input of the graph feature extraction neural network is $O^0 = \pi$, whereas the output is $G$'s spatiotemporal representation $R(G) = O^{L_G}$.

### 7.3.6 Encoder (Q) and Decoder (P)

Since graph feature extraction neural network captures spatiotemporal features of $\pi$ and stores them in R(G), one can view convolutional graph auto-encoder as a model estimating $P^*(V^* \mid R(G))$ instead of $P^*(V^* \mid \pi)$. In Section 7.3.3, (7.12) showed that Q can be viewed as an artificial neural network encoding input tensor $X$ into the latent vector $Z$ while $P$ is a decoding artificial neural network that maps $Z$ to $X$. As depicted in Figure 7.7, the input to the encoder $Q$ is $X = R(G)$. The encoder $Q$ is defined by a deep artificial neural network with $L_Q$ hidden layers and ReLU activations for each hidden layer are trained to encode $V^*$ into latent vector $Z \in \mathbf{Z}$, such that the resulting $Z$ can be decoded back to $V^*$. As discussed in (7.19) and also shown in Figure 7.7, the error function for the encoder $Q$ is defined by:

$$Err_Q = KL[Q(Z \mid \pi, V^*) \mid\mid N(0, 1)] = KL[Q(Z \mid R(G), V^*) \mid\mid N(0, 1)] \tag{7.23}$$

Similar to encoder $Q$, the decoder $P$ is implemented by a deep artificial neural network with $L_p$ hidden layers using ReLU activations to take the latent vector $Z$ learned by $Q$, as well as the graph representation $R(G)$, and decode them to generate an approximation of $V^*$, denoted by $\hat{V}$. To make the generated sample $\hat{V}(t')$ as close as possible to the real future value $V^*(t')$, we minimize the following reconstruction error for $P$:

$$Err_P = \left\| V^* \left( t' - \hat{V}(t') \right) \right\|^2 \tag{7.24}$$

Therefore, the total error optimized by the SGD method is $E = Err_Q + Err_P$.

### 7.3.7 Estimation of $P(V^*|\pi)$

As shown in Figure 7.8, during test time, $R(G)$ and $Z \sim N(0, I)$ are fed to the decoder artificial neural network and the estimation $\hat{V}(t')$ is obtained. No encoding is needed; hence, generating estimations $\hat{V}(t') \approx V^*(t')$ is dramatically fast. To generate a new sample $V^*(t')$, the test model makes a sample $Z \sim N(0, I)$ and runs a feedforward algorithm on the graph feature extraction neural network (to obtain $R(G)$) and the decoder artificial neural network (to obtain the desired result, that is, $\hat{V}(t')$). Following this approach, we generate $\rho$ number of samples $\hat{V} \sim P(V^*|\pi)$ to estimate $P(V^* | \pi)$ using the decoder. As a result, our decoder $P$ generates the probability distribution function of future solar radiation mapping $N(0, I)$ to $P(V^*|\pi)$.

## References

1  Jordan, M.I. and Mitchell, T.M. (2015). Machine learning: trends, perspectives, and prospects. *Science* 349 (6245): 255–260.

2  Papalexopoulos, A.D., Hao, S., and Peng, T.M. (1994). An implementation of a neural network based load forecasting model for the EMS. *IEEE Transactions On Power System* 9 (4): 1956–1962.

3  Voyant, C., Notton, G., Kalogirou, S. et al. (2017). Machine learning methods for solar radiation forecasting: a review. *Renewable Energy* 105: 569–582.

4  Wan, C., Xu, Z., Pinson, P. et al. (2014). Probabilistic forecasting of wind power generation using extreme learning machine. *IEEE Transactions on Power Systems* 29 (3): 1033–1044.

5  Pathiravasam, C. and Venayagamoorthy, G. (2018). Comparison of learning cellular computational networks with EKF and CPSO for multi-location wind speed prediction. *2018 IEEE Symposium Series on Computational Intelligence (SSCI)*, Bangalore, India, pp. 897–904. https://doi.org/10.1109/SSCI.2018.8628620.

6  Kamel, M., Wang, Y., Yuan, C. et al. (2020). A reinforcement learning approach for branch overload relief in power systems. *2020 IEEE Power & Energy Society General Meeting (PESGM)*, Montreal, QC, Canada, pp. 1–5. https://doi.org/10.1109/PESGM41954.2020.9281402.

7  McCalley, J.D. and Krause, B.A. (1995). Rapid transmission capacity margins determination for dynamic security assessment using artificial neural networks. *Electric Power Systems Research* 34: 37–45.

8  Park, Y.-M., Kim, G.-W., Cho, H.-S. et al. (1997). A new algorithm for Kohonen layer learning with application to power stability analysis. *IEEE Transactions on Systems, Man and Cybernetics, Part B-Cybernetics* 27 (6): 1030–1034.

9  Wu, L., Venayagamoorthy, G.K., and Gao, J. (2021). Online steady-state security awareness using cellular computation networks and fuzzy techniques. *Energies* 14 (1): 148.

10  Kamel, M., Dai, R., Wang, Y. et al. (2021). Data-driven and model-based hybrid reinforcement learning to reduce stress on power systems branches. *CSEE Journal of Power and Energy Systems* 7 (3): 433–442.

11  Fischetti, M. and Jo, J. (2017). Deep neural networks as 0-1 mixed integer linear programs: a feasibility study. *arXiv* 1712.06174 [cs.LG]: 1–13.

**12** Wu, L., Venayagamoorthy, G.K., and Gao, J. (2019). Cellular computational networks for distributed prediction of active power flow in power systems under contingency. *2019 IEEE PES Innovative Smart Grid Technologies Europe (ISGT-Europe)*, Bucharest, Romania, pp. 1–5. https://doi.org/10.1109/ISGTEurope.2019.8905569.

**13** Jiang, Y., Long, H., Zhang, Z., and Song, Z. (2017). Day-ahead prediction of bihourly solar radiance with a Markov switch approach. *IEEE Transactions on Sustainable Energy* 8 (4): 1536–1547.

**14** Wan, C., Zhao, J., Song, Y. et al. (2015). Photovoltaic and solar power forecasting for smart grid energy management. *CSEE Journal of Power and Energy Systems* 1 (4): 38–46.

**15** Hottel, H.C. (1976). A simple model for estimating the transmittance of direct solar radiation through clear atmospheres. *Solar Energy* 18 (2): 129–134.

**16** Pfenninger, S. and Staffell, I. (2016). Long-term patterns of European PV output using 30 years of validated hourly reanalysis and satellite data. *Energy* 114: 1251–1265.

**17** Larson, D.P., Nonnenmacher, L., and Coimbra, C.F. (2016). Day-ahead forecasting of solar power output from photovoltaic plants in the American southwest. *Renewable Energy* 91: 11–20.

**18** Nou, J., Chauvin, R., Thil, S. et al. (2015). Clear-sky irradiance model for real-time sky imager application. *Energy Procedia* 69: 1999–2008.

**19** Crisosto, C., Hofmann, M., Mubarak, R., and Seckmeyer, G. (2018). One-hour prediction of the global solar irradiance from all-sky images using artificial neural networks. *Energies* 11 (11): 2906.

**20** Jayawardene, I. and Venayagamoorthy, G.K. (2016). Spatial predictions of solar irradiance for photovoltaic plants. *2016 IEEE 43rd Photovoltaic Specialists Conference (PVSC)*, Portland, OR, USA, pp. 0267–0272. https://doi.org/10.1109/PVSC.2016.7749592.

**21** Wei, Y., Jayawardene, I., and Venayagamoorthy, G.K. (2015). Frequency prediction of synchronous generators in a multi-machine power system with a photovoltaic plant using a cellular computational network. *2015 IEEE Symposium Series on Computational Intelligence*, Cape Town, South Africa, pp. 673–678. https://doi.org/10.1109/SSCI.2015.103.

**22** Tiwari, S., Sabzchgar, R., and Rasouli, M. (2018). Short term solar irradiance forecast using numerical weather prediction (NWP) with gradient boost regression. *2018 9th IEEE International Symposium on Power Electronics for Distributed Generation Systems (PEDG)*, Charlotte, NC, USA. IEEE, pp. 1–8. https://doi.org/10.1109/PEDG.2018.8447751.

**23** Chen, Y., Wang, Y., Kirschen, D., and Zhang, B. (2018). Model-free renewable scenario generation using generative adversarial networks. *IEEE Transactions on Power Systems* 33 (3): 3265–3275.

**24** Bayindir, R., Yesilbudak, M., Colak, M., and Genc, N. (2017). A novel application of naive bayes classifier in photovoltaic energy prediction. *2017 16th IEEE International Conference on Machine Learning and Applications (ICMLA)*, Cancun, Mexico, pp. 523–527. https://doi.org/10.1109/ICMLA.2017.0-108.

**25** Chu, Y. and Coimbra, C.F. (2017). Short-term probabilistic forecasts for direct normal irradiance. *Renewable Energy* 101: 526–536.

**26** Girvan, M. and Newman, M.E. (2002). Community structure in social and biological networks. *Proceedings of the National Academy of Sciences* 99 (12): 7821–7826.

**27** Khodayar, M., Mohammadi, S., Khodayar, M.E. et al. (2020). Convolutional graph autoencoder: a generative deep neural network for probabilistic spatio-temporal solar irradiance forecasting. *IEEE Transactions on Sustainable Energy* 11 (2): 571–583.

**28** Joyce, J.M. (2011). Kullback-Leibler divergence. In: *International Encyclopedia of Statistical Science* (ed. M. Lovric), 102–235. Berlin, Heidelberg: Springer.

**29** Kingma, D.P. and Welling, M. (2013). Auto-encoding variational bayes. *arXiv preprint arXiv* 1312.6114: 1–14.

**30** Kipf, T.N. and Welling, M. (2016). Semi-supervised classification with graph convolutional networks. *arXiv preprint arXiv* 1609.02907: 1–14.

**31** Khodayar, M. and Wang, J. (2018). Spatio-temporal graph deep neural network for short-term wind speed forecasting. *IEEE Transactions on Sustainable Energy* 10: 670–681.

**Part II**

**Implementations and Applications**

# 8

# Power Systems Modeling

Conventionally, we use a relational database to organize data into tables for power system modeling. A Relational Database Management System (RDBMS) uses Structured Query Language (SQL) for querying and maintaining the database. The database stores structured records and their attributes in an equal-length table. Ideally, data relationships of arbitrary complexity can be presented by a relational database. However, the limitations of RDBMS on power system analysis are obvious when records have different lengths of attributes. A straightforward illustration can be observed in the case of a transmission line and a transformer, which both fall under the category of branches. However, they have different numbers and meanings of attributes. In conventional power system modeling, they are usually organized into two different tables by a relational database.

In contrast to the relational database, the graph database is concise when modeling different data structures, from simple array data to the more complex structures which store graphs and trees. Hence, the graph database is very accommodating of data interconnections within a data set that fits the characteristics of power systems since the power system is naturally modeled as a graph consisting of nodes and branches. Nodes are physically connected through branches as edges. The unconstructed parameters of the bus, generator, load, and branch, such as active and reactive power of a generator and load, resistance and reactance of lines and transformers, and so on are stored in the node or edge. The graph structure itself naturally represents the topology of an electric power grid.

## 8.1 Power System Graph Modeling

As discussed in Chapter 2, graphs are used to model relationships among entities. The graph uses vertices to store subject information, and edges to articulate the relationship between the vertices. A graph can be expressed as $G = (V, E)$, where $V$ denotes the set of vertices and $E$ denotes the set of edges.

To support power system analysis, the bus-branch model and the node-breaker model are commonly used to describe a power system network. In the bus-branch model, edges are usually adopted to represent both multi-terminal equipment (for example, transmission line and transformer) and connecting relationship (buses connected by transmission lines or transformers) to minimize the size of the graph while vertices are served as buses. Single-terminal devices such as generators, loads, and shunts are modeled as attributes of vertices. In the node-breaker

*Graph Database and Graph Computing for Power System Analysis*, First Edition. Renchang Dai and Guangyi Liu.
© 2024 The Institute of Electrical and Electronics Engineers, Inc. Published 2024 by John Wiley & Sons, Inc.

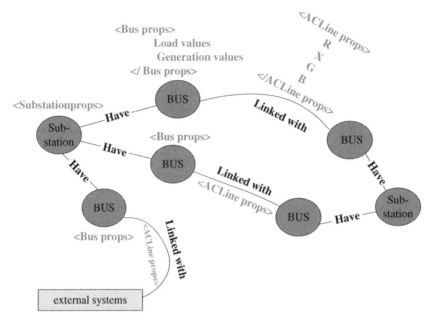

**Figure 8.1** Conceptual bus-branch graph model.

model, to facilitate topology analysis, all physical elements and devices of power systems, including transmission lines, transformers, generators, loads, and shunts, along with breakers, switches, disconnectors, and so on are modeled as vertices and they are connected to each other or connected to virtual vertices through virtual edges.

Figure 8.1 shows a conceptual bus-branch graph model, in which transmission lines are modeled as edges to connect buses at their two terminals.

Figure 8.2 shows a conceptual node-breaker graph model for a simple substation, in which circuit breakers, disconnectors, load, and generator are modeled as vertices connected by virtual edges (directly or through physical vertices or nodes). These physical vertices or nodes are named topology nodes or connectivity nodes in graph modeling terminology.

## 8.2 Physical Graph Model and Computing Graph Model

In power system graph modeling terminology, there are two models: the physical graph model and the computing graph model. The physical graph model describes the power system's physical connectivity and attributes. The physical connectivity is generalized to include both node-breaker connectivity and bus-branch connectivity which is reduced from node-breaker connectivity by topology processing based on breaker and switch statuses.

The physical attributes cover power system equipment and devices' physical characteristics and operational parameters in both steady-state and dynamics, such as resistance, reactance, capacitance, tap settings, nominal voltage, active power, reactive power, power limits, voltage limits, generation cost, inertia, control gains, time constant, dam elevation, and so on.

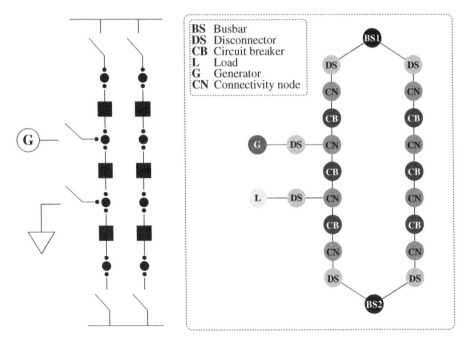

**Figure 8.2** Conceptual node-break graph model.

Bus-branch model is categorized into a physical graph model to align with the power system analysis practice in the industry since most power system analysis starts with a bus-branch model. In graph computing, the node-breaker model and the bus-branch model are defined by a set of vertices and edges in a graph.

The computing graph model is an abstract graph model defined specifically for graph computing purposes. For example, to solve power flow equations, admittance graph $G(A)$ is defined as an equivalent admittance matrix in which an element $a_{ij} \neq 0$ $(i \neq j)$ in admittance matrix $A$ is equivalent to vertices $v_i$ and $v_j$ and are connected; on the contrary, $a_{ij} = 0$ equivalent to the edge $e_{ij}$ does not exist, which means vertices $v_i$ and $v_j$ are not connected. The built admittance graph is a computing graph since there may be no physical device equivalent to the edge and vertex in the graph. The admittance graph preserves the sparsity of the admittance matrix and it is defined for computing purposes.

Other examples of computing graphs are filled graph $G^+(A)$, elimination tree graph T(A) discussed in Chapter 3, the graph for the simplex table $\begin{bmatrix} -c^T & 0 \\ A & b \end{bmatrix}$, the graph for the interior-point update equation matrix $\begin{bmatrix} A & 0 & 0 \\ 0 & A^T & I \\ S & 0 & X \end{bmatrix}$ developed in Chapter 6, and the graph for machine learning in Chapter 7. The vertices and edges in these graphs are abstract and have no explicit mapping to power system's physical equipment. They are highly used to implement graph computing for power system analysis.

## 8.3   Node-Breaker Model and Graph Representation

The node-breaker model represents the high granularity of a power system network which models field sectionalizing breakers and their status to show physical or actual connections between the network components. Node-breaker models are widely used to analyze power system network topology, protection and control scheme, circuit breaker power flow, and power system postcontingency security. The majority of power system online operation applications and part of offline planning studies use a node-breaker model to represent power system networks.

In the power industry, Common Information Model (CIM) is developed by Electric Power Research Institute (EPRI) as a standard to model power systems and interoperate data in CIM/XML format [1, 2]. The node-breaker graph model is established based on the CIM standard [3]. To demonstrate the approach of using a graph data structure to model a power system, CIM/E, a new version of the CIM model with higher efficiency is adopted [4, 5]. The CIM/E model simplifies CIM/XML model by ignoring terminals without compromising network topology with a much smaller size and simpler form.

In CIM/XML, the topology relationship is described by equipment, terminals, and connectivity nodes as shown in Figure 8.3 [6].

In the CIM/E model, the above topology is represented by equipment and connectivity nodes as shown in Figure 8.4.

By the CIM/E model, a typical substation consisting of busbar(s), circuit breaker(s), disconnector(s), generator(s), and load(s) is represented in Figure 8.5 in which all the electrical components are connected by connectivity nodes.

In the graph node-breaker model, all power system physical components are represented by vertices, and the connection relationship between physical components is represented by edges.

The node-breaker model standardized by CIM/E contains 15 tables, namely base voltage, substation, busbar, AC line, generator, transformer, load, `compensator_P`, `compensator_S`, converter, DC line, island, topological node, breaker, and disconnector tables. The attributes of each table are listed as follows:

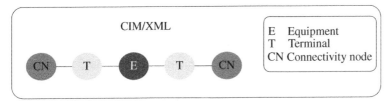

**Figure 8.3**   CIM/XML model.

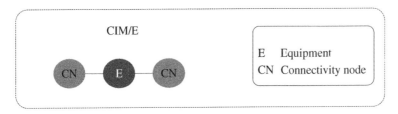

**Figure 8.4**   CIM/E model.

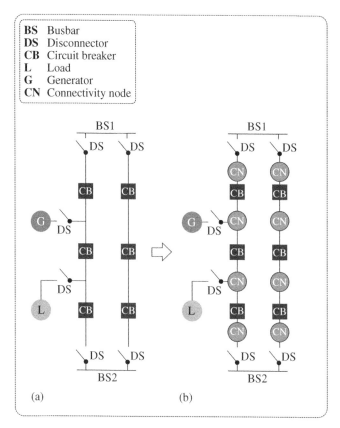

**Figure 8.5** Substation representation in one line diagram and CIM/E. (a) one line diagram; (b) CIM/E representation.

| | |
|---|---|
| **base value** | *id, name, value, unit* |
| **substation** | *name, volt, type, config, nodes, islands, island, dvname* |
| **busbar** | *id, name, volt, node, V, Ang, off, V_meas, Ang_meas, nd, bs, island, v_max, v_min* |
| **AC line** | *id, name, volt, Eq, R, X, B, I_node, J_node, I_P, I_Q, J_P, J_Q, I_off, J_off, Ih, Pi_meas, Qi_meas, Pj_meas, Qj_meas, I_nd, J_nd, I_bs, J_bs, I_island, J_island, R\*, X\*, B\** |
| **generator** | *id, name, Eq, position, V_rate, P_rate, volt_n, node, P, Q, Ue, Ang, off, P_meas, Q_meas, Ue_meas, Ang_meas, P_max, P_min, Q_max, Q_min, pf, nd, bs, island* |
| **transformer** | *id, name, type, I_Vol, K_Vol, J_Vol, I_S, K_S, J_S, Itap_H, Itap_L, Itap_E, Itap_C, Itap_V, Ktap_H, Ktap_L, Ktap_E, Ktap_C, Ktap_V, Jtap_V, Ri, Xi, Rk, Xk, Rj, Xj, I_node, K_node, J_node, I_P, I_Q, K_P, K_Q, J_P, J_Q, I_tap, K_tap, I_off, K_off, J_off, Pi_meas, Qi_meas, Pk_meas, Qk_meas, Pj_meas, Qj_meas, Ti_meas, Tk_meas, G, B, I_nd, K_nd, J_nd, I_bs, K_bs, J_bs, I_island, K_island, J_island, Ri\*, Xi\*, Rk\*, Xk\*, Rj\*, Xj\*, I_t, K_t, J_t, IBase_V, KBase_V, JBase_V* |

| | |
|---|---|
| **load** | *id, name, volt, Eq, position, node, P, Q, off, P_meas, Q_meas, nd, bs, island* |
| **compensator_P** | *id, name, volt, Q_rate, V_rate, position, node, P, Q, off, Q_meas, nd, bs, island)* |
| **compensator_S** | *id, name, volt, Q_rate, V_rate, I_node, J_node, Zk, off, Pi_meas, Qi_meas, Pj_meas, Qj_meas, Ih, Zk_max, Zk_min, Line, I_nd, J_nd, I_bs, J_bs, I_island, J_island, Zk\*)* |
| **converter** | *id, name, N, I_DCName, J_DCName, N_ACName, Vdc, Idc, Wdc, Pac, Qac, Mode, state, off, Ang, Vdc_meas, Idc_meas, Wdc_meas, Pac_meas, Qac_meas, Ang_meas, Wdc_rate, Ang_max, Ang_min, Wdc_Rate, Kt, I_dcnd, J_dcnd, N_acnd, I_dcbs, J_dcbs, N_acbs* |
| **DC line** | *id, name, R, I_node, J_node, Conduct, off, Ih, I_nd, J_nd, I_bs, J_bs, I_island, J_island* |
| **island** | *id, name, Ref_bus, off, F_sys* |
| **topological node** | *id, name, island, v, ang, vbase* |
| **breaker** | *id, name, volt, point, I_nd, J_nd, I_node, J_node* |
| **disconnector** | *id, name, volt, point, I_nd, J_nd, I_node, J_node* |

The annotations of the attributes of each table are listed in Appendix A.

To map the power system CIM/E model to a node-breaker graph, the following vertices and edges are defined:

```
CREATE VERTEX BV (PRIMARY_ID basevalueid int, id int, name string,
    bvvalue double, unit string)

CREATE VERTEX Substation (PRIMARY_ID subid int, id int, name string,
    volt double, subtype string, config string, nodes int, islands
    uint, island string, dvname string)

CREATE VERTEX BUS (PRIMARY_ID busid int, topoID int, id int, name
    string, volt double, node string, V double, Ang double, off uint,
    V_meas double, Ang_meas double, nd int, bs uint, island string,
    v_max double, v_min double, typename string)

CREATE VERTEX ACline (PRIMARY_ID AClineid int, id int, name string,
    volt double, Eq uint, R double, X double, B double, I_node string,
    J_node string, I_P double, I_Q double,J_P double, J_Q double,
    I_off uint, J_off uint, Ih double, Pi_meas double, Qi_meas double,
    Pj_meas double,Qj_meas double,I_nd int, J_nd int, I_bs uint, J_bs
    uint, I_island string, J_island string, line_R double, line_X
    double, line_B double, typename string)

CREATE VERTEX Unit (PRIMARY_ID unitid int, topoID int, id int, name
    string, Eq uint, position string, V_rate double, P_rate double,
    volt_n double, node string, control_bus_number int, desired_volts
    double, qUp double, qLower double, P double, Q double, Ue double,
    Ang double, off uint, P_meas double, Q_meas double, Ue_meas double,
```

Ang_meas double, P_max double, P_min double, Q_max double, Q_min double, pf double, nd int, bs uint, island string, typename string)

CREATE VERTEX Transformer (PRIMARY_ID transformerID int, topoID int, index int, name string, type string, volt double, S double, itapH double, itapL double, itapC double, R double, X double, I_node string, J_node string, I_P double, I_Q double, J_P double, J_Q double, off uint, Pi_meas double, Qi_meas double, Pj_meas double, Qj_meas double, I_nd int, J_nd int, I_bs uint, J_bs uint, Rstar double, Xstar double, t double, base double, typename string)

CREATE VERTEX Load (PRIMARY_ID loadid int, topoID int, id int, name string, volt double, Eq uint, position string, node string, P double, Q double, off uint, P_meas double, Q_meas double, nd int, bs uint, island string, typename string)

CREATE VERTEX C_P (PRIMARY_ID compensatorPid int, topoID int, id int, name string, volt double, Q_rate double, V_rate double, position string, node string, control_bus_number int, desired_volts double, qUp double, qLower double, P double, Q double, off uint, Q_meas double, nd int, bs uint, island string, typename string)

CREATE VERTEX C_S (PRIMARY_ID compensatorSid int, topoID int, id int, name string, volt double, Q_rate double, V_rate double, I_node string, J_node string, Zk double, off uint, Pi_meas double, Qi_meas double, Pj_meas double, Qj_meas double, Ih uint, Zk_max double, Zk_min double, Line string, I_nd int, J_nd int, I_bs uint, J_bs uint, I_island string, J_island string, cs_ZK double, typename string)

CREATE VERTEX Converter (PRIMARY_ID convertorid string, topoID uint, id uint, name string, N uint, I_DCNAME string, J_DCNAME string, N_ACNAME string, Vdc uint, Idc uint, Wdc uint, Pac uint, Qac double, Mode uint, state uint, off uint, Ang uint, Vdc_meas string, Idc_meas string, Wdc_meas string, Pac_meas string, Qac_meas string, Ang_meas string, Wdc_rate string, Ang_max uint, Ang_min uint, Wdc_Rate uint, Kt uint, I_dcnd uint, J_dcnd uint, N_acnd uint, I_dcbs uint, J_dcbs uint, N_acbs uint, typename string)

CREATE VERTEX DCline (PRIMARY_ID DClineid string, topoID uint, id uint, name string, R double, I_node string, J_node string, Conduct uint, off uint, Ih double, I_nd int, J_nd int, I_bs uint, J_bs uint, I_island string, typename string)

CREATE VERTEX ISLAND (PRIMARY_ID islandid string, id uint, name string, Ref_bus string, off uint, F_sys string, typename string)

```
CREATE VERTEX topoNd (PRIMARY_ID topoNdid int, topoID int, id uint,
    name string, v double, ang, double, vbase double)

CREATE VERTEX Breaker (PRIMARY_ID breakerid uint, topoID uint, id
    uint, name string, volt uint, point uint, I_nd uint, J_nd uint,
    I_node string, J_node string, typename string)

CREATE VERTEX Disconnector (PRIMARY_ID disconnectorid uint, topoID
    uint, id uint, name string, volt uint, point uint, I_nd uint, J_nd
    uint, I_node string, J_node string, typename string)

CREATE VERTEX CN (PRIMARY_ID cnid int, topoID int, CN_id int,
    processed int)
```

In the above definition, a two-winding transformer vertex is defined. Three-winding transformer is modeled as 3 two-winding transformers. Vertex CN is a connectivity node to connect devices in the systems.

In the node-breaker model, each device is modeled as a vertex and their attributes match the CIM/E device attributes with two additional attributes: PRIMARY_ID and topoID. PRIMARY_ID is unique in the whole database to identify the device. topoID represents the topological identification of the device in the graph connecting to other topoID through connectivity nodes.

On top of the vertex schema, edges are defined as follows to hold the device connectivity:

```
CREATE UNDIRECTED EDGE connected_Bus_CN(from BUS, to CN)

CREATE UNDIRECTED EDGE connected_ACline_CN(from ACline, to CN)

CREATE UNDIRECTED EDGE connected_Unit_CN(from Unit, to CN)

CREATE UNDIRECTED EDGE connected_Transformer_CN(from Transformer,
    to CN)

CREATE UNDIRECTED EDGE connected_Load_CN(from Load, to CN)

CREATE UNDIRECTED EDGE connected_Compensator_P_CN(from C_P, to CN)

CREATE UNDIRECTED EDGE connected_Compensator_S_CN(from C_S, to CN)

CREATE UNDIRECTED EDGE connected_Converter_CN(from Converter, to CN)

CREATE UNDIRECTED EDGE connected_DCline_CN(from DCline, to CN)

CREATE UNDIRECTED EDGE connected_Breaker_CN(from Breaker, to CN)

CREATE UNDIRECTED EDGE connected_Disconnector_CN(from Disconnector,
    to CN)
```

```
CREATE UNDIRECTED EDGE CN_CN(from CN, to CN, id string, name string,
    volt string, point uint, I_nd int, J_nd int, I_node string, J_node
    string)

CREATE DIRECTED EDGE connected_Bus_Sub(from BUS, to Substation) with
    reverse_edge = "connected_Sub_Bus"

CREATE DIRECTED EDGE connected_ACline_Sub(from ACline, to Substation)
    with reverse_edge = "connected_Sub_ACline"

CREATE DIRECTED EDGE connected_Unit_Sub(from unit, to Substation) with
    reverse_edge = "connected_Sub_Unit"

CREATE DIRECTED EDGE connected_Transformer_Sub(from transformer, to
    Substation) with reverse_edge = "connected_Sub_Transformer"

CREATE DIRECTED EDGE connected_Load_Sub(from load, to Substation) with
    reverse_edge = "connected_Sub_Load"

CREATE DIRECTED EDGE connected_Compensator_P_Sub(from C_P, to
    Substation) with reverse_edge = "connected_Sub_Compensator_P"

CREATE DIRECTED EDGE connected_Compensator_S_Sub(from C_S, to
    Substation) with reverse_edge = "connected_Sub_Compensator_S"

CREATE DIRECTED EDGE connected_Converter_Sub(from Converter, to
    Substation) with reverse_edge = "connected_Sub_Converter"

CREATE DIRECTED EDGE connected_DCline_Sub(from DCline, to Substation)
    with reverse_edge = "connected_Sub_DCline"

CREATE DIRECTED EDGE connected_Breaker_Sub(from Breaker , to
    Substation) with reverse_edge = "connected_Sub_Breaker"

CREATE DIRECTED EDGE connected_Disconnector_Sub(from Disconnector, to
    Substation) with reverse_edge = "connected_Sub_Disconnector"

CREATE UNDIRECTED EDGE CN_Sub(from CN, to Substation)

CREATE UNDIRECTED EDGE CN_Topo(from CN, to TopoND)

CREATE UNDIRECTED EDGE topo_Sub(from TopoND, to Substation)

CREATE UNDIRECTED EDGE topo_Bus(from TopoND, to BUS)

CREATE UNDIRECTED EDGE topo_ACline(from TopoND, to ACline)
```

```
CREATE UNDIRECTED EDGE topo_Unit(from TopoND, to Unit)

CREATE UNDIRECTED EDGE topo_Transformer(from TopoND, to Transformer)

CREATE UNDIRECTED EDGE topo_Load (from TopoND, to Load)

CREATE UNDIRECTED EDGE topo_Compensator_P (from TopoND, to C_P)

CREATE UNDIRECTED EDGE topo_Compensator_S (from TopoND, to C_S)

CREATE UNDIRECTED EDGE topo_Converter (from TopoND, to Converter)

CREATE UNDIRECTED EDGE topo_DCline (from TopoND, to DCline)

CREATE UNDIRECTED EDGE topo_Breaker (from TopoND, to Breaker)

CREATE UNDIRECTED EDGE topo_Disconnector (from TopoND, to
    Disconnector)
```

The edges are defined in three categories.

The first categorized edge *"connected_ < Component > _CN"* connects a component to a connectivity node as shown in Figure 8.2. <Component> can be replaced by any component of a unit, load, transformer, etc., and connectivity node as well.

The second categorized edge *"connected_<Component>_Sub"* defines the relationship that a substation has the component. The clause *"with reverse_edge"* defined the reverse edge by switching the from-vertex and the to-vertex to enable a path from a substation to a component reversely.

The third categorized edge *"Topo_<Component>"* connects TopoND to a component which is utilized to reduce the node-breaker model to the bus-branch model. Please note the vertex TopoND is not a topoNd defined in the node-breaker model schema in which the topoNd is a physical node. Vertex TopoND is a computing node which will be defined in the bus-branch model schema. TopoND is functioning as a bridge between the node-breaker graph representation and the bus-branch graph representation.

The node-breaker graph schema is then defined as:

```
CREATE GRAPH nodeBreakerGraph(
    //Vertex
    BV, Substation, BUS, ACline, Unit, Transformer, Load, C_P, C_S,
    Converter, DCline, ISLAND, topoNd, Breaker, Disconnector, CN,

    //Component to Connectivity Node Edge
    connected_Bus_CN, connected_ACline_CN, connected_Unit_CN,
    connected_Transformer_CN, connected_Load_CN,
    connected_Compensator_P_CN, connected_Compensator_S_CN,
    connected_Converter_CN, connected_DCline_CN, connected_Breaker_CN,
    connected_Disconnector_CN, connected_CN_CN,
```

```
//Component to Substation Edge
connected_Bus_Sub, connected_Unit_Sub, connected_ACline_Sub,
connected_Transformer_Sub, connected_Load_Sub,
connected_Compensator_P_Sub, connected_Compensator_S_Sub,
connected_Converter_Sub, connected_DCline_Sub,
connected_Breaker_Sub,
connected_Disconnector_Sub,connected_CN_Sub, connected_topo_Sub,

//Substation to Component Edge
connected_Sub_Bus, connected_Sub_Unit, connected_Sub_ACline,
connected_Sub_Transformer, connected_Sub_Load,
connected_Sub_Compensator_P, connected_Sub_Compensator_S,
connected_Sub_Converter, connected_Sub_DCline,
connected_Sub_Breaker, connected_Sub_Disconnector,

//Topo Node to Component Edge
topo_Bus, topo_ACline, topo_Unit, topo_Transformer,topo_Load,
topo_Compensator_P, topo_Compensator_S, topo_Converter,
topo_DCline, topo_Breaker, topo_Disconnector,

//Connectivity Node to Connectivity Node Edge
CN_CN,

//Connectivity Node to Topology Node Edge
CN_Topo)
```

Once the graph schema is defined with the desired structure and attributes placeholders, the actual data in the CIM/E tables are loaded to the graph and mapped values to the vertex and edge attributes as examples of the following loading jobs.

```
load "$sys.data_root/Substation.csv" TO VERTEX Substation
    values($"id",$"id",$"name",$"volt",$"type",$"config",$"nodes",
    $"islands",$"island",$"dvname")
    using Separator=",", Header="true";

load "$sys.data_root/Bus.csv" TO VERTEX BUS
    values($"id",$"id",$"id",$"name",$"volt",$"node",$"V",$"Ang",
    $"off",$"V_meas",$"Ang_meas",$"nd",$"bs",$"island",$"v_max",
    $"v_min", $"typename")
    using Separator=",", Header="true";

load "$sys.data_root/ACline.csv" TO VERTEX ACline values
    ($"id",$"id",$"name",$"volt",$"Eq",$"R",$"X",$"B",$"I_node",
    $"J_node",$"I_P",$"I_Q",$"J_P",$"J_Q",$"I_off",$"J_off",$"Ih",
    $"Pi_meas",$"Qi_meas",$"Pj_meas",$"Qj_meas",$"I_nd",$"J_nd",
    $"I_bs",$"J_bs",$"I_island",$"J_island",$"R*",$"X*",$"B*",
    $"typename")
    using Separator=",", Header="true";
```

```
load "$sys.data_root/Unit.csv" TO VERTEX Unit
    values($"id",$"id",$"id",$"name",$"Eq",$"position",$"V_rate",
    $"P_rate",$"volt_n",$"node",$"P",$"Q",$"Ue",$"Ang",$"off",
    $"P_meas",$"Q_meas",$"Ue_meas",$"Ang_meas",$"P_max",$"P_min",
    $"Q_max",$"Q_min",$"pf",$"nd",$"bs",$"island",$"typename")
    using Separator=",", Header="true";

load "$sys.data_root/Transformer.csv" TO VERTEX Transformer values
    ($"id",$"id",$"name",$"volt",$"S",$"itapH",$"itapL",$"itapC",$"R",
    $"X",$"I_node",$"J_node",$"I_P",$"I_Q",$"J_P",$"J_Q",$"off",
    $"Pi_meas",$"Qi_meas",$"Pj_meas",$"Qj_meas",$"I_nd",$"J_nd",
    $"I_bs",$"J_bs",$"island",$"R*",$"X*",$"t",$"typename")
    using Separator=",", Header="true";

load "$sys.data_root/Load.csv" TO VERTEX Load
    values($"id",$"id",$"id",$"name",$"volt",$"Eq",$"position",
    $"node",$"P",$"Q",$"off",$"P_meas",$"Q_meas",$"nd",$"bs",
    $"island",$"typename")
    using Separator=",", Header="true";

load "$sys.data_root/Compensator_P.csv" TO VERTEX C_P
    values($"id",$"id",$"id",$"name",$"volt",$"Q_rate",$"V_rate",
    $"position",$"node",$"P",$"Q",$"off",$"Q_meas",$"nd",$"bs",
    $"island",$"typename")
    using Separator=",", Header="true";

load "$sys.data_root/Compensator_S.csv" TO VERTEX C_S
    values($"id",$"id",$"id",$"name",$"volt",$"Q_rate",$"V_rate",
    $"I_node",$"J_node",$"Zk",$"off",$"Pi_meas",$"Qi_meas",
    $"Pj_meas",$"Qj_meas",$"Ih",$"Zk_max",$"Zk_min",$"off",$"I_nd",
    $"J_nd",$"I_bs",$"J_bs",$"I_island",$"J_island",$"Zk_star",,
    $"typename")
    using Separator=",", Header="true";

load "$sys.data_root/Converter.csv" TO VERTEX Converter
    values($"id",$"id",$"id",$"name",$"N",$"I_DCName",$"J_DCName",
    $"N_ACName",$"Vdc",$"Idc",$"Wdc",$"Pac",$"Qac",$"Mode",$"state",
    $"off",$"Ang",$"Vdc_meas",$"Idc_meas",$"Wdc_meas",$"Pac_meas",
    $"Qac_meas",$"Ang_meas",$"Wdc_rate",$"Ang_max",$"Ang_min",
    $"Wdc_Rate",$"Kt",$"I_dcnd",$"J_dcnd",$"N_acnd",$"I_dcbs",
    $"J_dcbs",$"N_acbs",$"typename")
    using Separator=",", Header="true";

load "$sys.data_root/Breaker.csv" TO VERTEX Breaker
    values($"id",$"id",$"id",$"name",$"volt",$"point",$"I_nd",
    $"J_nd",$"I_node",$"J_node",",$"typename")
    using Separator=",", Header="true";
```

```
load "$sys.data_root/Disconnector.csv" TO VERTEX Disconnector
    values($"id",$"id",$"id",$"name",$"volt",$"point",$"I_nd",
    $"J_nd",$"I_node",$"J_node",",",$"typename")
    using Separator=",", Header="true";
```

## 8.4 Bus-Branch Model and Graph Representation

In the bus-branch model, vertex `TopoND` and edge `TopoConnect` are defined as follows containing the information for basic power system applications:

```
CREATE VERTEX TopoND (PRIMARY_ID topoid int, TOPOID int, island int,
    bus_name string, area string, loss_zone uint, busType int, Vm
    double, Va double, M_Vm double, M_Va double, base_kV double,
    desired_volts double, control_bus_number int, up_V double, lo_V
    double, GenP double, GenQ double, M_Gen_P double, M_Gen_Q double,
    qUp double, qLower double, LdP double, LdQ double, P double,
    Q double, M_Load_P double, M_Load_Q double, G double, B double, Sub
    string, OV int, UV int, sumB double, sumG double, sumBi double,
    volt double, double, SE_Vm double, SE_Va double, M_C_P double)
```

```
CREATE DIRECTED EDGE TopoConnect (from TopoND, to TopoND, typename
    string, edge_name List<string>, area List<string>, zone
    List<string>, from_bus int, to_bus int, flag uint, R List<double>,
    X List<double>, hB_list List<double>, line_Q1 double, line_Q2
    double, line_Q3 double, control_bus int, e.K List<double>,
    shifting_transformer_angle List<double>, min_tap double, max_tap
    double, step_size double, min_volt List<double>, max_volt
    List<double>, M_P_BR List<double>, M_Q_BR List<double>,
    G List<double>, B List<double>, BIJ List<double>, circuit uint,
    P_BR double, Q_BR double, from_open List<int>, to_open List<int>)
```

The annotation of the vertex and edge attributes is listed in Appendix A. The vertex `TopoND` is a bus in the bus-branch model. However, `TopoND` is not necessarily a physical busbar. To differentiate the physical busbar, `TopoND` other than Bus is used in the bus-branch graph model. `TopoConnect` is a generic branch which can be an AC line, a DC line, a transformer, a phase-shifting transformer, a series compensator, and so on. To model line- and transformer-in-parallel in between two `TopoNDs`, a data structure `List` is used to list attributes of component-in-parallel. To simplify the illustration, the DC line and converter are not defined in the vertex and edge of the bus-branch graph.

The `TopoND` is a generic bus that supports the generator bus, load bus, shunt terminal bus, and connecting bus. The attributions of the `TopoND` vertex contain parameters of generator, load, and/or shunt if any. The `TopoConnect` is a generic branch that supports transmission lines, transformers, phase shifters, series compensators, and so on. The attributes of the `TopoConnect` edge contain parameters of the transmission line, transformer, phase shifter, and series compensator.

The attributes of the vertex and edge are extensible to accommodate different power system analyses. The attributes to support power system security-constrained unit commitment and transient analysis will be discussed in Chapters 13 and 14.

With the defined vertex and edge above, the bus-branch graph schema is then simply defined as:

```
CREATE GRAPH busBranchGraph(TopoND, TopoConnect)
```

The bus-branch graph schema defined the graph structure. The bus-branch graph will be formed by adding `TopoND` and `TopoConnect` through topology processing. When the raw data defined the bus-branch model available (for instance, vertex information file vertexInfo.csv and edge information file edgeInfo.csv), the loading job statement will load the data into the bus-branch graph.

```
CREATE LOADING JOB job1 FOR GRAPH busBranchGraph {
    DEFINE FILENAME vertexFile = "$sys.data_root/vertexInfo.csv";
    DEFINE FILENAME egdeFile = "$sys.data_root/edgeInfo.csv";
    LOAD vertexFile TO VERTEX TopoND VALUES ($0, $1, $2, ...);
    LOAD egdeFile TO EDGE TopoConnect VALUES ($0, $1, $2, ...);

}
```

## 8.5 Graph-Based Topology Analysis

In the network topology processing, the electrically connected nodes in the Node-Breaker model will be gathered to form vertex `TopoND` in the Bus-Branch model. They are connected and modeled as a bus in power system analysis. Edge is defined depending on the connected vertex attributes.

Briefly, the network topology analysis aims to build up a computing model consisting of buses and branches for the online analysis of a power system. Network topology processing is usually conducted in two levels, substation-level topology processing, and then system-level topology processing. Substation-level topology processing processes substation topology in parallel by checking the updated status of circuit breakers and disconnectors in the substation. The system-level topology analysis process goes through the whole network model of the power system to finally generate the bus-branch model.

### 8.5.1 Substation-Level Topology Analysis

The target of substation-level topology processing is to form the topology nodes `TopoNDs` which reflect the collection information of objects of a substation. And then inside the substation, the edges are built up in between the topology nodes.

The basic idea of parallel topology processing is to start from the busbar vertices of the substation and then travel through the un-processed connectivity nodes to find its neighbor vertices through the connected edges. In parallel substation-level topology analysis, all the busbars in a substation are searched simultaneously. When the neighboring circuit breaker or disconnector status is on, the topology processor assigns the busbar's identification to its neighbor vertex [7]. This process is continuing the search on the vertices that are not visited until all the vertexes of the substation have been assigned with busbar identification. The detailed algorithm is presented in the following Algorithm 8.1.

---

**Algorithm 8.1    CIM/E Oriented Parallel Substation Topology Processing**

---

```
1: for (all substations) //in parallel
2:       for (un-processed busbars and connectivity nodes in the
         station) //in parallel
3:           Start from busbar vertices of all substations
4:           while (not all vertices are processed) do
5:               Find neighbor vertices that are not processed
6:               if (vertices are circuit breakers or disconnectors)
7:               check the connection status
8:                   if (the status is connected)
9:                       pass the busbar id to the vertices
10:                  end
11:              end
12:              if (vertices are connectivity nodes)
13:                  pass the busbar id to the connectivity nodes
14:              end
15:          end
16:      end
17:      Insert topology nodes (TopoND) according to the IDs created
18: end
```

---

The substation-level topology process is an incremental process, which is only necessary to be performed on these substations having breaker or disconnector status changes. Algorithm 8.1 can be implemented by pseudo-code in graph query as follows:

```
SetAccum<double>    @base_kV = 0
SetAccum<double>    @baseType = 1
MaxAccum<double>    @up_V = 0
MinAccum<double>    @lo_v = 1
SetAccum<int>       @control_bus_number = 0
SetAccum<double>    @desired_volts = 0
SumAccum<double>    @GenP = 0
SumAccum<double>    @GenQ = 0
SumAccum<double>    @LdP = 0
SumAccum<double>    @LdQ = 0
MaxAccum<double>    @qUp = 0
MinAccum<double>    @qLower = 1
SumAccum<double>    @B = 0
CREATE QUERY stationLevelTopology(
    SET<VERTEX<Breaker>> breakerid
    SET<VERTEX<Disconnector>> disconnectorid
  ) FOR GRAPH nodeBreakerGraph and GRAPH busBranchGraph{
    Station = SELECT sub FROM {breakerid}:s -((breaker_subid):e) ->:sub |
              {disconnectorid }:s -(( disconnector_subid):e) ->:sub
    ACCUM
    CN = SELECT cn FROM {subid}:s -((sub_cnid):e) ->:cn
       WHERE Breaker_CN.point == 1 | WHERE Disconnector_CN.point == 1
```

```
TopoNDMapping = SELECT t FROM CN:s -
  ((connected_Bus_CN|
   connected_ACline_CN|
   connected_Unit_CN|
   connected_Transformer_CN|
   connected_Load_CN|
   connected_Compensator_P_CN|
   connected_Compensator_S_CN|
   connected_DCline_CN|
   connected_Breaker_CN|
   connected_Disconnector_CN):e) ->:t
  WHERE t.processed == 0
  ACCUM
    CASE
    WHEN t.typename == "Bus" THEN
       INSERT INTO topo_Bus VALUES(s.@topoID, t.id)
       t.@base_kV = t.volt
       t.@up_V    = t.v_max
       t.@lo_V    = t.v_min
    WHEN t.typename == "ACline" THEN
       INSERT INTO topo_ACline VALUES(s.@topoID, t.id)
    WHEN t.typename == "Unit" THEN
       INSERT INTO topo_Unit VALUES(s.@topoID, t.id)
       t.@busType = 2
       t.@base_kV = t.V_rate
       t.@control_bus_number = t.control_bus_number
       t.@desired_volts     = t.desired_volts
       t.@qUp        = t.qUp
       t.@qLower     = t.qLower
       t.@GenP       = t.P
       t.@GenQ       = t.Q

    WHEN t.typename == "Transformer" THEN
       INSERT INTO topo_Transformer VALUES(s.@topoID, t.id)
       double G_tr = -e.Rstar/(e.Rstar*e.Rstar + e.Xstar*e.Xstar)
       double B_tr = e.Xstar/(e.Rstar*e.Rstar + e.Xstar*e.Xstar)
       INSERT INTO TopoConnect VALUES(s.@topoID,t.@topoID,
          e.name,_,_,e.I_bs,e.J_bs,e.off,e.R,e.X,0,_,_,_,_,e.t,_,
          e.itapL,e.itapH ,e.itapC,_,_,e.Pi_meas,e.Qi_meas,G_tr,
          B_tr,_,_,_,_,_,_)

    WHEN t.typename == "Load" THEN
       INSERT INTO topo_Load VALUES(s.@topoID, t.id)
       t.@base_kV  = t.volt
       t.@LdP          += t.P
       t.@LdQ          += t.Q
```

```
      WHEN t.typename == "Compensator_P" THEN
          INSERT INTO topo_Compensator_P VALUES(s.@topoID, t.id)
          t.@base_kV = t.volt
          t.@control_bus_number = t.control_bus_number
          t.@desired_volts      = t.desired_volts
          t.@qUp          = t.qUp
          t.@qLower       = t.qLower
          t.@B            += 1/t.cs_ZK
      WHEN t.typename == "Compensator_S" THEN
          INSERT INTO topo_Compensator_S VALUES(s.@topoID, t.id)
      WHEN t.typename == "DCline" THEN
          INSERT INTO topo_DCline VALUES(s.@topoID, t.id)
      WHEN t.typename == "Breaker" THEN
          INSERT INTO topo_Breaker VALUES(s.@topoID, t.id)
      WHEN t.typename == "Disconnector" THEN
          INSERT INTO topo_Disconnector VALUES(s.@topoID, t.id)
      t.processed == 1

  POST-ACCUM

    IF(!FIND(CN.@topoID,TopoND)) THEN
        INSERT INTO TopoND VALUES(s.@topoID,s.@topoID,_,_,_,_,_
            t.@busType,_,_,_,_,t.@base_kV,t.@desired_volts,
            t.@control_bus_number,t.@up_V,t.@lo_V, t.@GenP, t.
            @GenQ,   _,_, t.@qUp, t.@qLower, t.@LdP, t.@LdQ,_,_,_
            _,_, t.@B,_,_,_,_,_,_)

        INSERT INTO CN_Topo VALUES(CN.@topoID, TopoND)
     End
}
```

In the query above, vertex-attached accumulator variables are defined starting with a single add-sign @.

```
SetAccum<double>  @base_kV = 0
SetAccum<double>  @baseType = 1
MaxAccum<double>  @up_V = 0
MinAccum<double>  @lo_v = 0
SetAccum<int>     @control_bus_number = 0
SetAccum<double>  @desired_volts = 0
SumAccum<double>  @GenP = 0
SumAccum<double>  @GenQ = 0
SumAccum<double>  @LdP = 0
SumAccum<double>  @LdQ = 0
MaxAccum<double>  @qUp = 0
MinAccum<double>  @qLower = 0
SumAccum<double>  @B = 0
```

There are different accumulator types in the graph query language. The accumulation operation will be different on how the accumulator <example>Accum is updated when the statement

<example>Accum = newVal is executed as explained as follow. <example> can be replaced by Set, Sum, Max, and Min in the context.

SetAccum updates the variable as the right operand or union of the right operand if an operand is not single. For example, statement @base_kV = t.volt set @@base_kV as t.volt.

SumAccum adds the right operand to the variable. For example, statement @GenP += t.P adds t.p to @GenP and updates @GenP with the sum.

MaxAccum Updates the value of MaxAccum to the greater between MaxAccum and the right operand. For example, statement @qUp=t.qUp check if t.qUp is greater than @qUp, @qUp will be updated as t.qUp, otherwise, keep @qUp unchanged.

MinAccum Updates the value of MinAccum to the lesser between MinAccum and the right operand. For example, statement @qLower=t. qLower check if t.qLower is less than @qLower, @qLower will be updated as t.qLower, otherwise, keep @qLoweer unchanged.

The accumulator variables are a unique and important feature in graph parallel computing. Accumulators are mutual exclusion object variables shared among all the graph computation threads exploring the graph within a given query. The graph processing engine employs multi-thread processing in graph parallel computing. Modification of accumulator variables attached to a vertex is coordinated at run-time across all threads. These parallel processed variables are utilized in POST-ACCUM clause to insert a new TopoND with updated attributes updated in ACCUM clause.

In the query, the inputs of the query are the breakers and disconnectors whose status are changed from the previous snapshot.

```
CREATE QUERY stationLevelTopology(
    SET<VERTEX<Breaker>> breakerid
    SET<VERTEX<Disconnector>> disconnectorid
)
```

The substations with these breakers and disconnectors are selected by traveling the breaker_sub and disconnector_sub edges. These substations' topology will be updated in parallel by using the keyword ACCUM.

```
Station = SELECT sub FROM {breakerid}:s -((breaker_subid):e) ->:sub |
              {disconnectorid }:s -(( disconnector_subid):e) ->:sub;
```

The connectivity nodes in each of the substations are traveled to neighboring connectivity nodes connected by a closed breaker in the conditions of (Breaker_CN.point == 1) and closed disconnector (Disconnector_CN.point == 1).

```
CN = SELECT cn FROM {subid}:s -((sub_cnid):e) ->:cn
   WHERE Breaker_CN.point == 1 | WHERE Disconnector_CN.point == 1;
```

Inside the substation, the components connecting to the selected connectivity nodes CN are selected by the following statement.

```
TopoNDMapping = SELECT t FROM CN:s -
   ((connected_Bus_CN|
    connected_ACline_CN|
    connected_Unit_CN|
    connected_Transformer_CN|
    connected_Load_CN|
    connected_Compensator_P_CN|
    connected_Compensator_S_CN|
    connected_DCline_CN|
    connected_Breaker_CN|
    connected_Disconnector_CN):e) ->:t
   WHERE t.processed == 0
```

Then topology node to component edges is inserted by the statement `INSERT INTO topo_<-Component> VALUES(s.@topoID, t.id)`. The first attribute of the edge is assigned as the ID of the first connectivity node in the connected neighbor-node group which ensures the first connectivity node ID is propagated to all connected neighbor nodes in the graph. Meanwhile, the second attribute of the edge is the component ID.

Finally, unique topology node `TopoNDs` are inserted into the bus-branch graph vertex in the `POST-ACCUM` clause. And the corresponding nodes are consolidated to the topology node by edge `CN_Topo`.

```
POST-ACCUM

   IF(!FIND(CN.@topoID,TopoND)) THEN

      INSERT INTO TopoND VALUES(s.@topoID,s.@topoID,_,_,_,_,_,_,
          busType,_,_,_,_,base_kV,desired_volts, control_bus_number,
          up_V,lo_V,GenP,GenQ,_,_,qUp,qLower,LdP,LdQ,_,_,_,_,_,_,B,_,
          _,_,_,_,_)

      INSERT INTO CN_Topo VALUES(CN.@topoID, TopoND)
```

The `TopoND` identification is set as the first connectivity node identification `s.@topoID` which is also assigned as the unique primary identification of the `TopoND`.

In the substation-level topology analysis, the transformer is a `TopoConnect` edge inside the substation. The transformer edge is inserted by the statement after the vertex type is checked as "transformer". Note that all components are modeled as a vertex in the node-branch graph model.

```
WHEN t.typename == "Transformer" THEN
   INSERT INTO topo_Transformer VALUES(s.@topoID, t.id)
   double G_tr = -e.Rstar/(e.Rstar*e.Rstar + e.Xstar*e
   .Xstar)
   double B_tr =  e.Xstar/(e.Rstar*e.Rstar + e.Xstar*e.
   Xstar)
   INSERT INTO TopoConnect VALUES(s.@topoID,t.@topoID,
      e.name,_,_,e.I_bs,e.J_bs,e.off,e.R,e.X,0,_,_,_,_,
      e.t,_,e.itapL,e.itapH ,e.itapC,_,_,e.Pi_meas,
      e.Qi_meas,G,B,_,_,_,_,_,_)
```

`TopoConnect` is a generic edge representing AC/DC line, transformer, series component, and all other branches in the bus-branch graph model. The attributes defined for the `TopoConnect` edge cover the parameters of all kinds of branches. Some of the attributes may not be applied to all types of branches. They are entered as "_" when it is inapplicable for a new edge to be inserted.

At the initialization, the substation-level topology is updated for all substations to create an edge of `topo_ACline` and `topo_Compensator_S` in the node-breaker graph model. Using edges `topo_ACline` and `topo_Compensator_S`, transmission line, and series compensator edges in between substations are inserted into the bus-branch graph model in the system-level topology analysis below.

### 8.5.2 System-Level Network Topology Analysis

Substation-level topology consolidates connected connectivity nodes to a unique topology node `TopoND`. The connectivity of components inside substations (such as units, loads, shunt compensator, and transformer) or connecting to substations (ACline and series compensator) are created by edge topo_<Compnent> in the substation-level topology. `TopoConnect` for a transformer as an edge is also inserted into the bus-branch graph in the substation-level topology analysis. However, `TopoConnect` for AC line and series compensator is not created in the substation-level topology analysis. They are inserted into the bus-branch graph in the system-level network topology analysis using the information stored in the edge `topo_ACline` and `topo_Compensator_S`.

From the perspective of graph computing, we generalize the process of system-level topology analysis for a power system as follows.

The power system is physically composed of electrical components such as a busbar, load, generator, circuit breaker, disconnector, transformer, transmission line, etc. All the components are connected through topology nodes. Thus, with all the components and topology nodes denoted in corresponding defined formats and their connection relationship, we can form an undirected graph representing the whole power system.

The network level of topology processing is to describe the connectivity information between different substations. Starting from the AC lines (and DC lines if the system has any) in parallel, the neighboring topology nodes (`TopoND`) are searched, and topology edges (`TopoConnect`) are inserted to connect the topology nodes. The topology nodes and topology edges are buses and branches respectively in the system bus-branch model.

The detailed algorithm is shown in Algorithm 8.2.

---

**Algorithm 8.2   CIM/E Oriented Parallel System-Level Topology Processing**

---

```
1:   for (all substations)
2:       while (not all topology nodes are processed) do
3:           Start from all AC/DC lines and series compensators
4:           if (both sides are not opened or the component is not
                offline)
5:               Find their topology nodes IDs generated in algorithm 8.1
                    and store them on their attributes
6:           end
7:           Insert topology edge simultaneously according to the IDs
                stored in
                AC/DC line and series compensator
8:       end
9: end
```

---

Algorithm 8.2 can be implemented by pseudo-code in graph query as follow:

```
CREATE QUERY systemLevelTopology(
     SET<VERTEX<substation>> subid
 ) FOR GRAPH nodeBreakerGraph and GRAPH busBranchGraph{
   ACline_selectedSub = SELECT t FROM subid:s-(sub_ACline:e)->:t
     ACCUM

      double G = -e.line_R / (e.line_R * e.line_R + e.line_X * e.line_X)
      double B = e.line_X / (e.line_R * e.line_R + e.line_X * e.line_X)
      INSERT INTO TopoConnect VALUES(e.I_node-(CN_Topo)->:I_topo.
       topoID, e.J_node-(CN_Topo)->:J_topo.topoID,e.name,_,_,e.I_bs,
        e.J_bs,_, e.R,e.X,e.B,_,_,_,_,_,_,_,_,_,_,_,e.Pi_meas,e.
        Qi_meas,G,B, 1/e.line_X,_,_,_,e.I_off,e.J_off)
      C_S_selectedSub = SELECT t FROM subid:s - (sub_Compensator_S:e)
      ->:t

       ACCUM

       INSERT INTO TopoConnect VALUES(e.I_node-(CN_Topo)->:I_topo.
         topoID, e.J_node-(CN_Topo)->:J_topo.topoID,e.name,_,_
         e.I_bs,e.J_bs, e.off,0,e.Zk,0,_,_,_,_,_,_,_,_,_,_,_,_,
         e.Pi_meas,e.Qi_meas,0, 1/e.line_X,_,_,_,_,_,_)
}
```

In the query definition, the inputs `subid` of the query `systemLevelTopology` is the substation set whose topology is updated in the substation-level topology analysis.

```
CREATE QUERY systemLevelTopology(
     SET<VERTEX<substation>> subid
 ) FOR GRAPH nodeBreakerGraph and busBranchGraph
```

The AC lines and series compensators connecting to these substations in `subid` are selected by traveling the `sub_ACline` and `sub_Compensator_S` edges. Under `ACCUM`, the corresponding edges are inserted into the bus-branch graph in parallel.

```
ACline_selectedSub = SELECT t FROM subid:s-(sub_ACline:e)->:t
  ACCUM
  double G = -e.line_R / (e.line_R * e.line_R + e.line_X * e.line_X)
  double B =  e.line_X / (e.line_R * e.line_R + e.line_X * e.line_X)

  INSERT INTO TopoConnect VALUES(e.I_node-(CN_Topo)->:I_topo.topoID,
    e.J_node-(CN_Topo)->:J_topo.topoID,e.name,_,_,e.I_bs,e.J_bs,_,
    e.R,e.X,e.B,_,_,_,_,_,_,_,_,_,_,_,e.Pi_meas,e.Qi_meas,G,B,1/
    e.line_X,_,_,_,e.I_off,e.J_off)

  C_S_selectedSub = SELECT t FROM subid:s - (sub_Compensator_S:e)->:t

   ACCUM

   INSERT INTO TopoConnect VALUES(e.I_node-(CN_Topo)->:I_topo.topoID,
     e.J_node-(CN_Topo)->:J_topo.topoID,e.name,_,_,e.I_bs,
     e.J_bs, e.off,0,e.Zk,0,_,_,_,_,_,_,_,_,_,_,_,_, e.Pi_meas,
     e.Qi_meas,0, 1/e.line_X,_,_,_,_,_,_)
```

At the initialization, the system-level topology is updated for all AC/DC lines and series compensator to create edge `TopoConnect`.

After the substation-level and system-level topology analysis, the power system connectivity represented by the bus-branch model is built by an undirected graph with `TopoND` vertices connecting with `TopoConnect` edges. Then the vertex attributes are calculated in the topology postprocessing.

## References

**1** Common Information Model (CIM) (2001). *CIM 10 Version*. Palo Alto, CA: EPRI.

**2** CIM (2008). Introduction to CIM and its role in the utility enterprise data preparation, exchange, integration and enterprise information management. *Tutorial, CIM User Group Meeting*, Vasteras, Sweden, June 2008.

**3** Ravikumar, G. and Khaparde, S.A. (2016). A common information model (CIM) oriented graph database framework for power systems. *IEEE Transactions on Power Systems* 99: 1–1.

**4** Yaozhong, X., Hongzhu, T., Yisong, L., and Junjie, S. (2006). E language for electric power system model description. *Automation of Electric Power System* 30 (10): 48–51.

**5** Weimin, M., Yaozhong, X., and Guodong, J. (2013). Comparative analysis of grid model exchange standard CIM/E and CIM/XML. *Power System Technology* 37 (4): 936–941.

**6** Zhou, Z., Yuan, C., Yao, Z. et al. (2018). CIM/E oriented graph database model architecture and parallel network topology processing. *2018 IEEE Power & Energy Society General Meeting (PESGM)*, Portland, OR, USA, pp. 1–5. https://doi.org/10.1109/PESGM.2018.8586367.

**7** Dai, J., Yao, Z., Zhang, G. et al. (2019). Graph computing-based real-time network topology analysis for power system. *2019 IEEE Power & Energy Society General Meeting (PESGM)*, Atlanta, GA, USA, pp. 1–5. https://doi.org/10.1109/PESGM40551.2019.8973614.

# 9

# State Estimation Graph Computing

State estimation is critical in power system real-time operation. Power system state estimation estimates system operation states based on SCADA real-time telemetry. The network topology connection discussed in Chapter 8 and estimated real-time operation states are functioning as the fundamental basis of other power system analyses, such as power flow, contingency analysis, optimal power flow, transient stability, and so on. The theory of power system state estimation was proposed by Professor Schweppe of MIT in the early 1970s [1]. Since then, state estimation has become a key function in the supervisory control and planning of electric power grids. By the 1980s, the basic algorithms and theories of state estimation were being well developed. Since the 1990s, state estimation has been adopted in the power system control centers widely. It serves as a key component to monitor power system states and support other applications in energy management systems (EMS) [2, 3].

For different study purposes, the traditional state estimate model is derived and extended. Reference [4] discusses the possibility of distributed state estimation. Parallel state estimation based on a fast decoupled method was illustrated in [5]. Artificial intelligence was used to implement a hybrid state estimator [6]. A decentralized robust state estimator was proposed in [7]. The matrix inversion lemma was used for parallel static state estimation in [8] and the block-partitioning algorithm was also proposed. Dynamic state estimation is proposed in [9–11]. To demonstrate graph computing, the conventional steady-state state estimation is refactorized and implemented in graph parallel in this chapter.

## 9.1 Power System State Estimation

The weighted least square (WLS) algorithm is the most widely used methodology of state estimators to minimize the weighted sum of the square of residuals between the actual measurements and estimations. The measurement model in power system state estimation is presented below [12].

$$z = h(x) + e \tag{9.1}$$

where $z$ is the measurement vector, $x$ is the system state vector, $h(x)$ is the nonlinear measurement function vector and $e$ is a vector of measurement errors, meeting Gaussian distribution with zero means. The state variables typically are bus voltage magnitudes and phase angles while the typical measurements are bus voltage magnitudes, bus power injections, and power flows on branches. For an $n$-bus power system, assuming it has $m$ branches, then there are $3n + 4m$ measurements, i.e. $n$-bus voltage magnitudes, $2n$ power injects at buses and $4m$ power flows at the two ends of branches. Regarding the state variables, since the phase angle at the slack bus is considered a reference value, there are $2n-1$ states to be estimated. Nonlinear WLS state estimation is then formulated as:

$$\textbf{Min } J(x) = [z - h(x)]^T \cdot R^{-1} \cdot [z - h(x)] \tag{9.2}$$

$$\textbf{s.t. } z = h(x) + e \tag{9.3}$$

The optimal solution is obtained by solving the following equation:

$$g(x) = \frac{\partial J(x)}{\partial x} = -H^T(x) \cdot R^{-1} \cdot [z - h(x)] = 0 \tag{9.4}$$

where $g(x)$ is the matrix of the derivative of the objective function $J(x)$, $H(x) = \dfrac{\partial h(x)}{\partial x}$ is the Jacobian matrix of $h(x)$ and $R^{-1}$ is the weight matrix. Substituting the first-order Taylor's expansion of $g(x)$ in Eq. (9.4), the following Eq. (9.5) is solved iteratively and the solution is converged to minimize $J(x)$.

$$\Delta x = [G(x^k)]^{-1} \cdot H^T(x^k) \cdot R^{-1} \cdot (z - h(x^k))$$
$$G(x^k) \cdot \Delta x = H^T(x^k) \cdot R^{-1} \cdot (z - h(x^k)) \tag{9.5}$$

where,

$$g(x^{k+1}) = g(x^k) + G(x^k) \cdot (x^{k+1} - x^k) = 0 \tag{9.6}$$

$$G(x^k) = \frac{\partial g(x^k)}{\partial x} = H^T(x^k) \cdot R^{-1} \cdot H(x^k) \tag{9.7}$$

$$x^{k+1} = x^k + \Delta x \tag{9.8}$$

Measurement Jacobian matrix $H$ and gain matrix $G$ are functions of state variables. They are updated in every iteration. And gain matrix is factorized in each iteration to solve the linear Eq. (9.5). The matrix formation and factorization are time-consuming. In typical power transmission systems, a high $R/X$ ratio supports a good convergence of the fast decoupled method in which the Jacobian matrix $H$ and gain matrix $G$ in (9.5) are converted to be constant matrices with no need for updating in each iteration, thus it saves time on matrix formulation and factorization in each iteration. Since the sensitivity of the active/reactive power equations to changes in the bus voltage magnitude/phase angle is lower, the numerical values of the off-diagonal blocks of Jacobian matrix $H$ are significantly smaller than the counterparts of the diagonal blocks under the assumption of large $R/X$ ratio, the non-diagonal blocks are ignored in the fast decoupled state estimation method. The Jacobian matrix $H$ and gain matrix $G$ are simplified as shown in (9.9) and (9.10), where the subscripts $A$ and $R$ represent the active and reactive components respectively.

$$H = \begin{bmatrix} H_{AA} & H_{AR} \\ H_{RA} & H_{RR} \end{bmatrix} = \begin{bmatrix} H_{AA} & \mathbf{0} \\ \mathbf{0} & H_{RR} \end{bmatrix} \tag{9.9}$$

$$G = \begin{bmatrix} G_{AA} & G_{AR} \\ G_{RA} & G_{RR} \end{bmatrix} = \begin{bmatrix} H_{AA}^T R_{AA}^{-1} H_{AA} & \mathbf{0} \\ \mathbf{0} & H_{RR}^T R_{RR}^{-1} H_{RR} \end{bmatrix} \tag{9.10}$$

where, $R^{-1} = diag\left(R_{AA}^{-1}, R_{RR}^{-1}\right)$. In (9.10), $G_{AA} \in \mathcal{M}(n-1, n-1)$ and $G_{RR} \in \mathcal{M}(n, n)$. Since the voltage angle at the swing bus is the reference, the derivatives of the measurements to the swing bus angle are not part of (9.9).

In the fast decoupled state estimation method, Eq. (9.5) is decoupled as:

$$G_{AA}\Delta\theta^k = H_{AA}^T R_{AA}^{-1} r_A\left(x^k\right)$$
$$G_{RR}\Delta|V|^k = H_{RR}^T R_{RR}^{-1} r_R\left(x^k\right) \tag{9.11}$$

The two size-reduced equations above in (9.11) are linear. They are solved by LU factorization and forward-backward substitutions.

## 9.2 Graph Computing-Based State Estimation

Although fast decoupled state estimation improved the estimation efficiency, the computational burden is still presented in right-hand side vector formation, gain matrix formation, LU factorization, and forward-backward substitutions for large-scale power system state estimation. To improve the computation efficiency, distributed, and decentralized calculation [13–17] and parallel computing [18–22] are developed to solve state estimation problems more effectively. Leveraging the high computation efficiency in solving large-scale linear equations illustrated in Chapter 3, we are motivated to use graph parallel computing to solve state estimation in this chapter.

The graph computing-based state estimation is based on the centralized WLS fast decoupled method by graph nodal parallel computing on forming the system-level gain matrix, updating the right-hand side vector in (9.11), and hierarchical parallel computing to factorize the gain matrix and solve the Eq. (9.11) by forward-backward substitutions. These graph nodal and hierarchical parallel computation methods are well addressed in Chapter 3. By investigating the characteristics of power system state estimation, the graph computing method is explored with more granularity to solve state estimation problems in parallel.

### 9.2.1 State Estimation Graph Computing Algorithm

#### 9.2.1.1 Build Node-Based State Estimation [23]
When a power network is modeled as a node-connected-by-branch graph, measurements can be grouped and assigned as the node's attributes. Taking node/bus $i$ as an example, its active power injection measurement $P_i$, as well as the active power flow measurements $P_{ij}$ of its adjacent

branch $i$–$j$ are gathered to form the bus $i$ measurement set $z_{A,\ i}$ which is a subset of the total measurement $z_A$ of the system.

The grouping process is represented by (9.12)–(9.14). In (9.12), the system measurement vector $z_A$ consists of an error-free vector $h_A(x)$ and an error vector $e_A$. $z_A$ is partitioned into $n$ parts upon its association. $z_{A,i}$ indicates the bus $i$'s measurement set. Without loss of generality, taking active power measurements as an example, in (9.13), $z_{A,i}$ consists of active power injection measurement $P_i$ at bus $i$ and its adjacent branch active power flow measurements $P_{ij}$. $h_{A,i}(x)$ and $e_{A,i}$ are the respective error-free sub-vector and error sub-vector of $z_{A,i}$. The error-free vector is accordingly divided into $n$ partitions in (9.14).

$$z_A = h_A(x) + e_A \text{ where } z_A = \left[ z_{A,1}^T \, z_{A,2}^T \cdots z_{A,i}^T \cdots z_{A,n}^T \right]^T \tag{9.12}$$

$$z_{A,i} = h_{A,i}(x) + e_{A,i} \text{ where } z_{A,i} = \left[ P_i \, P_{ij}^T \right]^T, i = 1, \cdots, n, j \in i \tag{9.13}$$

$$h_A(x) = \left[ h_{A,1}(x)^T \, h_{A,2}(x)^T \cdots h_{A,i}(x)^T \cdots h_{A,n}(x)^T \right]^T \tag{9.14}$$

Based on the knowledge of $H(x) = \dfrac{\partial h(x)}{\partial x}$, $G_{AA} = H_{AA}^T R_{AA}^{-1} H_{AA}$ and Eq. (9.14), the $H$ matrix and gain matrix of bus $i$ are:

$$H_{AA,i} = \frac{\partial h_{A,i}(x)}{\partial \theta}, \theta = [\theta_1 \, \theta_2 \cdots \theta_i \cdots \theta_{n-1}] \tag{9.15}$$

$$G_{AA,i} = H_{AA,i}^T R_{AA,i}^{-1} H_{AA,i} \tag{9.16}$$

where $R_{AA,i}^{-1}$ is the weight matrix for active power measurements related to node $i$ directly. $h_{A,i}(x)$ includes the active power measurements that are directly related to bus $i$. Obviously, $\dfrac{\partial h_{A,i}(x)}{\partial \theta_j}$ is a zero vector when bus $j$ is neither bus $i$ nor bus $i$'s 1-step neighboring bus. Thus, $H_{AA,i}$ is only related to bus $i$, bus $i$'s 1-step neighbor(s) and the branches in between. The same reasoning applies to (9.16) as well. $G_{AA,i}$ is only related to bus $i$, bus $i$'s 1-step neighbor(s) and the branches in between. Modeled by a graph, $H_{AA,i}$ and $G_{AA,i}$ can be calculated in parallel at bus $i$ with at most one-step graph traversal.

Once the node-based $H_{AA,i}$ and $G_{AA,i}$ are formed for all buses, the system-level Jacobian matrix-$H_{AA}$ and gain matrix $G_{AA}$ are formulated by assembling all node-based Jacobian matrices $H_{AA,i}$ and gain matrices $G_{AA,i}$ as represented in (9.17) and (9.18), where (9.17) is derived by (9.14), (9.15), and (9.18) is obtained based on (9.10), (9.16), and (9.17).

$$
\begin{aligned}
H_{AA} = \frac{\partial h_A(x)}{\partial \theta} &= \frac{\partial \left[ h_{A,1}(x)^T \, h_{A,2}(x)^T \cdots h_{A,i}(x)^T \cdots h_{A,n}(x)^T \right]^T}{\partial \theta} \\
&= \left[ \left( \frac{\partial h_{A,1}(x)}{\partial \theta} \right)^T \left( \frac{\partial h_{A,2}(x)}{\partial \theta} \right)^T \cdots \left( \frac{\partial h_{A,i}(x)}{\partial \theta} \right)^T \cdots \left( \frac{\partial h_{A,n}(x)}{\partial \theta} \right)^T \right]^T \\
&= \left[ H_{AA,1}^T \, H_{AA,2}^T \cdots H_{AA,i}^T \cdots H_{AA,n}^T \right]^T
\end{aligned}
\tag{9.17}
$$

$$G_{AA} = H_{AA}^{\ T} R_{AA}^{\ -1} H_{AA} = \begin{bmatrix} H_{AA,1}^T \ H_{AA,2}^T \cdots H_{AA,i}^T \cdots H_{AA,n}^T \end{bmatrix} \cdot R_{AA}^{-1}$$
$$\cdot \begin{bmatrix} H_{AA,1}^T \ H_{AA,2}^T \cdots H_{AA,i}^T \cdots H_{AA,n}^T \end{bmatrix}^T$$
$$= \sum_{i=1}^{n} H_{AA,i}^T \cdot R_{AA,i}^{-1} \cdot H_{AA,i} = \sum_{i=1}^{n} G_{AA,i} \tag{9.18}$$

where, $R_{AA}^{-1} = diag\left(R_{AA,1}^{-1}, R_{AA,2}^{-1}, \cdots, R_{AA,i}^{-1}, \cdots, R_{AA,n}^{-1}\right)$

The system-level right-hand side (RHS) vector update in (9.11) can also be calculated by summing up the node-based RHS vectors, as shown in (9.19). The node-based RHS vector can be updated in parallel on the graph model.

$$RHS = H_{AA}^T R_{AA}^{-1} r_A\left(x^k\right) = \begin{bmatrix} H_{AA,1}^T \ H_{AA,2}^T \cdots H_{AA,i}^T \cdots H_{AA,n}^T \end{bmatrix} \cdot$$
$$R_{AA}^{-1} \cdot \begin{bmatrix} r_{A,1}\left(x^k\right) \ r_{A,2}\left(x^k\right) \cdots r_{A,i}\left(x^k\right) \cdots r_{A,n}\left(x^k\right) \end{bmatrix}$$
$$= \sum_{i=1}^{n} H_{AA,i}^T R_{AA,i}^{-1} r_{A,i}\left(x^k\right) \tag{9.19}$$

#### 9.2.1.2 Graph-Based State Estimation Parallel Algorithm

In graph computing, a power system is modeled as an undirected graph $G = (V, E)$, where $V$ is the set of vertices (buses) and $E$ is the set of edges (brunches). The distance between two vertices in a graph is defined as:

Definition: The distance $d(u, v)$ between two vertices $u$ and $v$ in a graph is the number of edges on the shortest path connecting $u$ and $v$.

And vertex $v$ is called $k$-neighbor of vertex $u$ if $d(u, v) = k$.

The connectivity of the system can be described by using the adjacency matrix in graph theory. Adjacency matrix $A$ is a $n \times n$ matrix. Entries in the matrix indicate whether pairs of vertices are adjacent in the graph.

$$A_{ij} = \begin{cases} 1, d(i,j) = 1 \\ 0, d(i,j) \neq 1 \end{cases} \tag{9.20}$$

The system connectivity matrix can be defined as:

$$T = A + I_{n \times n} \tag{9.21}$$

Where $I_{n \times n}$ is a $n \times n$ unit matrix. The connectivity matrix also has the following relationship with the nodal admittance matrix $Y$:

$$T_{ij} = \begin{cases} 1, Y_{ij} \neq 0 \\ 0, Y_{ij} = 0 \end{cases} \tag{9.22}$$

Note that the connectivity matrix will also have the same matrix structure as the Laplacian matrix of the graph. Each row in the matrix represents a node in the graph. The column index of non-zero elements in a certain row includes the node itself and all of its 1-neighbor nodes.

The index of node $i$ and its 1-neighbor nodes can be put into a set:

$$\alpha_i : = \left\{ j : T_{ij} \neq 0 \right\} \tag{9.23}$$

And the index of node $i$, its 1-neighbor nodes, and its 2-neighbor nodes can be defined as:

$$\beta_i := \alpha_{\alpha_i} \tag{9.24}$$

The 1-neighbor nodes and 2-neighbor nodes of $i$ can be defined as:

$$\varphi_1(i) = \alpha_i - i \tag{9.25}$$

$$\varphi_2(i) = \beta_i - \alpha_i \tag{9.26}$$

Then, the system-level gain matrix is divided into row vectors, where each row vector is stored as an attribute of the corresponding node, as shown in (9.27). For example, $G_{AA}(i)$ represents the $i$th row vector of the system-level gain matrix $G_{AA}$, and it is modeled as an attribute of node $i$. The column indices of non-zero elements in $G_{AA}(i)$ belong to $\beta_i$, indicating it only relates to node $i$, node $i$'s 1-step neighbor(s), and node $i$'s 2-step neighbor(s).

$$G_{AA} = \left[ G_{AA}(1)^T G_{AA}(2)^T \cdots G_{AA}(i)^T \cdots G_{AA}(n)^T \right]^T \tag{9.27}$$

As shown in Figure 9.1, node $i$ is centered in the system graph model for the convenience of illustration. Its 1-step neighboring nodes are represented as $\{a^1, b^1, \cdots\}$ and its 2-step neighboring nodes are signified as $\{a^2, b^2, \cdots\}$. The superscripts indicate the shortest distance, i.e. the fewest number of edges, from the central node to the corresponding node.

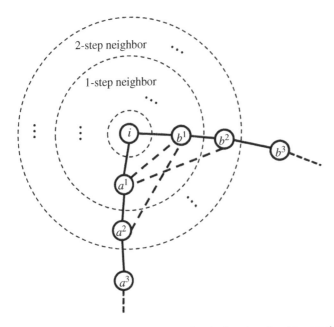

**Figure 9.1** A generalized structure of node-$i$-centered system graph.

Since node $i$ is only adjacent to its 1-step neighboring nodes, its nodal $H$ matrix is expressed by (9.28), where $H_{AA,ii} = \dfrac{\partial h_{A,i}(x)}{\partial \theta_i}$, $H_{AA,ia^1} = \dfrac{\partial h_{A,i}(x)}{\partial \theta_{a^1}}$, and $H_{AA,ib^1} = \dfrac{\partial h_{A,i}(x)}{\partial \theta_{b^1}}$ are the corresponding column vectors. To simplify the notation, $H_{AA,ii}$, $H_{AA,ia^1}$, and $H_{AA,ib^1}$ are denoted as $\mathcal{H}_{ii}$, $\mathcal{H}_{ia^1}$, and $\mathcal{H}_{ib^1}$, respectively.

$$H_{AA,i}^C = \begin{bmatrix} H_{AA,ii} & H_{AA,ia^1} & H_{AA,ib^1} \end{bmatrix} = \begin{bmatrix} \mathcal{H}_{ii} & \mathcal{H}_{ia^1} & \mathcal{H}_{ib^1} \end{bmatrix} \tag{9.28}$$

The corresponding node-based gain matrix is then expressed as:

$$G_{AA,i}^C = \begin{matrix} & i & a^1 & b^1 & \\ \begin{bmatrix} \mathcal{H}_{ii}^T \mathcal{R}_i^{-1} \mathcal{H}_{ii} & \mathcal{H}_{ii}^T \mathcal{R}_i^{-1} \mathcal{H}_{ia^1} & \mathcal{H}_{ii}^T \mathcal{R}_i^{-1} \mathcal{H}_{ib^1} \\ \mathcal{H}_{ia^1}^T \mathcal{R}_i^{-1} \mathcal{H}_{ii} & \mathcal{H}_{ia^1}^T \mathcal{R}_i^{-1} \mathcal{H}_{ia^1} & \mathcal{H}_{ia^1}^T \mathcal{R}_i^{-1} \mathcal{H}_{ib^1} \\ \mathcal{H}_{ib^1}^T \mathcal{R}_i^{-1} \mathcal{H}_{ii} & \mathcal{H}_{ib^1}^T \mathcal{R}_i^{-1} \mathcal{H}_{ia^1} & \mathcal{H}_{ib^1}^T \mathcal{R}_i^{-1} \mathcal{H}_{ib^1} \end{bmatrix} & \begin{matrix} i \\ a^1 \\ b^1 \end{matrix} \end{matrix} \tag{9.29}$$

In (9.29), $\mathcal{R}_i^{-1}$ is a simplified notation for $R_{AA,i}^{-1}$. The right-most column outside the matrix bracket is the row indices in the system-level gain matrix, and the row above the matrix is the column indices in the system-level gain matrix. For example, the first-row vector of $G_{AA,i}^C$ is a subset of the $i$th row vector in the system-level gain matrix $G_{AA}$, and it contributes to elements with column indices of $i$, $a^1$, $b^1$ in the $i$th row vector of $G_{AA}$.

In the system-level gain matrix, the $i$th row vector $G_{AA}(i)$ not only includes non-zero entries in the columns of node $i$ and its 1-step neighbors but also has non-zero elements contributed by the node $i$'s 2-step neighbors, as displayed in (9.30).

$$G_{AA}(i) = \begin{bmatrix} \mathcal{H}_{ii}^T \mathcal{R}_i^{-1} \mathcal{H}_{ii} + \mathcal{H}_{a^1 i}^T \mathcal{R}_{a^1}^{-1} \mathcal{H}_{a^1 i} + \mathcal{H}_{b^1 i}^T \mathcal{R}_{b^1}^{-1} \mathcal{H}_{b^1 i} \\ \mathcal{H}_{ii}^T \mathcal{R}_i^{-1} \mathcal{H}_{ia^1} + \mathcal{H}_{a^1 i}^T \mathcal{R}_{a^1}^{-1} \mathcal{H}_{a^1 a^1} + \mathcal{H}_{b^1 i}^T \mathcal{R}_{b^1}^{-1} \mathcal{H}_{b^1 a^1} \\ \mathcal{H}_{ii}^T \mathcal{R}_i^{-1} \mathcal{H}_{ib^1} + \mathcal{H}_{a^1 i}^T \mathcal{R}_{a^1}^{-1} \mathcal{H}_{a^1 b^1} + \mathcal{H}_{b^1 i}^T \mathcal{R}_{b^1}^{-1} \mathcal{H}_{b^1 b^1} \\ \mathcal{H}_{a^1 i}^T \mathcal{R}_{a^1}^{-1} \mathcal{H}_{a^1 a^2} + \mathcal{H}_{b^1 i}^T \mathcal{R}_{b^1}^{-1} \mathcal{H}_{b^1 a^2} \\ \mathcal{H}_{a^1 i}^T \mathcal{R}_{a^1}^{-1} \mathcal{H}_{a^1 b^2} + \mathcal{H}_{b^1 i}^T \mathcal{R}_{b^1}^{-1} \mathcal{H}_{b^1 b^2} \end{bmatrix}^T \begin{matrix} i \\ a^1 \\ b^1 \\ a^2 \\ b^2 \end{matrix} \tag{9.30}$$

The reasoning behind this is that node $i$ and its 2-step neighbors are one step away from their mutual 1-step neighbors. For example, as shown in Figure 9.2, $a^1$ and $a^2$ are node $i$'s 1-step and 2-step neighbors, and $a^1$ is located between node $i$ and $a^2$. So, in the node $a^1$'s assembled Jacobian matrix $H_{AA,a^1}^C$, non-zero derivatives $\dfrac{\partial P_{a^1}}{\partial \theta_i}$ and $\dfrac{\partial P_{a^1}}{\partial \theta_{a^2}}$ are elements of $\mathcal{H}_{a^1 i}$ and $\mathcal{H}_{a^1 a^2}$ respectively.

After the assembled Jacobian matrices $\mathcal{H}_{a^1 i}$ and $\mathcal{H}_{a^1 a^2}$ are formed, the system gain matrix element of $\mathcal{H}_{a^1 i}^T \mathcal{R}_{a^1}^{-1} \mathcal{H}_{a^1 a^2}$ calculation is straightforward. Note that $\mathcal{H}_{a^1 i}^T \mathcal{R}_{a^1}^{-1} \mathcal{H}_{a^1 a^2}$ sits in the $i$th row and $a^2$th column showing that node $a^2$, as the 2-step neighboring node of node $i$, contributes a non-zero element at the $i$th row and the $a^2$th column. Similarly, node $i$ contributes a non-zero element at the $a^2$th row and $i$th column. In general, the following lemmas are derived to highlight the process:

Lemma 1: For node $i$, only itself, its 1-step neighboring nodes, and the edges between them contribute non-zero in column $i$ of the system $H$ matrix $H_{AA}$.

Lemma 2: For node $i$, the node itself, its 1-step neighboring nodes, its 2-step neighboring nodes, and the edges between them contribute non-zero in row $i$ of the system gain matrix $G_{AA}$.

Investigating (9.30), the non-zeros in $G_{AA}(i)$ are categorized into three sets: diagonal entry = $\{G_{AA}(i, i)\}$, 1-step entry = $\{G_{AA}(i, a^1), G_{AA}(i, b^1), \cdots\}$, and 2-step entry=$\{G_{AA}(i, a^2), G_{AA}(i, b^2), \cdots\}$. Their contributions are illustrated graphically in Figures 9.2–9.4.

Reading (9.30) with Figures 9.2–9.4, we should understand that the value of the diagonal entry $G_{AA}(i, i)$ is determined by the node itself, its 1-step neighbors, and the edges between the node and its 1-step neighbors. Consistently, as demonstrated in Figure 9.2, the diagram shows that $\mathcal{H}_{ii}$, $\mathcal{H}_{a^1 i}$, $\mathcal{H}_{b^1 i}$, and the corresponding weights at nodes $i$, $a^1$, and $b^1$ determine $G_{AA}(i, i)$.

For the 1-step entry set, node $i$, its 1-step neighbors, and the edges between node $i$ and its 1-step neighboring nodes contribute to the elements in node $i$'s 1-step entry. As shown in Figure 9.3, the node $a^1$ is the node $i$'s 1-step neighbor. The highlighted black bold lines are the 1-step and 2-step edges between node $i$ and $a^1$, and they contribute to $G_{AA}(i, a^1)$ as presented as the $a^1$th element in (9.30). Similarly, elements in node $i$'s 2-step entry are determined by the 2-step edges between node $i$ and its 2-step neighbors.

For the 2-step entry set, as shown in Figure 9.4, $a^2$ is a 2-step neighbor of node $i$. Node $i$ travels through the two highlighted lines to $a^2$. So, $G_{AA}(i, a^2)$ contains information on these two edges, as shown in (9.30).

Based on (9.30), the $i$th row of the system-level gain matrix is sourced from the $i$th rows of the assembled node-based gain matrices of $i$, $a^1$, and $b^1$, as presented in (9.31)–(9.33).

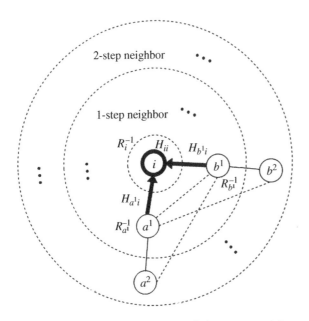

**Figure 9.2** Diagonal entry (taking node $i$ as an example).

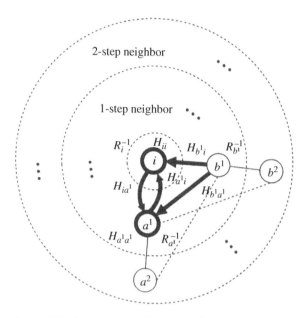

**Figure 9.3** 1-step entry (taking node $a^1$ as an example).

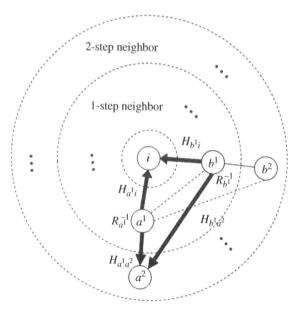

**Figure 9.4** 2-step entry (taking node $a^2$ as an example).

$$G_{AA,i}^C(i) = \begin{bmatrix} \left(\sum B_i^1\right)^2 \mathcal{R}_{ii}^{-1} + (B_{ia^1})^2 \mathcal{R}_{ia^1}^{-1} + (B_{ib^1})^2 \mathcal{R}_{ib^1}^{-1} \\[2mm] -B_{ia^1}\left(\left(\sum B_i^1\right)\mathcal{R}_{ii}^{-1} + B_{ia^1}\mathcal{R}_{ia^1}^{-1}\right) \\[2mm] -B_{ib^1}\left(\left(\sum B_i^1\right)\mathcal{R}_{ii}^{-1} + B_{ib^1}\mathcal{R}_{ib^1}^{-1}\right) \end{bmatrix}^T \begin{matrix} i \\[2mm] a^1 \\[2mm] b^1 \end{matrix} \tag{9.31}$$

$$G_{AA,a^1}^C(i) = \begin{bmatrix} (B_{a^1i})^2\left(\mathcal{R}_{a^1a^1}^{-1} + \mathcal{R}_{a^1i}^{-1}\right) \\[2mm] -B_{a^1i}\left(\left(\sum B_{a^1}^1\right)\mathcal{R}_{a^1a^1}^{-1} + B_{a^1i}\mathcal{R}_{a^1i}^{-1}\right) \\[2mm] B_{a^1i}B_{a^1b^1}\mathcal{R}_{a^1a^1}^{-1} \\[2mm] B_{a^1i}B_{a^1a^2}\mathcal{R}_{a^1a^1}^{-1} \\[2mm] B_{a^1i}B_{a^1b^2}\mathcal{R}_{a^1a^1}^{-1} \end{bmatrix}^T \begin{matrix} i \\[2mm] a^1 \\[2mm] b^1 \\[2mm] a^2 \\[2mm] b^2 \end{matrix} \tag{9.32}$$

$$G_{AA,b^1}^C(i) = \begin{bmatrix} (B_{b^1i})^2\left(\mathcal{R}_{b^1b^1}^{-1} + \mathcal{R}_{b^1i}^{-1}\right) \\[2mm] B_{b^1i}B_{b^1a^1}\mathcal{R}_{b^1b^1}^{-1} \\[2mm] -B_{b^1i}\left(\left(\sum B_{b^1}^1\right)\mathcal{R}_{b^1b^1}^{-1} + B_{b^1i}\mathcal{R}_{b^1i}^{-1}\right) \\[2mm] B_{b^1i}B_{b^1a^2}\mathcal{R}_{b^1b^1}^{-1} \\[2mm] B_{b^1i}B_{b^1b^2}\mathcal{R}_{b^1b^1}^{-1} \end{bmatrix}^T \begin{matrix} i \\[2mm] a^1 \\[2mm] b^1 \\[2mm] a^2 \\[2mm] b^2 \end{matrix} \tag{9.33}$$

where $\sum B_i^1, \sum B_{a^1}^1, \sum B_{b^1}^1$ is the total susceptance of branches that are directly connected to node $i, a^1, b^1$, which means the total susceptance between node $i$ and its 1-step neighbors(s), $B_{ia^1}, B_{ib^1}, B_{a^1i}, B_{b^1i}, B_{a^1b^1}, B_{a^1a^2}, B_{a^1b^2}, B_{b^1a^1}, B_{b^1a^2}, B_{b^1b^2}$ are branch susceptances.

In summary, the algorithm of the graph computing-based WLS fast decoupled state estimation is presented below.

---

**Algorithm 9.1   Graph-Based WLS Fast Decoupled State Estimation**

---

1: **Set** iteration index $k = 0$
2: **Initialize** the system state vector $x^0 = (\theta^0, |V|^0)$
3: **Form** nodal gain matrix $G_{AA,i}$ //in nodal parallel
4: **Form** nodal gain matrix $G_{RR,i}$ //in nodal parallel
5: **Assemble** gain matrices $G_{AA}$ and $G_{RR}$
6: **Factorize** gain matrices $G_{AA}$ and $G_{RR}$ //in hierarchical parallel
7: **While** $(\|\Delta\theta^k\| \geq \epsilon_\theta$ or $\|\Delta|V|^k\| \geq \epsilon_V)$ **do**
8:    **Update** RHS vector $H_{AA}^T R_{AA}^{-1} r_A(x^k)$ //in nodal parallel
9:    **Solve** $\Delta\theta^k$ in $G_{AA}\Delta\theta^k = H_{AA}^T R_{AA}^{-1} r_A(x^k)$ //in hierarchical parallel
10:    **Update** RHS vector $H_{RR}^T R_{RR}^{-1} r_R(x^k)$ //in nodal parallel
11:    Solve $\Delta|V|^k$ in $G_{RR}\Delta|V|^k = H_{RR}^T R_{RR}^{-1} r_R(x^k)$ //in hierarchical parallel
12:    $k = k+1$

---

In the Algorithm, steps 3 and 4 build a node-based gain matrix. Step 5 assembles the system-level gain matrix. Step 6 factorizes gain matrices and steps 9 and step 11 solve two linear equations by forward and backward substitution using graph hierarchical parallel computing as discussed in detail in Chapter 3.

### 9.2.2  Numerical Example

To illustrate the gain matrix assembling process, the IEEE-14 bus system as shown in Figure 9.5 is used. The system data and measurements are listed in Tables 9.1–9.4.

We assume that different type of measurements has different standard deviation, as such $\sigma_{V_i}^2 = 1 \times 10^{-5}$, $\sigma_{P_i}^2 = \sigma_{Q_i}^2 = 1 \times 10^{-4}$, $\sigma_{P_{ij}}^2 = \sigma_{Q_{ij}}^2 = 6.4 \times 10^{-5}$.

The graph is partitioned into 14 partitions based on nodes. Bus 9 is highlighted as the studied node. Its 1-neighbor nodes are $\varphi_1(9) = $ 4, 7, 10, 14 and its 2-neighbor nodes are $\varphi_2(9) = $ 2, 3, 5, 8, 11, 13. The nodal $H$ matrix and the nodal $G$ matrix for bus 9 are given by:

$$
H_9 = 
\begin{array}{ccccc}
\theta_4 & \theta_7 & \theta_9 & \theta_{10} & \theta_{14} \\
\end{array}
\begin{bmatrix}
0 & 0 & 0 & 0 & 0 \\
-2.001 & -10.049 & 26.628 & -11.299 & -3.279 \\
0.160 & 0.280 & -6.110 & 4.212 & 1.459 \\
0 & 0 & 11.299 & -11.299 & 0 \\
0 & 0 & 3.279 & 0 & -3.279 \\
-2.001 & 0 & 2.001 & 0 & 0 \\
0 & -10.049 & 10.049 & 0 & 0 \\
0 & 0 & -4.212 & 4.212 & 0 \\
0 & 0 & -1.459 & 0 & 1.459 \\
-0.160 & 0 & -0.160 & 0 & 0 \\
0 & 0.280 & -0.280 & 0 & 0 \\
\end{bmatrix}
\tag{9.34}
$$

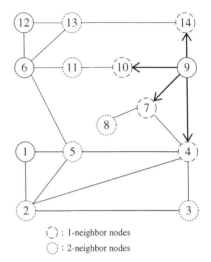

**Figure 9.5**  IEEE 14-bus system.

○ : 1-neighbor nodes

⋯ : 2-neighbor nodes

**Table 9.1** IEEE 14-bus test system branch data.

| Line number | From bus | To bus | Line impedance (p.u.) | | Line charging susceptance (p.u.) | Transformer turns ratio |
|---|---|---|---|---|---|---|
| | | | Resistance | Reactance | | |
| 1 | 1 | 2 | 0.0194 | 0.0592 | 0.0528 | 0.0000 |
| 2 | 1 | 5 | 0.0540 | 0.2230 | 0.0492 | 0.0000 |
| 3 | 2 | 3 | 0.0470 | 0.1980 | 0.0438 | 0.0000 |
| 4 | 2 | 4 | 0.0581 | 0.1763 | 0.0340 | 0.0000 |
| 5 | 2 | 5 | 0.0570 | 0.1739 | 0.0346 | 0.0000 |
| 6 | 3 | 4 | 0.0670 | 0.1710 | 0.0128 | 0.0000 |
| 7 | 4 | 5 | 0.0134 | 0.0421 | 0.0000 | 0.0000 |
| 8 | 4 | 7 | 0.0000 | 0.2091 | 0.0000 | 0.9780 |
| 9 | 4 | 9 | 0.0000 | 0.5562 | 0.0000 | 0.9690 |
| 10 | 5 | 6 | 0.0000 | 0.2520 | 0.0000 | 0.9320 |
| 11 | 6 | 11 | 0.0950 | 0.1989 | 0.0000 | 0.0000 |
| 12 | 6 | 12 | 0.1229 | 0.2558 | 0.0000 | 0.0000 |
| 13 | 6 | 13 | 0.0662 | 0.1303 | 0.0000 | 0.0000 |
| 14 | 7 | 8 | 0.0000 | 0.1762 | 0.0000 | 0.0000 |
| 15 | 7 | 9 | 0.0000 | 0.1100 | 0.0000 | 0.0000 |
| 16 | 9 | 10 | 0.0318 | 0.0845 | 0.0000 | 0.0000 |
| 17 | 9 | 14 | 0.1271 | 0.2704 | 0.0000 | 0.0000 |
| 18 | 10 | 11 | 0.0821 | 0.1921 | 0.0000 | 0.0000 |
| 19 | 12 | 13 | 0.2209 | 0.1999 | 0.0000 | 0.0000 |
| 20 | 13 | 14 | 0.1709 | 0.3480 | 0.0000 | 0.0000 |

**Table 9.2** IEEE 14-bus test system bus data.

| Bus number | Load_P (p.u.) | Load_Q (p.u.) | Unit_P (p.u.) | Unit_Q (p.u.) |
|---|---|---|---|---|
| 1 | 0.0000 | 0.0000 | 2.3240 | −0.1690 |
| 2 | 0.2170 | 0.1270 | 0.4000 | 0.4240 |
| 3 | 0.9420 | 0.1900 | 0.0000 | 0.2340 |
| 4 | 0.4780 | −0.0390 | 0.0000 | 0.0000 |
| 5 | 0.0760 | 0.0160 | 0.0000 | 0.0000 |
| 6 | 0.1120 | 0.0750 | 0.0000 | 0.1220 |
| 7 | 0.0000 | 0.0000 | 0.0000 | 0.0000 |
| 8 | 0.0000 | 0.0000 | 0.0000 | 0.1740 |
| 9 | 0.2950 | 0.1660 | 0.0000 | 0.0000 |
| 10 | 0.0900 | 0.0580 | 0.0000 | 0.0000 |
| 11 | 0.0350 | 0.0180 | 0.0000 | 0.0000 |
| 12 | 0.0610 | 0.0160 | 0.0000 | 0.0000 |
| 13 | 0.1350 | 0.0580 | 0.0000 | 0.0000 |
| 14 | 0.1490 | 0.0500 | 0.0000 | 0.0000 |

**Table 9.3** IEEE 14-bus test system branch measurements.

| Line number | From bus | To bus | M_P_TLPF | M_Q_TLPF | M_P_TLPF (reverse) | M_Q_TLPF (reverse) |
|---|---|---|---|---|---|---|
| 1 | 1 | 2 | 1.5680 | −0.2039 | −5.0000 | 0.2764 |
| 2 | 1 | 5 | 0.7549 | 0.0385 | −0.7273 | 0.0223 |
| 3 | 2 | 3 | 0.7321 | 0.0356 | −0.7089 | 0.0159 |
| 4 | 2 | 4 | 0.5612 | −0.0158 | −0.5445 | 0.0305 |
| 5 | 2 | 5 | 0.4151 | 0.0116 | −0.4061 | −0.0209 |
| 6 | 3 | 4 | −0.2327 | 0.0444 | 0.2364 | −0.0480 |
| 7 | 4 | 5 | −0.6113 | 0.1590 | 0.6164 | −0.1428 |
| 8 | 4 | 7 | 0.2810 | −0.0971 | −0.2810 | 0.1142 |
| 9 | 4 | 9 | 0.1610 | −0.0045 | −0.1610 | 0.0176 |
| 10 | 5 | 6 | 0.4412 | 0.1248 | −0.4412 | −0.0805 |
| 11 | 6 | 11 | 0.0738 | 0.0366 | −0.0732 | −0.0354 |
| 12 | 6 | 12 | 0.0786 | 0.0270 | −0.0779 | −0.0254 |
| 13 | 6 | 13 | 0.1785 | 0.0749 | −0.1764 | −0.0706 |
| 14 | 7 | 8 | 0.0000 | −0.1709 | 0.0000 | 0.1755 |
| 15 | 7 | 9 | 0.2811 | 0.0571 | −0.2811 | −0.0491 |
| 16 | 9 | 10 | 0.0525 | 0.0425 | −0.0523 | −0.0421 |
| 17 | 9 | 14 | 0.0944 | 0.0364 | −0.0932 | −0.0339 |
| 18 | 10 | 11 | −0.0373 | −0.0144 | 0.0374 | 0.0146 |
| 19 | 12 | 13 | 0.0157 | 0.0072 | −0.0157 | −0.0071 |
| 20 | 13 | 14 | 0.0559 | 0.0162 | −0.0554 | −0.0152 |

**Table 9.4** IEEE 14-bus test system bus measurements.

| Bus number | M_Va (degree) | M_Vm (p.u.) | M_P (p.u.) | M_Q (p.u.) |
|---|---|---|---|---|
| 1 | 0.0000 | 1.0600 | 2.3229 | −0.1653 |
| 2 | −4.9800 | 1.0450 | 0.1833 | 0.3079 |
| 3 | −12.7193 | 1.0100 | −0.9416 | 0.0603 |
| 4 | −10.3099 | 1.0177 | −0.4773 | 0.0398 |
| 5 | −8.7712 | 1.0195 | −0.0758 | −0.0167 |
| 6 | −14.2221 | 1.0700 | −0.1102 | 0.0579 |
| 7 | −13.3588 | 1.0616 | 0.0001 | 0.0003 |
| 8 | −13.3588 | 1.0900 | 0.0000 | 0.1755 |
| 9 | −14.9392 | 1.0561 | −0.2952 | −0.1645 |
| 10 | −15.0982 | 1.0512 | −0.0896 | −0.0565 |
| 11 | −14.7899 | 1.0567 | −0.0359 | −0.0208 |
| 12 | −15.0749 | 1.0546 | −0.0621 | −0.0182 |
| 13 | −15.1556 | 1.0500 | −0.1362 | −0.0615 |
| 14 | −16.0342 | 1.0356 | −0.1486 | −0.0490 |

$$
G_9 = 10^4 \times
\begin{array}{ccccc}
\theta_4 & \theta_7 & \theta_9 & \theta_{10} & \theta_{14}
\end{array}
\left[
\begin{array}{ccccc}
10.329 & 20.156 & -60.566 & 23.286 & 6.795 \\
20.156 & 258.95 & -427.18 & 114.72 & 33.356 \\
\mathbf{-60.566} & \mathbf{-427.18} & \mathbf{1157.9} & \mathbf{-553.80} & \mathbf{-116.35} \\
23.286 & 114.72 & -553.80 & 372.60 & 43.193 \\
6.7950 & 33.356 & -116.35 & 43.193 & 33.006
\end{array}
\right]
\tag{9.35}
$$

The $H$ matrices of its 1-neighbor nodes are as follows:

$$
H_4 =
\begin{array}{cccccc}
\theta_2 & \theta_3 & \theta_4 & \theta_5 & \theta_7 & \theta_9
\end{array}
\left[
\begin{array}{cccccc}
0 & 0 & 0 & 0 & 0 & 0 \\
-5.452 & -5.490 & 41.241 & -22.954 & -5.344 & -2.001 \\
2.348 & 1.889 & -11.737 & 7.936 & -0.277 & -0.160 \\
0 & 0 & 22.954 & -22.954 & 0 & 0 \\
0 & 0 & 5.344 & 0 & -5.344 & 0 \\
0 & 0 & 2.001 & 0 & 0 & -2.001 \\
-5.452 & 0 & 5.452 & 0 & 0 & 0 \\
0 & -5.490 & 5.490 & 0 & 0 & 0 \\
0 & 0 & -7.936 & 7.936 & 0 & 0 \\
0 & 0 & 0.277 & 0 & -0.277 & 0 \\
0 & 0 & 0.160 & 0 & 0 & -0.160 \\
2.348 & 0 & -2.348 & 0 & 0 & 0 \\
0 & 1.889 & 1.889 & 0 & 0 & 0
\end{array}
\right]
\tag{9.36}
$$

$$
H_7 =
\begin{array}{cccc}
\theta_4 & \theta_7 & \theta_8 & \theta_9
\end{array}
\left[
\begin{array}{cccc}
0 & 0 & 0 & 0 \\
-5.344 & 21.907 & -6.514 & -10.049 \\
0.277 & 0.003 & 0.001 & -0.280 \\
0 & 6.514 & -6.514 & 0 \\
0 & 10.049 & 0 & -10.049 \\
-5.344 & 5.344 & 0 & 0 \\
0 & 0.001 & 0.001 & 0 \\
0 & 0.280 & 0 & -0.280 \\
0.277 & -0.277 & 0 & 0
\end{array}
\right]
\tag{9.37}
$$

$$
H_{10} = \begin{array}{c}
\begin{array}{ccc} \theta_9 & \theta_{10} & \theta_{11} \end{array} \\
\begin{bmatrix}
0 & 0 & 0 \\
-11.271 & 16.042 & -4.770 \\
4.285 & -6.353 & 2.067 \\
0 & 4.770 & -4.770 \\
-11.271 & 11.271 & 0 \\
0 & -2.067 & 2.067 \\
4.285 & -4.285 & 0
\end{bmatrix}
\end{array}
\qquad (9.38)
$$

$$
H_{14} = \begin{array}{c}
\begin{array}{ccc} \theta_9 & \theta_{13} & \theta_{14} \end{array} \\
\begin{bmatrix}
0 & 0 & 0 \\
-3.216 & -2.432 & 5.647 \\
1.594 & 1.242 & -2.836 \\
-3.216 & 0 & 3.216 \\
0 & -2.432 & 2.432 \\
1.594 & 0 & -1.594 \\
0 & 1.242 & -1.242
\end{bmatrix}
\end{array}
\qquad (9.39)
$$

The corresponding $G$ matrices of its 1-neighbor nodes are as follows:

$$G_4 = 10^4 \times$$

$$
\begin{array}{c}
\begin{array}{cccccc} \theta_2 & \theta_3 & \theta_4 & \theta_5 & \theta_7 & \theta_9 \end{array} \\
\begin{bmatrix}
90.276 & 34.363 & -307.42 & 143.76 & 28.484 & 10.536 \\
34.363 & 86.374 & -301.24 & 141.01 & 28.815 & 10.686 \\
-307.42 & -301.24 & 2919.0 & -1961.4 & -261.89 & -86.965 \\
143.76 & 141.01 & -1961.4 & 1511.5 & 120.47 & 44.674 \\
28.484 & 28.815 & -261.89 & 120.47 & 73.378 & 10.740 \\
\mathbf{10.536} & \mathbf{10.686} & \mathbf{-86.965} & \mathbf{44.674} & \mathbf{10.740} & \mathbf{10.329}
\end{bmatrix}
\end{array}
\qquad (9.40)
$$

$$
G_7 = 10^4 \times \begin{array}{c}
\begin{array}{cccc} \theta_4 & \theta_7 & \theta_8 & \theta_9 \end{array} \\
\begin{bmatrix}
73.378 & -161.81 & 34.812 & 53.623 \\
-161.81 & 748.85 & -209.01 & -378.03 \\
34.812 & -209.01 & 108.74 & 65.458 \\
\mathbf{53.623} & \mathbf{-378.03} & \mathbf{65.458} & \mathbf{258.95}
\end{bmatrix}
\end{array}
\qquad (9.41)
$$

$$
G_{10} = 10^4 \times
\begin{array}{ccc}
\theta_9 & \theta_{10} & \theta_{11} \\
\left[\begin{array}{ccc}
\mathbf{372.60} & -\mathbf{435.23} & \mathbf{62.625} \\
-435.23 & 567.11 & -131.89 \\
62.625 & -131.89 & 69.261
\end{array}\right]
\end{array}
\tag{9.42}
$$

$$
G_{14} = 10^4 \times
\begin{array}{ccc}
\theta_9 & \theta_{13} & \theta_{14} \\
\left[\begin{array}{ccc}
\mathbf{33.006} & \mathbf{9.799} & -\mathbf{42.805} \\
9.799 & 19.107 & -28.906 \\
-42.805 & -28.906 & 71.711
\end{array}\right]
\end{array}
\tag{9.43}
$$

Then according to (9.40)–(9.43), we have:

$$
G(9) = 10^4 \times
\left[\begin{array}{c}
10.536 \\
10.686 \\
-60.566 - 86.965 + 53.623 \\
44.674 \\
-427.18 + 10.740 - 378.03 \\
65.458 \\
1157.9 + 10.329 + 258.95 + 372.60 + 33.006 \\
-553.80 - 435.23 \\
62.625 \\
9.799 \\
-116.35 - 42.805
\end{array}\right]^T
\begin{array}{c}
\theta_2 \\
\theta_3 \\
\theta_4 \\
\theta_5 \\
\theta_7 \\
\theta_8 \\
\theta_9 \\
\theta_{10} \\
\theta_{11} \\
\theta_{13} \\
\theta_{14}
\end{array}
$$

$$
= 10^4 \times
\left[\begin{array}{c}
10.536 \\
10.686 \\
-93.908 \\
44.674 \\
-794.47 \\
65.458 \\
1832.8 \\
-989.03 \\
62.625 \\
9.799 \\
-159.16
\end{array}\right]^T
\begin{array}{c}
\theta_2 \\
\theta_3 \\
\theta_4 \\
\theta_5 \\
\theta_7 \\
\theta_8 \\
\theta_9 \\
\theta_{10} \\
\theta_{11} \\
\theta_{13} \\
\theta_{14}
\end{array}
\tag{9.44}
$$

Each node in the system will conduct the above graph-based algorithm in parallel until the system gain matrix is formed.

## 9.2.3  Graph-Based State Estimation Implementation

To implement the graph-based state parallel estimation developed above, the defined bus-branch model graph schema for topology processing in Chapter 8 is extended to support state estimation graph computing.

### 9.2.3.1  Graph-Based State Estimation Graph Schema

In the bus-branch model, vertex `TopoND` and edge `TopoConnect` are defined. They are extended as follows to contain the information for state estimation:

```
CREATE VERTEX TopoND (PRIMARY_ID topoid int, TOPOID int, island int,
    bus_name string, area string, loss_zone uint, busType int, Vm
    double, Va double, M_Vm double, M_Va double, base_kV double,
    desired_volts double, control_bus_number int, up_V double, lo_V
    double, GenP double, GenQ double, M_Gen_P double, M_Gen_Q double,
    qUp double, qLower double, LdP double, LdQ double, P double,
    Q double, M_Load_P double, M_Load_Q double, G double, B double, Sub
    string, OV int, UV int, sumB double, sumG double, sumBi double,
    volt double, double, M_C_P double, SE_Vm double, SE_Va double,
    Ri_vP double, Ri_vQ double, Ri_eP double, Ri_eQ double, SE_Inj
    tuple, GainP tuple, GainQ tuple, H_r_P double, H_r_Q double)
```

```
CREATE DIRECTED EDGE TopoConnect (from TopoND, to TopoND, typename
    string, edge_name List<string>, area List<string>, zone
    List<string>, from_bus int, to_bus int, flag uint, R List<double>,
    X List<double>, hB_list List<double>, line_Q1_list List<double>,
    line_Q2_list List<double>, line_Q3_list List<double>, control_bus
    int, K List<double>, shifting_transformer_angle List<double>,
    min_tap double, max_tap double, step_size double, min_volt
    List<double>, max_volt List<double>, M_P_BR List<double>, M_Q_BR
    List<double>, G List<double>, B List<double>, BIJ List<double>,
    circuit uint, P_BR double, Q_BR double, from_open List<int>,
    to_open List<int>, SE_P List<double>, SE_Q List<double>, SE_I
    List<double>, SE_Tap List<double>)
```

In the vertex and edge definition above, measurement attributes have been defined to meet the CIM/E standard. Two new attributes `GainP` and `GainQ` are highlighted in black bold in the vertex `TopoND` definition. They are defined in the type of tuple to calculate the nodal Jacobian matrix and gain matrix at node $i$ in the graph queries when we implement graph-based state estimation.

```
typedef tuple<double GainP_i, double GainP_a1, double GainP_b1, double
    GainP_a2, double GainP_b2> GainP;
```

```
typedef tuple<double GainQ_i, double GainQ_a1, double GainQ_b1, double
    GainQ_a2, double GainQ_b2> GainQ;
```

A tuple is a data structure consisting of a sequence of baseType variables. Tuple types are created and named using a typedef statement. The two tuples are defined first before they are used in the vertex TopoND schema definition.

To store estimated results, SE_Vm and SE_Va in double are defined as attributes of vertex TopoND. In addition, tuple SE_Inj is added to store active power and/or reactive power injection estimate of a generator, load, and shunt. Ri_vP, Ri_vQ, Ri_eP, Ri_eQ are weights. H_r_P and H_r_Q are RHS.

```
typedef tuple<double Unit_P, double Unit_Q, double Load_P, double
    Load_Q, double Shunt_Q> SE_Inj;
```

In edge TopoConnect, additional SE_P, SE_Q, SE_I, and SE_Tap are defined to store branch active power, reactive power, current, and tap position estimates. They are defined as a List type to accommodate components in parallel.

After the vertex and edge definitions are updated, the graph will be recreated by the statement as follows and the bus-branch graph is not necessarily reformed since the adding attributes do not change the network topology and parameters.

```
CREATE GRAPH busBranchGraph(TopoND, TopoConnect)
```

Now, the graph is ready to perform graph-based state estimation.

### 9.2.3.2 Nodal Gain Matrix Formation

As shown in Algorithm: Graph-based WLS Fast Decoupled State Estimation, the graph-based state estimation is consistent with five key steps: (i) build node-based gain matrix, (ii) assemble the system-level gain matrix, (iii) factorizes gain matrices, (iv) form the RHS vector, and (v) solve linear equations. Where gain matrix factorization and solving linear equations by forward/backward substitution graph computing implementation can be leveraged by the approaches in Chapter 3. RHS vector formation, nodal gain matrix formation, and system-level matrix assembling are unique in graph-based state estimation which will be motivated and highlighted with details in this chapter.

The pseudo-code in the graph query to implement the nodal gain matrix formation is shown as follows:

```
SumAccum<double> @sumG = 0
SumAccum<double> @sumB = 0
SumAccum<double> @sumBedge = 0
SumAccum<double> @sumBi = 0
SumAccum<double> @sumGii_P = 0 //diagonal element in Gain matrix
SumAccum<double> @sumGij_P = 0 //off-diagonal element in Gain matrix
SumAccum<double> @sumGii_Q = 0 //diagonal element in Gain matrix
SumAccum<double> @sumGij_Q = 0 //off-diagonal element in Gain matrix
MapAccum<int, double> @neighbor_B //connected edges info (B)
MapAccum<int, double> @neighbor_BIJ //connected edges info (BIJ)
MapAccum<int, double> @neighbor_Pweight //weight of neighbor's injection
MapAccum<int, double> @neighbor_Qweight //weight of neighbor's injection
MapAccum<int, double> @G_offdiag_B //gain matrix offdiag_B
MapAccum<int, double> @G_offdiag_BIJ //gain matrix offdiag_BIJ
MapAccum<int, double> @G_2_neighbor_B //gain matrix offdiag_B
MapAccum<int, double> @G_2_neighbor_BIJ //gain matrix offdiag_BIJ
ListAccum<sort_id> @Gip
ListAccum<sort_id> @Giq
```

```
CREATE QUERY stateEstimationWeighedGain() FOR GRAPH busBranchGraph{
 T0 = {TopoND.*}
 T0 = SELECT s FROM T0:s-(connected:e)->:t
 ACCUM
    s.@sumG += e.G/e.K/e.K
    s.@sumB += -e.B/e.K/e.K+e.hB
    s.@sumBedge += e.B/e.K
    s.@sumBi += -e.BIJ
    s.@sumGii_Q += (e.B/e.K-2*e.hB)*(e.B/e.K-2*e.hB)*e.Ri_eQ
    t.@sumGii_Q += (e.B*e.B/e.K/e.K)*s.Ri_vQ+(e.B*e.B/e.K/e.K)*e.Ri_eQ
    s.@neighbor_B += (t.topoid->e.B/e.K)
    s.@sumGii_P += (e.BIJ*e.BIJ)*e.Ri_eP
    t.@sumGii_P += (e.BIJ*e.BIJ)*e.Ri_eP+(e.BIJ*e.BIJ)*s.Ri_vP
    s.@neighbor_BIJ += (t.topoid->e.BIJ)
    s.@neighbor_Pweight += (t.topoid->t.Ri_vP)
    s.@neighbor_Qweight += (t.topoid->t.Ri_vQ)
 POST-ACCUM
    s.@sumB += s.B
    s.@sumG += s.G
    s.@sumGii_P += (s.@sumBi*s.@sumBi)*s.Ri_vP
    s.@sumGii_Q += (-2*s.@sumB-s.@sumBedge)*
                   (-2*s.@sumB-s.@sumBedge)*s.Ri_vQ+s.Ri_V
    s.@Gip += sort_id (s.topoid, s.@sumGii_P)
    s.@Giq += sort_id (s.topoid, s.@sumGii_Q)

 T0 = SELECT s FROM T0:s-(connected:e)->:t
 ACCUM
    // Processing 1-step-neighbor
    s.@G_offdiag_BIJ += (t.topoid->s.@neighbor_BIJ*
                         t.@neighbor_BIJ*s.@neighbor_Pweight)
    s.@G_offdiag_BIJ += (t.topoid->s.@sumBi*e.BIJ*s.Ri_vP+
                         t.@sumBi*e.BIJ*t.Ri_vP-e.BIJ*e.BIJ*e.Ri_eP)
    t.@G_offdiag_BIJ += (s.topoid->(- e.BIJ*e.BIJ*e.Ri_eP))
    s.@G_offdiag_B += (t.topoid->s.@neighbor_B*
                       t.@neighbor_B*s.@neighbor_Qweight)
    s.@G_offdiag_B += (t.topoid->(2*s.@sumB+s.@sumBedge*e.B/e.K*s.
                       Ri_vQ+
    2*t.@sumB+t.@sumBedge*e.B/e.K*t.Ri_vQ-e.B/e.K-2*e.hB*e.B/e.K*e.
                       Ri_eQ))
    t.@G_offdiag_B += (s.topoid ->(-(e.B/e.K-2*e.hB)*e.B/e.K*e.Ri_eQ))

    // Processing 2-step-neighbor
    FOREACH (key,val) in t.@neighbor_BIJ DO
        key = SELECT s FROM key:s-(connected:e)->:t
            s.@G_2_neighbor_BIJ += (t.topoid->e.BIJ*val*t.Ri_vP)
    END
```

```
    FOREACH (key,val) in t.@neighbor_B DO
        key = SELECT s FROM key:s-(connected:e)->:t
            s.@G_2_neighbor_B += (t.topoid-> e.B*val/e.K*t.Ri_vQ)
    END
    POST-ACCUM
        Map2List(s.@G_offdiag_BIJ, s.@G_offdiag_B, s.@G_2_neighbor_BIJ,
                s.@G_2_neighbor_B, s.@Gip, s.@Giq)
        s.GainP += s.@Gip
        s.GainQ += s.@Giq
}
```

Before the query `stateEstimationWeighedGain()` is implemented, vertex-attached accumulator variables are defined with a single add-sign @ to store the nodal gain matrix elements and corresponding index. Note that two gain matrices $G_{AA}$(`GainP`) and $G_{RR}$(`GainQ`) are built to support fast decoupled state estimation.

In the query definition, all `TopoND` vertices in the graph are assigned to `T0` by the statement `T0 = {TopoND.*}` to be processed in parallel.

For all vertices, their edges e are selected starting from vertex s to the end vertex t by the following statement.

```
T0 = SELECT s FROM T0:s-(connected:e)->:t
```

The attributes of these edges are used to calculate the contribution of each edge to the diagonal terms in the nodal gain matrix in parallel by using the keyword `ACCUM`. In the `ACCUM` block, statement `s.@neighbor_B += (t.topoid->e.B/e.K)` and

`s.@neighbor_BIJ += (t.topoid->e.BIJ)` find 1-step-neighbor t of vertex s and save its topoid, *B*, and *BIJ* into vertex-attached MapAccum `@neighbor_B` and `@neighbor_BIJ` respectively. The corresponding weights for *P* and *Q* injection of the 1-step-neighbor are saved in the MapAccum `@neighbor_Pweight` and `@neighbor_Qweight` by the following statements in parallel.

```
s.@neighbor_Pweight += (t.topoid->t.Ri_vP)
s.@neighbor_Qweight += (t.topoid->t.Ri_vQ)
```

In the `POST-ACCUM`, shunt conductance and shunt susceptance are added to the corresponding diagonals.

```
s.@sumB += s.B
s.@sumG += s.G
s.@sumGii_P += (s.@sumBi*s.@sumBi)*s.Ri_vP
s.@sumGii_Q += (-2*s.@sumB-s.@sumBedge)*
            (-2*s.@sumB-s.@sumBedge)*s.Ri_vQ+s.Ri_V
```

To process 1-step-neighbor contributions to the gain matrix, all vertices are traveled from vertex s to their 1-step neighbors t by `T0 = SELECT s FROM T0:s-(connected:e)->:t`. Then corresponding off-diagonals are calculated.

To find a 2-step-neighbor, starting from a 1-step-neighbor stored in the MapAccum `@neighbor_B` and `@neighbor_BIJ`, their neighbors are found by the following two FOR loops and the second step traversal `key = SELECT s FROM key:s-(connected:e)->:t`.

```
FOREACH (key,val) in t.@neighbor_BIJ DO
FOREACH (key,val) in t.@neighbor_B DO
```

Finally, in the POST-ACCUM, the gain matrix is assembled in GainP and GainQ by:

```
Map2List(s.@G_offdiag_BIJ, s.@G_offdiag_B, s.@G_2_neighbor_BIJ,
         s.@G_2_neighbor_B, s.@Gip, s.@Giq)
s.GainP += s.@Gip
s.GainQ += s.@Giq
```

### 9.2.3.3  Build RHS

To solve the fast decoupled state estimation by (9.11), the RHS vectors $H_{AA}^T R_{AA}^{-1} r_A(x^k)$ and $H_{RR}^T R_{RR}^{-1} r_R(x^k)$ are updated in each iteration in the Graph-based WLS Fast Decoupled State Estimation algorithm in parallel. They are implemented by the following pseudo-code:

```
SumAccum<double> @deltaP = 0
SumAccum<double> @deltaQ = 0
SumAccum<double> @deltaVm = 0
CREATE QUERY buildRighHandSideSE() FOR GRAPH busBranchGraph{
 T0 = {TopoND.*}
 //Build H'*R*r for P measurements
 T0 = SELECT s FROM T0:s-(connected:e)->:t
  ACCUM
   s.@deltaP += s.@Vm*t.@Vm *
           (-e.G/e.K*cos(s.@Vs-t.@Vs)+(e.B/e.K*sin(s.@Vs-t.@Vs)))
   s.H_r_P += e.BIJ*(e.M_P_BR-(s.@Vm*s.@Vm*e.G/e.K/e.K-s.@Vm*t.@Vm*
      ((e.G/e.K)*cos(s.@Vs-t.@Vs)+(-e.B/e.K)*sin(s.@Vs-t.@Vs)))))*e.
Ri_eP
    t.H_r_P+=-e.BIJ*(e.M_P_BR-(s.@Vm*s.@Vm*e.G/e.K/e.K-s.@Vm*t.@Vm*
        ((e.G/e.K)*cos(s.@Vs-t.@Vs)+(-e.B/e.K)*sin(s.@Vs-t.@Vs)))))*e.
Ri_eP
POST-ACCUM
   s.@deltaP = s.@P - (s.@deltaP + s.@Vm*s.@Vm*s.@sumG)
   s.H_r_P += -s.@sumBi * s.@deltaP * s.Ri_vP
 T0 = SELECT s FROM T0:s-(connected:e)->:t
  ACCUM
    t.H_r_P +=  -e.BIJ * s.@deltaP * s.Ri_vP


 //Build H'*R*r for Q measurements
 T0 = SELECT s FROM T0:s-(connected:e)->:t
  ACCUM
   s.@deltaQ += s.@Vm*t.@Vm*
        (-1*e.G/e.K*sin(s.@Vs-t.@Vs)-(e.B/e.K*cos(s.@Vs-t.@Vs)))
   s.H_r_Q+=(e.B/e.K-2*e.hB)*(e.M_Q_BR-(-s.@Vm*s.@Vm*
      (-e.B/e.K/e.K+e.hB)-s.@Vm*t.@Vm*((e.G/e.K)*sin(s.@Vs-t.@Vs)-
      (-e.B/e.K)*cos(s.@Vs-t.@Vs)))))*e.Ri_eQ
   t.H_r_Q +=(-1)*(e.B/e.K)*(e.M_Q_BR-(-s.@Vm*s.@Vm*
      (-e.B/e.K/e.K+e.hB)-s.@Vm*t.@Vm*((e.G/e.K)*sin(s.@Vs-t.@Vs)-
      (-e.B/e.K)*cos(s.@Vs-t.@Vs)))))*e.Ri_eQ
```

```
POST-ACCUM
    s.@deltaQ = s.@Q - (s.@deltaQ - s.@Vm*s.@Vm*s.@sumB),
    s.@deltaVm = s.M_Vm - s.@Vm,
    s.H_r_Q += (-1) * (2*s.@sumB + s.@sumBedge)* s.@deltaQ * s.Ri_vQ +
        s.@deltaVm * s.Ri_V
TO = SELECT s FROM TO:s-(connected:e)->:t
  ACCUM
    t.H_r_Q += (-1) * (e.B/e.K) * s.@deltaQ * s.Ri_vQ
}
```

Without loss of generality, we assume branch active power measurement M_P_BR and reactive power measurement M_Q_BR are available to estimate system states. Before the query, vertex-attached SumAccum @deltaP, @deltaQ, @deltaVm are defined to store active power, reactive power, and voltage magnitude residual of each bus/vertex. Graph traversal statement TO = SELECT s FROM TO:s-(connected:e)->:t coupling with keyword ACCUM travels to all branches from vertex s to vertex t and updates RHS attributes H_r_P and H_r_Q in nodal parallel computing. In the POST-ACCUM block, contributions of active power and reactive power injections from generators and shunts are added to the RHS.

### 9.2.4 Graph-Based State Estimation Computation Efficiency

The graph-based state estimation computing method divides the entire system into $n$ sub-systems (nodes). The computation complexity is reduced, and computation efficiency is improved.

For illustration purposes, the indices of node $i$ and its 1-step neighbors are grouped into a set $\alpha_i$. For $H_{AA,i}$ in (9.17), the column indices of its non-zero entries are the elements in the set of $\alpha_i$.

$$\alpha_i : = \{j : Y_{ij} \neq 0\}, j = 1, 2, \cdots, n \tag{9.45}$$

$\beta_i$ is a set that consists of node $i$, its 1-step neighbors, and its 2-step neighbors' indices.

$$\beta_i : = \{j : Y_{ij} \neq 0\} \cup \{j : d(i,j) = 2\} = \alpha_i \cup \{j : d(i,j) = 2\} \tag{9.46}$$

Denote the number of branch power flow measurements by $n_{BFM}$ and the number of power injection measurements by $n_{IM}$. A row vector of $H_{AA}$ corresponding to a branch power flow measurement has two non-zeros. Let $d_i$ denote the number of degree(s) of node $i$. A row vector of $H_{AA}$ corresponding to a power injection measurement at node $i$ has $1 + d_i$ non-zeros.

The number of non-zeros in $H_{AA}$ is therefore

$$nz_{H_{AA}} = 2n_{BFM} + n_{IM} + \sum_{i \in \gamma} d_i \tag{9.47}$$

where $\gamma$ is the set of power injection measured buses.

The scaled $H_{AA}$ matrix is defined as

$$\overline{H}_{AA} = R_{AA}^{-1/2} H_{AA} \tag{9.48}$$

and the scaled active residual vector at iteration $k$ is

$$\overline{r}_A(x^k) = R_{AA}^{-1/2} r_A(x^k) \tag{9.49}$$

Then, at iteration $k$, the RHS vector is given as

$$RHS = H_{AA}^T R_{AA}^{-1} r_A(x^k) = \overline{H}_{AA}^T \overline{r}_A(x^k) \tag{9.50}$$

The number of floating-point operations (flops) required to form the system-level RHS vector $nflop_{RHS}$ is

$$nflop_{RHS} = nnz_{H_{AA}} \tag{9.51}$$

where $nnz_{H_{AA}}$ is the number of non-zeros in $H_{AA}$.

In (9.19), the system-level RHS vector is the summation of each node-based RHS vector. After matrix scaling, the vector is updated as:

$$RHS = \sum_{i=1}^{n} H_{AA,i}^T R_{AA,i}^{-1} r_{A,i}(x^k) = \sum_{i=1}^{n} \overline{H}_{AA,i}^T \overline{r}_{A,i}(x^k) \tag{9.52}$$

Then, the number of flops for each node-based RHS vector $nflop_{RHS,i}$ is

$$nflop_{RHS,i} = nnz_{H_{AA,i}} = 2n_{BFM,i} + 1 + d_i \tag{9.53}$$

where $nnz_{H_{AA,i}}$ indicates the number of non-zeros in the bus $i$'s Jacobian matrix $H_{AA,i}$, $n_{BFM,i}$ is the number of power flow measurements of branches connecting to bus $i$, and $d_i$ denotes the number of degree(s) of node $i$.

Different from the RHS vector, the formation of the gain matrix involves the multiplication of the scaled $H_{AA}$ matrix by its transpose:

$$G_{AA} = H_{AA}^T R_{AA}^{-1} H_{AA} = \overline{H}_{AA}^T \overline{H}_{AA} \tag{9.54}$$

Similar to (9.17), $\overline{H}_{AA}$ is written in terms of its rows:

$$\overline{H}_{AA} = \begin{pmatrix} \overline{H}_{AA}(1) \\ \overline{H}_{AA}(2) \\ \vdots \\ \overline{H}_{AA}(i) \\ \vdots \\ \overline{H}_{AA}(m_A) \end{pmatrix} \tag{9.55}$$

where $\overline{H}_{AA}(i)$ is the $i$th row vector of $\overline{H}_{AA}$ in the dimension of $(n-1)$, and $m_A$ is the total number of active power measurements. The product of $\overline{H}_{AA}^T \overline{H}_{AA}$ can be expressed as a sum of the outer product of the row vectors of $\overline{H}_{AA}$.

$$\overline{H}_{AA}^T \overline{H}_{AA} = \sum_{i=1}^{m_A} \left( \overline{H}_{AA}(i)^T \overline{H}_{AA}(i) \right) \tag{9.56}$$

For a branch power flow measurement, forming $\overline{H}_{AA}(i)^T \overline{H}_{AA}(i)$ requires four flops. Also, forming $\overline{H}_{AA}(i)^T \overline{H}_{AA}(i)$ for a power injection measurement at bus $i$ with $d_i$ connected branches requires $(1+d_i)^2$ flops.

Suppose the active power measurement vector $z_A$ contains $n_{BFM}$ branch power flow measurements and $n_{IM}$ power injection measurements, the total number of flops required to form the system-level gain matrix $nflop_{Gain}$ is

$$nflop_{Gain} = 4n_{BFM} + \sum_{i \in \gamma} (1 + d_i)^2 \tag{9.57}$$

Since the matrix is symmetric and only its upper (lower) triangular matrix needs to be computed and the number of flops is approximately half of $nflop_{Gain}$.

In (9.18), the system-level gain matrix is decomposed into the summation of node-based gain matrices. After matrix scaling, it becomes

$$G_{AA} = \sum_{i=1}^{n} H_{AA,i}^T R_{AA,i}^{-1} H_{AA,i} = \sum_{i=1}^{n} \overline{H}_{AA,i}^T \overline{H}_{AA,i} \tag{9.58}$$

For each node-based gain matrix, the number of flops $nflop_{Gain,i}$ is

$$nflop_{Gain,i} = 4n_{BFM,i} + (1 + d_i)^2 \tag{9.59}$$

where, $n_{BFM,i}$ is the number of branch power flow measurements starting from bus $i$, and $d_i$ denotes the number of degree(s) of node $i$.

Since the formulation of each node-based RHS vector and node-based gain matrix $G_{AA,i}$ is independent and can be implemented in parallel, the computation efficiency of the node-based graph computing approach is significantly improved.

The node-based $H$ matrices and gain matrices are sparse since the derivatives of a node/edge measurement to its indirectly connected nodes' states are zero. The graph model preserves the sparsity of the node-based $H$ matrices and gain matrices and stores them in the graph database. In other words, each node-based Jacobian matrix only stores information from the node itself, its directly connected nodes, and the edge between them, excluding zero elements introduced by indirectly connected nodes. Compared with $H_{AA,i}$, the size of the node-based Jacobian matrix $H_{AA,i}^C$ is reduced to $m_{A,i} \times n_i$, eliminating the columns of the indirectly connected nodes. $m_{A,i}$ is the size of $z_{A,i}$ and $\sum_{i=1}^{n} m_{A,i} = m_A$. $m_{A,i}$ is obviously smaller than the total number of active power measurements $m_A$ significantly. For the node-based gain matrix, taking node $i$ as an example, its size is reduced from $(n-1) \times (n-1)$ to $n_i \times n_i$, where $n_i$ is the size of $\alpha_i$.

To illustrate the efficiency, the IEEE standard 5-bus system in Figure 9.6 is taken as an example. The compressed node-based $H$ matrix for node 1 is $H_{AA,1}^C$ as presented in (9.60). It has non-zero column vectors only and its size is $3 \times 3$.

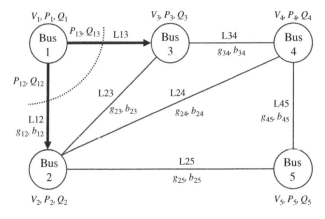

**Figure 9.6** Graph model of IEEE 5-bus system.

$$\boldsymbol{H}^C_{AA,1} = \begin{bmatrix} \dfrac{\partial P_1}{\partial\theta_1} & \dfrac{\partial P_1}{\partial\theta_2} & \dfrac{\partial P_1}{\partial\theta_3} \\[2mm] \dfrac{\partial P_{12}}{\partial\theta_1} & \dfrac{\partial P_{12}}{\partial\theta_2} & \dfrac{\partial P_{12}}{\partial\theta_3} \\[2mm] \dfrac{\partial P_{13}}{\partial\theta_1} & \dfrac{\partial P_{13}}{\partial\theta_2} & \dfrac{\partial P_{13}}{\partial\theta_3} \end{bmatrix} \in \mathcal{M}(3,3) \tag{9.60}$$

When the pre-compressed node-based $H$ matrix $\boldsymbol{H}_{AA,1}$ is used, as presented in (9.61), the size of the matrix $\boldsymbol{H}_{AA,1}$ is $3 \times 5$. As highlighted in shadow in (9.61), the elements in the last two columns are all-zero.

$$\boldsymbol{H}_{AA,1} = \begin{bmatrix} \dfrac{\partial P_1}{\partial\theta_1} & \dfrac{\partial P_1}{\partial\theta_2} & \dfrac{\partial P_1}{\partial\theta_3} & \dfrac{\partial P_1}{\partial\theta_4} & \dfrac{\partial P_1}{\partial\theta_5} \\[2mm] \dfrac{\partial P_{12}}{\partial\theta_1} & \dfrac{\partial P_{12}}{\partial\theta_2} & \dfrac{\partial P_{12}}{\partial\theta_3} & \dfrac{\partial P_{12}}{\partial\theta_4} & \dfrac{\partial P_{12}}{\partial\theta_5} \\[2mm] \dfrac{\partial P_{13}}{\partial\theta_1} & \dfrac{\partial P_{13}}{\partial\theta_2} & \dfrac{\partial P_{13}}{\partial\theta_3} & \dfrac{\partial P_{13}}{\partial\theta_4} & \dfrac{\partial P_{13}}{\partial\theta_5} \end{bmatrix} \in \mathcal{M}(3,5) \tag{9.61}$$

$$\boldsymbol{G}^C_{AA,1} = \boldsymbol{H}^{C\,T}_{AA,1} \cdot \boldsymbol{R}^{-1}_{A,1} \cdot \boldsymbol{H}^C_{AA,1} \in \mathcal{M}(3,3) \tag{9.62}$$

$$\boldsymbol{G}_{AA,1} = \boldsymbol{H}^T_{AA,1} \cdot \boldsymbol{R}^{-1}_{A,1} \cdot \boldsymbol{H}_{AA,1} \in \mathcal{M}(5,5) \tag{9.63}$$

In addition, as displayed in (9.62) and (9.63), the size of node $1'$s gain matrix is compressed from a $3 \times 5$ matrix to a $3 \times 3$ matrix. The efficiency improvement is more evident in large systems.

Note that there are zero entries (for example $\dfrac{\partial P_{12}}{\partial\theta_3}$) in the compressed matrix $H^C_{AA,i}$. By graph modeling, only non-zero entries in $H^C_{AA,i}$ are stored as attributes of nodes/edges in the graph, thus the storage efficiency of $H^C_{AA,i}$ is further improved. Taking $P_{12}$ as an example, since $P_{12}$ is the active power flow from node 1 to node 2, its derivation to node 3's state, i.e. $\dfrac{\partial P_{12}}{\partial\theta_3}$, is 0, while its derivations to $\theta_1$ and $\theta_2$ are saved as $H$ attributes as $L_{12}$ shown in Figure 9.6.

## 9.3 Bad Data Detection and Identification

In a real power system, measurements may contain errors due to the degree of accuracy of meters and telecommunications. Taking advantage of the sufficient redundancy of measurements, the WLS method can filter small measurement errors.

However, large measurement errors can also occur in the data acquisition system upon telecommunication system failures or malfunction of meters. Big measurement deviations will result in unexpected state estimates.

Some bad data can be detected and eliminated by a sanity check. For example, negative voltage magnitude, power flow multiple times higher or lower than its limits, tap position which is out of range, etc. Unfortunately, not all bad data are obviously detectable by a sanity check. A practicable state estimator should be equipped with an advanced function to identify and detect any type of bad data in measurements.

Classic bad data detection approaches are built on the WLS estimation method. The approaches essentially analyze the residuals and their probability distribution [24–28]. Researchers also developed other bad data identification and detection approaches based on robust state estimation [29, 30]. Because of its complexity, bad data detection based on robust state estimation is commonly computationally expensive. We will focus on classic bad data detection approaches based on WLS estimation.

### 9.3.1 Chi-Squares Test

Bad data detection refers to the determination of bad data in the measurement set. Identification is a procedure of finding specific bad data. Chi-squares test is commonly used to detect bad data. Then they are identified and eliminated to achieve an unbiased state estimate [31].

Consider a set of $n$ independent random variables $X_i$, $i = 1, 2, ..., n$ meet the standard normal distribution:

$$X_i \sim N(0, 1) \tag{9.64}$$

A new random variable $Y$ is defined as:

$$Y = \sum_{i=1}^{n} X_i^2 \tag{9.65}$$

Then $Y$ will have a Chi-squares $\chi^2$ distribution with $n$ degrees of freedom, i.e.

$$Y \sim \chi_n^2 \tag{9.66}$$

The degrees of freedom represent the number of independent variables in the sum of squares. In the bad data detection based on the WLS state estimation, the objective function $J(x)$ is used for the Chi-squares test for bad data approximately.

$$J(x) = [z - h(x)]^T \cdot R^{-1} \cdot [z - h(x)] = \sum_{i=1}^{m} \frac{(z_i - h_i(x))^2}{R_{ii}} \tag{9.67}$$

Assuming the degrees of freedom of $J(x)$ is $n$, for a given probability $p$ (typically 0.95) and degrees of freedom $n$, look up the value $\chi_{n,p}^2$ from the Chi-squares distribution table, e.g.

$$p = Pr\left(J(x) \leq \chi_{n,p}^2\right) \tag{9.68}$$

Test if $J(x) \geq \chi_{n,p}^2$, then bad data is detected.

### 9.3.2 Advanced Bad Data Detection

The approach used to detect and identify bad data is based on the core idea in the methods of residual search and estimation identification, i.e. the logic of iterative exploration and linear correction. In the suspicious data set, the largest residual is first selected to be identified. Residuals are corrected and updated after each bad data identification iteration. In the iterative estimate identification approach, residual sensitivity and the Jacobian matrix are utilized to correct residuals and system states.

Assuming the predicted residuals is updated after a measurement $i$ or a set of measurements $I$ is removed as expressed as (9.69) and (9.70).

$$r_\omega^{(i)} = z_\omega - h_\omega\left(\hat{x}^{(i)}\right) \tag{9.69}$$

$$r_\omega^{(I)} = z_\omega - h_\omega\left(\hat{x}^{(I)}\right) \tag{9.70}$$

where, $r_\omega^{(i)} \in R^m$ is the weighted residual vector after measurement data $i$ is removed; $z_\omega \in R^m$ is the weighted measurement vector, i.e. $R^{-1}z$; $\hat{x}^{(i)} \in R^n$ is the system state vector after the measurement data $i$ is removed; $r_\omega^{(I)} \in R^m$ is the weighted residual vector after measurement data set $I$ is removed; $\hat{x}^{(I)} \in R^n$ is the system state vector after the measurement data set $I$ is removed; $h_\omega(\cdot) = R^{-1}h(\cdot) \in R^m$ is the weighted measurement; $m$ is the number of measurements; and $n$ is the number of variables.

Then the system state correction equation is as follows.

$$\hat{x}^{(i)} = \hat{x} - \left(H_\omega^T H_\omega\right)^{-1} H_{\omega,i} W_{\omega,i,i}^{-1} r_{\omega,i} \tag{9.71}$$

$$\hat{x}^{(I)} = \hat{x} - \left(H_\omega^T H_\omega\right)^{-1} H_{\omega,I} W_{\omega,I}^{-1} r_{\omega,I} \tag{9.72}$$

where, $\hat{x}$ is the estimated system states before any measurement correction; $H_{\omega,i}$ is the $i$th row vector in the matrix $H_\omega$, and it's a one-dimensional vector in the size of $n$; $H_{\omega,I}$ is a row vector set, corresponding to the measurement data set $I$, in the matrix $H_\omega$ which is also a one-dimensional vector in the size of $n$; $H_\omega \in R^{m \times n}$ is the weighted Jacobian matrix, i.e. $R^{-1}H$; $W_\omega \in R^{m \times m}$ is the weighted residual sensitivity matrix; $W_{\omega,i,i}$ is the diagonal element of $W_\omega$, corresponding to the deleted measurement data $i$; $W_{\omega,I}$ is the submatrix of $W_\omega$, corresponding to the deleted measurement data set $I$; $r_{\omega,i}$ is the weighted residual corresponding to measurement data $i$ before measurement data $i$ is removed; $r_{\omega,I}$ is the weighted residual vector corresponding to measurement data $I$ before measurement data set $I$ is removed.

The weighted residuals after measurement data $i$ is removed are derived as (9.73)–(9.75).

$$r_\omega^{(i)} = r_\omega + H_\omega\left(H_\omega^T H_\omega\right)^{-1} H_{\omega,i} W_{\omega,i,i}^{-1} r_{\omega,i} \tag{9.73}$$

$$r_{\omega,j}^{(i)} = W_{\omega,i,i}^{-1} r_{\omega,i} \tag{9.74}$$

$$r_{\omega,j}^{(i)} = r_{\omega,i} - W_{\omega,i,j} W_{\omega,i,i}^{-1} r_{\omega,i} \ (j \in i, j \neq i) \tag{9.75}$$

The corresponding weighted residual variance is then expressed by (9.76) and (9.77).

$$var\left(r_{\omega,i}^{(i)}\right) = 1 \tag{9.76}$$

$$var\left(r_{\omega,j}^{(i)}\right) = \left(W_{\omega,i,i} W_{\omega,i,j} - W_{\omega,j,j}^2\right)/W_{\omega,i,i} \tag{9.77}$$

Similarly, the weighted residual calculation after measurement data set $I$ is removed is derived as (9.78)–(9.80).

$$r_\omega^{(I)} = r_\omega + H_\omega \left(H_\omega^T H_\omega\right)^{-1} H_{\omega,I} W_{\omega,I}^{-1} r_{\omega,I} \tag{9.78}$$

$$r_{\omega,I}^{(I)} = W_{\omega,I}^{-1} r_{\omega,I} \tag{9.79}$$

$$r_{\omega,M-I}^{(I)} = r_{\omega,M-I} - W_{\omega,M-I} W_{\omega,I}^{-1} r_{\omega,I} \tag{9.80}$$

The corresponding weighted residual covariance is then expressed by (9.81) and (9.82).

$$\text{cov}\left(r_{\omega,I}^{(I)}\right) = I \tag{9.81}$$

$$\text{cov}\left(r_{\omega,M-I}^{(I)}\right) = W_{\omega,M-I} - W_{\omega,M-I,I} W_{\omega,I}^{-1} W_{\mu,M-I,I}^T \tag{9.82}$$

The matrix $W_{\omega,I+1}^{-1}$ is calculated by

$$W_{\omega,I+1}^{-1} = \begin{bmatrix} W_\omega & W_{\omega,i+1} \\ W_{\omega,i+1}^T & W_{\omega,i+1,i+1} \end{bmatrix}^{-1} = \begin{bmatrix} W_{\omega,I}^{-1} + \dfrac{1}{C} bb^T & -\dfrac{1}{C} b \\ -\dfrac{1}{C} b^T & \dfrac{1}{C} \end{bmatrix} \tag{9.83}$$

where,

$$b = W_{\omega,I}^{-1} W_{i+1} \tag{9.84}$$

$$C = W_{\omega,i+1,i+1} - W_{\omega,i+1,i+1}^T W_{\omega,I}^{-1} W_{\omega,i+1} \tag{9.85}$$

Then the system state $\hat{x}^{(I+1)}$ is given by (9.86).

$$\hat{x}^{(I+1)} = \hat{x} - \left(H_\omega^T H_\omega\right)^{-1} H_{\omega,I+1}^T W_{\omega,I+1}^{-1} r_{\omega,I+1} \tag{9.86}$$

$$= \hat{x} - \left(H_\omega^T H_\omega\right)^{-1} \begin{bmatrix} H_{\omega,I} \\ H_{\omega,i+1} \end{bmatrix}^T \begin{bmatrix} W_{\omega,I}^{-1} + \dfrac{1}{C} bb^T & -\dfrac{1}{C} b \\ -\dfrac{1}{C} b^T & \dfrac{1}{C} \end{bmatrix} \begin{bmatrix} r_{\omega,I} \\ r_{\omega,i+1} \end{bmatrix}$$

$$= \hat{x} - \left(H_\omega^T H_\omega\right)^{-1} H_{\omega,I}^T \left(bb^T r_{\omega,I} - b r_{\omega,i+1}\right)/C - \left(H_\omega^T H_\omega\right)^{-1} H_{\omega,i+1}^T \left(r_{\omega,i+1} - b^T r_{\omega,I}\right)/C$$

Let

$$f = \left(r_{\omega,i+1} - b^T r_{\omega,I}\right)/C \tag{9.87}$$

Then

$$\hat{x}^{(I+1)} = \hat{x}^{(I)} - \left(H_\omega^T H_\omega\right)^{-1} H_{\omega,I+1}^T \begin{bmatrix} -b \\ 1 \end{bmatrix} \cdot f \tag{9.88}$$

The weighted residual $r_{\omega,i+1}^{(I+1)}$ is given by

$$r_{\omega,i+1}^{(I+1)} = \begin{bmatrix} W_{\omega,I}^{-1} + \dfrac{1}{C}bb^T & -\dfrac{1}{C}b \\ -\dfrac{1}{C}b^T & \dfrac{1}{C} \end{bmatrix} \begin{bmatrix} r_{\omega,I} \\ r_{\omega,i+1} \end{bmatrix}$$

$$= \begin{bmatrix} W_{\omega,I}^{-1}r_{\omega,I} + b(b^T r_{\omega,I} - r_{\omega,i+1})/C \\ (r_{\omega,i+1} - b^T r_{\omega,I})/C \end{bmatrix}$$

$$= -\begin{bmatrix} r_{\omega}^{(I)} - bf \\ f \end{bmatrix} \tag{9.89}$$

And the weighted residual $r_{\omega,M-I-1}^{(I+1)}$ is calculated by

$$r_{\omega,M-I-1}^{(I+1)} = r_{\omega,M-I-1} + H_{\omega,M-I-1}(H_{\omega}^T H_{\omega})^{-1}(H_{\omega,I}^T H_{\omega,i+1}^T)$$

$$\times \begin{bmatrix} W_{\omega,I}^{-1} + \dfrac{1}{C}bb^T & -\dfrac{1}{C}b \\ -\dfrac{1}{C}b^T & \dfrac{1}{C} \end{bmatrix} \begin{bmatrix} r_{\omega,I} \\ r_{\omega,i+1} \end{bmatrix}$$

$$= r_{\omega,M-I-1} + H_{\omega,M-I-1}(H_{\omega}^T H_{\omega})^{-1}f\begin{bmatrix} -b \\ 1 \end{bmatrix} \tag{9.90}$$

The weighted residual covariance $cov\left(r_{\omega,M-I-1}^{(I+1)}\right)$ is updated by

$$cov\left(r_{\omega,M-I-1}^{(I+1)}\right) = cov\left(r_{\omega,M-I+1}^{(I)}\right) - W_{\omega,M-I-1,I+1}\begin{bmatrix} b^{-1} & -1 \end{bmatrix}\begin{bmatrix} -b \\ 1 \end{bmatrix}W_{\omega,M-I,I-1}^T \tag{9.91}$$

Its $j$th term is

$$D\left(r_{\omega,j}^{(I+1)}\right) = D\left(r_{\omega,j}^{(I)}\right) - \left(\sum_{k=1}^{I} W_{\omega,j,k-b_k} - W_{\omega,j,i+1}\right) \tag{9.92}$$

In summary, the steps of the iterative estimate identification method are as follows:

1) Using weighted residuals and measurement change detection approach to determine and locate the initial suspicious measurement set $S_0$, and let $I = 0$;
2) Select the largest suspicious measurement data $i$, if $|r_{\omega,i}| < \varepsilon_\omega$, go to step (4); if $|r_{\omega,i}| \geq \varepsilon_\omega$, $I = I + 1$, go to step (3);
3) If $I = 1$, calculate $r_{\omega,j}^{(i)}$ ($j \in S_0$) by Eqs. (9.74) and (9.75); If $I \neq 1$, correct $r_{\omega,I}^{(I)}, r_{\omega,M-I}^{(I)}$ by Eqs. (9.79) and (9.80); In the correction process, if C is smaller than a predefined threshold, the measurement is identified and replaced by its prediction. Go to step (2);
4) Calculate system states $x$ by Eq. (9.72).

### 9.3.3 Bad Data Identification

When bad data are detected in the measurement set, they can be identified by further processing on their normalized residuals. The largest normalized residual for bad data is used to identify and eliminate the bad data.

#### 9.3.3.1 Normalized Residual

To calculate normalized residual, the measurement Eq. (9.1) is linearized as:

$$\Delta z = H\Delta x + e \tag{9.93}$$

Assuming that measurement errors are not correlated, then $E(e) = 0$ and $cov(e) = R$ is a diagonal matrix.

The WLS state estimation of the linearized state vector will be given by:

$$\Delta \hat{x} = \left(H^T R^{-1} H\right)^{-1} H^T R^{-1} \Delta z = G^{-1} H^T R^{-1} \Delta z \tag{9.94}$$

and the estimated value of $\Delta z$ is:

$$\Delta \hat{z} = H\Delta \hat{x} = HG^{-1} H^T R^{-1} \Delta z = K\Delta z \tag{9.95}$$

where $K = HG^{-1} H^T R^{-1}$ is called the hat matrix. The measurement residual can be expressed in terms of $K$:

$$r = \Delta z - \Delta \hat{z} = (I - K)\Delta z = (I - K)(H\Delta x + e) = (I - K)e \tag{9.96}$$

Note that $(I - K)H = 0$. And we define $S = (I - K)$ as the residual sensitivity matrix representing the sensitivity of the measurement residuals $r$ to the measurement errors $e$.

Applying the measurement sensitivity Eq. (9.96) and the assumption $E(e) = 0$, the mean value and covariance of the measurement residual can be obtained as follows:

$$E(r) = E[(I - K)e] = E(S \cdot e) = S \cdot E(e) = 0 \tag{9.97}$$

$$Cov(r) = E\left[rr^T\right] = S \cdot E\left[ee^T\right] \cdot S^T = SRS^T = SR = \Omega \tag{9.98}$$

Therefore, measurement residuals meet the following unbiased normal distribution:

$$r \sim N(0, \Omega) \tag{9.99}$$

The residual covariance matrix $\Omega$ is used to identify bad data.

#### 9.3.3.2   Largest Normalized Residual for Bad Data Identification

The algorithm of the largest normalized residual test for bad data identification is presented below.

---

**Algorithm 9.2   Largest Normalized Residual Test**

---

1: **Solve** the WLS estimation and calculate the measurement residuals

$$r^{(i)} = z^{(i)} - h\left(\hat{x}^{(i)}\right) \quad i = 1, 2, \ldots, m$$

2: **Compute** the normalized residuals $r_N^{(i)} = \frac{|r^{(i)}|}{\sqrt{\Omega_{ii}}} \quad i = 1, 2, \ldots, m$

3: **Find** the largest normalized residual $r_N^{(k)} = max\left(r_N^{(i)}\right) \quad i = 1, 2, \ldots, m$

4: **If** $r_N^{(k)} > c$ //c is a predefined threshold

5:        **Identify** whether the $k$-th measurement is bad data

6:        **Eliminate** the $k$-th measurement

7:        **Go to** Step 1

8: **Else**

9:        **Stop**

---

## 9.4   Graph-Based Bad Data Detection Implementation

The pseudo-code in graph query to implement the Chi-squares bad data detection is shown as follows:

```
SumAccum<double> @deltaP_bdt = 0
SumAccum<double> @deltaQ_bdt = 0
SumAccum<double> @@error = 0
SumAccum<double> @@NumofMeas = 0
double n = 0.0
double freedom_degree = 0
double BadDataIndex = 0.0
double p = 0.95
CREATE QUERY badDataDetection () FOR GRAPH busBranchGraph{
 T0 = {TopoND.*}
 T0 = SELECT s FROM T0:s-(connected:e)->:t
ACCUM
    s.@deltaP_bdt+=s.@Vm*t.@Vm*(-1*e.G/e.K*cos(s.@Vs-t.@Vs)+
                (e.B/e.K*sin(s.@Vs-t.@Vs))),
    s.@deltaQ_bdt+=s.@Vm*t.@Vm*(-1*e.G/e.K*sin(s.@Vs-t.@Vs)-
                (e.B/e.K*cos(s.@Vs-t.@Vs)))
    @@error+=(e.M_P_BR-(s.@Vm*s.@Vm*e.G-s.@Vm*t.@Vm*((e.G/e.K)*cos(s.
        @Vs-t.@Vs)+(-e.B/e.K)*sin(s.@Vs-t.@Vs))))*(e.M_P_BR-(s.@Vm*s.
        @Vm*e.G-s.@Vm*t.@Vm*((e.G/e.K)*cos(s.@Vs-t.@Vs)+(-e.B/e.K)
        *sin(s.@Vs-t.@Vs))))*e.Ri_eP*e.Ri_eP
```

```
    @@error+=(e.M_Q_BR-(-s.@Vm*s.@Vm*(-e.B+0.5*e.hB)-s.@Vm*t.@Vm*((e.
    G/e.K)*sin(s.@Vs-t.@Vs)-(-e.B/e.K)*cos(s.@Vs-t.@Vs))))*(e.
    M_Q_BR-(-s.@Vm*s.@Vm*(-e.B+0.5*e.hB)-s.@Vm*t.@Vm*((e.G/e.K)*sin
    (s.@Vs-t.@Vs)-(-e.B/e.K)*cos(s.@Vs-t.@Vs))))*e.Ri_eQ*e.Ri_eQ

    IF (e.Ri_eP > 0) THEN @@NumofMeas += 1 END
    IF (e.Ri_eQ > 0) THEN @@NumofMeas += 1 END

POST-ACCUM
    IF (s.Ri_vP > 0) THEN @@NumofMeas += 1 END
    IF (s.Ri_vQ > 0) THEN @@NumofMeas += 1 END
    s.@deltaP_bdt = s.@P - (s.@deltaP_bdt + s.@Vm*s.@Vm*s.@sumG)
    s.@deltaQ_bdt = s.@Q - (s.@deltaQ_bdt - s.@Vm*s.@Vm*s.@sumB)
    @@error += (s.M_Vm - s.@Vm) * (s.M_Vm - s.@Vm) * s.Ri_V * s.Ri_V
    @@error += s.@deltaP_bdt * s.@deltaP_bdt * s.Ri_vP * s.Ri_vP
    @@error += s.@deltaQ_bdt * s.@deltaQ_bdt * s.Ri_vQ * s.Ri_vQ

    n = T0.size()*2 - 1;
    freedom_degree = @@NumofMeas - n;
    BadDataIndex = chi_square_index(p, freedom_degree)

    IF (@@error >= BadDataIndex) THEN print "Bad Data Detected"
        ELSE print "No Bad Data Detected"
    END
}
```

Before the query `badDataDetection ()` is implemented, vertex-attached accumulator variables are defined with a single add-sign @ to store $J(x)$ and the number of measures.

In the query definition, all `TopoND` vertices in the graph are assigned to `T0` by the statement `T0 = {TopoND.*}` to be processed in parallel.

For all vertices, their edges e are selected starting from vertex s to the end vertex t by the following statement.

```
    T0 = SELECT s FROM T0:s-(connected:e)->:t
```

The attributes of these edges are used to calculate the sum of squares of the measurement errors in parallel by using the keyword `ACCUM`. In the `ACCUM` block, the number of measurements is also accumulated in parallel.

In the `POST-ACCUM` block, the sum of squares of the voltage measurement and zero injection errors are calculated. The degree of freedom is also calculated by the number of measurements minus the number of variables.

In the `POST-ACCUM` b, function `chi_square_index(p, freedom_degree)` is a look-up table returning the value $\chi^2_{m-n,p}$ with probability $p$ and degrees of freedom $m - n$. Statement `IF (@@error >= BadDataIndex)` THEN `print` "Bad Data Detected" tests if $J(x) \geq \chi^2_{m-n,p}$. If the statement `(@@error >= BadDataIndex)` is true, then bad data is detected.

# References

**1** Schweppe, F.C. and Rom, D.B. (1970). Power system static estimation part I,II, III. *IEEE Transactions on Power Apparatus and Systems* PAS-89 (1): 130–135.

**2** Wu, F.F. (1990). Power system state estimation: a survey. *International Journal of Electrical Power & Energy Systems* 12 (2): 80–87.

**3** Monticelli, A. (2000). Electric power system state estimation. *Proceedings of the IEEE* 88 (2): 262–282.

**4** Korres, G.N. (2011). A distributed multiarea state estimation. *IEEE Transactions on Power Apparatus and Systems* 26 (1): 73–84.

**5** Xie, L., Choi, D.H., Kar, S., and Poor, H.V. (2012). Fully distributed state estimation for wide-area monitoring systems. *IEEE Transactions on Smart Grid* 3 (3): 1154–1169.

**6** Rahman, M.A. and Venayagamoorthy, G.K. (2017). A hybrid method for power system state estimation using cellular computational network. *Engineering Applications of Artificial Intelligence* 64: 140–151.

**7** Kekatos, V. and Giannakis, G.B. (2013). Distributed robust power system state estimation. *IEEE Transactions on Power Apparatus and Systems* 28 (2): 1617–1626.

**8** Sasaki, H., Aoki, K., and Yokoyama, R. (1987). A parallel computation algorithm for static state estimation by means of matrix inversion lemma. *IEEE Power Engineering Review* PER-7 (8): 40–41.

**9** Rahman, M.A. and Venayagamoorthy, G.K. (2016). Power system distributed dynamic state prediction. *2016 IEEE Symposium Series on Computational Intelligence (SSCI)*, Athens, Greece, pp. 1–7. https://doi.org/10.1109/SSCI.2016.7849847.

**10** Karimipour, H. and Dinavahi, V. (2016). Parallel relaxation-based joint dynamic state estimation of large-scale power systems. *IET Generation Transmission and Distribution* 10 (2): 452–459.

**11** Dai, R., Liu, G., Yuan, C. et al. Systems and methods for hybrid dynamic state estimation. US Patent, US 11,119,462 B2.

**12** Abur, A. and Exposito, A.G. (2004). *Power System State Estimation: Theory and Implementation*. CRC Press.

**13** Brice, C.W. and Cavin, R.K. (1982). Multiprocessor static state estimation. *IEEE Transactions on Power Apparatus and Systems* PAS-101 (2): 302–308.

**14** Korres, G.N. (2011). A distributed multiarea state estimation. *IEEE Transactions on Power Apparatus and Systems* 26 (1): 73–84.

**15** Xie, L., Choi, D.H., Kar, S., and Poor, H.V. (2012). Fully distributed state estimation for wide-area monitoring systems. *IEEE Transactions on Smart Grid* 3 (3): 1154–1169.

**16** Abur, A. and Tapadiya, P. (1990). Parallel state estimation using multiprocessors. *Electric Power Systems Research* 18 (1): 67–73.

**17** Kekatos, V. and Giannakis, G.B. (2013). Distributed robust power system state estimation. *IEEE Transactions on Power Apparatus and Systems* 28 (2): 1617–1626.

**18** Sasaki, H., Aoki, K., and Yokoyama, R. (1987). A parallel computation algorithm for static state estimation by means of matrix inversion lemma. *IEEE Power Engineering Review* PER-7 (8): 40–41.

**19** Chen, Y., Jin, S., Rice, M., and Huang, Z. (2013). Parallel state estimation assessment with practical data. *IEEE Power and Energy Society General Meeting* Vancouver, BC, Canada, pp. 1–5. https://doi.org/10.1109/PESMG.2013.6672742.

**20** Minot, A., Lu, Y.M., and Li, N. (2016). A distributed Gauss-Newton method for power system state estimation. *IEEE Transactions on Power Apparatus and Systems* 31 (5): 3804–3815.

**21** Seshadri Sravan Kumar, V. and Dhadbanjan, T. (2013). State estimation in power systems using linear model infinity norm-based trust region approach. *IET Generation Transmission and Distribution* 7 (5): 500–510.

**22** Karimipour, H. and Dinavahi, V. (2016). Parallel relaxation-based joint dynamic state estimation of large-scale power systems. *IET Generation Transmission and Distribution* 10 (2): 452–459.

**23** Yuan, C., Zhou, Y., Liu, G. et al. (2020). Graph computing-based WLS fast decoupled state estimation. *IEEE Transactions on Smart Grid* 11 (3): 2440–2451.

**24** Slutsker, I.W. (1989). Bad data identification in power system state estimation based on measurement compensation and linear residual calculation. *IEEE Transactions on Power Systems* 4 (1): 53–60.

**25** Clements, K.A. and Davis, P.W. (1986). Multiple bad data detectability and identifiability, a geometric approach. *IEEE Power Engineering Review* PER-6 (7): 73–73.

**26** Abur, A. (1990). A bad data identification method for linear programming state estimation. *IEEE Transactions on Power Systems* 5 (3): 894–901.

**27** Abur, A. and Exposito, A.G. (1993). Observability and bad data identification when using ampere measurements in state estimation. *1993 IEEE International Symposium on Circuits and Systems*, Chicago, IL, USA, vol. 4, pp. 2668–2671836. https://doi.org/10.1109/ISCAS.1993.394315.

**28** Abur, A. and Gomez Exposito, A. (1997). Bad data identification when using ampere measurements. *IEEE Transactions on Power Systems* 12 (2): 831–836.

**29** Baldick, R., Clements, K.A., Pinjo-Dzigal, Z., and Davis, P.W. (1997). Implementing nonquadratic objective functions for state estimation and bad data rejection. *IEEE Transactions on Power Systems* 12 (1): 376–382.

**30** Rousseeuw, P.J. and Leroy, A.M. (1987). *Robust Regression and Outlier Detection*, Wiley Series in Probability and Mathematical Statistics. Wiley.

**31** Abur, A. and Exposito, A.G. (2004). *Power System State Estimation: Theory and Implementation*. New York, NY: Marcel Dekker.

# 10

# Power Flow Graph Computing

Power flow calculation is one critical underlying problem for power system analysis. The solution of power flow shows the voltage at every bus and real and reactive power flows on branches in the system. The intention of power flow calculation is to obtain bus voltage magnitude and angle information. Once the voltage information is known, the active power and reactive power flow on each branch can be analytically determined.

Power flow calculation is a well-known application in power system analysis. It takes a key role in power system long-term planning, operational planning, and real-time monitoring and control. Although many researchers have been working on achieving fast and accurate power flow calculation, the high computation performance of power flow calculation is still a challenge for modern highly interconnected, large, and complex power systems. In this chapter, power flow graph computing is discussed and implemented on classic Gauss–Seidel, Newton–Raphson, fast decoupled, and conjugate gradient methods in parallel to improve the power flow calculation efficiency.

## 10.1 Power Flow Mathematical Model

Generally, for a power system with $n$ independent buses, we can write the node voltage equation as follow:

$$\begin{bmatrix} \dot{I}_1 \\ \dot{I}_2 \\ \vdots \\ \dot{I}_n \end{bmatrix} = \begin{bmatrix} \dot{Y}_{11} \dot{Y}_{12} \cdots \dot{Y}_{1n} \\ \dot{Y}_{21} \dot{Y}_{22} \cdots \dot{Y}_{2n} \\ \vdots \ \vdots \ \ddots \ \vdots \\ \dot{Y}_{n1} \dot{Y}_{n2} \cdots \dot{Y}_{nn} \end{bmatrix} \cdot \begin{bmatrix} \dot{V}_1 \\ \dot{V}_2 \\ \vdots \\ \dot{V}_n \end{bmatrix} \tag{10.1}$$

The injection current at the bus $i$ can be expressed as follows:

$$\dot{I}_i = \sum_{j=1}^{n} \dot{Y}_{ij} \cdot \dot{V}_j \tag{10.2}$$

In a power system, usually injection power other than injection current at a bus is modeled. To meet power system convention, the complex power injection at bus $i$ is expressed in terms of the bus voltage and the bus injection current as follows:

$$\dot{S}_i = P_i + jQ_i = \dot{V}_i \cdot \dot{I}_i^* \tag{10.3}$$

where, $P_i$ and $Q_i$ are the active power and reactive power injected into bus $i$, respectively.

*Graph Database and Graph Computing for Power System Analysis*, First Edition. Renchang Dai and Guangyi Liu.
© 2024 The Institute of Electrical and Electronics Engineers, Inc. Published 2024 by John Wiley & Sons, Inc.

In order to solve the power flow problem, by substituting injection current (10.2) into (10.3), we build the following bus voltage equation to represent the relationship among bus power, voltage, and admittance matrix.

$$\dot{S}_i^* = P_i - jQ_i = \dot{V}_i^* \cdot \sum_{j=1}^{n}\left(Y_{ij} \cdot \dot{V}_j\right) \tag{10.4}$$

$$\dot{S}_i^* = \dot{V}_i^* \cdot \left[Y_{i1} \cdot \dot{V}_1 + Y_{i2} \cdot \dot{V}_2 + \cdots + Y_{ii} \cdot \dot{V}_i + \cdots + Y_{in} \cdot \dot{V}_n\right] \tag{10.5}$$

$$\dot{V}_i = \frac{1}{Y_{ii}} \cdot \left[\frac{P_i - jQ_i}{\dot{V}_i^*} - \sum_{\substack{j=1 \\ j \neq i}}^{n} Y_{ij} \cdot \dot{V}_j\right] \tag{10.6}$$

Equation (10.6) is the basic equation for power flow calculation. In the power system, the variables representing the operation status of each bus are the voltage phasor and the complex power at the corresponding bus, which means each bus has four variables that represent the operating status of the bus. Therefore, the $n$-bus power system has $4n$ variables. Since Eq. (10.6) is a complex equation, it can be separated into real and imaginary parts. The $n$ complex equations are equivalent to $2n$ real number equations. To solve the power flow equations, two of the four variables at each bus should be known. According to the known variables, buses are typically categorized into the following three types:

PV Bus: The active power $P$ and the voltage magnitude $V$ are known as the PV bus. Reactive power $Q$ and voltage angle $\theta$ are unknown. Generally, PV buses are generator buses and others are voltage-controlled buses.

PQ Bus: The active power $P$ and the reactive power $Q$ are known as the PQ bus. Voltage magnitude $V$ and voltage angle $\theta$ are unknown. Generally, PQ buses are load buses and other buses are those whose voltage is not directly under control.

Slack Bus: The slack bus is also called the swing bus or the reference bus. In power flow calculation, at least one bus must be the slack bus to balance the system's active power and reactive power. Generally, the slack bus is a generator bus whose voltage magnitude $V$ and voltage angle $\theta$ are known as references. Its active power $P$ and reactive power $Q$ are unknown variables. A practical power flow application usually supports multiple slack buses in which selected slack buses are used to balance the power mismatch weighted on their participation factors or other defined rules. Without loss of generality, we will investigate a single slack bus case only in power flow graph computing.

In power flow calculation, bus voltage magnitude and voltage angle are defined as dependent variables. Since voltage magnitude and voltage angle at the slack bus are given, the remained $n-1$ bus voltage magnitude and voltage angle need to be calculated.

## 10.2 Gauss–Seidel Method

The Gauss–Seidel method is an iterative method to solve a system of linear or nonlinear equations. Applying the Gauss–Seidel method, a mathematical model of the power flow problem (10.6) is solved iteratively by the following Eq. (10.7).

$$\dot{V}_i^{(k+1)} = \frac{1}{Y_{ii}} \cdot \left[ \frac{P_i - jQ_i^{(k)}}{\dot{V}_i^{(k)*}} - \sum_{\substack{j=1 \\ j \neq i}}^{n} Y_{ij} \cdot \dot{V}_j^{(k)} \right] \tag{10.7}$$

For the PQ bus, the active power and reactive power are known. The voltage can be updated by (10.7). For the PV bus, the bus's active power and the voltage magnitude are known. The bus reactive power is calculated in each iteration by:

$$Q_i^{(k)} = \mathrm{Im}\left[ \dot{V}_i^{(k)} \dot{I}_i^{(k)*} \right] = \mathrm{Im}\left[ \dot{V}_i^{(k)} \sum_{\substack{j=1 \\ j \neq i}}^{n} Y_{ij}^* \cdot \dot{V}_j^{(k)*} \right] \tag{10.8}$$

Assuming bus $n$ is a slack bus. Its voltage $\dot{V}_n$ is given. In (10.7), the values of the bus $i$ voltage $\dot{V}_i$ at the $k$-th iteration are substituted into the right side of the iteration equation to get the updated values of the voltage $\dot{V}_i$ at the $(k+1)$-th iteration until the following convergence conditions are satisfied for all variables:

$$\left| \dot{V}_i^{(k+1)} - \dot{V}_i^{(k)} \right| < \varepsilon, i = 1, 2, ..., n-1 \tag{10.9}$$

At the $(k+1)$-th iteration, since all voltage $\dot{V}_i^{(k)}$ are known, the iteration Eq. (10.7) can be simply calculated in parallel. And the graph-based Gauss–Seidel power flow calculation algorithm is presented as follows:

---

**Algorithm 10.1   Graph-Based Gauss–Seidel Power Flow Calculation**

---

1:  **Build a** power system graph
2:  **Initialize** bus voltage $\dot{V}_i^{(0)}$, k = 0
3:  **While** $\left( \left| \dot{V}_i^{(k+1)} - \dot{V}_i^{(k)} \right| > \varepsilon \right)$ **do**
4:      **Update** $\dot{V}_i^{(k+1)}$ by (10.7) //in nodal parallel
5:      k = k + 1

---

To implement the graph-based Gauss–Seidel power flow calculation, the defined bus-branch model graph schema in Chapter 8 is used to support power flow graph computing.

In the bus-branch model, vertex TopoND and edge TopoConnect are defined as follows containing the information for power flow calculation:

```
CREATE VERTEX TopoND (PRIMARY_ID topoid int, TOPOID int, island int,
     bus_name string, area string, loss_zone uint, busType int, Vm
     double, Va double, M_Vm double, M_Va double, base_kV double,
     desired_volts double, control_bus_number int, up_V double, lo_V
     double, GenP double, GenQ double, M_Gen_P double, M_Gen_Q double,
     qUp double, qLower double, LdP double, LdQ double, P double,
     Q double, M_Load_P double, M_Load_Q double, G double, B double,
```

```
    Sub string, OV int, UV int, sumB double, sumG double, sumBi
    double, volt double, double, SE_Vm double, SE_Va double, M_C_P
    double)
```

```
CREATE DIRECTED EDGE TopoConnect (from TopoND, to TopoND, typename
    string, edge_name List<string>, area List<string>, zone
    List<string>, from_bus int, to_bus int, flag uint, R List<double>,
    X List<double>, hB_list List<double>, line_Q1_list List<double>,
    line_Q2_list List<double>, line_Q3_list List<double>, control_bus
    int, transformer_turns_ratio List<double>,
    shifting_transformer_angle List<double>, min_tap double, max_tap
    double, step_size double, min_volt List<double>, max_volt
    List<double>, M_P_BR List<double>, M_Q_BR List<double>,
    G List<double>, B List<double>, BIJ List<double>, circuit uint,
    P_BR double, Q_BR double, from_open List<int>, to_open List<int>)
```

The annotation of the vertex and edge attributes are listed in Appendix A. With the defined vertex and edge above, the bus-branch graph schema is then simply defined as:

```
CREATE GRAPH busBranchGraph(TopoND, TopoConnect)
```

The bus-branch graph schema defined the graph structure. The bus-branch graph will be formed by adding `TopoND` and `TopoConnect` through topology processing or by reading available raw data files which define the bus-branch model. Taking the IEEE-14 bus system for instance, the bus information file busInfo.csv and the branch information file branchInfo.csv are given in Tables 10.1 and 10.2.

All values in Tables 10.1 and 10.2 are in per unit and the bus voltage angle is in degrees. Buses in type 0 or 1 are PQ buses. Buses in type 2 are PV buses. Type 3 bus is the slack bus. The following loading job statement will load the data into the bus-branch graph.

```
CREATE LOADING JOB loadingVertexandEdgeInfo FOR GRAPH busBranchGraph {
    DEFINE FILENAME BusInfo = "/data/ busInfo.csv";
    DEFINE FILENAME BranchInfo = "/data/ branchInfo.csv";
    LOAD BusInfo TO VERTEX TopoND VALUES(_,$"Bus ID",_,_,
        $"Ares",$"Loss Zone",$"Type",$"Voltage",$"Angle",_,_,_,
        $"Desired Voltage",_,_"_",$"Unit_P",$"Unit_Q",_,_,
        $"MAX Q",$"MIN Q",$"Load_P",$"Load_Q",_,_,_,_,$"G",$"B",
        _,_,_,_,_,_,_,_,_);
    LOAD BranchInfo TO EDGE TopoConnect VALUES($"From Bus",$"To Bus",_,
        $"Ares",$"Zone",$"From Bus",$"To Bus",_, $"R",$"X",$"B",
        $"Line Q1",$"Line Q2",$"Line Q3",_,$"Turns Ratio",_,_,_,_,_,_,
        _,_,_,_,_,_,_,_,_);
}
```

In the LOAD statement, the values are read from the CSV file and added to the vertex or edge attributes in the sequence when they are defined in the schema. When the CVS file is read, the values are matched by the table header in the format of $"table header", and the default value is applied at the position of "_".

**Table 10.1** IEEE-14 bus system bus table.

| Bus ID | Area | Loss zone | Type | Voltage | Angle | Load_P | Load_Q | Unit_P | Unit_Q | Desired voltage | MAX Q | MIN Q | G | B |
|--------|------|-----------|------|---------|--------|--------|--------|--------|--------|-----------------|-------|-------|---|---|
| 1 | 1 | 1 | 3 | 1.060 | 0 | 0 | 0 | 2.324 | −0.169 | 1.06 | 0 | 0 | 0 | 0 |
| 2 | 1 | 1 | 2 | 1.045 | −4.98 | 0.217 | 0.127 | 0.4 | 0.424 | 1.045 | 0.50 | −0.4 | 0 | 0 |
| 3 | 1 | 1 | 2 | 1.010 | −12.72 | 0.942 | 0.190 | 0 | 0.234 | 1.01 | 0.40 | 0 | 0 | 0 |
| 4 | 1 | 1 | 0 | 1.019 | −10.33 | 0.478 | −0.039 | 0 | 0 | 0 | 0 | 0 | 0 | 0 |
| 5 | 1 | 1 | 0 | 1.020 | −8.78 | 0.076 | 0.016 | 0 | 0 | 0 | 0 | 0 | 0 | 0 |
| 6 | 1 | 1 | 2 | 1.070 | −14.22 | 0.112 | 0.075 | 0 | 0.122 | 1.07 | 0.24 | −0.6 | 0 | 0 |
| 7 | 1 | 1 | 0 | 1.062 | −13.37 | 0 | 0 | 0 | 0 | 0 | 0 | 0 | 0 | 0 |
| 8 | 1 | 1 | 2 | 1.090 | −13.36 | 0 | 0 | 0 | 0.174 | 1.09 | 0.24 | −0.6 | 0 | 0 |
| 9 | 1 | 1 | 0 | 1.056 | −14.94 | 0.295 | 0.166 | 0 | 0 | 0 | 0 | 0 | 0 | 0.19 |
| 10 | 1 | 1 | 0 | 1.051 | −15.10 | 0.090 | 0.058 | 0 | 0 | 0 | 0 | 0 | 0 | 0 |
| 11 | 1 | 1 | 0 | 1.057 | −14.79 | 0.035 | 0.018 | 0 | 0 | 0 | 0 | 0 | 0 | 0 |
| 12 | 1 | 1 | 0 | 1.055 | −15.07 | 0.061 | 0.016 | 0 | 0 | 0 | 0 | 0 | 0 | 0 |
| 13 | 1 | 1 | 0 | 1.050 | −15.16 | 0.135 | 0.058 | 0 | 0 | 0 | 0 | 0 | 0 | 0 |
| 14 | 1 | 1 | 0 | 1.036 | −16.04 | 0.149 | 0.050 | 0 | 0 | 0 | 0 | 0 | 0 | 0 |

**Table 10.2** IEEE-14 bus system branch table.

| From bus | To bus | Area | Zone | Circuit | R | X | B | Line Q1 | Line Q2 | Line Q3 | Turns ratio |
|---|---|---|---|---|---|---|---|---|---|---|---|
| 1 | 2 | 1 | 1 | 1 | 0.01938 | 0.05917 | 0.0528 | 120 | 120 | 120 | 0 |
| 1 | 5 | 1 | 1 | 1 | 0.05403 | 0.22304 | 0.0492 | 65 | 65 | 65 | 0 |
| 2 | 3 | 1 | 1 | 1 | 0.04699 | 0.19797 | 0.0438 | 36 | 36 | 36 | 0 |
| 2 | 4 | 1 | 1 | 1 | 0.05811 | 0.17632 | 0.034 | 65 | 65 | 65 | 0 |
| 2 | 5 | 1 | 1 | 1 | 0.05695 | 0.17388 | 0.0346 | 50 | 50 | 50 | 0 |
| 3 | 4 | 1 | 1 | 1 | 0.06701 | 0.17103 | 0.0128 | 65 | 65 | 65 | 0 |
| 4 | 5 | 1 | 1 | 1 | 0.01335 | 0.04211 | 0 | 45 | 45 | 45 | 0 |
| 4 | 7 | 1 | 1 | 1 | 0 | 0.20912 | 0 | 55 | 55 | 55 | 0.978 |
| 4 | 9 | 1 | 1 | 1 | 0 | 0.55618 | 0 | 32 | 32 | 32 | 0.969 |
| 5 | 6 | 1 | 1 | 1 | 0 | 0.25202 | 0 | 45 | 45 | 45 | 0.932 |
| 6 | 11 | 1 | 1 | 1 | 0.09498 | 0.1989 | 0 | 18 | 18 | 18 | 0 |
| 6 | 12 | 1 | 1 | 1 | 0.12291 | 0.25581 | 0 | 32 | 32 | 32 | 0 |
| 6 | 13 | 1 | 1 | 1 | 0.06615 | 0.13027 | 0 | 32 | 32 | 32 | 0 |
| 7 | 8 | 1 | 1 | 1 | 0 | 0.17615 | 0 | 32 | 32 | 32 | 0 |
| 7 | 9 | 1 | 1 | 1 | 0 | 0.11001 | 0 | 32 | 32 | 32 | 0 |
| 9 | 10 | 1 | 1 | 1 | 0.03181 | 0.0845 | 0 | 32 | 32 | 32 | 0 |
| 9 | 14 | 1 | 1 | 1 | 0.12711 | 0.27038 | 0 | 32 | 32 | 32 | 0 |
| 10 | 11 | 1 | 1 | 1 | 0.08205 | 0.19207 | 0 | 12 | 12 | 12 | 0 |
| 12 | 13 | 1 | 1 | 1 | 0.22092 | 0.19988 | 0 | 12 | 12 | 12 | 0 |
| 13 | 14 | 1 | 1 | 1 | 0.17093 | 0.34802 | 0 | 12 | 12 | 12 | 0 |

When the power system graph is created and the raw data are loaded to the vertex and edge attributes, the pseudo-code in graph query can be implemented to form the nodal admittance as shown as follow:

```
CREATE QUERY formAdmittanceMatrix(List<double> initial_Vm,
List<double> initial_Va
    ) FOR GRAPH busBranchGraph{
SumAccum<double> @sumG = 0
SumAccum<double> @sumB = 0

T0 = {TopoND.*}
T0.Vm  = initial_Vm
T0.Va  = initial_Va
T1 = SELECT v FROM T0:v
 ACCUM
   v.P = (v.GenP - v.LdP)
   v.Q = (v.GenQ - v.LdQ)
   v.@sumG = v.G
   v.@sumB = v.B
```

```
T2 = SELECT v FROM T1:v-(connected:e)->:t
   ACCUM
     IF (e.K == 0) THEN
        v.@sumG += e.G
        v.@sumB += -1*e.B + 0.5*e.hB
     ELSE
          v.@sumG += 1/(e.K*e.K)* e.G
          v.@sumB += 1/(e.K*e.K)*(-1*e.B) + 0.5*e.hB
     END
}
```

In the query `formAdmittanceMatrix`, initial voltage magnitude and angle `initial_Vm` and `initial_Va` are passed. Vertex-attached `SumAccum` `@sumG`, `@sumB` are defined to store nodal conductance and susceptance (equivalent to diagonal elements in the admittance matrix).

In the query definition, graph traversal statement `T1 = SELECT v FROM T0:v` coupling with keyword `ACCUM` travels all vertices `v`, calculate the total injection active power `v.P` and reactive `v.Q`.

For all vertices, their edges `e` are selected starting from vertex `v` to the end vertex `t` by the following statement.

```
T2 = SELECT v FROM T1:v-(connected:e)->:t
```

The attributes of these edges are used to calculate the contribution of each edge to the diagonal terms in the nodal admittance matrix in parallel by using the keyword `ACCUM`. Nested in `ACCUM`, statements for transmission lines

```
v.@sumG += e.G
v.@sumB += -1*e.B + 0.5*e.hB
```

and for transformers

```
v.@sumG += 1/(e.K*e.K)* e.G
v.@sumB += 1/(e.K*e.K)*(-1*e.B) + 0.5*e.hB
```

calculate contribution to the diagonal conductance and susceptance.

When diagonal conductance and susceptance are calculated for all vertices, the Gauss–Seidel iterations are implemented by the following pseudo-code.

```
CREATE QUERY powerFlowbyGaussSeidel(int MaxIteration, double Tolerance
     ) FOR GRAPH busBranchGraph{
MaxAccum<double> @@maxDeltVx = -INF
MaxAccum<double> @@maxDeltVy = -INF
MaxAccum<double> @@PI = 3.1415926535898
SumAccum<double> @Vreal = 0
SumAccum<double> @Vimag = 0
SumAccum<double> @Vreal0 = 0
SumAccum<double> @Vimag0 = 0
SumAccum<double> @sum_Ix = 0
SumAccum<double> @sum_Iy = 0
```

```
WHILE (abs(@@maxDeltVx)>Tolerance or abs(@@maxDeltVy)>Tolerance)
    LIMIT MaxIteration DO
    T0 = {TopoND.*}
    T1 = SELECT s FROM T0:s -(connected:e)->:t
      ACCUM
        s.@Vreal = s.Vm*cos(s.Va*@@PI/180)
        s.@Vimag = s.Vm*sin(s.Va*@@PI/180)
        IF ( e.K == 0 ) THEN e.K == 1 END
        s.@sum_Ix += e.G/e.K * s.@Vreal - e.B/e.K * s.@Vimag
        s.@sum_Iy += e.G/e.K * s.@Vimag + e.B/e.K * s.@Vreal


      POST-ACCUM
        s.@Vreal0 = s.@Vreal
        s.@Vimag0 = s.@Vimag
        IF (s.busType == 0 or s.busType == 1) THEN
          double sumIx = (s.P*s.@Vreal+s.Q*s.@Vimag)/(s.Vm*s.Vm)-s.@sum_Ix
          double sumIy = (s.P*s.@Vimag-s.Q*s.@Vreal)/(s.Vm*s.Vm)-s.@sum_Iy
          s.@Vreal = (sumIx*s.@sumG+sumIy*s.@sumB)/
                     (s.@sumG*s.@sumG+s.@sumB*s.@sumB)
          s.@Vimag = (sumIy*s.@sumG-sumIx*s.@sumB)/
                     (s.@sumG*s.@sumG+s.@sumB*s.@sumB)
        ELSE IF (s.busType == 2) THEN
          double sumIx = (s.@Vreal*s.@sumG+s.@Vimag*s.@sumB)/
                         (s.@sumG*s.@sumG+s.@sumB*s.@sumB)-s.@sum_Ix
          double sumIy = (s.@Vimag*s.@sumG-s.@Vreal*s.@sumB)/
                         (s.@sumG*s.@sumG+s.@sumB*s.@sumB)-s.@sum_Iy
          s.Q = s.@Vimag * sumIx - s.@Vreal * sumIy
          IF (s.Q > s.qUp) THEN
            s.Q = s.qUp
            s.busType = 1
          ELSE IF (s.Q < s.qLower) THEN
            s.Q = s.qLower
            s.busType = 1
          END
        double sumIx = (s.P*s.@Vreal+s.Q*s.@Vimag)/(s.Vm*s.Vm)-s.@sum_Ix
        double sumIy = (s.P*s.@Vimag-s.Q*s.@Vreal)/(s.Vm*s.Vm)-s.@sum_Iy
        s.@Vreal = (sumIx*s.@sumG+sumIy*s.@sumB)/
                   (s.@sumG*s.@sumG+s.@sumB*s.@sumB)
        s.@Vimag = (sumIy*s.@sumG-sumIx*s.@sumB)/
                   (s.@sumG*s.@sumG+s.@sumB*s.@sumB)
      END
    END
    @@maxDeltVx = max(abs(s.@Vreal - s.@Vreal0))
    @@maxDeltVy = max(abs(s.@Vimag - s.@Vimag0))
    s.@sum_Ix = 0
    s.@sum_Iy = 0
END
```

```
T1 = SELECT s FROM T0
 ACCUM
   s.Vm = s.@Vreal * s.@Vreal + s.@Vimag * s.@Vimag
   s.Va = atan2(s.@Vimag, s.@Vreal) * 180/@@PI
}
```

In the query `powerFlowbyGaussSeidel`, maximum iteration number `MaxIteration` and voltage tolerance `Tolerance` are passed through. Global variables `@@maxDeltVx` and `@@maxDeltVy` are defined to store maximum voltage updates in each iteration. `@@PI` is defined as a constant. Vertex-attached variables `SumAccum @Vreal, @Vimag` are defined to store bus voltage in rectangular coordinates. Vertex-attached variables are associated with vertex functioning as vertex attributes in the graph database temporarily. Their memory allocations are released after the query execution to save memory. An alternative way to define vertex-attached variables is to define them as vertex attributes stored in the graph for other application use at the cost of memory. Similarly, the vertex-attached `SumAccum @Vreal0, @Vimag0` are defined to store the bus voltage in the previous iteration to check the voltage convergence by comparing the voltage updates against the tolerance. `SumAccum @sum_Ix, @sum_Iy` are temporary vertex-attached variables defined to store the total branch current injections at the bus.

The `WHILE` loop statement is straightforward which is checking the maximum voltage updates against `Tolerance` with the `LIMIT` clause which counts the iterations in the `WHILE` loop against the maximum iteration number `MaxIteration`.

```
WHILE (@@maxDeltVx>Tolerance or @@maxDeltVy>Tolerance)
      LIMIT MaxIteration DO
```

In the query definition, graph traversal statement `T1 = SELECT t FROM T0:s-(connected: e)->:t` coupling with keyword `ACCUM` travels all edges `e` from vertex `s` to the end vertex `t`, calculates the total injection current `s.@sum_Ix` and `s.@sum_Iy` in nodal parallel.

In the `POST-ACCUM`, current bus voltages are saved by

```
s.@Vreal0 = s.@Vreal
s.@Vimag0 = s.@Vimag
```

And PQ bus voltage is updated by

```
double sumIx = (s.P*s.@Vreal+s.Q*s.@Vimag)/(s.Vm*s.Vm)-
               s.@sum_Ix
double sumIy = (s.P*s.@Vimag-s.Q*s.@Vreal)/(s.Vm*s.Vm)-
               s.@sum_Iy
s.@Vreal = (sumIx*s.@sumG+sumIy*s.@sumB)/
           (s.@sumG*s.@sumG+s.@sumB*s.@sumB)
s.@Vimag = (sumIy*s.@sumG-sumIx*s.@sumB)/
           (s.@sumG*s.@sumG+s.@sumB*s.@sumB)
```

PV bus reactive power injection is calculated by

```
double sumIx = (s.@Vreal*s.@sumG+s.@Vimag*s.@sumB)/
               (s.@sumG*s.@sumG+s.@sumB*s.@sumB)-s.@sum_Ix
double sumIy = (s.@Vimag*s.@sumG-s.@Vreal*s.@sumB)/
               (s.@sumG*s.@sumG+s.@sumB*s.@sumB)-s.@sum_Iy
s.Q = s.@Vimag * sumIx - s.@Vreal*sumIy
```

PV bus reactive power injection is compared against the reactive power limits. When the reactive power injection is over the limit, we switch the type of the bus.

```
IF (s.Q > s.qUp) THEN
    s.Q = s.qUp
    s.busType = 1
ELSE IF (s.Q < s.qLower) THEN
    s.Q = s.qLower
    s.busType = 1
END
```

At the end of the `WHILE` loop, the maximum voltage updates of the iteration are calculated:

```
@@maxDeltVx = max(abs(s.@Vreal - s.@Vreal0))
@@maxDeltVy = max(abs(s.@Vimag - s.@Vimag0))
```

After the iteration is converged or reached the maximum iteration number, the bus voltage is converted from rectangular coordinates to polar coordinates and saved as bus attributes in nodal parallel by using the keyword `ACCUM`:

```
T1 = SELECT s FROM T0
    ACCUM
        s.Vm = sqrt(s.@Vreal * s.@Vreal + s.@Vimag * s.@Vimag)
        s.Va = atan2(s.@Vimag, s.@Vreal) * 180/@@PI
```

## 10.3  Newton–Raphson Method

In the polar coordinate system, the bus voltage, active and reactive power are written as follows:

$$\dot{V}_i = V_i(\cos\theta_i + j\sin\theta_i) \tag{10.10}$$

$$P_i = V_i \sum_{j=1}^{n} V_j \left( G_{ij}\cos\theta_{ij} + B_{ij}\sin\theta_{ij} \right) \tag{10.11}$$

$$Q_i = V_i \sum_{j=1}^{n} V_j \left( G_{ij}\sin\theta_{ij} - B_{ij}\cos\theta_{ij} \right) \tag{10.12}$$

where $V_i$ and $V_j$ are voltage magnitude at bus $i$ and bus $j$; $\theta_{ij} = \theta_i - \theta_j$ is the voltage angle difference between the $i$th and $j$th buses; $G_{ij}$ and $B_{ij}$ are the real and imaginary parts of the element in the bus admittance matrix $Y_{BUS}$ corresponding to the $i$th row and $j$th column.

Without loss of generality, assuming in the system, the first $m$ buses are PQ buses, the $n$th bus is the slack bus, and the reminding buses are PV buses. Active power and reactive power at each bus are balanced. In mathematics, for each PV and PQ bus, the active power mismatch equation is shown as follows:

$$\Delta P_i = P_i - V_i \sum_{j=1}^{n} V_j \left( G_{ij}\cos\theta_{ij} + B_{ij}\sin\theta_{ij} \right) \tag{10.13}$$

For each PQ bus, we have the following reactive power mismatch equation:

$$\Delta Q_i = Q_i - V_i \sum_{j=1}^{n} V_j \left( G_{ij} \sin \theta_{ij} - B_{ij} \cos \theta_{ij} \right) \tag{10.14}$$

where $P_i$ and $Q_i$ are the net active and reactive power injections at bus $i$.

There are several different methods to solve the nonlinear equations. The well-known Newton–Raphson method linearizes Eqs. (10.13) and (10.14) using a Taylor Series and keeping first-order approximation as:

$$\begin{bmatrix} \Delta P \\ \Delta Q \end{bmatrix} = -J \begin{bmatrix} \Delta \theta \\ \Delta V/_V \end{bmatrix} \tag{10.15}$$

$$J = \begin{bmatrix} \dfrac{\partial \Delta P}{\partial \theta} & V \dfrac{\partial \Delta P}{\partial V} \\ \dfrac{\partial \Delta Q}{\partial \theta} & V \dfrac{\partial \Delta Q}{\partial V} \end{bmatrix} = \begin{bmatrix} H & N \\ K & L \end{bmatrix} \tag{10.16}$$

Where, $J$ is the Jacobian matrix.

$$\Delta Q = \begin{bmatrix} \Delta Q_1 & \Delta Q_2 & \cdots & \Delta Q_m \end{bmatrix}^T \tag{10.17}$$

$$\Delta \theta = \begin{bmatrix} \Delta \theta_1 & \Delta \theta_2 & \cdots & \Delta \theta_{n-1} \end{bmatrix}^T \tag{10.18}$$

$$\Delta V = \begin{bmatrix} \Delta V_1 & \Delta V_2 & \cdots & \Delta V_m \end{bmatrix}^T \tag{10.19}$$

$$H = \begin{bmatrix} H_{1,1} & \cdots & H_{1,n-1} \\ \vdots & \ddots & \vdots \\ H_{n-1,1} & \cdots & H_{n-1,n-1} \end{bmatrix} \tag{10.20}$$

$$N = \begin{bmatrix} N_{1,1} & \cdots & N_{1,m} \\ \vdots & \ddots & \vdots \\ N_{n-1,1} & \cdots & N_{n-1,m} \end{bmatrix} \tag{10.21}$$

$$K = \begin{bmatrix} K_{1,1} & \cdots & K_{1,n-1} \\ \vdots & \ddots & \vdots \\ K_{m,1} & \cdots & K_{m,n-1} \end{bmatrix} \tag{10.22}$$

$$L = \begin{bmatrix} L_{1,1} & \cdots & L_{1,m} \\ \vdots & \ddots & \vdots \\ L_{m,1} & \cdots & L_{m,m} \end{bmatrix} \tag{10.23}$$

When $i \neq j$,

$$H_{ij} = \frac{\partial \Delta P_i}{\partial \theta_j} = -V_i V_j \left( G_{ij} \sin \theta_{ij} - B_{ij} \cos \theta_{ij} \right) \tag{10.24}$$

$$N_{ij} = V_j \frac{\partial \Delta P_i}{\partial V_j} = -V_i V_j \left( G_{ij} \cos \theta_{ij} + B_{ij} \sin \theta_{ij} \right) \tag{10.25}$$

$$K_{ij} = \frac{\partial \Delta Q_i}{\partial \theta_j} = V_i V_j \left( G_{ij} \cos \theta_{ij} + B_{ij} \sin \theta_{ij} \right) \tag{10.26}$$

$$L_{ij} = V_j \frac{\partial \Delta Q_i}{\partial V_j} = -V_i V_j \left( G_{ij} \sin \theta_{ij} - B_{ij} \cos \theta_{ij} \right) \tag{10.27}$$

When $i = j$,

$$H_{ii} = \frac{\partial \Delta P_i}{\partial \theta_i} = V_i \sum_{\substack{j=1 \\ j \neq i}}^{n} V_j \left( G_{ij} \sin \theta_{ij} - B_{ij} \cos \theta_{ij} \right) = V_i^2 B_{ii} + Q_i \tag{10.28}$$

$$N_{ii} = V_j \frac{\partial \Delta P_i}{\partial V_i} = -V_i \sum_{\substack{j=1 \\ j \neq i}}^{n} V_j \left( G_{ij} \cos \theta_{ij} + B_{ij} \sin \theta_{ij} \right) - 2V_i^2 G_{ii} = -V_i^2 G_{ii} - P_i \tag{10.29}$$

$$K_{ii} = \frac{\partial \Delta Q_i}{\partial \theta_i} = -V_i \sum_{\substack{j=1 \\ j \neq i}}^{n} V_j \left( G_{ij} \cos \theta_{ij} + B_{ij} \sin \theta_{ij} \right) = V_i^2 G_{ii} - P_i \tag{10.30}$$

$$L_{ii} = V_i \frac{\partial \Delta Q_i}{\partial V_i} = -V_i \sum_{\substack{j=1 \\ j \neq i}}^{n} V_j \left( G_{ij} \sin \theta_{ij} - B_{ij} \cos \theta_{ij} \right) + 2V_i^2 G_{ii} = V_i^2 G_{ii} - Q_i \tag{10.31}$$

Based on graph database and graph computing, the power flow calculation algorithm by the Newton–Raphson method is presented below.

---

**Algorithm 10.2   Graph-Based Power Flow Calculation by the Newton–Raphson**

---

```
1:  Set iteration index k = 0
2:  Initialize the system state vector x⁰ = (θ⁰, |V|⁰) //in nodal
    parallel
3:  Form admittance matrix //in nodal parallel
4:  While (‖Δθᵏ‖ ≥ ϵθ or ‖Δvᵏ‖ ≥ ϵᵥ) do
5:      Build Jacobian graph by (10.16) //in nodal parallel
6:      Factorize Jacobian matrix //in hicrarchical parallel
7:      Update active power and reactive power mismatch by (10.13) and
        (10.14)
8:      Solve Δθᵏ and Δvᵏ in (10.15) by substitution//in hierarchical
        parallel
9:      k = k+1
10: Calculate branch flow //in nodal parallel
```

---

In the algorithm, bus voltage initialization, admittance formation, Jacobian graph creation, and branch flow calculation can be performed by nodal parallel computing, e.g. calculation on each bus and branch does not rely on other bus and branch calculations. Jacobian matrix factorization and forward/backward substitution mathematically are nodal dependent, e.g. the calculation of a row is dependent only on the columns at the same level in the elimination tree as addressed in Chapter 3. Calculations for Jacobian matrix factorization and forward/backward substitution can be implemented by hierarchical graph parallel computing [1, 2].

To implement the above graph-based power flow calculation by the Newton–Raphson method, the bus-branch model graph schema is defined as the same as that for the Gauss–Seidel method in Section 10.2. And the query `formAdmittanceMatrix` defined in Section 10.2 is called to form an admittance matrix in nodal parallel.

## 10.3.1 Build Jacobian Graph

The Jacobian graph is a graph representation of a Jacobian matrix. The Jacobian graph is defined as a set of (vertex, edge) pairs. Each pair of (vertex, edge) represents a nonzero element in the Jacobian matrix.

To build a Jacobian graph, the vertex JacobianND and edge JacobianConnect are defined as follows containing the nonzero value of the four matrix blocks, $H$, $N$, $K$, and $L$ by $aij$ and $aii$ in double:

```
CREATE VERTEX JacobianND (PRIMARY_ID JacobianID int, aii double)
CREATE DIRECTED EDGE JacobianConnect (fromJacobianND, toJacobianND,
aij double)
```

The vertex `JacobianND` and edge `JacobianConnect` are virtual vertices and virtual edges. They are not physical nodes or branches in power systems. In the Jacobian graph structure, the counterparts of the edges between vertices are nonzero off-diagonal elements in the Jacobian matrix. The diagonal elements in the Jacobian matrix are modeled as attributes of a vertex. Zero (absent) elements in the Jacobian matrix indicate that no direct connections between the vertices exist in the Jacobian graph.

With the defined vertex and edge above, the Jacobian graph schema is then simply defined as:

```
CREATE GRAPH JacobianGraph(JacobianND, JacobianConnect)
```

The Jacobian graph schema defines the graph structure. The Jacobian graph will be built by adding `JacobianND` and `JacobianConnect` by the following pseudo-query code.

```
CREATE QUERY buildJacobianGraph(int nBus
) FOR GRAPH busBranchGraph and JacobianGraph{
SumAccum<double> @Pcal = 0
SumAccum<double> @Qcal = 0
MaxAccum<double> @@PI = 3.1415926535898
double H, N, K, L

T0 = {TopoND.*}
T1 = SELECT s FROM T0:s-(connected:e)->:t
  ACCUM
    double dDeg = (s.Va-t.Va)*@@PI/180
    s.@Pcal += s.Vm*t.Vm*(e.G*cos(dDeg) + e.B*sin(dDeg))
    s.@Qcal += s.Vm*t.Vm*(e.G*sin(dDeg) - e.B*cos(dDeg))
```

```
POST-ACCUM
    s.@Pcal += s.Vm*s.Vm*e.sumG
    s.@Qcal += -s.Vm*s.Vm*e.sumB

    IF (t.busType == 0 or t.busType == 1) THEN
      H =  e.B*s.Vm*s.Vm + s.@Qcal
      N = -e.G*s.Vm*s.Vm - s.@Pcal
      K =  e.G*s.Vm*s.Vm - s.@Pcal
      L =  e.B*s.Vm*s.Vm - s.@Qcal
      INSERT INTO JacobianND(s.TOPOID) VAULES(H)
      INSERT INTO JacobianND(s.TOPOID + nBus-1) VAULES(L)
      INSERT INTO JacobianConnect(s.TOPOID, s.TOPOID + nBus-1) VALUES(N)
      INSERT INTO JacobianConnect(s.TOPOID + nBus-1, s.TOPOID) VALUES(K)
    ELSE IF (t.busType == 2) THEN
      H =  e.B*s.Vm*s.Vm + s.@Qcal
      N = -e.G*s.Vm*s.Vm - s.@Pcal
      K =  0
      L =  s.Vm
      INSERT INTO JacobianND(s.TOPOID) VAULES(H)
      INSERT INTO JacobianND(s.TOPOID + nBus-1) VAULES(L)
      INSERT INTO JacobianConnect(s.TOPOID, s.TOPOID + nBus-1) VALUES(N)
    ELSE IF (t.busType == 3) THEN
      H =  s.Vm
      N =  0
      K =  0
      L =  s.Vm
      INSERT INTO JacobianND(s.TOPOID) VAULES(H)
      INSERT INTO JacobianND(s.TOPOID + nBus-1) VAULES(L)
    END

T1 = SELECT s FROM T0:s-(connected:e)->:t
  ACCUM
    double dDeg = (s.Va-t.Va)*@@PI/180
    IF (t.busType == 0 or t.busType == 1) THEN
      H = -s.Vm*t.Vm*(e.G*sin(dDeg) - e.B*cos(dDeg))
      N = -s.Vm*t.Vm*(e.G*cos(dDeg) + e.B*sin(dDeg))
      K =  s.Vm*t.Vm*(e.G*cos(dDeg) + e.B*sin(dDeg))
      L = -s.Vm*t.Vm*(e.G*sin(dDeg) - e.B*cos(dDeg))
      INSERT INTO JacobianConnect(s.TOPOID, t.TOPOID) VALUES(H)
      INSERT INTO JacobianConnect(s.TOPOID, t.TOPOID + nBus-1) VALUES(N)
      INSERT INTO JacobianConnect(s.TOPOID + nBus-1, t.TOPOID) VALUES(K)
      INSERT INTO JacobianConnect(s.TOPOID + nBus-1, t.TOPOID + nBus-1)
              VALUES(L)
    ELSE IF (t.busType == 2) THEN
      H = -s.Vm*t.Vm*(e.G*sin(dDeg) - e.B*cos(dDeg))
      N = -s.Vm*t.Vm*(e.G*cos(dDeg) + e.B*sin(dDeg))
```

```
      K = 0
      L = 0
      INSERT INTO JacobianConnect(s.TOPOID, t.TOPOID) VALUES(H)
      INSERT INTO JacobianConnect(s.TOPOID, t.TOPOID + nBus-1)
VALUES(N)
 }
```

In the query `buildJacobianGraph` arguments, the number of bus `nBus` is passed in to locate the position of matrix blocks, *H, N, K,* and *L.*

In the query, the vertex-attached variables `SumAccum @Pcal, @Qcal` are defined to store bus active and reactive power injections.

In the query definition, graph traversal statement `T1 = SELECT s FROM T0:s-(connected: e)->:t` travels all edges from vertex s in the vertex set `T0 = {TopoND.*}`. Nested in the `ACCUM`, bus active and reactive power injections `@Pcal, @Qcal` are calculated in nodal parallel.

In the `POST-ACCUM`, diagonal elements in the matrix blocks *H, N, K,* and *L* are calculated for PQ, PV, and slack buses. Each calculated nonzero diagonal element is inserted as either a vertex or an edge into the Jacobian graph by:

```
INSERT INTO JacobianND(s.TOPOID) VAULES(H)
INSERT INTO JacobianND(s.TOPOID + nBus-1) VAULES(L)
INSERT INTO JacobianConnect(s.TOPOID, s.TOPOID + nBus-1) VALUES(N)
INSERT INTO JacobianConnect(s.TOPOID + nBus-1, s.TOPOID) VALUES(K)
```

Note that *N* and *K* are diagonal elements of matrix blocks, but nondiagonal elements of the Jacobian matrix. They are inserted as edges in the Jacobian graph.

In the second graph traversal, `T1 = SELECT s FROM T0:s-(connected:e)->:t`, nondiagonal elements in the matrix blocks *H, N, K,* and *L* are calculated for PQ and PV buses. Each calculated nonzero diagonal element is inserted as an edge into the Jacobian graph. The nondiagonal elements for the slack bus are zero, they are excluded in the Jacobian graph.

### 10.3.2 Graph-Based Symbolic Factorization

Symbolic factorization is the process to determine fill-ins. The Jacobian graph is extended with fill-in edges. Each fill-in edge represents a fill-in element in the Gaussian elimination. As discussed in Section 3.3.1, the graph-based symbolic factorization algorithm is rewritten as follows:

---
**Algorithm 10.3   Graph-Based Symbolic Factorization**
---
```
1: CREATE QUERY formFilledGraph() FOR GRAPH JacobianGraph
2:    FOR NODE i IN V OF JacobianGraph
3:       SELECT j AND k FROM V WHERE j∈i AND k∈i
4:          ACCUM
5:                DELETE E(i,j)
6:                DELETE E(i,k)
7:                INSERT INTO E+ (j, k)      //fill-in
```
---

The pseudo-code to implement the graph-based symbolic factorization is presented as follows:

```
CREATE QUERY formFilledGraph(
) FOR GRAPH JacobianGraph{
TYPEDEF TUPLE <int v, int w> FILL_IN
ListAccum<FILL_IN> @@fillIns = <EMPTY>
HeapAccum<int JacobianID, double aii> (JacobianID ASC) @@orderedND
E0 = {JacobianConnection.*}
T0 = {JacobianND.*}
T1 = SELECT s FROM T0:s
 ACCUM
  @@orderedND += s
FOREACH v IN @@orderedND DO
  T1 = SELECT s FROM v:s-(connected:e)->:s.neighbors()
  ACCUM
    DELETE e FROM T0
    FOREACH j IN s.neighbors() DO
      FOREACH k IN s.neighbors() WHERE NOT j=k DO
          @@fillIns += FILL_IN<j,k>
          INSERT INTO JacobianConnect(j,k)
        END
      END
END
JacobianConnection = E0
FOREACH fillIn IN @@fillIns DO
  INSERT INTO JacobianConnect(fillIn.v,fillIn.w)
END
}
```

Symbolic factorization is a dynamic process in which adjacent edges of a vertex are temporarily removed. Thus, statement E0 = {JacobianConnection.*} is taken to save the original edges which will be restored after the fill-in process is completed.

In the query, ListAccum @@fillIns is defined to list all fill-ins from vertex and to vertex saved in a TUPLE. HeapAccum @@orderedND is defined with the keyword ASC to order JacobianND in ascending by

```
T0 = {JacobianND.*}
T1 = SELECT s FROM T0:s
  ACCUM
    @@orderedND += s
```

HeapAccum is an accumulator type that maintains a sorted collection of tuples. The output of a HeapAccum is a sorted collection of tuple elements. When a vertex s is added to the HeapAccum @@orderedND, it is sorted by JacobianID as defined in the declaration of HeapAccum<int. JacobianID, double aii > (JacobianID ASC) @@orderedND.

Looping through vertex v in HeapAccum @@orderedND, statement T1 = SELECT s FROM v : s-(connected:e)->:s.neighbors() travels adjacent edges from each vertex s to its neighbors by calling the build-in function s.neighbors().

The adjacent edges are temporarily removed by DELETE e in nodal parallel nested in ACCUM. The fill-ins are inserted into the graph to determine follow-up fill-ins. After all, fill-ins are determined, the edges are restored by JacobianConnection = E0. And fill-ins are inserted to build the extended Jacobian graph by

```
INSERT INTO JacobianConnect(fillIn.v,fillIn.w)
```

### 10.3.3   Graph-Based Elimination Tree Creation and Node Partition

The elimination tree defines the column dependencies in the Gaussian elimination. To solve the iteration equations by the Gaussian elimination method, matrix factorization and forward/ backward substitution are performed by column for each row. The calculation of a row is dependent only on the columns at the same level in the elimination tree. The nodes at the same level are partitioned. The information on the partition will be saved as attributes in the Jacobian graph.

To save the partition information, the Jacobian graph schema is extended by adding the attribute hLevel to store the hierarchical level where the vertex is at in the elimination tree. The extended Jacobian graph is designed as follows:

```
CREATE VERTEX JacobianND (PRIMARY_ID JacobianID int, aii double,
hLevel int)
```

And the graph schema for the elimination tree is defined as follows:

```
CREATE VERTEX eTreeND (PRIMARY_ID eTreeID int)
CREATE DIRECTED EDGE eTreeConnection (fromeTreeND, toeTreeND)
```

With the defined vertex and edge above, the elimination tree graph schema is then simply defined as:

```
CREATE GRAPH eTreeGraph(eTreeND, eTreeConnection)
```

The algorithm to create an elimination tree is shown as follows.

---

**Algorithm 10.4   Create an Elimination Tree**

---

```
1: CREATE QUERY createEliminationTree() FOR JacobianGraph
2:    SELECT i IN V FROM JacobianGraph
3:      ACCUM
4:        SELECT j FROM V WHERE j∈i
5:              v.parent = MIN(j>i)
6:          INSERT E (i.parent, i) INTO eTreeGraph
```

---

The pseudo-code in graph query can be implemented to create the elimination tree as shown as follows:

```
CREATE QUERY createEliminationTree(
) FOR GRAPH JacobianGraph and GRAPH eTreeGraph{
SetAccum<int> @parent

eTreeND.eTreeID =  JacobianND.JacobianID

T0= {JacobianND.*}
T1 = SELECT v FROM T0
```

```
ACCUM
   v.@parent = MIN(v.neighbors().JacobianID)
               WHERE v.neighbors().JacobianID > v.JacobianID
   INSERT INTO eTreeConnect(v.@parent, v.JacobianID)
}
```

In the query, the vertex-attached variable SetAccum @parent is defined to store the vertex parent. Since the @parent is a temporary variable, it is unnecessary to define it as an attribute of the graph vertex.

The elimination tree graph shares the same vertices as the Jacobian graph. The following code copies the Jacobian graph vertex JacobianID to the elimination tree graph eTreeID.

```
eTreeND.eTreeID = JacobianND.JacobianID
```

Then statement T1 = SELECT v FROM T0 selects all vertices and searches their neighbors JacobianID by v.neighbors().JacobianID in nodal parallel using the keyword ACCUM. The parent is found as the smallest JacobianID in all neighbors while the neighbor's JacobianID is larger than the vertex's JacobianID by the following code:

```
v.@parent = MIN(v.neighbors().JacobianID)
            WHERE v.neighbors().JacobianID > v.JacobianID
```

Meanwhile, the new directed edge is inserted into the elimination tree graph from the parent to the child by the statement

```
INSERT INTO eTreeConnect(v.@parent, v.JacobianID)
```

The directed edge supports to determine the leaves on the tree. Leaves have zero out-degree.

Based on the elimination tree graph, the vertex in the Jacobian graph is partitioned hierarchically by the following algorithm.

---

**Algorithm 10.5   Node Hierarchical Partition**

---

```
1:      LEVEL = 1
2:      CREATE QUERY formPartition(T(A), LEVEL) FOR JacobianGraph
3:           AND eTreeGraph
4:        WHILE T(A) != EMPTY
5:          formPartition(T(A), LEVEL)
6:          SELECT i FROM V OF T(A) WHERE LEAVE(i) == TRUE
7:             ACCUM
8:                i.hLevel = LEVEL
9:                DELETE i //remove node i from T(A)
10:     LEVEL ++
```

---

The pseudo-query code to partition nodes based on the elimination tree is implemented as follows:

```
CREATE QUERY nodePartition( int LEVEL
) FOR GRAPH JacobianGraph and GRAPH eTreeGraph{
Tj = {JacobianND.*}
Te = {eTreeND.*}
```

```
WHILE (Te != EMPTY) DO
   T1 = SELECT ve FROM Te WHERE ve.outdegree() == 0
      ACCUM
         T2 = SELECT vj FROM Tj WHERE vj.JacobianID == ve.eTreeID
            ACCUM
               vj.hLevel = LEVEL
         DELETE ve FROM Te
   END
   LEVEL += 1
   nodePartition(LEVEL)
}
```

The query `nodePartition` is called recursively starting with `nodePartition(1)` till all nodes are processed. The graph partition starts from level 1 (LEVEL = 1). The leave nodes are selected from the elimination tree graph vertex set `Te` by `T1 = SELECT ve FROM Te WHERE ve.outdegree() == 0`, where `outdegree()` is a build-in function to check if the node `ve` is a leave node. All leave nodes are deleted from the elimination tree `Te` and hierarchical level `LEVEL` is assigned as an attribute `hLevel` of the extended Jacobian graph vertex which will be used for hierarchical parallel factorization and substitutions. When the first-level leave nodes are removed, the second-level nodes are new leaves on the elimination tree. The query `nodePartition` is called with a new LEVEL till no node is left by checking the condition (`Te != EMPTY`) in the `WHILE` loop.

### 10.3.4  Graph Numerical Factorization

The extended Jacobian graph stores the information of the Jacobian matrix $J$ and the hierarchical level of each vertex. Using the node partitioning information, the pseudo-code for the numerical factorization hierarchical parallel computing approach is:

---
**Algorithm 10.6   Factorize Matrix $J$**
---

```
1:    CREATE QUERY factorizeJacobian(nLevel) FOR GRAPH JacobianGraph
2:     FOR l in Range(nLevel)
3:        SELECT NODE i FROM V WHERE i.hLevel ==l
4:           ACCUM
5:             SELECT j FROM V WHERE j∈i
```
$$6: \qquad e_{ij}.a^{(n)} = \frac{e_{ij}.a^{(n-1)}}{v_i.a^{(n-1)}}$$
$$7: \qquad e_{kj}.\ a^{(n)} = e_{kj}.\ a^{(n-1)} - e_{ki}.\ a^{(n-1)} \cdot e_{ij}.\ a^{(n)}$$
```
8:     POST-ACCUM
9:        SELECT NODE i FROM V
10:          ACCUM
```
$$11: \qquad v_i.l = v_i.\ a$$
```
12:          SELECT NODE j FROM V
13:          WHERE j∈i AND j
```
$$14: \qquad e_{ij}.l = e_{ij}.\ a$$

---

where $v_i.\,a$ is the attribute of the vertex $v_i$ in the Jacobian graph representing the diagonal element $a_{ii}$ and $e_{ij}.\,a$ is the attribute of the edge $e_{ij}$ in the Jacobian graph representing the nondiagonal element $a_{ij}$ ($i \neq j$) of the Jacobian matrix. Step 3 selects all nodes within the same level, and steps 5–7 nested in ACCUM perform the factorization for the nodes selected in step 3 in parallel.

After the matrix is formed by the query factorize Jacobian, in the POST-ACCUM clause, the elements are saved to attributes $L$ in the nested parallel ACCUM clause.

To implement graph-based numerical factorization, the Jacobian graph is extended as follows to store attributes $L$:

```
CREATE VERTEX JacobianND (PRIMARY_ID JacobianID int, aii double,
hLevel int, lii double)
CREATE DIRECTED EDGE JacobianConnect (fromJacobianND, toJacobianND,
aij double, lij double)
```

The pseudo-code in the graph query for graph numerical factorization is presented as follows:

```
CREATE QUERY factorizeJacobian( int nLevel
) FOR GRAPH JacobianGraph{
  T0 = {JacobianND.*}
  FOREACH level IN RANGE[1,nLevel] DO
  T1 = SELECT v FROM T0:v WHERE v.hLevel == level
   ACCUM
     T2 = SELECT vi FROM T1:vi-(connected:eij)->:vj
      ACCUM
       eij.aij = eij.aij/vi.aii
       T3 = SELECT vk FROM vi.neighbors():vk-(connected:ekj)->:vj
        eki = SELECT e FROM eij
              WHERE e.fromJacobianND == vk.JacobianID
              AND e.toJacobianND == vi.JacobianID
         ekj.aij = ekj.aij - eki.aij*eij.aij
  POST-ACCUM
   T2 = SELECT vi FROM T0:vi-(connected:eij)->:vj
    ACCUM
       vi.lii  = vi.aii
       eij.lij = eij.aij
  END
}
```

In the query `factorizeJacobian` argument, the number of hierarchical level `nLevel` is passed. For vertex set `T1` at each of the hierarchical levels selected by `T1 = SELECT v FROM T0:v WHERE v.hLevel == level`, the attribute `aij` of edges `eij` are normalized by `eij.aij = eij.aij/vi.aii`. For each adjacent vertex `vj` of `vi`'s neighbor `vk`, the attribute `aij` is calculated by `ekj.aij = ekj.aij - eki.aij*eij.aij`. After the Gaussian elimination is performed, the attributes `lii` and `lij` are assigned to attributes of the corresponding vertex and edge by `vi.lii = vi.aii` and `eij.lij = eij.aij`.

## 10.3.5 Build Right-Hand Side

To solve the power flow update Eq. (10.15), the right-hand-side vector is updated at each iteration in the graph-based Newton–Raphson power flow calculation algorithm in parallel.

The right-hand-side vector is stored in the Jacobian graph. To save the right-hand-side value, the Jacobian graph schema is extended by adding attributes `bii`. The extended Jacobian graph is designed as follows:

```
CREATE VERTEX JacobianND (PRIMARY_ID JacobianID int, aii double,
hLevel int, bii double)
```

They are implemented by the following pseudo-code.

```
SumAccum<double> @deltaP = 0
SumAccum<double> @deltaQ = 0

 CREATE QUERY buildRighHandSidePF(int nBus) FOR GRAPH busBranchGraph
AND GRAPH JacobianGraph{
  T0 = {busBranchGraph.TopoND.*}
  Tj = {JacobianGraph.JacobianND.*}

  T1 = SELECT s FROM T0:s-(connected:e)->:t
   ACCUM
     IF (e.K == 0) THEN e.K = 1 END
     s.@deltaP += s.Vm*t.Vm * (-e.G/e.K*cos(s.Va-t.Va)+
                 e.B/e.K*sin(s.Va-t.Va))
     IF (s.busType == 2) THEN
        s.@deltaQ += s.Vm*s.Vm * (-e.G/e.K*sin(s.Va-t.Va)-
                 e.B/e.K*cos(s.Va-t.Va))
   END

 POST-ACCUM
     s.@deltaP = s.GenP - s.LdP - (s.@deltaP + s.Vm*s.Vm*s.sumG)
     UPDATE j_nd FROM Tj WHERE j_nd.JacobianID == s.TOPOID
     SET j_nd.bii = s.@deltaP
     IF (s.busType == 2) THEN
        s.@deltaQ = s.GenQ - s.LdQ - (s.@deltaQ + s.Vm*s.Vm*s.sumB)
        UPDATE j_nd FROM Tj WHERE j_nd.JacobianID == s.TOPOID
        SET (j_nd+nBus).bii = s.@deltaQ
     END
}
```

Before the query `buildRighHandSidePF()` is implemented, vertex-attached accumulator variables `SumAccum @deltaP`, `@deltaQ` are defined with a single add-sign @ to store the active power and reactive power mismatch.

In the query, the number of buses `nBus` is passed through to allocate the voltage magnitude update vector at the right vertex of the Jacobian graph. In the query definition, graph traversal statement `T1 = SELECT t FROM T0:s-(connected:e)->:t` coupling with keyword `ACCUM` travels all edges e from vertex s to the end vertex t in the bus-branch graph busBranchGraph and calculates the total branch power flow to the vertex s in nodal parallel.

In the POST-ACCUM, active power and reactive power mismatches are calculated by adding the active power and reactive power injections from generators, loads, and shunts. The mismatches are assigned to the Jacobian graph attribute bii.

### 10.3.6 Graph Forward and Backward Substitution

The last step to solve the power flow update Eq. (10.15) is forward and backward substitution. The temporary and final solutions are stored in the Jacobian graph. To save the solution value, the Jacobian graph schema is extended by adding attributes xi. The extended Jacobian graph is designed as follows:

```
CREATE VERTEX JacobianND (PRIMARY_ID JacobianID int, aii double,
hLevel int, bii double, xi double)
```

They are implemented by the following pseudo-code in hierarchical parallel.

```
CREATE QUERY Substitution(int nLevel, int nBus) FOR GRAPH
JacobianGraph and GRAPH busBranchGraph {
    T0 = {JacobianGraph.JacobianND.*}
    Tb = {busBranchGraph.TopoND.*}

    FOREACH level IN RANGE[1,nLevel] DO
        T1 = SELECT v FROM T0:v WHERE v.hLevel == level
            ACCUM
              v.xi = v.bii
              T2 = SELECT vi FROM T1:vi-(connected:eij)->:vj
                  ACCUM
                     v.xi -= eij.lij*v.xi

        POST-ACCUM
            v.xi  = v.xi/v.aii
    END

    FOREACH level IN RANGE[1,nLevel] DO
        T1 = SELECT v FROM T0:v WHERE v.hLevel == level
            ACCUM
                T2 = SELECT vi FROM T1:vi-(connected:eij)->:vj
                    ACCUM
                        v.xi -= eij.lij*v.xi
    END

    T1 = SELECT v FROM T0
        ACCUM
          T2 = SELECT topo_nd FROM Tb
              ACCUM
```

```
        UPDATE topo_nd FROM T2 WHERE topo_nd.TOPOID==v.JacobianID
        SET topo_nd.Va   = topo_nd.Va + v.xi
        IF (topo_nd.busType == 2) THEN
             SET topo_nd.Vm   = topo_nd.Vm + v.xi/topo_nd.Vm
        END
}
```

In the query arguments, the number of hierarchical levels nLevel and the number of buses are passed through. For vertex set T1 at each of the hierarchical levels selected by T1 = SELECT v FROM T0:v WHERE v.hLevel == level, the vertex attribute xi is forward-substituted by the following statements in parallel.

```
    T2 = SELECT vi FROM T1:vi-(connected:eij)->:vj
       ACCUM
         v.xi -= eij.lij*v.xi
```

Then vertex attribute is normalized by v.xi = v.xi/v.aii. The backward substitution is performed similarly in hierarchical parallel for each of the hierarchical levels.

At the end of the query, the voltage angle and voltage magnitude are updated to the vertex attributes Va and Vm in the bus-branch graph busBranchGraph.

### 10.3.7  Graph-Based Newton–Raphson Power Flow Calculation

The queries to support the Newton–Raphson method for power flow calculations are integrated to solve power flow problems. The procedure is straightforward by the following pseudo-code.

```
CREAT QUERY newtonRaphsonPowerFlow(int MaxIteration, double
Tolerance)
FOR GRAPH busBranchGraph and GRAPH JacobianGraph and GRAPH eTreeGraph{

MaxAccum<double> @@xi = -INF

T0 = {busBranchGraph.TopoND.*}
Tj = {JacobianGraph.JacobianND.*}

T1 = SELECT v FROM T0:v-(connected:e)->:t
   ACCUM
       e.P_BR = 0
       e.Q_BR = 0
       IF (e.K == 0) THEN
            v.sumG += e.G
            v.sumB += -e.B + 0.5*e.hB
       ELSE
        v.sumG += 1/(e.K*e.K)* e.G
        v.sumB += 1/(e.K*e.K)*(-e.B) + 0.5*e.hB
       END
```

```
    POST-ACCUM
        v.sumG += v.G
        v.sumB += v.B

int nBus = busBranchGraph.getVertexCount()
WHILE( abs(@@xi) > Tolerance ) LIMIT MaxIteration DO
    buildJacobianGraph(int nBus)
                            FOR GRAPH busBranchGraph and GRAPH
                            JacobianGraph
    formFilledGraph() FOR GRAPH JacobianGraph
    createEliminationTree() FOR GRAPH JacobianGraph and GRAPH
    eTreeGraph
    int nLevel = nodePartition(1)
                            FOR GRAPH JacobianGraph and GRAPH
                            eTreeGraph
    factorizeJacobian(int nLevel) FOR GRAPH JacobianGraph
    buildRighHandSidePF(int nBus)
                            FOR GRAPH busBranchGraph AND GRAPH
                            JacobianGraph
    Substitution(int nLevel, int nBus)
                            FOR GRAPH JacobianGraph and GRAPH
                            busBranchGraph
    T1 = SELECT v FROM Tj
        ACCUM
            @@xi = max(abs(v.xi))

T1 = SELECT s FROM T0:s-(connected:e)->:t
    IF (e.K == 0) THEN e.K = 1 END
    e.P_BR += s.Vm*t.Vm * (e.G/e.K*cos(s.Va-t.Va)+e.B/e.K*sin(s.Va-t.Va))
    e.Q_BR += s.Vm*s.Vm * (e.G/e.K*sin(s.Va-t.Va)-e.B/e.K*cos(s.Va-t.Va))
}
```

In the query `newtonRaphsonPowerFlow`, maximum iteration number `MaxIteration` and tolerance `Tolerance` are passed through. The global variable `MaxAccum` `@@xi` is defined to store the maximum voltage magnitude or angle update which will be used to compare against the tolerance `Tolerance` to check the convergence. Vertex attributes `sumG` and `sumB` are calculated first in the query.

In the while loop, in each of the iterations, the Jacobian graph is built and factorized. Active power and reactive power mismatch are updated by `buildRighHandSidePF`. The power flow update equation is solved and bus voltage is updated by query `Substitution`. At the end of the while loop, `@@xi` is calculated in nodal parallel.

After the bus voltage is solved or the maximum iteration number is reached, the branch power flow is calculated by traversal statement `T1 = SELECT s FROM T0:s-(connected:e)->:t` which travels all edges e from vertex s to the end vertex t in the bus-branch graph busBranch-Graph. With the keyword ACCUM, the branch active power `e.P_BR` and reactive power `e.Q_BR` are calculated in nodal parallel.

## 10.4 Fast Decoupled Power Flow Calculation

The Newton–Raphson method is robust which demonstrates great convergence in solving power flow problems. However, the Newton–Raphson method requires recalculating the Jacobian matrix (10.16) in each iteration. Power flow calculation study on real power systems shows Jacobian matrix formation and factorization are the most time-consuming tasks in the Newton–Raphson method. In a practical power transmission system, the reactance of a branch is generally far greater than the resistance of the branch. And angle separation $\theta_{ij}$ between the two terminal buses of a branch is relatively small in a stable power system. Thus

$$N_{ij} = V_j \frac{\partial \Delta P_i}{\partial V_j} = -V_i V_j \left( G_{ij} \cos \theta_{ij} + B_{ij} \sin \theta_{ij} \right) \approx 0 \tag{10.32}$$

$$K_{ij} = \frac{\partial \Delta Q_i}{\partial \theta_j} = V_i V_j \left( G_{ij} \cos \theta_{ij} + B_{ij} \sin \theta_{ij} \right) \approx 0 \tag{10.33}$$

Equation (10.15) is simplified to:

$$\begin{bmatrix} \Delta P \\ \Delta Q \end{bmatrix} = - \begin{bmatrix} H & 0 \\ 0 & L \end{bmatrix} \begin{bmatrix} \Delta \theta \\ \Delta V / V \end{bmatrix} \tag{10.34}$$

Since the nondiagonal blocks of the matrix $\begin{bmatrix} H & 0 \\ 0 & L \end{bmatrix}$ are zero, the equation can be decoupled as:

$$\Delta P = -H \Delta \theta \tag{10.35}$$

$$\Delta Q = -L \frac{\Delta V}{V} \tag{10.36}$$

The two simplified equations are iteratively calculated to update the bus voltage angle by active power mismatch only, and the bus voltage magnitude by reactive power mismatch only. This simplified method is called the fast decoupled power flow method. In the fast decoupled power flow method, with assumptions of $G_{ij} \sin \theta_{ij} \ll B_{ij} \cos \theta_{ij}$ and $\cos \theta_{ij} \approx 1$, elements in the matrices $H$ and $L$ are further simplified as:

$$H_{ij} = \frac{\partial \Delta P_i}{\partial \theta_j} = -V_i V_j \left( G_{ij} \sin \theta_{ij} - B_{ij} \cos \theta_{ij} \right) \approx V_i V_j B_{ij} \tag{10.37}$$

$$L_{ij} = V_j \frac{\partial \Delta Q_i}{\partial V_j} = -V_i V_j \left( G_{ij} \sin \theta_{ij} - B_{ij} \cos \theta_{ij} \right) \approx V_i V_j B_{ij} \tag{10.38}$$

The matrices $H$ and $L$ are rewritten as:

$$H = \begin{bmatrix} H_{1,1} & \cdots & H_{1,n-1} \\ \vdots & \ddots & \vdots \\ H_{n-1,1} & \cdots & H_{n-1,n-1} \end{bmatrix} = \begin{bmatrix} V_1 & & \\ & \ddots & \\ & & V_{n-1} \end{bmatrix} \begin{bmatrix} B_{1,1} & \cdots & B_{1,n-1} \\ \vdots & \ddots & \vdots \\ B_{n-1,1} & \cdots & B_{n-1,n-1} \end{bmatrix}$$

$$\begin{bmatrix} V_1 & & \\ & \ddots & \\ & & V_{n-1} \end{bmatrix} = V B' V$$

$$\tag{10.39}$$

$$
L =
\begin{bmatrix}
H_{1,1} & \cdots & H_{1,n-1} \\
\vdots & \ddots & \vdots \\
H_{n-1,1} & \cdots & H_{n-1,n-1}
\end{bmatrix}
=
\begin{bmatrix}
V_1 & & \\
& \ddots & \\
& & V_m
\end{bmatrix}
\begin{bmatrix}
B_{1,1} & \cdots & B_{1,m} \\
\vdots & \ddots & \vdots \\
B_{m,1} & \cdots & B_{m,m}
\end{bmatrix}
\begin{bmatrix}
V_1 & & \\
& \ddots & \\
& & V_m
\end{bmatrix}
= VB''V
$$

$$(10.40)$$

Substituting $H = VB'V$ and $L = VB''V$ into (10.35) and (10.36), we get

$$
\frac{\Delta P}{V} = -B'V\Delta\theta \tag{10.41}
$$

$$
\frac{\Delta Q}{V} = -B''\Delta V \tag{10.42}
$$

In order to make the coefficient matrix at the right-hand side of the above Eq. (10.41) to be constant, in practical decoupled power flow calculation, the voltage magnitudes on the right-hand side of Eq. (10.41) are assumed as 1.0. Thus, Eq. (10.41) is further simplified as

$$
\frac{\Delta P}{V} = -B'\Delta\theta \tag{10.43}
$$

In the decoupled power flow method, matrices $B'$ and $B''$ are constant. In each iteration, $B'$ and $B''$ do not need to be rebuilt and refactorized. Although the decoupled power flow method takes a few more iterations than the Newton–Raphson method to converge, each iteration takes much less time. For reactance-dominated transmission networks, the decoupled power flow method outperforms the Newton–Raphson method in computation efficiency. The cost is the approximation on the Jacobian matrices by the decoupled power flow method deteriorates power flow convergence.

Based on graph database and graph computing, the power flow calculation algorithm by the fast decoupled method is presented below.

---

**Algorithm 10.7   Graph-Based Power Flow Calculation by Fast Decoupled Method**

---

1: **Set** iteration index $k$ = 0
2: **Initialize** the system state vector $\mathbf{x}^0$ = $(\boldsymbol{\theta}^0, |\mathbf{v}|^0)$ //in nodal parallel
3: **Create g**raphs B_P and B_PP for $B'$ and $B''$ //in nodal parallel
4: **Factorize** $B'$ and $B''$ on graphs B_P and B_PP//in hierarchical parallel
5: **While** ($\|\Delta\boldsymbol{\theta}^k\| \geq \epsilon_\theta$ or $\|\Delta\mathbf{v}^k\| \geq \epsilon_v$) **do**
6:     **Update** active power and reactive power mismatch by (10.13) and (10.14)
7:     **Solve** $\Delta\boldsymbol{\theta}^k$ and $\Delta\mathbf{v}^k$ in (10.42) and (10.43) //in hierarchical parallel
8:     $k$ = $k+1$
9: **Calculate** branch flow //in nodal parallel

---

Similar to the Newton–Raphson method, in the graph-based fast decoupled power flow calculation algorithm, bus voltage initialization, $B'$ and $B''$ graph creation, and branch flow calculation can be performed by nodal parallel computing, e.g. calculation on each bus and branch does not rely on other bus and branch's calculations. $B'$ and $B''$ matrix factorization and forward/backward substitution mathematically are nodal dependent, e.g. the calculation of a row is dependent only on the columns at the same level in the elimination tree as addressed in Chapter 3. Calculations for $B'$ and

$B''$ matrix factorization and forward/backward substitution can be implemented by hierarchical graph parallel computing [3].

To implement the above graph-based power flow calculation by fast decoupled method, the bus-branch model graph schema is defined as the same as that for the Gauss–Seidel and Newton–Raphson methods.

### 10.4.1   Build B_P and B_PP Graphs

The B_P and B_PP graphs represent $B'$ and $B''$ matrices respectively by graphs. The B_P and B_PP graphs are defined as a set of (vertex, edge) pairs. Each pair of (vertex, edge) represents a nonzero element in the $B'$ and $B''$ matrices.

To build B_P and B_PP graphs, the vertices B_pND, B_ppND and edges B_pConnect and B_ppConnect are defined as follows containing the nonzero value of the $B'$ and $B''$ matrices by $b_{ij}$ and $b_{ii}$ in double:

```
CREATE VERTEX B_pND (PRIMARY_ID B_pID int, bii double)
CREATE VERTEX B_ppND (PRIMARY_ID B_ppID int, bii double)
CREATE DIRECTED EDGE B_pConnect (from B_pND, to B_pND, bij double)
CREATE DIRECTED EDGE B_ppConnect (from B_ppND, to B_ppND, bij double)
```

The vertices B_pND, B_ppND, and edges B_pConnect, B_ppConnect are virtual vertices and virtual edges. They are not physical nodes or branches in power systems. In the B_P and B_PP graph structure, the counterparts of the edges between vertices are nonzero off-diagonal elements in the $B'$ and $B''$ matrices. The diagonal elements in the $B'$ and $B''$ matrices are modeled as attributes of the vertex. Zero (absent) elements in the $B'$ and $B''$ matrices indicate that no direct connections between the vertices exist in the B_P and B_PP graphs.

With the defined vertex and edge above, the B_P and B_PP graph schema is then simply defined as:

```
CREATE GRAPH B_pGraph(B_pND, B_pConnect)
CREATE GRAPH B_ppGraph(B_ppND, B_ppConnect)
```

The B_P and B_PP graph schema defined the graph structure. And the B_P and B_PP graphs will be built by adding vertices B_pND, B_ppND, and edges B_pConnect, B_ppConnect by the following pseudo-query code.

```
CREATE QUERY buildB_pAndB_ppGraph(
) FOR GRAPH busBranchGraph and GRAPH B_pGraph and GRAPH B_ppGraph{

SumAccum<double> @sumB  = 0
SumAccum<double> @sumBi = 0

T0= {TopoND.*}
T1 = SELECT s FROM T0:s-(connected:e)->:t
  IF (t.busType!= 3) THEN
    ACCUM
      s.@sumB  += 1.0/e.X
    INSERT INTO B_pND(s.TOPOID)
    INSERT INTO B_pND(t.TOPOID)
      INSERT INTO B_pConnect(s.TOPOID, t.TOPOID) VALUES(-1.0/e.X)
    POST-ACCUM
```

```
        UPDATE b_p FROM B_pND WHERE b_p.B_pID == s.TOPOID
            SET b_p.bii = s.@sumB
     END

  T1 = SELECT s FROM T0:s-(connected:e)->:t
     IF (t.busType == 0 or t.busType == 1) THEN
        ACCUM
           s.@sumBi += e.X/(e.R*e.R + e.X*e.X)
           INSERT INTO B_ppND(s.TOPOID)
           INSERT INTO B_ppND(t.TOPOID)
           INSERT INTO B_ppConnect(s.TOPOID, t.TOPOID)
                       VALUES(e.X/(e.R*e.R + e.X*e.X))
        POST-ACCUM
          UPDATE b_pp FROM B_ppND WHERE b_pp.B_ppID == s.TOPOID
          SET b_pp.bii = s.@sumB
     END
  }
```

In the query, vertex-attached variables SumAccum @sumB, @sumBi are defined to store the diagonal elements in the $B'$ and $B''$ matrices which will be assigned to be vertex attributes of in the B_P and B_PP graphs.

In the query definition, graph traversal statement T1 = SELECT s FROM T0:s-(connected: e)->:t travels all edges from vertex s in the vertex set T0 = {TopoND.*}. Nested in the ACCUM, @sumB is calculated in nodal parallel. And new edges are added to B_pConnect and B_ppConnect by:

```
INSERT INTO B_pConnect(s.TOPOID, t.TOPOID) VALUES(-1.0/e.X)
INSERT INTO B_ppConnect(s.TOPOID, t.TOPOID)
            VALUES(e.X/(e.R*e.R + e.X*e.X))
```

New vertices are added to B_pND and B_ppND before new edges are added in between the vertices by:

```
INSERT INTO B_pND(s.TOPOID)
INSERT INTO B_pND(t.TOPOID)
INSERT INTO B_ppND(s.TOPOID)
INSERT INTO B_ppND(t.TOPOID)
```

In the POST-ACCUM, diagonal elements in the $B'$ and $B''$ matrices are calculated for all $n-1$ buses and PQ buses respectively. Each calculated diagonal element is assigned to the vertex attribute by the two UPDATE statements:

```
UPDATE b_p FROM B_pND WHERE b_p.B_pID == s.TOPOID
  SET b_p.bii = s.@sumB
UPDATE b_pp FROM B_ppND WHERE b_pp.B_ppID == s.TOPOID
  SET b_pp.bii = s.@sumB
```

After the B_P and B_PP graphs are created, the graph symbolic factorization, elimination tree creation, node partition, and graph numerical factorization are the same as we process power flow calculation by the Newton–Raphson method. The only change is to apply these graph-based algorithms to graphs `B_pGraph` and `B_ppGraph` other than graph `JacobianGraph`.

## 10.5 Ill-Conditioned Power Flow Problem Solution

### 10.5.1 Introduction

Finding a solution to the power flow problem is imperative for many power system applications and several iterative approaches are employed to achieve this objective. However, because of the non-convex feasibility region of this problem, the possibility of finding a solution is dependent on the choice of the initial point. As the feasibility region of the power flow problem is nonconvex, and rank deficiency in power flow equations is possible, where the traditional Jacobian-based approaches fail to procure a feasible solution for ill-conditioned problems.

To solve an ill-conditioned power flow problem, a noniterative approach based on convex relaxation is employed to verify the existence of a feasible solution and then find a solution. To ensure the scalability of the convex relaxation, the sparse semi-definite programming relaxation corresponding to each maximal clique within the graph associated with the network is formulated. Perturbation and network reconfiguration schemes are employed to improve the tightness of the convex relaxation and verify a feasible solution for the original nonconvex problem.

Power flow is one critical underlying problem for power system analysis. Integration of intermittent renewable energy resources and possible network contingencies further highlight the merit of providing an efficient framework to solve this nonlinear problem efficiently. The complexity of modern power systems further underlines the essence of having reliable tools for power flow analysis. The power flow is formulated as a set of nonlinear equations. Several iterative approaches including the Gauss–Seidel method and the Newton–Raphson method are adopted by the industry sector to solve this problem.

However, the convergence and stability of these approaches could not always be guaranteed. The Newton–Raphson method is the most popular approach to solve this system of equations as it provides a better convergence compared to other techniques. Although the Newton–Raphson method provides an improved convergence, it may be unable to procure the solution as a result of an inappropriate initial point or rank deficiency in the system of power flow equations that renders a singular Jacobean matrix. Employing the Newton–Raphson method to solve the power flow problem could lead to the following scenarios:

1) A unique solution exists and could be found regardless of the initial guess,
2) Multiple solutions exist, and one solution is returned based on an initial guess,
3) No solution exists,
4) A unique or multiple solutions exist while a solution cannot be procured because of an improper initial point that renders a singular Jacobean matrix in the iterative process. Such a scenario corresponds to ill-conditioned power flow.

A framework is developed to overcome the numerical issues raised by the iterative solution frameworks for the power flow problem. The first step is to quickly find out the feasibility of the power flow problem. If the power flow is infeasible, one can change the set points and provide corrective actions to make the problem feasible.

To tackle the issues with iterative approaches for solving the ill-conditioned power flow problems, convex relaxation approaches are used to check the feasibility of the power flow problem.

For this step, a sparse perturbed semi-definite programming formulation is applied as a noniterative methodology to procure a feasible solution for the power flow problem in polynomial time. The second step is to determine a feasible solution for the power flow problem. Once it is ensured that the power flow problem has a feasible solution, a mathematical framework is developed to procure a unique solution that provides minimum power loss and is independent of the choice of the initial point. For this step, an efficient convex relaxation approach is developed to find a feasible solution for the power flow problem. The tightness of the presented convex relaxation is ensured using appropriate perturbation matrices as well as network reconfiguration schemes.

### 10.5.2 Determine the Feasibility of the Power Flow

Although many research works address the feasibility of the power flow problem [4, 5], iterative approaches such as the Newton–Raphson and the Gauss–Seidel may work well to solve large-scale problems, but they are incapable of handling the ill-conditioned power flow problems as addressed numerically in [6, 7].

Similar to the optimal power flow problem, the power flow problem has a nonconvex feasibility region as the power flow constraints/equations are the same in both problems. Several approaches were proposed to solve the optimal power flow problem using convex relaxation techniques [8–11]. These relaxations and convex approximations provide a solution to a broader class of power system problems compared to traditional methods.

While finding the global optimal solution is the ultimate goal for the optimal power flow problem, we are focusing on checking if the power flow problem has any feasible solution. An efficient algorithm is required to procure a tight convex relaxation for the power flow problem. It is shown in [12] that the semi-definite programming relaxation presented in [9] for the optimal power flow problem is unable to render a feasible solution for the optimal power flow problem.

Semi-definite programming relaxation is a special case of the hierarchy of moment relaxation with order one in the hierarchy. The computation burden of the moment relaxation approach is very large and this challenge makes it impractical for large-scale applications.

However, exploiting the sparsity of the electricity network enables the attainment of the solution for mid-size optimal power flow problems. An alternative convex relaxation approach is proposed in [13] in which the quadratic convex relaxation approach along with a bound-tightening algorithm is presented to procure the global solution for the optimal power flow problem. Another approach is to employ the second-order cone programming (SOCP) relaxation as proposed in [11] where several convex envelopes are formulated to procure a tight convex relaxation of the original nonconvex optimal power flow problem.

The approaches used for solving the optimal power flow problem can be used to check for the feasibility of the power flow problem. The moment relaxation for the nonconvex optimal power flow problem is considered tight if the rank of the relaxation matrix is equal to one. As a lower-order moment relaxation renders a lower bound for the global solution of the optimal power flow problem, the procured solution may not be feasible for the original nonconvex problem. One approach is to employ higher-order moment relaxation that increases the number of monomials and computation burden. The other approach to procuring a solution that is feasible for the original nonconvex problem is to perturb the moment relaxation to reduce its rank to 1 [13–15].

Although reducing the rank of the moment relaxation matrix to 1 renders a feasible solution for the original nonconvex problem, it may not procure the global solution to the original problem.

Since the objective is to check the feasibility of the power flow, the perturbation techniques could be used to check the feasibility of the power flow without focusing on the global solution for this problem.

A rank minimization approach is proposed in [16] in which the hidden solution for the convex relaxation of the optimal power flow problem is procured using the proposed linearization-minimization procedure. We employ the semi-definite programming relaxation of the power flow problem with perturbation to determine the feasibility of this problem. As the sparse formulation for the semi-definite programming relaxation is employed the computation burden of the problem is not considerable and this approach can be applied to large-scale systems.

### 10.5.3  Problem Formulation for Determining the Feasibility of Power Flow

The power flow problem is formulated as given in (10.44)–(10.51), where power system buses are divided into three categories. If a power network has $n$ buses, there are $2n$ known parameters and $2n$ unknown variables in the power flow equations.

The magnitude and angle of the voltage for the slack/reference bus, the voltage magnitude and real power injection at voltage-controlled buses, and real and reactive power injection at load buses are the known parameters for the power flow problem. The unknown variables are the real and reactive power injection for the slack bus, voltage angles and reactive power injection for the voltage-controlled (PV) buses, and the voltage magnitude and voltage angle for the load (PQ) buses. Real and reactive power injections are not usually enforced for the slack bus as it is supposed to compensate for the real and reactive power mismatch in the network. The slack bus is considered a reference bus to calculate the voltage angle for the load and voltage-controlled buses.

The limits on the real power generation of the slack bus are shown in (10.44), where $P_i^{G,\min}$ and $P_i^{G,\max}$ are the minimum and maximum power generation on bus $i$, $P_i^D$ is the real power demand on bus $i$, and $f_{ij}$ is a function of the real and imaginary part of the voltage phasor ($V_{(.)}^d$ and $V_{(.)}^q$) on buses $i$ and $j$ from the total number of buses in the network.

The limits on the reactive power generation on slack buses and PV buses are given in (10.45), where $Q_i^{G,\min}$ and $Q_i^{G,\max}$ are the minimum and maximum reactive power generation on bus $i$, $Q_i^D$ is the reactive power demand on bus $i$, and $g_{ij}$ is a function of the real and imaginary part of the voltage phasor on buses $i$ and $j$.

The real power balance equation for PV and PQ buses is given in (10.46), where $P_i^G$ is the given real power generation. The reactive power balance equation for PQ is given in (10.47), where $Q_i^G$ is the given reactive power generation.

The limits on the magnitude of the voltage on each bus are given in (10.48), where $V_i^{\min}$ and $V_i^{\max}$ are the minimum and maximum limits for the voltage magnitude. The voltage magnitude for PV buses is provided in (10.49).

The functions for the real and reactive power utilized in (10.44)–(10.47) are introduced in (10.50) and (10.51), respectively, where $G_{ij}$ and $B_{ij}$ are the elements of the conductance and susceptance matrices of the network.

$$P_i^{G,\min} \leq P_i^D + \sum_{j=1}^{n} f_{ij}\left(V_i^d, V_j^d, V_i^q, V_j^q\right) \leq P_i^{G,\max} \, \forall i \in S \tag{10.44}$$

$$Q_i^{G,\min} \leq Q_i^D + \sum_{j=1}^{n} g_{ij}\left(V_i^d, V_j^d, V_i^q, V_j^q\right) \leq Q_i^{G,\max} \, \forall i \in \{PV\} \tag{10.45}$$

$$P_i^G - P_i^D = \sum_{j=1}^{n} f_{ij}\left(V_i^d, V_j^d, V_i^q, V_j^q\right) \forall i \in \{PV, PQ\} \tag{10.46}$$

$$Q_i^G - Q_i^D = \sum_{j=1}^{n} g_{ij}\left(V_i^d, V_j^d, V_i^q, V_j^q\right) \forall i \in \{PQ\} \tag{10.47}$$

$$V_i^{\min} \leq \sqrt{\left(V_i^d\right)^2 + \left(V_i^q\right)^2} \leq V_i^{\max} \forall i \in PQ \tag{10.48}$$

$$\sqrt{\left(V_i^d\right)^2 + \left(V_i^q\right)^2} = \left|V_i'\right| \forall i \in PV \tag{10.49}$$

$$f_{ij}\left(V_i^d, V_j^d, V_i^q, V_j^q\right) = G_{ij}\left(V_i^d V_j^d + V_i^q V_j^q\right) - B_{ij}\left(V_i^d V_j^q + V_i^q V_j^d\right) \tag{10.50}$$

$$g_{ij}\left(V_i^d, V_j^d, V_i^q, V_j^q\right) = -B_{ij}\left(V_i^d V_j^d + V_i^q V_j^q\right) - G_{ij}\left(V_i^d V_j^q + V_i^q V_j^d\right) \tag{10.51}$$

Solving the problem presented in (10.44)–(10.51) is challenging and iterative approaches may not lead to a feasible solution. Thus, a noniterative solution methodology is needed.

### 10.5.4 Power Flow Feasibility Verification

A tractable algorithm is presented in this section to procure a feasible solution for the power flow problem formulated in (10.44)–(10.51). The semi-definite programming relaxation is the first order for Lasserre's hierarchy of moment relaxation, where convergence to a feasible solution for the original nonconvex problem is guaranteed once the order of moment relaxation goes to infinity.

However, employing higher orders of the moment relaxation to procure such a feasible solution is computationally expensive and not desirable. Thus, a new methodology based on semi-definite programming relaxation is presented here. If the rank of the semi-definite programming relaxation for the power flow problem is one, the relaxation is exact, and the power flow problem has a feasible solution. The problem given in (10.44)–(10.51) is reformulated as sparse semi-definite programming in which a first-order moment relaxation matrix is presented for each maximal clique within the network.

In a clique, all buses are adjacent to each other, and the maximal set of cliques does not share any power line with each other. Thus, the semi-definite programming matrices associated with each maximal clique are dense. Although the average rank of each sparse semi-definite programming relaxation matrix is lower than the single nonsparse semi-definite programming relaxation matrix, it will not necessarily be one as desired. The original power flow problem is feasible when the rank of all of the sparse semi-definite programming matrices is one.

To address this challenge, a rank minimization approach is required to determine the rank-1 semi-definite programming relaxation matrix. The rank minimization problem is NP-hard and the semi-definite programming relaxation would not provide the desired rank-1 solution for the power flow problem. To procure the desired rank-1 solution, a perturbation is needed for the semi-definite programming relaxation formulation of the power flow problem [17]. The perturbed relaxed problem is tighter than the relaxed problem while the procured solution is feasible for the original nonconvex problem. The solution procured by the perturbed relaxed problem is desirable as it is a feasible solution for the original power flow problem.

The perturbed convex relaxation formulation of the power flow problem is given in (10.52)–(10.60). The perturbation matrices ($\zeta_c$) associated with each maximal clique ($c$) are chosen in the objective function as given in (10.52) to perturb the presented semi-definite programming relaxation. The sparse semi-definite programming matrix for each maximal clique is given in (10.60), where $\gamma_{(.)}$ represents the lifting variables associated with the nonlinear terms for voltages within each maximal clique and $|c|$ is the size of the associated maximal clique. For example, $V_i^q V_j^d$ is a nonlinear term in (10.50) and (10.51) which is replaced by the lifting variable $\gamma_{V_i^q V_j^d}$ in (10.57) and (10.58) and semi-definite programming matrix in (10.60). Here, the inequalities associated with real and reactive power injections at each bus are rewritten in (10.53)–(10.56). The limits on the magnitude of voltage in each bus are enforced in (10.57). Utilizing the presented lifting variables, the functions for active and reactive power injections that are utilized in (10.53)–(10.56) are given in (10.58) and (10.59), respectively.

$$\min \sum_c Tr(\zeta_c X_c) \tag{10.52}$$

Subject to

$$P_i^{G,\min} \leq P_i^D + \sum_{j=1}^n f'_{ij}\left(\gamma_{V_i^d},\gamma_{V_j^d},\gamma_{V_i^q},\gamma_{V_j^q}\right) \leq P_i^{G,\max} \forall i \in S \tag{10.53}$$

$$Q_i^{G,\min} \leq Q_i^D + \sum_{j=1}^n g'_{ij}\left(\gamma_{V_i^d},\gamma_{V_j^d},\gamma_{V_i^q},\gamma_{V_j^q}\right) \leq Q_i^{G,\max} \forall i \in \{S,PV\} \tag{10.54}$$

$$P_i^G - P_i^D = \sum_{j=1}^n f'_{ij}\left(\gamma_{V_i^d},\gamma_{V_j^d},\gamma_{V_i^q},\gamma_{V_j^q}\right) \forall i \in \{PV,PQ\} \tag{10.55}$$

$$Q_i^G - Q_i^D = \sum_{j=1}^n g'_{ij}\left(\gamma_{V_i^d},\gamma_{V_j^d},\gamma_{V_i^q},\gamma_{V_j^q}\right) \forall i \in \{PQ\} \tag{10.56}$$

$$\left(V_i^{\min}\right)^2 \leq \gamma_{\left(V_i^d\right)^2} + \gamma_{\left(V_i^q\right)^2} \leq \left(V_i^{\max}\right)^2 \forall i \tag{10.57}$$

$$f'_{ij}\left(\gamma_{V_i^d},\gamma_{V_j^d},\gamma_{V_i^q},\gamma_{V_j^q}\right) = G_{ij}\left(\gamma_{V_i^d V_j^d} + \gamma_{V_i^q V_j^q}\right) - B_{ij}\left(\gamma_{V_i^d V_j^q} + \gamma_{V_i^q V_j^d}\right) \tag{10.58}$$

$$g'_{ij}\left(\gamma_{V_i^d},\gamma_{V_j^d},\gamma_{V_i^q},\gamma_{V_j^q}\right) = -B_{ij}\left(\gamma_{V_i^d V_j^d} + \gamma_{V_i^q V_j^q}\right) - G_{ij}\left(\gamma_{V_i^d V_j^q} + \gamma_{V_i^q V_j^d}\right) \tag{10.59}$$

$$X_c = \begin{pmatrix} 1 & \gamma_{V_i^d} & \cdots & \gamma_{V_{|c|}^d} & \gamma_{V_i^q} & \cdots & \gamma_{V_{|c|}^q} \\ \gamma_{V_i^d} & \gamma_{\left(V_i^d\right)^2} & \cdots & \gamma_{V_i^d V_{|c|}^d} & \gamma_{V_i^d V_i^q} & \cdots & \gamma_{V_i^d V_{|c|}^q} \\ \vdots & \vdots & \ddots & \vdots & \vdots & \ddots & \vdots \\ \gamma_{V_{|c|}^d} & \gamma_{V_{|c|}^d V_i^d} & \cdots & \gamma_{\left(V_{|c|}^d\right)^2} & \gamma_{V_{|c|}^d V_i^q} & \cdots & \gamma_{V_{|c|}^d V_{|c|}^q} \\ \gamma_{V_i^q} & \gamma_{V_i^q V_i^d} & \cdots & \gamma_{V_i^q V_{|c|}^d} & \gamma_{\left(V_i^q\right)^2} & \cdots & \gamma_{V_i^q V_{|c|}^q} \\ \vdots & \vdots & \ddots & \vdots & \vdots & \ddots & \vdots \\ \gamma_{V_{|c|}^q} & \gamma_{V_{|c|}^q V_i^d} & \cdots & \gamma_{V_{|c|}^q V_{|c|}^d} & \gamma_{V_{|c|}^q V_i^q} & \cdots & \gamma_{\left(V_{|c|}^q\right)^2} \end{pmatrix} \succeq 0 \tag{10.60}$$

Finding a feasible solution to the power flow problem using perturbed convex relaxation is computationally inexpensive. To ensure the feasibility of the procured solution for the original power flow problem, the rank of all semi-definite programming matrices given in (10.60) should be one. The rank of a matrix is equal to the number of its nonzero eigenvalues. A matrix is considered as rank-1 when its second-largest eigenvalue is significantly smaller than its largest eigenvalue.

Thus, the ratio of the largest and second-largest eigenvalues is captured as a measure of tightness. If this ratio is significantly large, the rank of that matrix is one and the perturbed convex relaxation is tight. The presented perturbation matrix enforces the rank-1 solution as given in (10.61).

$$
\zeta_c = \begin{pmatrix} 1 & -\alpha^T & \alpha^T \\ -\alpha & -\sqrt{\alpha}\sqrt{\alpha}^T & 0 \\ \alpha & 0 & \sqrt{\alpha}\sqrt{\alpha}^T \end{pmatrix}
\tag{10.61}
$$

The choice of the perturbation matrix has no impact on the feasibility of the relaxed problem. However, the proper choice of perturbation matrix would enforce the desired rank-1 solution which is feasible for the original power flow problem. The proper choice of the perturbation matrix is not necessarily unique, where various perturbation matrices may lead to various solutions to the power flow problem in case of the existence of multiple solutions.

As the focus of this approach is to quickly find at least a solution for the power flow problem, in the proposed perturbation matrix, all elements of $\alpha$, which is a vector of size equal to $|c|$, are equal to each other e.g. 1e-2. If the perturbed relaxed problem is infeasible or procuring a rank-1 solution is not possible, the power flow problem has no feasible solution.

### 10.5.5 Find a Feasible Solution for the Power Flow Problem

The nonlinear terms in (10.44)–(10.51) are represented by respective lifting variables in the semi-definite programming relaxation matrix as formulated for the relaxed problem in (10.62)–(10.75). If all sparse semi-definite programming relaxation matrices are rank-1, the presented relaxation in (10.62)–(10.75) is tight, and a feasible solution for the power flow problem in (10.44)–(10.51) is procured.

The objective of the perturbed convex relaxation is given in (10.62), where the lifting terms associated with the square of real and imaginary parts of voltage on each bus are employed. The choice of perturbation plays a significant role in procuring a rank-1 solution. This choice of perturbation not only contributes to the rank minimization but also enforces the real and imaginary parts of the voltage on each bus to be closest to 1 and 0 p.u., respectively. The choice of perturbation is not unique; however, it should lead to a rank-1 solution.

The objective is to procure a feasible solution for the power flow problem; however, power flow problems may have multiple solutions and other arrangements for the choice of perturbation could facilitate reaching those solutions. The nonlinear terms in (10.44)–(10.51) are presented by their associated lifting variables in (10.62)–(10.75), and the semi-definite programming matrices associated with each maximal clique given in (10.75) contain the lifting variables. For example, $V_i^q V_j^d$ is a nonlinear term in (10.44)–(10.51) which is replaced by a lifting variable $\gamma_{V_i^q V_j^d}$ as defined in the semi-definite constraint (10.75).

The voltage for the slack/reference bus is enforced by (10.63) where $S$ is the slack bus set. Although, the real power balance for the slack bus is ignored in (10.44)–(10.51), considering the generation capacity limits for the slack bus will ensure the feasibility of the solution procured by solving the relaxed problem. Here, the real and reactive generation capacity of the generation unit connected to the slack bus is enforced by (10.64) and (10.65), respectively.

The real power balance for the voltage-controlled and load buses is presented in (10.66). The reactive power balance for the load buses is shown in (10.67). The voltage limits for load buses are not usually considered for the power flow problem. However, to ensure the feasibility of the procured solution, the voltage limits for the load buses are presented in (10.68).

For the voltage-controlled bus, the voltage magnitude is enforced by (10.69) and (10.70). Once the reactive power of the generation units connected to a voltage-controlled bus reaches its limits, the voltage-controlled bus will change into a load bus and its reactive power is fixed while the voltage magnitude will be unknown. In the convex relaxation problem, this condition is captured by two auxiliary binary variables for each voltage-controlled bus to check if any of the upper and lower limits for the reactive power generation of the generation units is reached.

The reactive power generation of the units connected to voltage-controlled buses is enforced by (10.71)–(10.74). Here once the reactive power generation reaches the upper or lower limits, the auxiliary binary variable becomes 1.

$$\min_{X_c} \sum_i \left( \gamma_{(V_i^d)^2} + \gamma_{(V_i^q)^2} - 2V_i^d \right) \tag{10.62}$$

Subject to

$$\gamma_{(V_i^d)^2} = \left( V_{ref}^d \right)^2, \gamma_{(V_i^q)^2} = \left( V_{ref}^q \right)^2 \forall i \in S \tag{10.63}$$

$$P_i^{G,\min} \le P_i^D + \sum_{j=1}^n \left( G_{ij} \left( \gamma_{V_i^d V_j^d} + \gamma_{V_i^q V_j^q} \right) - B_{ij} \left( \gamma_{V_i^d V_j^q} - \gamma_{V_i^q V_j^d} \right) \right) \le P_i^{G,\max} \quad \forall i \in S \tag{10.64}$$

$$Q_i^{G,\min} \le Q_i^D + \sum_{j=1}^n \left( -B_{ij} \left( \gamma_{V_i^d V_j^d} + \gamma_{V_i^q V_j^q} \right) - G_{ij} \left( \gamma_{V_i^d V_j^q} - \gamma_{V_i^q V_j^d} \right) \right) \le Q_i^{G,\max} \quad \forall i \in S \tag{10.65}$$

$$P_i'^G - P_i^D = \sum_{j=1}^n \left( G_{ij} \left( \gamma_{V_i^d V_j^d} + \gamma_{V_i^q V_j^q} \right) - B_{ij} \left( \gamma_{V_i^d V_j^q} - \gamma_{V_i^q V_j^d} \right) \right)$$

$$\forall i \in \{PV, PQ\} \tag{10.66}$$

$$Q_i'^G - Q_i^D = \sum_{j=1}^n \left( -B_{ij} \left( \gamma_{V_i^d V_j^d} + \gamma_{V_i^q V_j^q} \right) - G_{ij} \left( \gamma_{V_i^d V_j^q} - \gamma_{V_i^q V_j^d} \right) \right)$$

$$\forall i \in PQ \tag{10.67}$$

$$\left( V_i^{\min} \right)^2 \le \gamma_{(V_i^d)^2} + \gamma_{(V_i^q)^2} \le \left( V_i^{\max} \right)^2 \forall i \in PQ \tag{10.68}$$

$$|V_i'|^2 - \left( u_i^{Q,\max} + u_i^{Q,\min} \right) \left( \left( V_i^{\max} \right)^2 - \left( V_i^{\min} \right)^2 \right) \le \gamma_{(V_i^d)^2} + \gamma_{(V_i^q)^2}$$

$$\forall i \in PV \tag{10.69}$$

$$\gamma_{(V_i^d)^2} + \gamma_{(V_i^q)^2} \le |V_i'|^2 + \left( u_i^{Q,\min} + u_i^{Q,\max} \right) \left( \left( V_i^{\max} \right)^2 - \left( V_i^{\min} \right)^2 \right)$$

$$\forall i \in PV \tag{10.70}$$

$$Q_i^D + \sum_{j=1}^n \left( -B_{ij} \left( \gamma_{V_i^d V_j^d} + \gamma_{V_i^q V_j^q} \right) - G_{ij} \left( \gamma_{V_i^d V_j^q} - \gamma_{V_i^q V_j^d} \right) \right) - \varepsilon u_i^{Q,\max} \le Q_i^{G,\max} - \varepsilon \quad \forall i \in PV \tag{10.71}$$

$$Q_i^D + \sum_{j=1}^n \left( -B_{ij} \left( \gamma_{V_i^d V_j^d} + \gamma_{V_i^q V_j^q} \right) - G_{ij} \left( \gamma_{V_i^d V_j^q} - \gamma_{V_i^q V_j^d} \right) \right)$$

$$+ (2 + \varepsilon) Q_i^{G,\max} u_i^{Q,\max} \ge Q_i^{G,\max} (-1 - \varepsilon) \quad \forall i \in PV \tag{10.72}$$

$$Q_i^D + \sum_{j=1}^{n} \left( -B_{ij} \left( \gamma_{V_i^d V_j^d} + \gamma_{V_i^q V_j^q} \right) - G_{ij} \left( \gamma_{V_i^d V_j^q} - \gamma_{V_i^q V_j^d} \right) \right) + \varepsilon u_i^{Q,\min} \geq Q_i^{G,\min} + \varepsilon \forall i \in PV$$

(10.73)

$$Q_i^D + \sum_{j=1}^{n} \left( -B_{ij} \left( \gamma_{V_i^d V_j^d} + \gamma_{V_i^q V_j^q} \right) - G_{ij} \left( \gamma_{V_i^d V_j^q} - \gamma_{V_i^q V_j^d} \right) \right)$$

$$- (2 + \varepsilon) Q_i^{G,\min} u_i^{Q,\min} \leq Q_i^{G,\min} (-1 - \varepsilon) \ \forall i \in PV$$

(10.74)

$$X_c = \begin{pmatrix} \gamma_{(V_i^d)^2} & \cdots & \gamma_{V_i^d V_{|c|}^d} & \gamma_{V_i^d V_i^q} & \cdots & \gamma_{V_i^d V_{|c|}^q} \\ \vdots & \ddots & \vdots & \vdots & \ddots & \vdots \\ \gamma_{V_{|c|}^d V_i^d} & \cdots & \gamma_{(V_{|c|}^d)^2} & \gamma_{V_i^d V_i^q} & \cdots & \gamma_{V_{|c|}^d V_{|c|}^q} \\ \gamma_{V_i^q V_i^d} & \cdots & \gamma_{V_i^q V_{|c|}^d} & \gamma_{(V_i^q)^2} & \cdots & \gamma_{V_i^q V_{|c|}^q} \\ \vdots & \ddots & \vdots & \vdots & \ddots & \vdots \\ \gamma_{V_{|c|}^q V_i^d} & \cdots & \gamma_{V_{|c|}^q V_{|c|}^d} & \gamma_{V_{|c|}^q V_i^q} & \cdots & \gamma_{(V_{|c|}^q)^2} \end{pmatrix} \succeq 0$$

(10.75)

Considering the auxiliary binary variable for voltage-controlled buses, the convex relaxation problem is formulated as a mixed-integer semi-definite programming problem. To solve this problem, a branch and bound algorithm could be employed. The sub-problem in each node of the branch and bound algorithm is a semi-definite programming problem with sparse semi-definite programming relaxation matrices.

The solution procured from the problem presented in (10.62)–(10.75) is feasible for the power flow problem when the convex relaxation is tight, and the rank of all semi-definite programming relaxation matrices is one. The measure for the tightness of the solution procured (10.62)–(10.75) is the ratio of the two largest eigenvalues of the procured moment matrices as shown in (10.76), where $TR_c$ is the tightness measure for maximal clique $c$.

$$TR_c = log \left( \frac{\lambda_{|c|}^c}{\lambda_{|c|-1}^c} \right)$$

(10.76)

If the ratio is a substantial number, the rank of the moment relaxation matrix is one. Therefore, the relaxation is tight and the voltages for the original problem can be procured using the Cholesky decomposition of the sparse semi-definite programming matrices. The vector of real and imaginary parts of voltages within a clique is procured by (10.77).

$$\left[ V_i^d, ..., V_{|c|}^d, V_i^q, ..., V_{|c|}^q \right] = \lambda_{|c|}^c q_{|c|}^{c \ T}$$

(10.77)

Formulating the convex relaxation for the power flow problem in sparse form along with the perturbation will provide a feasible solution if it exists. However, if the procured tightness measure is small and solving the relaxed problem would render a higher than rank-1 solution, a network reconfiguration method is applied to avoid employing the higher-order moment relaxation.

A number of buses exist in the power network to facilitate the connection between the generation and load buses while they have zero power injections. On many occasions, the outcome of the relaxed problem is not rank-1 if the maximal cliques have load buses with zero real and reactive power injections. The reason is that the zero injection buses increase the degree of freedom for the set of power flow equations by introducing (10.78) and (10.79). This leads to a higher than rank-1 moment matrix corresponding to the maximal clique containing these buses.

$$0 = \sum_{j=1}^{NB} G_{ij} \left( \gamma_{V_i^d V_j^d} + \gamma_{V_i^q V_j^q} \right) - B_{ij} \left( \gamma_{V_i^d V_j^q} - \gamma_{V_i^q V_j^d} \right) \tag{10.78}$$

$$0 = \sum_{j=1}^{NB} - B_{ij} \left( \gamma_{V_i^d V_j^d} + \gamma_{V_i^q V_j^q} \right) - G_{ij} \left( \gamma_{V_i^d V_j^q} - \gamma_{V_i^q V_j^d} \right) \tag{10.79}$$

To tackle this challenge, network reconfiguration is employed to eliminate the load buses with zero injection from the network. The procured topology is equivalent to the original network topology; however, the degree of freedom for the relaxed power flow problem is reduced.

The following cases in which the degree of the load bus with zero injection in the graph associated with the power network is equal to 1, 2, and 3 are considered. If the degree of the zero injection bus is 1 as shown in Figure 10.1, the bus can be removed from the network. The flow of the line connected to this bus is zero, and the voltage of this bus is equal to the bus connected to it.

If the degree of the zero injection bus is 2 as shown in Figure 10.2, the load bus with zero injection can be removed from the network while the two lines connected to this bus will merge into a single line in the reconfigured topology as shown in Figure 10.3. Here, $Z_{ik} = Z_{ij} + Z_{jk}$ and the Y matrix of the configured network are further adjusted. The voltage $V_j^d + j V_j^q$ of this bus $j$ can be further recovered by solving Eqs. (10.80) and (10.81). Here, two unknowns i.e. $V_j^d$ and $V_j^q$, could be found once $\gamma_{V_i^d}$ and $\gamma_{V_i^q}$ are procured from the power flow solution of the reconfigured network.

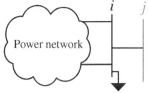

**Figure 10.1** A zero injection load bus with degree 1.

$$\sum_i \left( G_{ij} \left( \gamma_{V_i^d} V_j^d + \gamma_{V_i^q} V_j^q \right) + B_{ij} \left( \gamma_{V_i^d} V_j^q - \gamma_{V_i^q} V_j^d \right) \right) = 0 \tag{10.80}$$

$$\sum_i \left( - B_{ij} \left( \gamma_{V_i^d} V_j^d + \gamma_{V_i^q} V_j^q \right) + G_{ij} \left( \gamma_{V_i^d} V_j^q - \gamma_{V_i^q} V_j^d \right) \right) = 0 \tag{10.81}$$

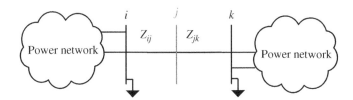

**Figure 10.2** A zero injection load bus with degree 2.

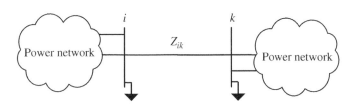

**Figure 10.3** The reconfigured network for zero injection load bus with degree 2.

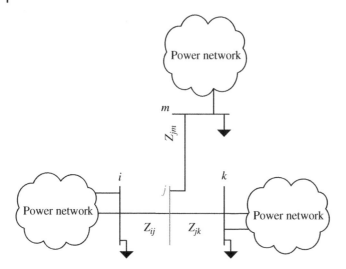

**Figure 10.4** A zero injection load bus with degree 3.

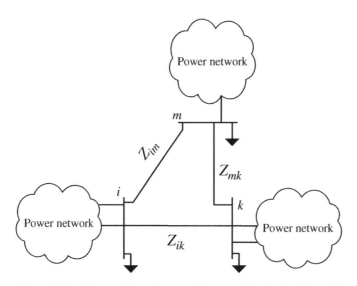

**Figure 10.5** The reconfigured network for zero injection load bus with degree 3.

If the degree of the zero injection bus is 3 as shown in Figure 10.4, the load bus with zero injection can be removed from the network by changing the network topology to its equivalent shown in Figure 10.5. The branch impedances are procured using Wye-delta conversion shown in (10.82) and the $Y$ matrix of the reconfigured network is constructed accordingly.

$$
\begin{cases}
Z_{im} = \left(Z_{ij}Z_{jm}\right)\left(Z_{ij} + Z_{jk} + Z_{jm}\right)^{-1} \\
Z_{ik} = \left(Z_{ij}Z_{jk}\right)\left(Z_{ij} + Z_{jk} + Z_{jm}\right)^{-1} \\
Z_{km} = \left(Z_{kj}Z_{jm}\right)\left(Z_{ij} + Z_{jk} + Z_{jm}\right)^{-1}
\end{cases}
\tag{10.82}
$$

The voltage $V_j^d + jV_j^q$ of this bus $j$ can be further recovered by solving Eqs. (10.80) and (10.81).

# References

1 Shi, J., Liu, G., Dai, R. et al. (2018). Graph based power flow calculation for energy management system. *2018 IEEE Power & Energy Society General Meeting (PESGM)*, Portland, OR, USA, pp. 1–5. https://doi.org/10.1109/PESGM.2018.8586233.

2 Dai, R., Liu, G., Wang, Z. et al. (2020). A novel graph-based energy management system. *IEEE Transactions on Smart Grid* 11 (3): 1845–1853.

3 Yuan, C., Lu, Y., Feng, W. et al. (2019). Graph computing based distributed fast decoupled power flow analysis. *2019 IEEE Power & Energy Society General Meeting (PESGM)*, Atlanta, GA, USA, pp. 1–5. https://doi.org/10.1109/PESGM40551.2019.8973870.

4 Makarov, Y.V., Dong, Z.Y., and Hill, D.J. (2008). On convexity of power flow feasibility boundary. *IEEE Transactions on Power Systems* 23 (2): 811–813.

5 Ferris, L.L. and Sasson, A.M. (1968). Investigation of the load flow problem. *IEE Proceedings* 115 (10): 1459–1470.

6 Tripathy, S.C., Prasad, G.D., Malik, O.P., and Hope, G.S. (1982). Load-flow solutions for ill-conditioned power systems by a Newton-like method. *IEEE Transactions on Power Apparatus and Systems* PAS-101 (10): 3648–3657.

7 Milano, F. (2009). Continuous Newton's method for power flow analysis. *IEEE Transactions on Power Apparatus and Systems* 24 (1): 50–57.

8 Molzahn, D. (2017). Computing the feasible spaces of optimal power flow problems. *IEEE Transactions on Power Apparatus and Systems* 32 (6): 4752–4763.

9 Low, S.H. (2014). Convex relaxation of optimal power flow—part I: formulations and equivalence. *IEEE Transactions on Control of Network Systems* 1 (1): 15–27.

10 Madani, R., Sojoudi, S., and Lavaei, J. (2015). Convex relaxation for optimal power flow problem: mesh networks. *IEEE Transactions on Power Apparatus and Systems* 30 (1): 199–211.

11 Coffrin, C., Hijazi, H.L., and Van Hentenryck, P. (2015). The QC relaxation: a theoretical and computational study on optimal power flow. *IEEE Transactions on Power Apparatus and Systems* 31 (4): 3008–3018.

12 Kocuk, B., Dey, S.S., and Sun, X.A. (2016). Strong SOCP relaxations for the optimal power flow problem. *Operations Research* 64 (6): 1177–1196.

13 Molzahn, D.K., Holzer, J.T., Lesieutre, B.C., and DeMarco, C.L. (2013). Implementation of a large-scale optimal power flow solver based on semidefinite programming. *IEEE Transactions on Power Apparatus and Systems* 28 (4): 3987–3998.

14 Sojoudi, S., Madani, R., and Lavaei, J. (2013). Low-rank solution of convex relaxation for optimal power flow problem. *2013 IEEE International Conference on Smart Grid Communications (SmartGridComm)*, Vancouver, BC, Canada, pp. 636–641. https://doi.org/10.1109/SmartGridComm.2013.6688030.

15 Chen, C., Atamtürk, A., and Oren, S.S. (2016). Bound tightening for the alternating current optimal power flow problem. *IEEE Transactions on Power Apparatus and Systems* 31 (5): 3729–3736.

16 Louca, R., Seiler, P., and Bitar, E. (2013). A rank minimization algorithm to enhance semidefinite relaxations of optimal power flow. *2013 51st Annual Allerton Conference on Communication, Control, and Computing (Allerton)*, Monticello, IL, USA, pp. 1010–1020, https://doi.org/10.1109/Allerton.2013.6736636.

17 Manshadi, S.D., Liu, G., Khodayar, M.E. et al. (2020). A distributed convex relaxation approach to solve the power flow problem. *IEEE Systems Journal* 14 (1): 803–812.

# 11

# Contingency Analysis Graph Computing

It is a challenge and a goal to operate a large-scale, complex, and dynamic power grid with safety and cost-effectiveness. Contingency analysis is one of the applications to secure power systems operating with no violation. Contingency analysis uses base case power flow driven from state estimation to assess the security of power systems under the contingency of a single equipment outage and their combinations. It is usually based on online power grid analysis, combined with the operator's experience, to figure out weak points and security risks of the power grid and issue an alarm when the system is running at risk. It facilitates operators to deal with potential operation issues in time to prevent cascading events and blackouts.

In traditional contingency analysis applications, it is usually running periodically every 5–10 minutes, making it impossible to accommodate the current complex and variable grid operation conditions in a timely manner.

Real-time performance is the fundamental requirement of contingency analysis and situational awareness. Due to the limitation of the computing performance of conventional contingency analysis applications, it is challenging to achieve real-time contingency analysis and look ahead situational awareness for operators in the control center.

To address the concerns from the issues discussed above, researchers have proposed multiple solutions [1–3]. The commonly used approach is automatic contingency screening. The approach is being adopted to identify and rank severe outages which potentially will cause power flow and/or voltage violations.

The capability of Artificial Neural networks on nonlinear adaptive filtering has been applied to perform contingency screening as well [4–6]. Pattern recognition technology is also used for power system security analysis [7].

Besides the fast screening approaches and machine learning methods, parallel computing is promising to accelerate the contingency analysis process. Analyzing each contingency in a given contingency is naturally independent. They can be easily and straightforwardly performed by graph computing in parallel. With more granularity, the postcontingency power flow can be solved by graph parallel computing as addressed in Chapter 10.

## 11.1 DC Power Flow

The DC power flow is used in contingency analysis screening when the requirement of calculation efficiency is of most concern other than the requirement of calculation accuracy, particularly for a large-scale power system. The DC power flow is an approximation version of the AC power flow in

*Graph Database and Graph Computing for Power System Analysis*, First Edition. Renchang Dai and Guangyi Liu.
© 2024 The Institute of Electrical and Electronics Engineers, Inc. Published 2024 by John Wiley & Sons, Inc.

the format of linear equations. It can be solved directly without iteration. The merit makes the DC power flow attractive when computation efficiency is a higher priority than accuracy.

In the DC power flow model, the voltage magnitude update equations in the fast decoupled power flow model are dropped. Only the voltage angle update equation is kept correcting the bus voltage angle by the active power mismatch.

$$\frac{\Delta P}{V} = -B'V\Delta\theta \tag{11.1}$$

DC power flow is further simplified by assuming all the voltage magnitudes are equal to 1.0. Thus, the following DC power flow equations can be obtained:

$$\Delta P = -B'\Delta\theta \tag{11.2}$$

$$P_{ij} = -B'_{ij}\left(\theta_i - \theta_j\right) \tag{11.3}$$

where, $B'_{ij} = 1/x_{ij}$, $x_{ij}$ is the reactance of branch $i-j$.

Applying the above equation to all buses, the following compact form equation is obtained:

$$P = B'\theta \tag{11.4}$$

where, $B = \left[B'_{ij}\right] \in \mathbb{R}^{n \times n}$, $P = [P_i] \in \mathbb{R}^n$, $\theta = [\theta_i] \in \mathbb{R}^n$, $B'_{ii} = \sum_{j=1}^{n} B'_{ij}$.

The DC power flow linear equation can be effectively solved by the Gaussian elimination involving a matrix formation, factorization, and forward/backward substitutions. Based on graph database and graph computing, the DC power flow calculation algorithm is presented below.

---

**Algorithm 11.1   Graph-Based DC Power Flow Calculation**

---

```
1:   Create graph B_P for B' //in nodal parallel
2:   Factorize B' on graph B_P //in hierarchical parallel
3:   Solve θ in (11.4) //in hierarchical parallel
4:   Calculate branch flow by (11.3) //in nodal parallel
```

---

Similar to the fast decoupled method, in the graph-based DC power flow calculation algorithm, $B'$ graph creation and branch flow calculation can be performed by nodal parallel computing, e.g. calculation on each bus and branch does not rely on other bus and branch's calculations. $B'$ matrix factorization and forward/backward substitution mathematically are nodal dependent, e.g. the calculation of a row is dependent only on the columns at the same level in the elimination tree as addressed in Chapter 3. Calculations for $B'$ matrix factorization and forward/backward substitution can be implemented by hierarchical graph parallel computing.

To implement the above graph-based DC power flow calculation, the bus-branch model graph schema is defined as the same as that for the fast decoupled power flow methods. And B_P graph is built to calculate DC power flow by graph parallel computing.

The B_P graph represents $B'$ matrix by a graph. The B_P graph is defined as a set of (vertex, edge) pairs. Each pair of (vertex, edge) represents a nonzero element in the $B'$ matrix.

To build a B_P graph, the vertex B_pND and edge B_pConnect are defined as follows containing the nonzero value of the $B'$ matrix by $b_{ij}$ and $b_{ii}$ in double:

```
CREATE VERTEX B_pND (PRIMARY_ID B_pID int, bii double)
CREATE DIRECTED EDGE B_pConnect (from B_pND, to B_pND, bij double)
```

The vertex B_pND and edge B_pConnect are virtual vertices and virtual edges. They are not physical nodes or branches in power systems. In the B_P graph structure, the counterparts of the edges between vertices are nonzero off-diagonal elements in the $B'$ matrix. The diagonal elements in the $B'$ matrix are modeled as attributes of the vertex. Zero (absent) elements in the $B'$ matrix indicate that no direct connections between the vertices exist in the B_P graph.

With the defined vertex and edge above, the B_P graph schema is then simply defined as:

```
CREATE GRAPH B_pGraph(B_pND, B_pConnect)
```

The B_P graph schema defined the graph structure. The B_P graph will be built by adding vertex B_pND and edge B_pConnect by the following pseudo-query code.

```
CREATE QUERY buildB_pGraph(
) FOR GRAPH busBranchGraph and GRAPH B_pGraph {

SumAccum<double> @sumB   = 0

T0= {TopoND.*}
T1 = SELECT s FROM T0:s-(connected:e)->:t
  IF (s.busType != 3) THEN
    ACCUM
      s.@sumB   += 1.0/e.X
     INSERT INTO B_pND(s.TOPOID)
     INSERT INTO B_pND(t.TOPOID)
     INSERT INTO B_pConnect(s.TOPOID, t.TOPOID) VALUES(-1.0/e.X)
    POST-ACCUM
     UPDATE b_p FROM B_pND WHERE b_p.B_pID = = s.TOPOID
       SET b_p.bii = s.@sumB
  END
}
```

In the query, the vertex-attached variable SumAccum @sumB is defined to store the diagonal elements in the $B'$ matrix which will be assigned to be vertex attributes of in the B_P graph.

In the query definition, graph traversal statement T1 = SELECT s FROM T0:s-(connected: e)->:t travels all edges from vertex s in the vertex set T0 = {TopoND.*}. Nested in the ACCUM, @sumB is calculated in nodal parallel. And new edges are added to B_pConnect by:

```
        INSERT INTO B_pConnect(s.TOPOID, t.TOPOID) VALUES(-1.0/e.X)
```

New vertices are added to B_pND before new edges are added in between the vertices by:

```
        INSERT INTO B_pND(s.TOPOID)
        INSERT INTO B_pND(t.TOPOID)
```

In the POST-ACCUM, diagonal elements in the $B'$ matrix are calculated for all $n - 1$ buses. Each calculated diagonal element is assigned to the vertex attribute by the following UPDATE statement:

```
        UPDATE b_p FROM B_pND WHERE b_p.B_pID = = s.TOPOID
          SET b_p.bii = s.@sumB
```

After the B_P graph is created, the graph symbolic factorization, elimination tree creation, node partition, and graph numerical factorization are the same as we process power flow calculation by the Newton method. The only change is to apply these graph-based algorithms to graph B_pGraph other than graph JacobianGraph.

## 11.2   Bridge Search

Contingency analysis calculates postcontingency power flow, compares the branch power flow and bus voltage against predefined limits, and reports violations if any. Contingency analysis also monitors power flow on important and desired flow gates or interfaces which is defined as a summation of power flow on multiple branches. Bridges at high voltage levels represent vulnerabilities in a power system. Operators are concerned about islanding and system separation caused by high voltage-level element outages. Monitoring power flow on bridges at high voltage levels is a critical and practical task of contingency analysis to operate a reliable network in which bridges are dynamically detected and identified for base cases and contingencies.

Graph theory supports bridge identification by the depth-first algorithm or the breadth-first algorithm which can be implemented in parallel by graph computing. To develop graph parallel computing-based bridge detection and identification, the following terminologies are defined [8].

**Definition 11.1**   an undirected graph $G = (V, E)$ is **connected**, if $\forall v \in V$ and $\forall w \in V$, there exists at least one path connecting vertices $v$ and $w$.

**Definition 11.2**   a **connected component** is a subgraph of a given graph. Any two vertices in the connected component can be connected by at least one path but cannot be connected to any other vertices in the given undirected graph other than the connected component.

For example, in Figure 11.1, the whole graph is a connected component. If the edge between vertices 1 and 6 is removed, then there will be two connected components consisting of vertices {1,2,3,4,5} and {6,7,8,9,10}.

**Definition 11.3**   a **parent** is the vertex $u$ when we travel an unvisited vertex $v$ through edge $u - v$. Vertex $u$ is the parent of vertex $v$. Vertex $v$ is the **child** of vertex $u$. All vertices visited before vertex $u$ are the **ancestor** of vertex $v$.

**Definition 11.4**   a **bridge** is an edge when it is removed, the number of connected components will be increased by one. Obviously, the edge 1–6 is a bridge in Figure 11.1.

The simplest approach to find all bridges in a given graph is to remove all edges one by one and check if the removal of an edge causes connected components to increase by a depth-first or breadth-first search. The time complexity of this simple approach is $O(E \times (V + E))$ which is

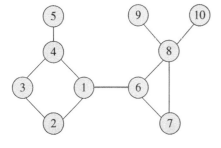

**Figure 11.1**   Example of a connected undirected graph.

**Figure 11.2** Tarjan's algorithm illustration (a).

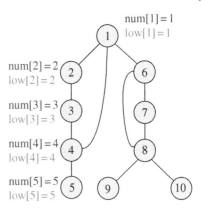

impractical for a large-scale real network. Tarjan's algorithm is the first linear computing algorithm for finding bridges in a given graph based on depth-first search (DFS) developed by Robert E. Tarjan [9]. The time complexity of Tarjan's algorithm is $O(V + E)$.

This algorithm introduces and maintains two record arrays: array "low" and array "num" during the graph traversal by DFS. The element of array *num*, num[$i$], denotes the sequential order of each vertex when it is initially visited during the DFS traversal. The element of the array *low*, low[$i$], is the lowest reachable ancestor vertex visited order number *num* without going through its parent vertex for vertex $i$.

For example, to search bridge in the graph shown in Figure 11.1 by DFS starting from vertex 1, Figure 11.2 is redrawn for a better illustration.

When vertex 5 is reached, no more unvisited vertices are connected to vertex 5. DFS will backtrack to vertex 4. For vertex 4, edge 4–1 is not traveled. DFS travels edge 4–1 to its ancestor vertex 1 since vertex 1 is not the parent of the vertex 4 and the num[1] = 1 is lower than vertex 4's current low[4] = 4, the Tarjan's algorithm update the low[4] = 1, the lowest reachable ancestor vertex visited order number *num* without going through its parent vertex shown in Figure 11.3.

When all edges connecting to vertex 4 are traveled, DFS will backtrack recursively to vertices 3 and 2 till vertex 1 and update low[3] = 1 and low[2] = 1 since vertex 1 is the lowest ordered vertex they can reach without going through their parent. Then the graph is updated in Figure 11.4.

After the left branch is processed, DFS continues to travel edges 1–6, 6–7, 7–8, and 8–9 on the right branch till it reaches vertex 9. The graph is then updated as Figure 11.5.

**Figure 11.3** Tarjan's algorithm illustration (b).

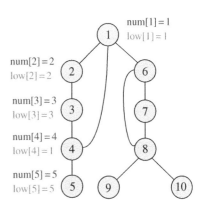

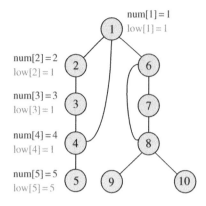

num[1]=1
low[1]=1

num[2]=2
low[2]=1

num[3]=3
low[3]=1

num[4]=4
low[4]=1

num[5]=5
low[5]=5

**Figure 11.4** Tarjan's algorithm illustration (c).

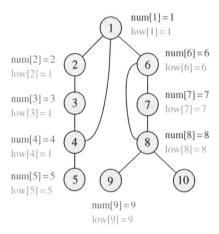

num[1]=1
low[1]=1

num[2]=2
low[2]=1

num[3]=3
low[3]=1

num[4]=4
low[4]=1

num[5]=5
low[5]=5

num[6]=6
low[6]=6

num[7]=7
low[7]=7

num[8]=8
low[8]=8

num[9]=9
low[9]=9

**Figure 11.5** Tarjan's algorithm illustration (d).

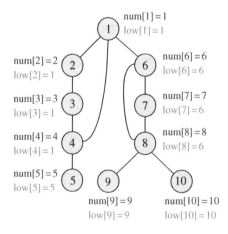

num[1]=1
low[1]=1

num[2]=2
low[2]=1

num[3]=3
low[3]=1

num[4]=4
low[4]=1

num[5]=5
low[5]=5

num[6]=6
low[6]=6

num[7]=7
low[7]=6

num[8]=8
low[8]=6

num[9]=9
low[9]=9

num[10]=10
low[10]=10

**Figure 11.6** Tarjan's algorithm illustration (f).

When vertex 9 is reached, no more unvisited vertices are –connected to vertex 9. DFS will back-track to vertex 8, then visit vertex 10 through edges 8–10 as shown in Figure 11.6.

DFS takes the final step to travel back to vertex 1. All vertices are visited, and all edges are traveled. The DFS is completed.

For a given edge from *parent* to *child* (*parent* is visited before the *child*), if the lowest reachable ancestor vertex of the child low[child] is greater than its parent's visited order number num[parent], e.g. low[child] > num[parent], the edge (*parent, child*) is a bridge. In other words, without going through the edge (*parent, child*), the child cannot reach the ancestor vertices which have ordered the visited number lower than num[parent].

With this condition, the edges 1–6 (low[6] = 6 > num[1] = 1), 4–5 (low[5] = 5 > num[4] = 1), 8–9 (low[9] = 9 > num[8] = 8), 8–10 (low[10] = 9 > num[8] = 8) are bridges.

The pseudo-code of the Tarjan algorithm [10] in determining whether a given edge is a bridge or not is listed below.

---

**Algorithm 11.2  Finding Bridges by Tarjan's Algorithm**

---

```
1:   Input: G = <V, E>
2:   Initialize visited, num, low, parent, time
3:   For node v in V of G(V,E)
4:       If visited[v] == False
5:           Call bridge(v, visited, parent, num, low, time)
```

---

Where time is the discovery time for visited vertices. The variable visited, num, low, parent, and time is initialized to be False, Infinity, Infinity, −1, and 1, respectively. For each vertex, if it is not visited, call the subroutine bridge recursively by DFS. The pseudo-code of the subroutine bridge is shown as follows:

---

**Algorithm 11.3  Subroutine Bridge**

---

```
1:   Input: v, visited, parent, num, low, time
2:   Update visited[v] = True
3:   Update num[v] = time
4:   Update low[v] = time
5:   Update time = time + 1
6:   For u FROM V WHERE u∈v
7:       If visited[u] == False
8:           Update parent[u] = v
9:           Call bridge(u, visited, parent, low, num, time)
10:          Update low[v] = min(low[v], low[u])
11:          If low[u] > num[v]
12:              Set edge<v,u> = 'Bridge'
13:      Elseif u! = parent[v]
14:          Update low[v] = min(low[v], low[u])
```

---

In the subroutine, lines 2–5 update the variables visited, num, low, and time. Line 6 loops through all the vertices $u$ adjacent to the vertex $v$. If the vertex $u$ is not visited, update its parent and call the subroutine bridge recursively till a leave is reached. Then the lowest reachable ancestor vertex visited order number *low* is updated and the bridge is identified by checking the condition low[$v$] > num[$u$].

To implement the graph-based Tarjan's algorithm, the defined bus-branch model graph schema in Chapter 10 is extended to support bridge searching.

In the bus-branch model, vertex `TopoND` and edge `TopoConnect` are defined. They are extended as follows to record vertex visited flag, parent, num, and low as vertex's attributes and bridge as edge's attribute:

```
CREATE VERTEX TopoND (PRIMARY_ID topoid int, TOPOID int, island int,
    bus_name string, area string, loss_zone uint, busType int, Vm
    double, Va double, M_Vm double, M_Va double, base_kV double,
    desired_volts double, control_bus_number int, up_V double, lo_V
    double, GenP double, GenQ double, M_Gen_P double, M_Gen_Q double,
    qUp double, qLower double, LdP double, LdQ double, P double,
    Q double, M_Load_P double, M_Load_Q double, G double, B double,
    Sub string, OV int, UV int, sumB double, sumG double, sumBi double,
    volt double, double, SE_Vm double, SE_Va double, M_C_P double,
    visited boolean, parent int, num int, low int)
```

```
CREATE DIRECTED EDGE TopoConnect (from TopoND, to TopoND, typename
    string, edge_name List<string>, area List<string>, zone
    List<string>, from_bus int, to_bus int, flag uint, R List<double>,
    X List<double>, hB_list List<double>, line_Q1_list List<double>,
    line_Q2_list List<double>, line_Q3_list List<double>, control_bus
    int, K List<double>, shifting_transformer_angle List<double>,
    min_tap double, max_tap double, step_size double, min_volt
    List<double>, max_volt List<double>, M_P_BR List<double>, M_Q_BR
    List<double>, G List<double>, B List<double>, BIJ List<double>,
    circuit uint, P_BR double, Q_BR double, from_open List<int>,
    to_open List<int>, bridge boolean)
```

After the vertex and edge definitions are updated, the graph will be recreated by the statement as follow and the bus-branch graph is not necessarily reformed since the adding attributes do not change the network topology and parameters.

```
CREATE GRAPH busBranchGraph(TopoND, TopoConnect)
```

Now, the graph is ready to perform a bridge search.

The pseudo-code in the graph query to implement Tarjan's algorithm is shown as follows:

```
SetAccum<int> @@time

CREATE QUERY TarjanAlgorithm(
) FOR GRAPH busBranchGraph{

@@time = 1
T0= {TopoND.*}

T1 = SELECT s FROM T0:s-(connected:e)->:t
  ACCUM
    s.visited = False
    s.sum = Inf
```

```
    s.low = Inf
    s.parent = -1

FOREACH v in T0
  IF (v.visited == False) THEN
    CALL bridge(v)
  END

T1 = SELECT s FROM T0:s-(connected:e)->:t
  ACCUM
    IF t.low > s.num THEN
      e.bridge = True
  END
}
```

Before the query `TarjanAlgorithm()` is implemented, global variable time is defined with a double add-sign @@ to store the discovery time for visited vertices.

In the query definition, all `TopoND` vertices in the graph are assigned to `T0` by the statement `T0 = {TopoND.*}`.

All vertices are selected by the following statement.

```
        T0 = SELECT s FROM T0:s-(connected:e)->:t
```

The attributes of the vertices are updated in parallel by using the keyword ACCUM. Nested in ACCUM, statements `s.visited = False, s.sum = Inf, s.low = Inf,` and `s.parent = -1` initialize the vertex attributions visited, num, low, and parent. Then, for each vertex, if it is not visited, call the subroutine bridge by the following statements:

```
FOREACH v in T0
  IF (v.visited == False) THEN
    CALL bridge(v)
  END
```

At the end of the query, the edge e from `TopoND` s to `TopoND` t is checked as a bridge or not by the following code in parallel:

```
T1 = SELECT s FROM T0:s-(connected:e)->:t
  ACCUM
    IF t.low > s.num THEN
      e.bridge = True
END
```

In the query `TarjanAlgorithm()`, query `bridge()` is called. The pseudo-code in the graph query to implement the function `bridge` is shown as follows:

```
SetAccum<int> @@time

CREATE QUERY bridge( TopoND v
) FOR GRAPH busBranchGraph{

  v.visited = True
  v.num = @@time
```

```
v.low = @@time
@@time += 1

FOREACH u in v.neighbor()
  IF (u.visited == False) THEN
    u.parent = v.topoid
    CALL bridge(u)
    v.low = min(v.low, u.low)
  ELSE IF u.topoid != v.parent
    v.low = min(v.low, u.low)
  END
END
}
```

In the query `bridge` arguments, the currently visited vertex v in the query `TarjanAlgorithm` is passed. It's `visited`, `num` and `low` attributes are updated, and the visiting time stamp is updated as well.

Statement `FOREACH u in v.neighbor()` loops through adjacent vertices u of the vertex v by a built-in function `v.neighbor()`. Inside the loop, the parent of the unvisited neighbors is assigned, and the query `bridge` is called recursively by the DFS method till no more downstream vertex is unvisited, in other words, a leave is reached. Then the lowest reachable ancestor vertex visited order number *low* is updated. For visited neighbors, if the neighbor is not the parent, update the lowest reachable ancestor vertex visited order number *low*.

## 11.3 Conjugate Gradient for Postcontingency Power Flow Calculation

The conjugate gradient (CG) algorithm is an iterative technique to solve sparse linear systems that are symmetric and positive definite. The CG is feasible to solve power flow when it's converted to be a symmetric positive definite system, for example, power flow equations by the fast decoupled method, and linearized equations by the Newton–Raphson method.

The CG algorithm calculates the search direction and the search step size in each iteration to approach the converged solution which involves intensive computing. Typically, in series computing, the CG is less effective than the Newton–Raphson method and the fast decoupled method. However, in the preconditioned CG algorithm, the steps of residual, searching direction, and searching step size are great candidates for parallel computing, and selecting the preconditioner as base case $A$ matrix for $N-1$ contingency analysis is straightforward. It makes sense to use graph parallel computing onto CG algorithm and Incomplete LU (ILU) preconditioner for $N-1$ contingency analysis.

The CG algorithm is a node-updated based iteration method. It is different from the Gauss–Seidel Methods, in which the update of each component $x_i^{(k)}$ not only depends upon previous iteration results $x_j^{(k-1)}$, but also depends on the results obtained from the current iteration $x_j^{(k)}$. By CG algorithm, $x_i^{(k)}$ are updated independently.

The pseudo-code of the preconditioned CG algorithm in solving linear system $Ax = b$ is listed below [11].

---

**Algorithm 11.4  Preconditioned CG Algorithm**

---

1:   **Initialize** $x_0 = 0$, $r_0 := b - A \cdot x_0$, $z_0 := M^{-1} {}^* r_0$, $p_0 := z_0$,

2:   **While** $\|r_k\| < \varepsilon$

3:      $k = k + 1$

4:      $\alpha_k = r_{k-1}^T z_{k-1} / p_{k-1}^T A p_{k-1}$

5:      $x_k = x_{k-1} + \alpha_k \cdot p_{k-1}$

6:      $r_k = r_{k-1} - \alpha_k \cdot A \cdot p_{k-1}$

7:      $z_k = M^{-1} \cdot r_k$

8:      $\beta_k = z_k^T r_k / z_{k-1}^T r_{k-1}$

9:      $p_k = z_k + \beta_k \cdot p_{k-1}$

---

In the method of CG, $M$ is the preconditioner. The new approximate solution $x_k$ is calculated by its value $x_{k-1}$ in the previous iteration. The new residual $r_k$ of each node is orthogonal to its previous residuals $r_{k-1}$ and the search directions $p_{k-1}$ respectively. Each new search direction is constructed (from the residual) to be T-orthogonal to its previous residuals $r_k$ and the search directions $p_{k-1}$.

In postcontingency power flow calculation, a standard ILU approach is to use a nonsingular matrix $M$ for preconditioning, and then the linear system $A'x = b$ is rewritten as

$$M^{-1} \cdot A' \cdot x = M^{-1} \cdot b \tag{11.5}$$

where $A'$ is the postcontingency $A$ matrix. The preconditioner $M$ is carefully chosen to make $\hat{A} = M^{-1} \cdot A$ a better-conditioned matrix for the CG. The best choice of preconditioner $M$ obviously is $M = A'$ which makes $M^{-1} \cdot A' = I$. Then the solution $x$ simply equates to $M^{-1} \cdot b$. Usually, we use LU decomposition and forward/backward substitution to calculate $M^{-1} \cdot b$. However, LU decomposition on $M$ takes a lot of computational effort. For the $N - 1$ contingency analysis, since in each contingency, only one element is out, which changes only four nonzero elements in the base case power flow coefficient matrix $A$, two nondiagonals, and two diagonals as illustrated in Figure 11.7 for a 14-bus system when edge 4–9 is out.

When the edge 4–9 is disconnected, only two off-diagonal nonzero elements $b_{4,9}$, $b_{9,4}$ and two diagonal elements $b_{4,4}$, $b_{9,9}$ will change. For a large-scale power network matrix $A$, the 4-element changes should not have significant impacts on the LU factorization of the postcontingency $A'$.

**Figure 11.7** Postcontingency matrix changes.

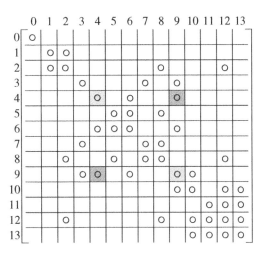

With this assumption, preconditioner $M = A$ to perform postcontingency power flow by preconditioned CG is a straightforward choice.

Taking the voltage angle update equation in the fast decoupled power flow method as an example:

$$\frac{\Delta P}{V} = -B'\Delta\theta \tag{11.6}$$

$$
\begin{bmatrix}
\dfrac{\Delta P_1}{V_1} \\
\dfrac{\Delta P_2}{V_2} \\
\vdots \\
\dfrac{\Delta P_{n-1}}{V_{n-1}}
\end{bmatrix}
=
\begin{bmatrix}
-B_{1,1} & -B_{2,1} & \cdots & -B_{1,n-1} \\
-B_{1,2} & -B_{2,2} & \cdots & -B_{2,n-1} \\
\vdots & \vdots & \vdots & \vdots \\
-B_{n-1,1} & -B_{n-1,2} & \cdots & -B_{n-1,n-1}
\end{bmatrix}
\begin{bmatrix}
\Delta\theta_1 \\
\Delta\theta_2 \\
\vdots \\
\Delta\theta_{n-1}
\end{bmatrix}
\tag{11.7}
$$

We assume Eq. (11.7) is used to update the postcontingency power flow voltage angle in the linear system standard form $A'x = b$, where $b = \left[\dfrac{\Delta P_1}{V_1}, \dfrac{\Delta P_2}{V_2}, ..., \dfrac{\Delta P_{n-1}}{V_{n-1}}\right]^T$,

$$
A' =
\begin{bmatrix}
-B_{1,1} & -B_{1,2} & \cdots & -B_{1,n-1} \\
-B_{2,1} & -B_{2,2} & \cdots & -B_{2,n-1} \\
\vdots & \vdots & \vdots & \vdots \\
-B_{n-1,1} & -B_{n-1,2} & \cdots & -B_{n-1,n-1}
\end{bmatrix}
, \text{ and } x = [\Delta\theta_1, \Delta\theta_2, ..., \Delta\theta_{n-1}]^T. \text{ To solve the}
$$

Eq. (11.7) by preconditioned CG algorithm, select preconditioner $M = A$. The pseudo-code of the graph-based preconditioned CG algorithm in solving postcontingency voltage angle update equation $\frac{\Delta P}{V} = -B'\Delta\theta$ is given as follows:

---

**Algorithm 11.5    Graph-Based Preconditioned CG Algorithm**

---

```
 1:  Initialize x₀ = 0, r₀ := b - A·x₀, z₀ := M⁻¹*r₀, p₀ := z₀,
 2:  Solve x₀ = 0, r₀ := b - A·x₀, z₀ := M⁻¹*r₀, p₀ := z₀, by Mzₖ = rₖ for all
 3:  vertices
 4:  Initialize pₖ = zₖ for all vertices
 5:  While ‖rₖ‖ < ε
 6:       Select edges e from vertex s to vertex t
 7:           PAP += t. Pₖ₋₁*e.BIJ*t.Pₖ₋₁
 8:           s.TempAP += e.BIJ*t.Pₖ₋₁
 9:       Foreach vertex s
10:           PAP += t.Pₖ₋₁*e.BIJ*t.Pₖ₋₁
11:           s.TempAP += e.BIJ*t.Pₖ₋₁
12:           αₖ₊₁ = s.rₖ* s.zₖ/PAP
13:           t.xₖ = t.xₖ₋₁ + αₖ*t.Pₖ₋₁
14:           t.rₖ = t.rₖ₋₁ - αₖ*t.TempAP
15:       Solve x₀ = 0, r₀ := b - A·x₀, z₀ := M⁻¹*r₀, p₀ := z₀, by Mzₖ₊₁ = rₖ₊₁
16:  for all vertices
17:       Foreach vertex s
18:           βₖ₊₁ = s.rₖ₊₁* s.zₖ₊₁/(s.rₖ* s.zₖ)
19:           s.pₖ₊₁ = s.zₖ₊₁ + βₖ₊₁* s.pₖ
20:       Update Δθₖ = Δθₖ₊₁, rₖ = rₖ₊₁, zₖ = zₖ₊₁, pₖ = pₖ₊₁
```

In the graph-based preconditioned CG algorithm, $\Delta\theta_k$, $\Delta\theta_{k+1}$, $r_k$, $r_{k+1}$, $z_k$, $z_{k+1}$, $p_k$, $p_{k+1}$, $AP$ are defined as vertex-attached variables. $PAP$, $\alpha_k$, $\alpha_{k+1}$, $\beta_k$, and $\beta_{k+1}$ are local variables. Since preconditioner $M$ is selected as unchanged base case matrix $A$, for all contingencies, it's unnecessary to refactorize $M$ to solve $Mz_k = r_k$ in line 2 and $Mz_{k+1} = r_{k+1}$ in line 15. Forward and backward substitution by graph-based hierarchical parallel computing is applied to solve the two equations. However, forward/backward substitution involved in solving the two linear equations above for large-scale systems is time-consuming even with hierarchical parallel computing. To form nodal parallel computing, preconditioner $M$ is defined as $diag[-B_{11}, -B_{22}, ..., -B_{n-1,n-1}]$. The pseudo-code of the graph-based preconditioned CG algorithm is changed as follows:

---

**Algorithm 11.6   Graph-Based Preconditioned CG Algorithm**

---

```
1:    Initialize x₀ = 0, r₀ := b - A·x₀, z₀ := M⁻¹*r₀, p₀ := z₀,
2:    Calculate zₖ = rₖ/Bii for all vertices
3:    Initialize pₖ = zₖ for all vertices
4:    While ‖rₖ‖ < ε
5:        Foreach vertex s
6:            PAP += t.Pₖ₋₁*e.BIJ*t.Pₖ₋₁
7:            s.TempAP += e.BIJ*t.Pₖ₋₁
8:            αₖ₊₁ = s.rₖ × s.zₖ/PAP
9:            t.xₖ = t.xₖ₋₁ + αₖ*t.Pₖ₋₁
10:           t.rₖ = t.rₖ₋₁ - αₖ*t.TempAP
11:           s.zₖ₊₁ = s.rₖ₊₁/s.Bii
12:           βₖ₊₁ = s.rₖ₊₁ × s.zₖ₊₁/(s.rₖ × s.zₖ)
13:           s.pₖ₊₁ = s.zₖ₊₁ + βₖ₊₁ × s.pₖ
14:       Update Δθₖ = Δθₖ₊₁, rₖ = rₖ₊₁, zₖ = zₖ₊₁, pₖ = pₖ₊₁
```

---

In Algorithm 11.6 above, forward and backward substitution is eliminated and replaced simply by divisions in nodal parallel which improves the computation efficiency in each iteration but at a cost of more iterations required. When solving a linear system with a diagonally dominant matrix, Algorithm 11.6 is proper. By applying Algorithm 11.5, since the preconditioner $M$ is selected as unchanged base case matrix $A$ and the postcontingency matrix $A'$ has minor changes from matrix $A$, the solution will be converged in a few iterations. Matrix $A$ corresponding to a real power system typically is highly sparse which supports great computation efficiency by using hierarchical parallel graph computing. These characteristics are factors driving us to select a proper preconditioner $M$ in practice.

The following simple example is taken to illustrate the CG approach and the preconditioned CG with different preconditioners.

$$\begin{bmatrix} 4 & 1 \\ 1 & 3 \end{bmatrix} \begin{bmatrix} x_1 \\ x_2 \end{bmatrix} = \begin{bmatrix} 1 \\ 2 \end{bmatrix} \tag{11.8}$$

The CG approach is performed beginning with the initial point $x_0$.

$$x_0 = \begin{bmatrix} 0 \\ 0.5 \end{bmatrix} \tag{11.9}$$

$$r_0 = b - Ax_0 = \begin{bmatrix} 1 \\ 2 \end{bmatrix} - \begin{bmatrix} 4 & 1 \\ 1 & 3 \end{bmatrix} \begin{bmatrix} 0 \\ 0.5 \end{bmatrix} = \begin{bmatrix} 0.5 \\ 0.5 \end{bmatrix} = p_0 \tag{11.10}$$

In the initialization, the search direction $p_0$ is set as the residual vector $r_0$.

In the first iteration, the search step size $\alpha_1$ is calculated by (11.11).

$$\alpha_1 = \frac{r_0^T r_0}{p_0^T A p_0} = \frac{[0.5 \quad 0.5]\begin{bmatrix} 0.5 \\ 0.5 \end{bmatrix}}{[0.5 \quad 0.5]\begin{bmatrix} 4 & 1 \\ 1 & 3 \end{bmatrix}\begin{bmatrix} 0.5 \\ 0.5 \end{bmatrix}} = \frac{2}{9} \tag{11.11}$$

The solution and residual are updated by

$$x_1 = x_0 + \alpha_1 p_0 = \begin{bmatrix} 0 \\ 0.5 \end{bmatrix} + \frac{2}{9} \times \begin{bmatrix} 0.5 \\ 0.5 \end{bmatrix} = \begin{bmatrix} \frac{1}{9} \\ \frac{11}{18} \end{bmatrix} \tag{11.12}$$

$$r_1 = r_0 - \alpha_1 A p_0 = \begin{bmatrix} 0.5 \\ 0.5 \end{bmatrix} + \frac{2}{9} \times \begin{bmatrix} 4 & 1 \\ 1 & 3 \end{bmatrix}\begin{bmatrix} 0.5 \\ 0.5 \end{bmatrix} = \begin{bmatrix} \frac{-5}{90} \\ \frac{5}{90} \end{bmatrix} \tag{11.13}$$

The scalar $\beta_1$ is calculated by (11.14)

$$\beta_1 = \frac{r_1^T r_1}{r_0^T r_0} = \frac{\begin{bmatrix} \frac{-5}{90} & \frac{5}{90} \end{bmatrix}\begin{bmatrix} \frac{-5}{90} \\ \frac{5}{90} \end{bmatrix}}{[0.5 \quad 0.5]\begin{bmatrix} 0.5 \\ 0.5 \end{bmatrix}} = \frac{1}{81} \tag{11.14}$$

The new search direction $p_1$ is calculated as

$$p_1 = r_1 + \beta_1 p_0 = \begin{bmatrix} \frac{-5}{90} \\ \frac{5}{90} \end{bmatrix} + \frac{1}{81} \times \begin{bmatrix} 0.5 \\ 0.5 \end{bmatrix} = \begin{bmatrix} \frac{-4}{81} \\ \frac{5}{81} \end{bmatrix} \tag{11.15}$$

In the second iteration, the search step size $\alpha_2$ is calculated by (11.16).

$$\alpha_2 = \frac{r_1^T r_1}{p_1^T A p_1} = \frac{\begin{bmatrix} \frac{-5}{90} & \frac{5}{90} \end{bmatrix}\begin{bmatrix} \frac{-5}{90} \\ \frac{5}{90} \end{bmatrix}}{\begin{bmatrix} \frac{-4}{81} & \frac{5}{81} \end{bmatrix}\begin{bmatrix} 4 & 1 \\ 1 & 3 \end{bmatrix}\begin{bmatrix} \frac{-4}{81} \\ \frac{5}{81} \end{bmatrix}} = \frac{9}{22} \tag{11.16}$$

The solution and residual are updated by

$$x_2 = x_1 + \alpha_2 p_1 = \begin{bmatrix} \frac{1}{9} \\ \frac{11}{18} \end{bmatrix} + \frac{9}{22} \times \begin{bmatrix} \frac{-4}{81} \\ \frac{5}{81} \end{bmatrix} = \begin{bmatrix} \frac{1}{11} \\ \frac{7}{11} \end{bmatrix} \tag{11.17}$$

$$r_2 = r_1 - \alpha_2 A p_1 = \begin{bmatrix} \frac{-5}{90} \\ \frac{5}{90} \end{bmatrix} - \frac{9}{22} \times \begin{bmatrix} 4 & 1 \\ 1 & 3 \end{bmatrix}\begin{bmatrix} \frac{-4}{81} \\ \frac{5}{81} \end{bmatrix} = \begin{bmatrix} 0 \\ 0 \end{bmatrix} \tag{11.18}$$

$x_2 = \begin{bmatrix} \dfrac{1}{11} & \dfrac{7}{11} \end{bmatrix}^T$ is the exact solution.

In preconditioned CG approach, assuming the base case equation is

$$Ax = b \Leftrightarrow \begin{bmatrix} 5 & 2 \\ 2 & 4 \end{bmatrix} \begin{bmatrix} x_1 \\ x_2 \end{bmatrix} = \begin{bmatrix} 1 \\ 2 \end{bmatrix} \tag{11.19}$$

And the postcontingency equation is

$$A'x = b \Leftrightarrow \begin{bmatrix} 4 & 1 \\ 1 & 3 \end{bmatrix} \begin{bmatrix} x_1 \\ x_2 \end{bmatrix} = \begin{bmatrix} 1 \\ 2 \end{bmatrix} \tag{11.20}$$

which is the same as (11.8). To solve the same equation but by preconditioned CG, we first select preconditioner $M = A = \begin{bmatrix} 5 & 2 \\ 2 & 4 \end{bmatrix}$, then $M = \begin{bmatrix} 5 & 0 \\ 0 & 4 \end{bmatrix}$.

When $M = A = \begin{bmatrix} 5 & 2 \\ 2 & 4 \end{bmatrix}$, the preconditioned CG approach is performed beginning with the same initial point $x_0$ as (11.9).

$$x_0 = \begin{bmatrix} 0 \\ 0.5 \end{bmatrix} \tag{11.21}$$

The initial residual is calculated by

$$r_0 = b - A'x_0 = \begin{bmatrix} 1 \\ 2 \end{bmatrix} - \begin{bmatrix} 4 & 1 \\ 1 & 3 \end{bmatrix} \begin{bmatrix} 0 \\ 0.5 \end{bmatrix} = \begin{bmatrix} 0.5 \\ 0.5 \end{bmatrix} \tag{11.22}$$

Different from the general CG approach, in the initialization of the preconditioned CG, the search direction $p_0$ is calculated as:

$$p_0 = M^{-1}r_0 = \begin{bmatrix} 5 & 2 \\ 2 & 4 \end{bmatrix}^{-1} \begin{bmatrix} 0.5 \\ 0.5 \end{bmatrix} = \begin{bmatrix} \dfrac{1}{16} \\ \dfrac{3}{32} \end{bmatrix} = z_0 \tag{11.23}$$

In the first iteration, the search step size $\alpha_1$ is calculated by (11.24).

$$\alpha_1 = \frac{r_0^T z_0}{p_0^T A' p_0} = \frac{\begin{bmatrix} 0.5 & 0.5 \end{bmatrix} \begin{bmatrix} \dfrac{1}{16} \\ \dfrac{3}{32} \end{bmatrix}}{\begin{bmatrix} \dfrac{1}{16} & \dfrac{3}{32} \end{bmatrix} \begin{bmatrix} 4 & 1 \\ 1 & 3 \end{bmatrix} \begin{bmatrix} \dfrac{1}{16} \\ \dfrac{3}{32} \end{bmatrix}} = \frac{16}{11} \tag{11.24}$$

The solution and residual are updated by

$$x_1 = x_0 + \alpha_1 p_0 = \begin{bmatrix} 0 \\ 0.5 \end{bmatrix} + \frac{16}{11} \times \begin{bmatrix} \dfrac{1}{16} \\ \dfrac{3}{32} \end{bmatrix} = \begin{bmatrix} \dfrac{1}{11} \\ \dfrac{7}{11} \end{bmatrix} \tag{11.25}$$

$$r_1 = r_0 - \alpha_1 A' p_0 = \begin{bmatrix} 0.5 \\ 0.5 \end{bmatrix} - \frac{16}{11} \times \begin{bmatrix} 4 & 1 \\ 1 & 3 \end{bmatrix} \begin{bmatrix} \dfrac{1}{16} \\ \dfrac{3}{32} \end{bmatrix} = \begin{bmatrix} 0 \\ 0 \end{bmatrix} \tag{11.26}$$

Taking one iteration, the exact solution $\left[\dfrac{1}{11} \quad \dfrac{7}{11}\right]^T$ is reached out at the cost of solving $z_k$ in the equation $M z_k = r_k$ by keeping the same base case matrix $M$ factorization but performing forward and backward substitution.

To save the time of performing forward and backward substitution, the preconditioner $M$ is selected as $M = \begin{bmatrix} 5 & 0 \\ 0 & 4 \end{bmatrix}$. Then the same processing of conducting preconditioned CG is as follows with the same initial condition and initial residual:

$$x_0 = \begin{bmatrix} 0 \\ 0.5 \end{bmatrix} \tag{11.27}$$

The initial residual is calculated by

$$r_0 = b - A' x_0 = \begin{bmatrix} 1 \\ 2 \end{bmatrix} - \begin{bmatrix} 4 & 1 \\ 1 & 3 \end{bmatrix} \begin{bmatrix} 0 \\ 0.5 \end{bmatrix} = \begin{bmatrix} 0.5 \\ 0.5 \end{bmatrix} \tag{11.28}$$

The search direction $p_0$ is calculated as:

$$p_0 = M^{-1} r_0 = \begin{bmatrix} 5 & 0 \\ 0 & 4 \end{bmatrix}^{-1} \begin{bmatrix} 0.5 \\ 0.5 \end{bmatrix} = \begin{bmatrix} \dfrac{1}{5} & 0 \\ 0 & \dfrac{1}{4} \end{bmatrix} \begin{bmatrix} 0.5 \\ 0.5 \end{bmatrix} = \begin{bmatrix} \dfrac{1}{10} \\ \dfrac{1}{8} \end{bmatrix} = z_0 \tag{11.29}$$

In the first iteration, the search step size $\alpha_1$ is calculated by (11.30).

$$\alpha_1 = \frac{r_0^T z_0}{p_0^T A' p_0} = \frac{[0.5 \quad 0.5] \begin{bmatrix} \dfrac{1}{10} \\ \dfrac{1}{8} \end{bmatrix}}{\begin{bmatrix} \dfrac{1}{10} & \dfrac{1}{8} \end{bmatrix} \begin{bmatrix} 4 & 1 \\ 1 & 3 \end{bmatrix} \begin{bmatrix} \dfrac{1}{10} \\ \dfrac{1}{8} \end{bmatrix}} = \frac{180}{179} \tag{11.30}$$

The solution and residual are updated by

$$x_1 = x_0 + \alpha_1 p_0 = \begin{bmatrix} 0 \\ 0.5 \end{bmatrix} + \frac{180}{179} \times \begin{bmatrix} \dfrac{1}{10} \\ \dfrac{1}{8} \end{bmatrix} = \begin{bmatrix} \dfrac{18}{179} \\ \dfrac{112}{179} \end{bmatrix} \tag{11.31}$$

$$r_1 = r_0 - \alpha_1 A' p_0 = \begin{bmatrix} 0.5 \\ 0.5 \end{bmatrix} - \frac{180}{179} \times \begin{bmatrix} 4 & 1 \\ 1 & 3 \end{bmatrix} \begin{bmatrix} \dfrac{1}{10} \\ \dfrac{1}{8} \end{bmatrix} = \begin{bmatrix} \dfrac{-5}{179} \\ \dfrac{4}{179} \end{bmatrix} \tag{11.32}$$

The scalar $\beta_1$ is calculated by:

$$z_1 = M^{-1}r_1 = \begin{bmatrix} 5 & 0 \\ 0 & 4 \end{bmatrix}^{-1} \begin{bmatrix} \dfrac{-5}{179} \\ \dfrac{4}{179} \end{bmatrix} = \begin{bmatrix} \dfrac{1}{5} & 0 \\ 0 & \dfrac{1}{4} \end{bmatrix} \begin{bmatrix} \dfrac{-5}{179} \\ \dfrac{4}{179} \end{bmatrix} = \begin{bmatrix} \dfrac{-1}{179} \\ \dfrac{1}{179} \end{bmatrix} \tag{11.33}$$

$$\beta_1 = \frac{r_1^T z_1}{r_0^T z_0} = \frac{\begin{bmatrix} \dfrac{-5}{179} & \dfrac{4}{179} \end{bmatrix} \begin{bmatrix} \dfrac{-1}{179} \\ \dfrac{1}{179} \end{bmatrix}}{\begin{bmatrix} 0.5 & 0.5 \end{bmatrix} \begin{bmatrix} \dfrac{1}{10} \\ \dfrac{1}{8} \end{bmatrix}} = \frac{80}{32041} \tag{11.34}$$

The new search direction $p_1$ is calculated as

$$p_1 = z_1 + \beta_1 p_0 = \begin{bmatrix} \dfrac{-1}{179} \\ \dfrac{1}{179} \end{bmatrix} + \frac{80}{32041} \times \begin{bmatrix} \dfrac{1}{10} \\ \dfrac{1}{8} \end{bmatrix} = \begin{bmatrix} \dfrac{-171}{32041} \\ \dfrac{189}{32041} \end{bmatrix} \tag{11.35}$$

In the second iteration, the search step size $\alpha_2$ is calculated by (11.36).

$$\alpha_2 = \frac{r_1^T z_1}{p_1^T A' p_1} = \frac{\begin{bmatrix} \dfrac{-5}{179} & \dfrac{4}{179} \end{bmatrix} \begin{bmatrix} \dfrac{-1}{179} \\ \dfrac{1}{179} \end{bmatrix}}{\begin{bmatrix} \dfrac{-171}{32041} & \dfrac{189}{32041} \end{bmatrix} \begin{bmatrix} 4 & 1 \\ 1 & 3 \end{bmatrix} \begin{bmatrix} \dfrac{-171}{32041} \\ \dfrac{189}{32041} \end{bmatrix}} = \frac{179}{99} \tag{11.36}$$

The solution and residual are updated by

$$x_2 = x_1 + \alpha_2 p_1 = \begin{bmatrix} \dfrac{18}{179} \\ \dfrac{112}{179} \end{bmatrix} + \frac{179}{99} \times \begin{bmatrix} \dfrac{-171}{32041} \\ \dfrac{189}{32041} \end{bmatrix} = \begin{bmatrix} \dfrac{1}{11} \\ \dfrac{7}{11} \end{bmatrix} \tag{11.37}$$

$$r_2 = r_1 - \alpha_2 A p_1 = \begin{bmatrix} \dfrac{-5}{90} \\ \dfrac{5}{90} \end{bmatrix} + \frac{9}{22} \times \begin{bmatrix} 4 & 1 \\ 1 & 3 \end{bmatrix} \begin{bmatrix} \dfrac{-4}{81} \\ \dfrac{5}{81} \end{bmatrix} = \begin{bmatrix} 0 \\ 0 \end{bmatrix} \tag{11.38}$$

$x_2 = \begin{bmatrix} \dfrac{1}{11} & \dfrac{7}{11} \end{bmatrix}^T$ is the exact solution.

When the preconditioner $M$ is selected as a diagonal matrix $M = \begin{bmatrix} 5 & 0 \\ 0 & 4 \end{bmatrix}$, $z_k$ in the equation $Mz_k = r_k$ is solved simply by division. However, it takes more iterations to reach the exact solution.

The CG approach produces the exact solution after a number of iterations less than the size of the matrix. A proper selection of preconditioner $M$ in contingency analysis supports fast convergence and parallel computing.

To implement the contingency analysis by graph-based preconditioned CG approach, B_P and B_PP graphs defined in Chapter 10 are used to represent $B'$ and $B''$ matrices, respectively by a graph. They are extended as follows to contain the information for preconditioned CG contingency analysis:

```
CREATE VERTEX B_pND (PRIMARY_ID B_pID int, bii double, bii0 double,
    deltaThetak double)
CREATE VERTEX B_ppND (PRIMARY_ID B_ppID int, bii double, bii0 double,
    deltaVk double)
```

In the extended vertex definition above, the new attribute bii0 is highlighted in black bold in the vertex B_pND and B_ppND definition to represent the diagonal elements in the base case $B'$ and $B''$ matrices. New attributes **deltaThetak** and **deltaVk** store voltage angle and voltage magnitude updates. And bii is the diagonal elements in the postcontingency case $B'$ and $B''$ matrices.

With the graphs B_pGraph and B_ppGraph, the pseudo-code in graph query to implement the preconditioned CG approach is shown as follows:

```
SumAccum<double> @deltaP = 0
SumAccum<double> @deltaQ = 0
SumAccum<double> @rk = 0
SumAccum<double> @rk1 = 0
SumAccum<double> @zk = 0
SumAccum<double> @zk1 = 0
SumAccum<double> @pk = 0
SumAccum<double> @AP = 0

SumAccum<double> @@PAP = 0
SumAccum<double> @@rkzk = 0
SumAccum<double> @@rk1zk1 = 0
SumAccum<double> @@alphak = 0
SumAccum<double> @@betak = 0
MaxAccum<double> @@maxrk =-INF


CREATE QUERY preconditoinedConjugateGradient(int MaxIteration, double
Tolerance
) FOR GRAPH busBranchGraph and GRAPH B_pGraph and GRAPH B_ppGraph{

T0= {busBranchGraph.TopoND.*}
Tp= {B_pGraph.TopoND.*}
Tp.@va = T0.Va

T1 = SELECT s FROM Tp:s-(connected:e)->:t WHERE (t.busType != 3)
  ACCUM
    IF ( e.K == 0 ) THEN e.K == 1 END
    s.@deltaP += s.Vm*t.Vm * (-1*e.G/e.K*cos(s.Va - t.Va) +
                e.B/e.K * sin(s.Va - t.Va))
```

```
  POST-ACCUM
    s.@deltaP = s.GenP -s.LdP  - (s.@deltaP + s.Vm*s.Vm*s.@sumG)
    s.@deltaP = s.@deltaP/s.Vm
    s.@rk = s.@deltaP
    s.@zk = s.@rk/s.bii0
    s.@pk = s.@zk
    s.@AP += s.@pk*e.bii
    @@PAP += s.@pk*s.@AP

WHILE( abs(@@maxrk) > Tolerance ) LIMIT MaxIteration DO
  T1 = SELECT s FROM Tp:s -(connected:e)->:t WHERE (s.busType  != 3)
    ACCUM
      s.@AP += s.@pk*e.bij
      @@PAP += s.@pk*s.@AP

  T1 = SELECT s FROM Tp WHERE (s.busType  != 3)
    ACCUM
      @@rkzk += s.@rk*s.@zk
    POST-ACCUM
      @@alphak = @@rkzk/@@PAP

  T1 = SELECT s FROM Tp WHERE (s.busType  != 3)
    ACCUM
      s.deltaThetak = s.deltaThetak + @@alphak*s.@pk
      s.@rk1 = s.@rk-@@alphak*s.@AP
      s.@zk1 = s.@rk1/s.bii0

  T1 = SELECT s FROM Tp WHERE (s.busType  != 3)
    ACCUM
      @@rk1zk1 += s.@rk1*s.@zk1
    POST-ACCUM
      @@betak = @@rkzk/@@rk1zk1

  T1 = SELECT s FROM Tp WHERE (s.busType  != 3)
    ACCUM
      s.@pk = s.@zk1+@@betak*s.@pk
      s.Va = s.Va + s.deltaThetak

      s.@rk = s.@rk1
      s.@zk = s.@zk1

    POST-ACCUM
      @@maxrk = max(abs(s.@rk))
      s.@AP = 0
      @@PAP = 0
      @@rkzk = 0
      @@rk1zk1 = 0
```

```
Tpp= {B_ppGraph.TopoND.*}
Tpp.@vm = T0.Vm

T1 = SELECT s FROM Tpp:s -(connected:e)->:t
     WHERE (s.busType == 0 or s.busType == 1)
   ACCUM
     IF ( e.K == 0 ) THEN e.K == 1 END
     s.@deltaQ += s.Vm*s.Vm * (-1*e.G/e.K*sin(s.Va - t.Va) -
                   e.B/e.K * cos(s.Va - t.Va))

   POST-ACCUM
     s.@deltaQ = s.GenQ - s.LdQ - (s.@deltaQ + s.Vm*s.Vm*s.@sumB)
     s.@deltaQ = s.@deltaQ/s.Vm
     s.@rk = s.@deltaQ
     s.@zk = s.@rk/s.bii0
     s.@pk = s.@zk
     s.@AP += s.@pk*e.bii
     @@PAP += s.@pk*s.@AP

WHILE( @@maxrk > Tolerance ) LIMIT MaxIteration DO
   T1 = SELECT s FROM Tpp:s-(connected:e)->:t
        WHERE (s.busType == 0 or s.busType == 1)
      ACCUM
        s.@AP += s.@pk*e.bij
        @@PAP += s.@pk*s.@AP

   T1 = SELECT s FROM Tpp WHERE (s.busType == 0 or s.busType == 1)
      ACCUM
        @@rkzk += s.@rk*s.@zk
      POST-ACCUM
        @@alphak = @@rkzk/@@PAP

   T1 = SELECT s FROM Tpp WHERE (s.busType == 0 or s.busType == 1)
      ACCUM
        s.deltaVk = s.deltaVk + @@alphak*s.@pk
        s.@rk1 = s.@rk-@@alphak*s.@AP
        s.@zk1 = s.@rk1/s.bii0

   T1 = SELECT s FROM Tpp WHERE (s.busType == 0 or s.busType == 1)
      ACCUM
        @@rk1zk1 += s.@rk1*s.@zk1
      POST-ACCUM
        @@betak = @@rkzk/@@rk1zk1

   T1 = SELECT s FROM Tpp WHERE (s.busType == 0 or s.busType == 1)
      ACCUM
        s.@pk = s.@zk1+@@betak*s.@pk
        s.Vm = s.Vm + s.deltaVk
```

```
    s.@rk = s.@rk1
    s.@zk = s.@zk1

POST-ACCUM
    @@maxrk = max(abs(s.@rk))
    s.@AP = 0
    @@PAP = 0
    @@rkzk = 0
    @@rk1zk1 = 0
}
```

In the query `preconditoinedConjugateGradient`, maximum iteration number `MaxI-teration` and tolerance `Tolerance` are passed through.

Before the definition of the query, vertex-attached varaibles `SumAccum @deltaP`, `@deltaQ`, `@rk`, `@rk1`, `@zk`, `@zk1`, `@pk`, `@AP` are defined to store active power and reactive power mismatch (`@deltaP`, `@deltaQ`), residuals (`@rk`, `@rk1`), preconditioned residuals (`@zk`, `@zk1`), search direction (`@pk`), and $A \times P$ (`@AP`).

Global variables `@@PAP`, `@@rkzk`, `@@rk1zk1`, `@@alphak`, and `@@betak`, and `@@maxrk` are defined to store $P^T AP$, $r_k^T z_k$, $r_{k+1}^T z_{k+1}$, scalar $\alpha_k$ and scalar $\beta_k$.

In the query, the base case voltage angle is copied to the postcontingency voltage angle first by `Tp.@va = T0.Va`, where `Tp` is a postcontingency nonslack bus set. And then the active power mismatch is calculated by the following pseudo-code.

```
T1 = SELECT s FROM Tp:s -(connected:e)->:t WHERE (s.busType != 3)
   ACCUM
      IF ( e.K == 0 ) THEN e.K == 1 END
      s.@deltaP += s.Vm*s.Vm * (-1*e.G/e.K*cos(s.Va - t.Va) + e.B/e.K *
sin(s.Va - t.Va))

   POST-ACCUM
      s.@deltaP = s.GenP -s.LdP   - (s.@deltaP + s.Vm*s.Vm*s.@sumG)
```

Graph traversal statement `T1 = SELECT s FROM Tp:s-(connected:e)->:t WHERE (s.busType! = 3)` travels all edges `e` from vertex `s` to the end vertex `t`, and clause `WHERE (s.busType! = 3)` excludes slack buses. The keyword `ACCUM` ensures the active power of the branch adjacent to vertex `s` is calculated and added into active power mismatch in parallel. Nested in `POST-ACCUM`, the active power injections at the vertex `s` are added to get the final active power mismatch.

Nested in `POST-ACCUM`, residual, preconditioned residual, and search direction are initialized by:

```
    s.@deltaP = s.@deltaP/s.Vm
    s.@rk = s.@deltaP
    s.@zk = s.@rk/s.bii0
    s.@pk = s.@zk
```

Please note that `s.bii0` represents the element in the preconditioner $M = diag(b_{11}^0, b_{22}^0, \ldots)$.

Statement `s.@AP += s.@pk*s.bii` calculate the matrix A diagonal element contribution to *AP*, so as to `@@PAP += s.@pk*s.@AP`. The matrix A nondiagonal elements contribution is calculated in nodal parallel as follows:

```
T1 = SELECT s FROM Tp:s -(connected:e)->:t WHERE (s.busType != 3)
    ACCUM
        s.@AP += s.@pk*e.bij
        @@PAP += s.@pk*s.@AP
```

Then $r_k^T z_k$, the product of two vectors, are performed in parallel is implemented by:

```
T1 = SELECT s FROM Tp WHERE (s.busType != 3)
    ACCUM
        @@rkzk += s.@rk*s.@zk
```

In graph parallel computing, vector multiplication is ideal for nodal parallel computing since the multiplication of each element is independent. They are nested in `ACCUM` in the graph query.

After the `@@rkzk` and `@@PAP` are calculated, the scalar is calculated by `@@alphak = @@rkzk/@@PAP` which is a global variable applied to $\Delta\theta_{k+1}$ update for each vertex in nodal parallel by the following statements:

```
T1 = SELECT s FROM Tp WHERE (s.busType != 3)
    ACCUM
        s.deltaThetak = s.deltaThetak + @@alphak*s.@pk
```

Meanwhile the residual and preconditioned residual are updated in nodal parallel as well by:

```
        s.@rk1 = s.@rk-@@alphak*s.@AP
        s.@zk1 = s.@rk1/s.bii0
```

The following code in the query is straightforward to calculate the multiplication $z_{k+1}^T r_{k+1}$, scalar $\beta_k$, search direction $p_k$, voltage angle $\theta$ iteratively till the maximum residual is within tolerance.

The voltage magnitude calculation follows the same process as the voltage angle calculation above but applies to $B''$ matrix and reactive power mismatch.

## 11.4 Contingency Analysis Using Convolutional Neural Networks

Contingency analysis is a critical task for an power system online operation. In a large-scale power system, contingency analysis investigates hundreds and thousands of contingencies challenging the computation efficiency of state-of-art approaches. In a large number of contingencies, most contingencies are not harmful to the power system operation, e.g. the contingency does not result in any violation. Contingency prescreening is functioning as a filter to exclude these nonharmful contingencies in the detailed power flow calculation.

Using DC power flow is an option for contingency screening. The simplification and assumption improve the DC power flow calculation efficiency but at the cost of loss of calculation accuracy and misreporting voltage violations.

In the screening process, we care about whether the system is secure or insecure under a given contingency. Secure cases are filtered out and only insecure cases are evaluated by AC power flow with details. So, contingency screening is a classification problem. Convolutional neural networks

(CNN) as designed to solve classification problems, will be a fast alternative to the classical AC power flow to retrieve potential harmful cases from a contingency set which supports both branch and voltage violation checking with high accuracy.

### 11.4.1 Convolutional Neural Network

In general, a typical feedforward neural network as shown in Figure 11.8 maps input features to a given set of outputs. The selection of features depends on the application that the feedforward neural network would be utilized. Typically, the domain experts would decide on what can be used as an input feature for the problem under consideration. In other words, the input features are engineered. Power flow is determined by generation injections, load injections, and system topology. They are typically designed as input features in feedforward neural networks to tell whether the system is secure or insecure. However, the system topology is hard to be represented in the feedforward neural network input features.

In CNN, input features are not engineered but rather extracted using convolutional filters on a set of input data. The extracted features are then fed to a fully connected neural network. In other words, as shown in Figure 11.9, CNN are made up of two parts:

1) Feature Learning: convolution and pooling layers are used to extract useful features from the input data.
2) Classification or Regression: the extracted features are fed to a fully connected neural network which learns the relation between the input and the output.

The convolutional neural network is known to work well on structured data. This is due to the fact that useful features can be identified based on local connectivity within the input data set. That is why CNN have seen a lot of success in the domain of image-based data which has a clearly structured topology.

Power systems are similar to images in that they also have a well-defined structure. Power systems are physical networks with a given structure/topology in which nodes/buses are connected to one another through branches. Therefore, power system operational information, such as bus voltages, branch power flows, etc. can be presented in a way that reflects the physical topology of the system. To be specific, the information of a given $N$-bus power system can be presented as an $N \times N$ matrix that is in accordance with the adjacency matrix of the system.

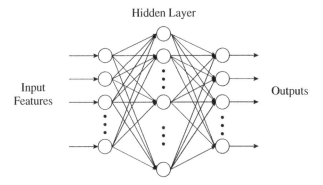

**Figure 11.8**   Feedforward neural network general structure.

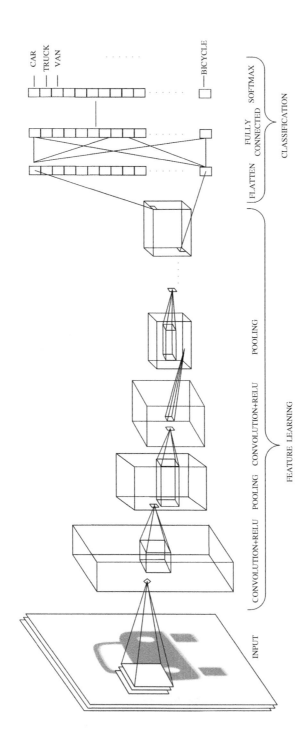

**Figure 11.9** Convolutional neural network general structure.

The feature allows for the use of convolutional filters over the power system data which in turn, would help extract better features for the power system. The schematics make the convolutional neural network a suitable tool for performing various classification and regression tasks for power systems contingency screening.

## 11.4.2 Convolutional Neural Network Components

### 11.4.2.1 Convolutional Neural Network Input

Given a general $N$-bus power system where $N$ is the number of system buses, $L$ is the set of branches, $V_i$ is the voltage magnitude of bus-$i$, and $S_{ij}$ is the MVA power flow of the branch connected between buses $i$ and $j$. Then, the information of a given operating point can be represented as a two-channel image of $N \times N$ pixels. The first channel contains the information on bus voltages on the diagonal. The second channel carries information on MVA branch flows of off-diagonal elements and MVA bus injections of diagonal elements.

For each bus, the value of its voltage magnitude $V_i$ is normalized. For a given bus, the limits on its voltage $V_{i-\max}$ and $V_{i-\min}$ can be used to normalize its voltage as follows:

$$V_{i-normalized} = \frac{V_i - V_{i-\min}}{V_{i-\max} - V_{i-\min}} \tag{11.39}$$

Similarly, for each branch, the value of its MVA flow $S_{ij}$ is normalized as follows:

$$\vec{S}_{ij-normalized} = \frac{\vec{S}_{ij}}{\left| S_{ij-\max} \right|} \tag{11.40}$$

where $S_{ij-\max}$ is the maximum MVA limit of lines or transformers. Please note that a bus MVA injection is calculated as the sum of all normalized flows of the lines and transformers connected to that bus.

### 11.4.2.2 Convolutional Neural Network Output

The output of a convolutional neural network is a vector whose elements correspond to the set (or subset) of all possible contingencies, i.e. output $k$ represents the system status under contingency $k$.

Without loss of generality, we define the system is secure if all branches are loaded within their limits and all bus voltages are operating within their limits, marginal insecure if one or more branches are loaded in the range of 100–120% of their limit, or insecure if any of the system branches is loaded at more than 120% of its limit or any bus voltage is running over limits. Accordingly, each contingency would be labeled as follows:

$$y_k = \begin{cases} -1 & \text{if system is secure} \\ 0 & \text{if system is marginally insecure} \\ 1 & \text{if system is insecure} \end{cases} \tag{11.41}$$

### 11.4.2.3 Convolutional Neural Network Convolutional Layer

The input to the convolutional neural network is a highly sparse $N \times N$ matrix. Information is carried on a small percentage of the image pixels: (a) diagonal pixels corresponding to bus information,

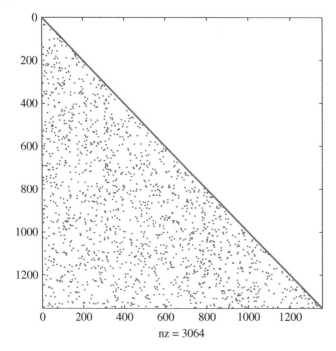

**Figure 11.10** 1354-bus system adjacency lower triangular matrix.

such as voltage and MVA injection while off-diagonal pixels corresponding to branch information, such as MVA flow between buses $i$ and $j$.

The input of standard convolutional neural network models is supposed to be dense. When the input to the network is sparse, the convolution filter and pooling will unnecessarily operate on zero pixels and result in zero pixels as well.

Taking a 1354-bus system as an example, the 1354-bus system represents the size and complexity of the European high voltage transmission network [12, 13]. The network contains 1354 buses, 260 generators, and 1991 branches and it operates at 380 and 220 kV. Please note that the data are fictitious and do not correspond to real-world data. Its input matrix is shown in Figure 11.10.

The $1354 \times 1345$ matrix has 3064 nonzeros. For a typical $3 \times 3$ convolution filter, most convolution operations will meet empty $3 \times 3$ pixels. Different from image recognition, moving nonzeros together by ordering buses does not change the topology and status of a power system. Reverse Cuthill–McKee (RCM), an ordering algorithm is utilized to bring connected buses closer to one another in the adjacency matrix as shown in Figure 11.11, the features extracted are expected to offer the same insight into the power system.

With the original ordering of buses, the number of extracted features is 12,620. This number is reduced to 4788 when RCM is used.

### 11.4.2.4 CNN Pooling Layer

Pooling layers are used to reduce the spatial size of the input feature maps. The types of pooling utilized for the convolutional neural network are average pooling and max pooling.

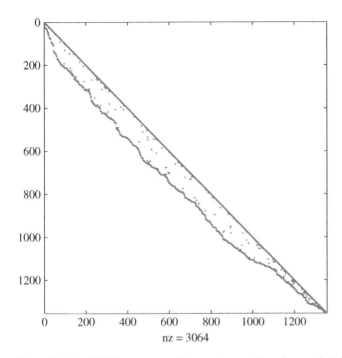

nz = 3064

**Figure 11.11** 1354-bus system adjacency lower triangular matrix (after RCM re-ordering).

#### 11.4.2.5 CNN Fully Connected Layer

As mentioned previously, each contingency would be labeled as −1 (secure), 0 (marginal insecure), or 1 (insecure). This multi-output for multi-contingencies, 3-label classification of secure, marginally insecure, and insecure problems can be implemented using a regression approach as follows:

First, a hyperbolic tangent, the tanh activation function is used at the output layer of the neural network. A tanh activation outputs a continuous value in the range of [−1, 1]. Second, since it is a regression problem, a mean squared error function is used as the loss function. Therefore, when minimizing the loss, the network will learn to generate an output as close as possible to the actual label (−1, 0, or 1). At the end of the process, the output is rounded to the nearest integer.

### 11.4.3 Evaluation Metrics

Three measures are used to evaluate the performance of the neural networks, those are accuracy, precision, and recall.

#### 11.4.3.1 Accuracy

Accuracy for a given instance $A_i$ i.e. an operating point is calculated as follows:

$$A_i = \frac{\text{number of correctly predicted outputs}}{\text{total number of outputs}} \tag{11.42}$$

The overall accuracy is the average across all instances.

#### 11.4.3.2 Precision

Precision is a measure of how much of the predicted data is relevant. For a given instance $i$, and for a given label $l$ (secure, marginal insecure, or insecure), precision is calculated as follows:

$$P_{i,l} = \frac{\text{number of correct predictions of label } l}{\text{number of predictions of label } l} \tag{11.43}$$

The overall precision is the average across all instances.

#### 11.4.3.3 Recall

Recall is a measure of how much of the relevant data is predicted. It is calculated as follows:

$$R_{i,l} = \frac{\text{number of correct predictions of label } l}{\text{number of outputs actually labeled } l} \tag{11.44}$$

The overall recall is the average across all instances.

### 11.4.4 Implementation of Convolutional Neural Network

Graph analytic platform, as addressed in Chapter 2, provides REST API to interface other programming languages, such as C/C++, Java, Python, etc. To demonstrate how to integrate Python code to
graph query, the convolutional neural network function is implemented by Tensorflow [14] in
Python and called in graph query, the pseudo-code to train the convolutional neural network model
is given as follows:

```
import tflearn as tf
import tflearn.layers as tfly

def trainCNN(x_train, y_train)
    convnet = tfly.core.input_data(shape=[None, h, w, d], name='X')
    convnet = tfly.conv.conv2d(convnet, 3, 8, 1, padding="SAME")
    convnet = tfly.conv.max_pool(convnet, 3, 3, padding="VALID")
    convnet = tfly.conv.conv2d(convnet, 3, 4, 1, padding="SAME")
    convnet = tfly.conv.max_pool(convnet, 7, 7, padding="VALID")
    convnet = tfly.core.fully_connected(convnet,1024,activation='relu')
    convnet = tfly.estimator.regression(convnet, optimizer='adam',
              learning_rate=0.001)
    model   = tf.models.dnn(convnet)

    model.fit(x_train,y_train,n_epoch=30)
    model.save('SecurityPredictor')
```

The convolutional neural network is designed with two convolution layers and two max pooling
layers along with a fully connected neural network. The network is optimized by the adaptive
moment estimation method 'adam'.

After being trained by the training set stored in x_train, y_train, the model is saved as
'SecurityPredictor'.

To apply the trained convolutional neural network, the pseudo-code of the function SecurityPredictor is shown as follows:

```
import tflearn as tf
import tflearn.layers as tfly

def testCNN(x_test)
    convnet = tfly.core.input_data(shape=[None, h, w, d], name='X')
    convnet = tfly.conv.conv2d(convnet, 3, 8, 1, padding="SAME")
    convnet = tfly.conv.max_pool(convnet, 3, 3, padding="VALID")
    convnet = tfly.conv.conv2d(convnet, 3, 4, 1, padding="SAME")
    convnet = tfly.conv.max_pool(convnet, 7, 7, padding="VALID")
    convnet = tfly.core.fully_connected(convnet,1024,
               activation='relu')
    convnet = tfly.estimator.regression(convnet, optimizer='adam',
               learning_rate=0.001)
    model   = tf.models.dnn(convnet)

    model.load('SecurityPredictor')
    y_predict = model.predict(x_test)
    return y_predict
```

So far graph database management system has been concentrated on making graph queries callable from Python. But calling Python functions from graph query needs a walk-around through expression function of C since nowadays, graph query supports calling C/C++ expression functions and a C interface can make use of callbacks to call Python functions [15].

In graph queries, user-defined functions are supplemented in the language of C/C++ and are called in queries to perform a set of defined actions and return a value like the built-in functions. The function definition follows the keyword inline. For example,

```
#include <Python.h>
inline int SecurityPredict(ArrayAccum<SumAccum<double>>& x_test){
  PyArg_ParseTuple(args, "testCNN", &x_test);
  ArrayAccum<int>& y_predict;
  y_predict = testCNN (x_test)
  return y_predict;
}
```

The `PyArg_ParseTuple()` function is declared to pass the Python function "testCNN" and argument `&x_test` from Python to the C inline function to be called in the C inline function. And the inline function `SecurityPredict` is callable in graph query.

To test the prediction accuracy and computation efficiency of the trained convolutional neural network for the 1354-bus system, 1000 random contingencies are tested on the trained convolutional neural network. The study results are shown in Table 11.1.

The convolutional neural network correctly predicts all insecure and marginal secure cases by 1.35 seconds. The decoupled power flow method takes 94.20 seconds to calculate the 1000 contingency cases. The 76 insecure and marginal secure cases are filtered into the detailed AC power flow calculation to report violations.

**Table 11.1** Test results for the 1354-bus system.

| Method | Computation time (s) | Unsecure case | Marginal secure case |
|---|---|---|---|
| Decoupled power flow | 94.20 | 72 | 4 |
| CNN | 1.35 | 72 | 4 |

## 11.5 Contingency Analysis Graph Computing Implementation

Contingency analysis implementation includes three major tasks: contingency screening, full power flow calculation, and violation report. To achieve the high computation efficiency of contingency analysis for large-scale power systems, the convolutional neural network is applied to perform contingency screening, the potential harmful contingencies are further studied by full power flow calculation using the preconditioned CG approach considering the merit of its parallelism. The high-level approach for the graph computing-based contingency analysis is presented below.

---

**Algorithm 11.7   Graph-Based Contingency Analysis**

---

```
1: Prescreen contingencies by convolutional neural network
2: For each contingency in the potentially harmful contingency set
3:    Calculate post-contingency power flow by preconditioned CG
4:    Check post-contingency power flow against limits to report
      violations
```

---

To implement the graph-based parallel contingency analysis, a new contingency analysis graph is defined by the following schema.

```
CREATE VERTEX CTGS (PRIMARY_ID ctgid int, ctgType string, from_bus
    int, to_bus int, harmful int)

CREATE VERTEX VIOLATIONS (PRIMARY_ID violationid int, violationType
    string, violationValue double, violationLimit double)
```

Vertex CTGS defined the contingency unique identification ctgid, contingency type (bus outage, line outage, transformer outage, generator outage, shunt capacitor outage etc), the outage element from end bus from_bus, and to end bus to_bus. When the outage element is a one-terminal element, the to_bus attribute is void.

Vertex VIOLATIONS defined the violation unique identification violationid, violation type (voltage limit violation, thermal limit violation, island, or unsolvable contingency), the actual value of the voltage and MVA flow violationValue, and the corresponding limit violationLimit. When the violation is in the type of island and unsolvable, the attributes violationValue and violationLimit are void.

On top of the vertex schema, edges are defined as follows to hold the contingency and violation connectivity:

```
CREATE UNDIRECTED EDGE connected_CTGS_VIOLATIONS (from CTGS, to
    VIOLATIONS) with reverse_edge = "connected_VIOLATIONS_CTGS"
```

Since contingency and violation are many-to-many relationships, undirected edge `connected_CTGS_VIOLATIONS` with `reverse_edge connected_VIOLATIONS_CTGS` are defined to map contingency to violation and violation to contingency.

With the vertex and edge defined as above, the contingency analysis graph schema is then defined as:

```
CREATE GRAPH contingencyAnalysisGraph(CTGS, VIOLATIONS,
    connected_CTGS_VIOLATIONS, connected_VIOLATIONS_CTGS)
```

Once the graph schema is defined with the desired structure and attribute placeholders, assuming the contingencies are defined in a CSV file `ContingencySet.csv`, the contingencies are loaded to the graph and mapped values to the vertex as examples as the following loading jobs.

```
load "$sys.data_root/ ContingencySet.csv" TO VERTEX CTGS
    values($"id",$"type",$"from_bus",$"to_bus",$"harmful")
    using Separator=",", Header="true";
```

To temporarily store the post-contingency branch power flow, two attributes `P_BR_CTG` and `Q_BR_CTG` are added to edge `TopoConnect`.

```
CREATE DIRECTED EDGE TopoConnect (from TopoND, to TopoND, typename
    string, edge_name List<string>, area List<string>, zone
    List<string>, from_bus int, to_bus int, flag uint, R List<double>,
    X List<double>, hB_list List<double>, line_Q1_list List<double>,
    line_Q2_list List<double>, line_Q3_list List<double>, control_bus
    int, transformer_turns_ratio List<double>,
    shifting_transformer_angle List<double>, min_tap double, max_tap
    double, step_size double, min_volt List<double>, max_volt
    List<double>, M_P_BR List<double>, M_Q_BR List<double>,
    G List<double>, B List<double>, BIJ List<double>, circuit uint,
    P_BR double, Q_BR double, P_BR_CTG double, Q_BR_CTG double,
    from_open List<int>, to_open List<int>)
```

The pseudo-code in graph query to implement the graph-based parallel contingency analysis is shown as follows:

```
CREAT QUERY contingencyAnalysis(int MaxIteration, double Tolerance)
FOR GRAPH busBranchGraph and GRAPH contingencyAnalysisGraph {

ArrayAccum<SumAccum<double>> @@x_test
SetAccum<double> @@MVA_BASE = 100
MaxAccum<double> @@maxVm = -INF
MaxAccum<double> @@maxVa = -INF
SetAccum<double> @@branchMVAFlow = 0
SetAccum<double> @@violationID = 0

T0 = {busBranchGraph.TopoND.*}
int nBus = busBranchGraph.getVertexCount()
```

```
T1 = SELECT s FROM T0:s-(connected:e)->:t
  ACCUM
    x_test[e.from_bus][e.to_bus] =
              sqrt(e.P_BR*e.P_BR + e.Q_BR*e.Q_BR)/@@MVA_BASE
  POST-ACCUM
    x_test[s.TOPOID][s.TOPOID] = (s.GenP-s.LdP)/@@MVA_BASE
    x_test[s.TOPOID][nBus+1] = (s.GenQ-s.LdQ)/@@MVA_BASE
    x_test[s.TOPOID][nBus+2] = s.Vm
    x_test[s.TOPOID][nBus+3] = s.Va

T0 = {contingencyAnalysisGraph.CTGS.*}
T0.harmful = SecurityPredict(ArrayAccum<SumAccum<double>>& x_test)

T1 = SELECT ctg FROM T0 WHERE (ctg.harmful!= 0)
  ACCUM
    buildB_pAndB_ppGraph() FOR GRAPH busBranchGraph
         and GRAPH B_pGraph and GRAPH B_ppGraph

    WHILE( abs(@@maxVm) > Tolerance OR abs(@@maxVa) > Tolerance )
         LIMIT MaxIteration DO
      preconditoinedConjugateGradient(int MaxIteration, double Tolerance)
        FOR GRAPH busBranchGraph and GRAPH B_pGraph and GRAPH B_ppGraph

      Tp = {B_pGraph.TopoND.*}
      T2 = SELECT v FROM Tp
        ACCUM
          @@maxVa = max(abs(v.deltaThetak))

      Tpp= {B_ppGraph.TopoND.*}
      T2 = SELECT v FROM Tpp
        ACCUM
          @@maxVm = max(abs(v.deltaVk))

    T1 = SELECT s FROM T0:s-(connected:e)->:t
      ACCUM
        IF (e.K == 0) THEN e.K = 1 END
        e.P_BR_CTG += s.Vm*t.Vm*(e.G/e.K*cos(s.Vs-t.Vs)
                    +e.B/e.K*sin(s.Vs-t.Vs))
        e.Q_BR_CTG += s.Vm*s.Vm*(e.G/e.K*sin(s.Va-t.Va)
                     -e.B/e.K*cos(s.Va-t.Va))
      POST-ACCUM
        @@branchMVAFlow= sqrt(e.P_BR_CTG*e.P_BR_CTG+e.Q_BR_CTG*e.
        Q_BR_CTG)
        e.P_BR_CTG = 0
        e.Q_BR_CTG = 0
        IF ( @@branchMVAFlow > e.line_Q3 ) THEN
          @@violationID += 1
```

```
        INSERT INTO VIOLATIONS VALUES(@@violationID,
            "Power Flow Load Shedding Limit Violation",
            @@branchMVAFlow, e.line_Q3)
        INSERT INTO connected_CTGS_VIOLATIONS VALUES(ctg.ctgid,
                @@violationID)
       INSERT INTO connected_VIOLATIONS_CTGS(@@violationID,ctg.ctgid)
      ELSE IF ( @@branchMVAFlow > e.line_Q2 ) THEN
        @@violationID += 1
        INSERT INTO VIOLATIONS VALUES(@@violationID,
            "Power Flow Emergency Limit Violation",
            @@branchMVAFlow, e.line_Q2)
        INSERT INTO connected_CTGS_VIOLATIONS VALUES(ctg.ctgid,
                @@violationID)
       INSERT INTO connected_VIOLATIONS_CTGS(@@violationID,ctg.ctgid)
      ELSE IF ( @@branchMVAFlow > e.line_Q3 ) THEN
        @@violationID += 1
        INSERT INTO VIOLATIONS VALUES(@@violationID,
            "Power Flow Normal Limit Violation",
            @@branchMVAFlow, e.line_Q1)
        INSERT INTO connected_CTGS_VIOLATIONS VALUES(ctg.ctgid,
            @@violationID)
       INSERT INTO connected_VIOLATIONS_CTGS( @@violationID,ctg.ctgid)
      END
      IF ( s.Vm < s.min_volt ) THEN
        @@violationID += 1
        INSERT INTO VIOLATIONS VALUES(@@violationID,
            "Low Voltage Violation", s.Vm, s.min_volt)
        INSERT INTO connected_CTGS_VIOLATIONS VALUES(ctg.ctgid,
                @@violationID)
       INSERT INTO connected_VIOLATIONS_CTGS( @@violationID,ctg.ctgid)
      ELSE IF ( s.Vm > s.max_volt ) THEN
        @@violationID += 1
        INSERT INTO VIOLATIONS VALUES(@@violationID,
            "High Voltage Violation", s.Vm, s.max_volt)
        INSERT INTO connected_CTGS_VIOLATIONS VALUES(ctg.ctgid,
            @@violationID)
       INSERT INTO connected_VIOLATIONS_CTGS( @@violationID,ctg.ctgid)
      END
}
```

The query `contingencyAnalysis` has two arguments maximum iteration number `MaxIteration` and tolerance `Tolerance` to control the iterations. The global variable `ArrayAccum<SumAccum<double>> @@x_test` is an array which is the convolutional neural network input matrix to predict if each contingency in the contingency set is secure ($-1$), marginal insecure (0), or insecure (1). Global variables `MaxAccum @@maxVm` and `@@maxVa` store the maximum voltage magnitude or angle update which will be used to compare against the tolerance `Tolerance` to check the convergence.

In the query, MVA flows in per unit are assigned to off-diagonal pixels in nodal parallel by:

```
T1 = SELECT s FROM T0:s-(connected:e)->:t
  ACCUM
    x_test[e.from_bus][e.to_bus] =
                    sqrt(e.P_BR*e.P_BR + e.Q_BR*e.Q_BR)/@@MVA_BASE
```

Nested in POST-ACCUM, active power injection in per unit is carried on diagonal pixels. In a typical convolutional neural network, the reactive power injection and bus voltage magnitude and phase angle are carried as the diagonal pixels on multiple additional layers using a three-dimensional array. Since the bus injection and voltage are carried on a small percentage of the image pixels on the additional layers, to simplify the input image, the reactive power injection and voltage magnitude and phase angle are saved in the extend columns of the ArrayAccum @@x_test by:

```
POST-ACCUM
    x_test[s.TOPOID][s.TOPOID] = (s.GenP-s.LdP)/@@MVA_BASE
    x_test[s.TOPOID][nBus+1]   = (s.GenQ-s.LdQ)/@@MVA_BASE
    x_test[s.TOPOID][nBus+2]   = s.Vm
    x_test[s.TOPOID][nBus+3]   = s.Va
```

Where nBus = busBranchGraph.getVertexCount() is the number of vertices in the bus-branch graph.

By calling the user-defined function SecurityPredict as follow, the convolutional neural network predicts the security of the contingency under the given base case power flow defined by the pixel image x_test.

```
    T0.harmful = SecurityPredict(ArrayAccum<SumAccum<double>>& x_test)
```

The marginal insecure and insecure contingencies are filtered into the fast decoupled power flow calculation using preconditioned CG approach.

The query firstly selects all marginal insecure and insecure cases by T1 = SELECT ctg FROM T0 WHERE (ctg.harmful! = 0). They are processed in nodal parallel.

The graphs B_pGraph and B_ppGraph are built nested in ACCUM first. Then in the while loop, preconditoinedConjugateGradient is called to solve the voltage angle and magnitude equations till the bus voltage is solved or the maximum iteration number is reached.

The postcontingency power flow is calculated in nodal parallel by traveling through all edges using T1 = SELECT s FROM T0:s-(connected:e)->:t. After the postcontingency bus voltage and branch power are determined, in POST-ACCUM, violations are reported if any. Note that for each violation, the bi-directed connections connected_CTGS_VIOLATIONS and connected_VIOLATIONS_CTGS between the contingency and violation are created to facilitate the searching of violations under a given contingency or different contingencies that result in the same violation.

## References

**1** Roy, A. and Jain, S. (2011). *Contingency Analysis in Power System: Thesis of Master of Engineering in Power Systems & Electric Drives*. Patiala: Thapar University 10.13140/RG.2.1.4481.4240.

**2** Turitsyn, K.S. and Kaplunovich, P.A. (2013). Fast algorithm for N-2 contingency problem. *2013 46th Hawaii International Conference on System Sciences*, Wailea, Maui, HI, USA, pp. 2161–2166. https://doi.org/10.1109/HICSS.2013.233.

**3** Vamshi Ram, B., Upadhyay, P., and Radhika, G. (2018). N-2 contingency analysis of an IEEE 14 bus system using MI power. *International Journal of Engineering and Technology* 7 (4.22): 211–214.

**4** Warwick, K., Ekwue, A., and Aggrawal, R. (1997). *Artificial Intelligence Techniques in Power Systems*. London, UK: The Institution of Electrical Engineers.

**5** La, S.M., Trovato, M., and Torelli, F. (1996). A neural network based method for voltage security monitoring. *IEEE Transactions on Power Systems* 11 (3): 1332–1341.

**6** Zhou, D.Q. and Annakage, U.D. (2010). Online monitoring of voltage stability margin using artificial neural network. *IEEE Transactions on Power Systems* 25 (3): 1566–1574.

**7** Pang, C.K., Prabhakara, F.S., El-Abiad, A.H., and Koivo, A.J. (1974). Security evaluation in power systems using pattern recognition. *IEEE Transaction on Power System Apparatus and Systems* PAS-93 (2): 969–976.

**8** Zhu, Y., Shi, L., Dai, R., and Liu, G. (2019). Fast grid splitting detection for N-1 contingency analysis by graph computing. *IEEE Innovative Smart Grid Technologies-Asia (ISGT Asia)* 2019: 673–677.

**9** Tarjan, R.E. (1974). A note on finding the bridges of a graph. *Information Processing Letters* 2: 160–161.

**10** Tarjan, R.E. and Vishkin, U. (1985). An efficient parallel biconnectivity algorithm. *SIAM Journal on Computing* 14 (4): 862–874.

**11** Zhao, Y., Yuan, C., Liu, G., and Grinberg, I. (2018). Graph-based preconditioning conjugate gradient algorithm for "N-1" contingency analysis. *2018 IEEE Power & Energy Society General Meeting (PESGM)*, Portland, OR, USA, pp. 1–5, https://doi.org/10.1109/PESGM.2018.8586214.

**12** Josz, C., Fliscounakis, S., Maeght, J., and Panciatici, P. (2016). AC power flow data in MATPOWER and QCQP format: iTesla, RTE snapshots, and PEGASE. *arXiv preprint arXiv* 1603.01533. https://arxiv.org/pdf/1603.01533.pdf.

**13** Fliscounakis, S., Panciatici, P., Capitanescu, F., and Wehenkel, L. (2013). Contingency ranking with respect to overloads in very large power systems taking into account uncertainty, preventive and corrective actions. *IEEE Transactions on Power Systems* 28 (4): 4909–4917.

**14** Abadi, M., Agarwal, A., Barham, P. et al. (2016). Tensorflow: large-scale machine learning on heterogeneous distributed systems. *arXiv* 1603.04467. https://arxiv.org/pdf/1603.04467.pdf.

**15** Python Software Foundation (2022). Extending python with C or C++. https://docs.python.org/3/extending/extending.html# (accessed 20 June 2022).

# 12

## Economic Dispatch and Unit Commitment

In power system operation, generation active power is dispatched to balance load demand. Economic dispatch is applied to determine the active power output of generation units at the lowest fuel cost subjected to meet the system load demands.

## 12.1 Classic Economic Dispatch

In classic economic dispatch, network security constraints are ignored. Generators are dispatched based on thermal unit fuel cost, hydro unit water consumption constraints, and system load demands. The equal incremental rate approach is the classic approach to solve the classic economic dispatch problems.

### 12.1.1 Thermal Unit Economic Dispatch

The classic economic dispatch optimizes generation power outputs to meet load demand and minimize the generation cost without considering network security constraints. The classic economic dispatch problem is formulated as follows:

$$\min F = \sum_{i=1}^{n} F_i(P_{Gi}) \tag{12.1}$$

Subject to

$$\sum_{i=1}^{n} P_{Gi} = P_D \tag{12.2}$$

$$P_{Gi}^{\min} \leq P_{Gi} \leq P_{Gi}^{\max}, \quad i = 1, ..., n \tag{12.3}$$

where $n$ is the number of generators in the system, $P_{Gi}$ is the active power output of the $i$-th generator, $F_i(P_{Gi})$ is the generation fuel cost function, $P_D$ is the system load demand, $P_{Gi}^{\min}$ and $P_{Gi}^{\max}$ are the generator's minimum and maximum active power outputs. In this model, network losses are not modeled.

Let us ignore the inequality constraint (12.3) temporarily. To solve the classic economic dispatch optimization problem defined by (12.1)–(12.2), Lagrange multiplier $\lambda$ is introduced to convert the equality constraints constrained optimization problem into an unconstrained problem as the following Lagrangian function:

$$\mathcal{L}(P_{Gi}, \lambda) = \sum_{i=1}^{n} F_i(P_{Gi}) + \lambda \left( P_D - \sum_{i=1}^{n} P_{Gi} \right) \tag{12.4}$$

*Graph Database and Graph Computing for Power System Analysis*, First Edition. Renchang Dai and Guangyi Liu.
© 2024 The Institute of Electrical and Electronics Engineers, Inc. Published 2024 by John Wiley & Sons, Inc.

To find the solution of (12.4), all partial derivatives of the Lagrangian function with respect to each of the unknown variables $P_{G1}$, $P_{G2}$, ..., $\lambda$ are equal to zero

$$\frac{\partial \mathcal{L}(P_{Gi}, \lambda)}{\partial P_{Gi}} = \frac{\partial F_i(P_{Gi})}{\partial P_{Gi}} - \lambda = 0, \quad i = 1, ..., n \tag{12.5}$$

$$\sum_{i=1}^{n} P_{Gi} - P_D = 0 \tag{12.6}$$

Where $\dfrac{\partial F_i(P_{Gi})}{\partial P_{Gi}}$ is the incremental fuel rate of the *i*-th generator. Assuming the fuel cost curve is quadratic, the incremental fuel rate curve is derived as

$$\frac{\partial \mathcal{L}(P_{Gi}, \lambda)}{\partial P_{Gi}} = \frac{\partial \left(a_i P_{Gi}^2 + b_i P_{Gi} + c_i\right)}{\partial P_{Gi}} - \lambda = 2a_i P_{Gi} + b_i - \lambda = 0 \quad i = 1, ..., n \tag{12.7}$$

$$2a_1 P_{G1} + b_1 = 2a_2 P_{G2} + b_2 = \cdots = 2a_n P_{Gn} + b_n = \lambda \tag{12.8}$$

Equation (12.8) implies the total fuel consumption *F* will be minimal if the incremental fuel rates of all generators are equal. That is how the name of the equal incremental rate approach comes from [1]. Taking account of the generator power output inequality constraint (12.3), the equal incremental rate algorithm is shown as follows:

---

**Algorithm 12.1  Equal Incremental Rate**

---

```
1:   k = 0, λ_min = min_i (2a_i P_Gi^min + b_i), λ_max = max_i (2a_i P_Gi^max + b_i)
2:   Initialize P_Gi^(0)
3:   While (|∑_{i=1}^n P_Gi^(k) − P_D| > ε)
4:       k = k + 1
5:       Calculate λ^(k) = ½(λ_min + λ_max)
6:       Calculate P_Gi^(k) = (λ^(k) − b_i) / 2a_i
7:       If P_Gi^(k) > P_Gi^max then P_Gi^(k) = P_Gi^max
8:       If P_Gi^(k) < P_Gi^min then P_Gi^(k) = P_Gi^min
9:       If ∑_{i=1}^n P_Gi^(k) > P_D then
10:          λ_max = λ^(k)
11:      Else
12:          λ_min = λ^(k)
```

---

Where $\varepsilon$ is the power balance tolerance which states the maximum power mismatch accepted for a converged solution.

To implement the graph-based equal incremental rate algorithm for solving the economic dispatch optimization problem, the bus-branch graph defined for power flow calculation and state estimation is extended as follows to contain the unit information for economic dispatch:

```
CREATE VERTEX TopoND (PRIMARY_ID topoid int, TOPOID int, island int,
    bus_name string, area string, loss_zone uint, busType int, Vm
    double, Va double, M_Vm double, M_Va double, base_kV double,
    desired_volts double, control_bus_number int, up_V double, lo_V
    double, GenP double, GenQ double, M_Gen_P double, M_Gen_Q double,
```

```
qUp double, qLower double, LdP double, LdQ double, P double,
Q double, M_Load_P double, M_Load_Q double, G double, B double, Sub
string, OV int, UV int, sumB double, sumG double, sumBi double,
volt double, double, M_C_P double, SE_Vm double, SE_Va double,
Ri_vP double, Ri_vQ double, Ri_eP double, Ri_eQ double, SE_Inj
tuple, GainP tuple, GainQ tuple, H_r_P double, H_r_Q double, Pmax
double, Pmin double, a double, b double, c double)
```

There is no update in the edge definition. In the vertex extension above, new attributes generation upper and lower limits **Pmax**, **Pmin**, and fuel cost coefficients **a**, **b**, and **c** highlighted in black bold are added in the vertex TopoND definition. After the vertex definition is updated, the graph will be recreated by the statement as follow.

```
CREATE GRAPH busBranchGraph(TopoND, TopoConnect)
```

With the new graph structure, the raw data file defining the generator fuel cost curve and maximum and minimum generation output (for instance, unit information file unitInfo.csv) will be read to the bus-branch graph by the following query `readUnitInfo`.

```
CREATE TABLE unitInfoTable (BusID String, UnitName String, Pmax
double, Pmin double, a double, b double, c double)
```

```
CREATE QUERY readUnitInfo() FOR GRAPH busBranchGraph{
BULK INSERT unitInfoTable
FROM '$sys.data_root/unitInfo.csv'
WITH(FIRSTROW = 2,FIELDTERMINATOR = ',',ROWTERMINATOR = '\n',TABLOCK)
T0= {TopoND.*}
T1 = SELECT s FROM T0
 ACCUM
   SELECT unit FROM unitInfoTable WHERE unit.BusID == s.TOPOID
     s.Pmax = unit.Pmax
     s.Pmin = unit.Pmin
     s.a       = unit.a
     s.b       = unit.b
     s.c       = unit.c
}
```

The table header of the unit information file unitInfo.csv saved at "`$sys.data_root`" is given in Table 12.1.

To read the unit information data to the busBranchGraph, a unitInfoTable is defined as

```
CREATE TABLE unitInfoTable (BusID String, UnitName String, Pmax
double, Pmin double, a double, b double, c double)
```

Statement BULK INSERT reads data from unitInfo.csv. The control parameters defined by a WITH clause WITH(FIRSTROW = 2, FIELDTERMINATOR = ',', ROWTERMINATOR = '\n', TABLOCK) tells the first row is a table header, the field delimiter is a comma, the row terminator is '\n', and the table is locked when it is read.

**Table 12.1** Unit information table header.

| Bus ID | Unit name | Pmax | Pmin | a | b | c |
| --- | --- | --- | --- | --- | --- | --- |

When the CSV file is read into the `unitInfoTable`, the data are assigned to `busBranch-Graph` attributes through matched unit `BusID` by the statement

```
SELECT unit FROM unitInfoTable WHERE unit.BusID == s.TOPOID
```

After the extended `busBranchGraph` is created and data are loaded into the graph, the following pseudocode in graph query is developed to implement classic economic dispatch in parallel.

```
CREATE QUERY classicEconomicDispatch(double Pload, int MaxIteration,
double Tolerance) FOR GRAPH busBranchGraph {
  MaxAccum<double> @@lamdaMax =-INF
  MinAccum<double> @@lamdaMin = INF
  SetAccum<double> @@lamda
  MaxAccum<double> @@Pmismatch =-INF

  T0 = {TopoND.*}
  WHILE (abs(@@Pmismatch)>Tolerance) LIMIT MaxIteration DO
    T1 = SELECT s FROM T0 WHERE s.busType == 2 or s.busType == 3
      ACCUM
        @@lamdaMax = 2*s.a*s.Pmax + s.b
        @@lamdaMin = 2*s.a*s.Pmin + s.b
      POST-ACCUM
        @@lamda = (@@lamdaMax + @@lamdaMin)/2
    @@Pmismatch = 0
    T1 = SELECT s FROM T0 WHERE s.busType == 2 or s.busType == 3
      ACCUM
        s.P = (@@lamda - s.b)/(2*s.a)
        IF (s.P > s.Pmax) THEN s.P = s.Pmax
        IF (s.P > s.Pmin) THEN s.P = s.Pmin
        @@Pmismatch + = s.P
      POST_ACCUM
        @@Pmismatch -= Pload
        IF (@@Pmismatch > 0) THEN
          @@lamdaMax = @@lamda
        ELSE
          @@lamdaMin = @@lamda
}
```

In the query `classicEconomicDispatch`, system load `Pload`, maximum iteration number `MaxIteration`, and active power mismatch tolerance `Tolerance` are passed through. Global variables `@@lamdaMax`, `@@lamdaMin`, `@@lamda`, and `@@Pmismatch` are defined to store updated $\lambda_{max}$, $\lambda_{min}$, $\lambda$, and active power mismatch.

The WHILE loop statement checks the active power mismatch absolute value against the Tolerance with the LIMIT clause which counts the iterations in the WHILE loop against the maximum iteration number `MaxIteration`.

```
WHILE (abs(@@Pmismatch)>Tolerance) LIMIT MaxIteration DO
```

In the query definition, statement `T1 = SELECT s FROM T0 WHERE s.busType == 2 or s.busType == 3` coupling with the keyword `ACCUM` selects all unit vertices s from vertex set `T0 = {TopoND.*}`, calculates the maximum and minimum incremental fuel rate $\lambda_{\max}$ and $\lambda_{\min}$ in nodal parallel by

```
@@lamdaMax = 2*s.a*s.Pmax + s.b
@@lamdaMin = 2*s.a*s.Pmin + s.b
```

Global variable `@@lamdaMax` is defined as a maximum accumulator by the keyword `MaxAccum`, while variable `@@lamdaMin` is defined as a minimum accumulator by key word `MinAccum`.

```
MaxAccum Updates the value of MaxAccum to the greater between
MaxAccum and the right operand. For example, statement @@lamdaMax =
2*s.a*s.Pmax + s.b checks (2*s.a*s.Pmax + s.b). If (2*s.a*s.Pmax + s.
b) is greater than @@lamdaMax, @@lamdaMax will be updated as (2*s.a*s.
Pmax + s.b), otherwise, keep @@lamdaMax unchanged.
MinAccum Updates the value of MinAccum to the lesser between
MinAccum and the right operand. For example, statement @@lamdaMin =
2*s.a*s.Pmin + s.b checks (2*s.a*s.Pmin + s.b). If (2*s.a*s.Pmin + s.
b) is less than @@lamdaMin, @@lamdaMin will be updated as (2*s.a*s.
Pmin + s.b), otherwise, keep @@lamdaMin unchanged.
```

In the `POST-ACCUM`, the current incremental fuel rate is updated by a binary search approach as follows:

```
@@lamda = (@@lamdaMax + @@lamdaMin)/2
```

With the updated incremental fuel rate, generator active power output is calculated and capped by corresponding upper and lower limits in nodal parallel.

```
s.P = (@@lamda - s.b)/(2*s.a)
IF (s.P > s.Pmax) THEN s.P = s.Pmax
IF (s.P > s.Pmin) THEN s.P = s.Pmin
```

Then the total active power output is also calculated by `@@Pmismatch + = s.P` in nodal parallel.
In the `POST-ACCUM` after, the active power mismatch is calculated and `@@lamdaMax` or `@@lamdaMin` are updated by a binary search.

```
@@Pmismatch -= Pload
IF (@@Pmismatch > 0) THEN
  @@lamdaMax = @@lamda
ELSE
  @@lamdaMin = @@lamda
```

Taking the IEEE-14 bus system as an example, the unit information is given in Table 12.2 [2].
The active power in per unit at each load bus is listed in Table 12.3 (the same as Table 10.1 in Chapter 10. The MW base is 100 MW). The total load is 259 MW.
Initially, the maximum and minimum incremental fuel rates $\lambda_{\max}$ and $\lambda_{\min}$ are calculated to set the range for binary searching to find the optimal incremental fuel rate.

$$\lambda_{\max} = \max_i \left(2a_i P_{Gi}^{\max} + b_i\right) = 14.0$$

$$\lambda_{\max} = \max_i \left(2a_i P_{Gi}^{\max} + b_i\right) = 7.7$$

**Table 12.2** Unit information.

| Bus ID | Unit name | Pmax (MW) | Pmin (MW) | a | b | c |
|--------|-----------|-----------|-----------|---|---|---|
| 1 | G1 | 500 | 50 | 0.0070 | 7.0 | 240 |
| 2 | G2 | 200 | 20 | 0.0095 | 10.0 | 200 |
| 3 | G3 | 300 | 20 | 0.0090 | 8.5 | 220 |
| 6 | G4 | 150 | 20 | 0.0090 | 11.0 | 200 |
| 8 | G5 | 200 | 20 | 0.0080 | 10.5 | 220 |

**Table 12.3** IEEE-14 bus system bus table.

| Bus ID | Area | Loss zone | Type | Voltage | Angle | Load_P | Load_Q | Unit_P | Unit_Q | Desired voltage | MAXQ | MINQ | G | B |
|--------|------|-----------|------|---------|-------|--------|--------|--------|--------|-----------------|------|------|---|---|
| 1 | 1 | 1 | 3 | 1.060 | 0 | 0 | 0 | 2.324 | −0.169 | 1.060 | 0 | 0 | 0 | 0 |
| 2 | 1 | 1 | 2 | 1.045 | −4.98 | 0.217 | 0.127 | 0.400 | 0.424 | 1.045 | 0.50 | −0.4 | 0 | 0 |
| 3 | 1 | 1 | 2 | 1.010 | −12.72 | 0.942 | 0.190 | 0 | 0.234 | 1.010 | 0.40 | 0 | 0 | 0 |
| 4 | 1 | 1 | 0 | 1.019 | −10.33 | 0.478 | −0.039 | 0 | 0 | 0 | 0 | 0 | 0 | 0 |
| 5 | 1 | 1 | 0 | 1.020 | −8.78 | 0.076 | 0.016 | 0 | 0 | 0 | 0 | 0 | 0 | 0 |
| 6 | 1 | 1 | 2 | 1.070 | −14.22 | 0.112 | 0.075 | 0 | 0.122 | 1.070 | 0.24 | −0.6 | 0 | 0 |
| 7 | 1 | 1 | 0 | 1.062 | −13.37 | 0 | 0 | 0 | 0 | 0 | 0 | 0 | 0 | 0 |
| 8 | 1 | 1 | 2 | 1.090 | −13.36 | 0 | 0 | 0 | 0.174 | 1.090 | 0.24 | −0.6 | 0 | 0 |
| 9 | 1 | 1 | 0 | 1.056 | −14.94 | 0.295 | 0.166 | 0 | 0 | 0 | 0 | 0 | 0 | 0.19 |
| 10 | 1 | 1 | 0 | 1.051 | −15.10 | 0.090 | 0.058 | 0 | 0 | 0 | 0 | 0 | 0 | 0 |
| 11 | 1 | 1 | 0 | 1.057 | −14.79 | 0.035 | 0.018 | 0 | 0 | 0 | 0 | 0 | 0 | 0 |
| 12 | 1 | 1 | 0 | 1.055 | −15.07 | 0.061 | 0.016 | 0 | 0 | 0 | 0 | 0 | 0 | 0 |
| 13 | 1 | 1 | 0 | 1.050 | −15.16 | 0.135 | 0.058 | 0 | 0 | 0 | 0 | 0 | 0 | 0 |
| 14 | 1 | 1 | 0 | 1.036 | −16.04 | 0.149 | 0.050 | 0 | 0 | 0 | 0 | 0 | 0 | 0 |

The incremental fuel rate is initialized as $\lambda^{(0)} = \frac{1}{2}(\lambda_{\min} + \lambda_{\max}) = 10.85$. The active power generation of each unit is calculated by $P_{Gi}^{(k)} = \frac{(\lambda^{(k)} - b_i)}{2a_i}$ and limited by its upper and lower limits. The total generation in the first iteration is $\sum_{i=1}^{n} P_{Gi}^{(k)} = 492.17$ MW which is greater than the system load of 259 MW. The maximum incremental fuel rate $\lambda_{\max}$ is updated to the current incremental fuel rate $\lambda^{(k)}$ by binary searching. Assuming the tolerance is 1 MW, after six iterations, the optimal solution is found as shown in Table 12.4.

**Table 12.4**  Thermal unit economic dispatch iteration.

| Iteration | $\lambda^{(k)}$ | PG1 (MW) | PG2 (MW) | PG3 (MW) | PG4 (MW) | PG5 (MW) | Total generation (MW) | System load (MW) | Total fuel cost ($) |
|---|---|---|---|---|---|---|---|---|---|
| 0 | 10.85 | 275.00 | 44.74 | 130.56 | 20.00 | 21.88 | 492.17 | 259 | 5,721.00 |
| 1 | 9.28 | 162.50 | 20.00 | 43.06 | 20.00 | 20.00 | 265.56 | 259 | 3,425.60 |
| 2 | 8.49 | 106.25 | 20.00 | 20.00 | 20.00 | 20.00 | 186.25 | 259 | 2,716.97 |
| 3 | 8.88 | 134.38 | 20.00 | 21.18 | 20.00 | 20.00 | 215.56 | 259 | 2,971.69 |
| 4 | 9.08 | 148.44 | 20.00 | 32.12 | 20.00 | 20.00 | 240.56 | 259 | 3,196.19 |
| 5 | 9.18 | 155.47 | 20.00 | 37.59 | 20.00 | 20.00 | 253.06 | 259 | 3,310.28 |
| 6 | 9.23 | 158.98 | 20.00 | 40.32 | 20.00 | 20.00 | 259.31 | 259 | 3,367.79 |

## 12.1.2  Hydrothermal Power Generation System Economic Dispatch

The above approach solves thermal unit economic dispatch at the lowest fuel cost to meet a given system load demand. In thermal and hydroelectric power generation systems, the hydro unit water consumption budget for a period is modeled as a constraint. The water consumption budget could be hourly, daily, and/or monthly budget depending on hydro unit operation flexibility. For a thermal and hydroelectric power generation system, besides the thermal unit economic dispatch model (12.1)–(12.3), water consumption constraint is added and the hydrothermal power generation system economic dispatch model is expressed as:

$$\min F = \sum_{i=1}^{n} \int_{0}^{\tau} F_i(P_{Ti}(t))dt \tag{12.9}$$

Subject to

$$P_{Ti}^{\min} \le P_{Ti}(t) \le P_{Ti}^{\max}, \quad i = 1, ..., n \tag{12.10}$$

$$\sum_{i=1}^{n} P_{Ti}(t) + \sum_{i=1}^{m} P_{Hi}(t) = P_D(t), t \in [0, \tau] \tag{12.11}$$

$$\int_{0}^{\tau} W_i(P_{Hi}(t))dt = C_i, i = 1, 2..., m \tag{12.12}$$

where, $W_i(P_{Hi}(t))$ is the water consumption function. $C_i$ is the water consumption budget. $m$ is the number of hydropower units. The generation outputs of the thermal unit and hydro unit are differentiated as $P_T$ and $P_H$ in the model. Assuming the thermal unit and hydro unit power output as well as the system load are constant for each time period $\Delta t_k$, $\left( \tau = \sum_{k=1}^{t_n} \Delta t_k \right)$, the model (12.9)–(12.12) is discretized as

$$\min F = \sum_{i=1}^{n} \sum_{k=1}^{t_n} F_i(P_{Ti,k})\Delta t_k \tag{12.13}$$

Subject to

$$\sum_{i=1}^{n} P_{Ti,k} + \sum_{i=1}^{m} P_{Hi,k} = P_{D,k}, k = 1, ..., t_n \tag{12.14}$$

$$\sum_{k=1}^{t_n} W_i(P_{Hi,k})\Delta t_k = C_i, i = 1, 2..., m \tag{12.15}$$

$$P_{Ti}^{\min} \leq P_{Ti,k} \leq P_{Ti}^{\max}, \quad i = 1, ..., n, k = 1, ..., t_n \tag{12.16}$$

Ignore the inequality constraint (12.16) temporarily. To solve the thermal and hydroelectric power generation system economic dispatch optimization problem defined by (12.13)–(12.15), Lagrange multipliers are introduced to convert the equality constraints constrained optimization problem into an unconstrained problem as the following Lagrangian function

$$
\begin{aligned}
\mathcal{L}(P_{Ti,k}, P_{Hi,k}, \lambda_k, \gamma_i) = {} & \sum_{i=1}^{n} \sum_{k=1}^{t_n} F_i(P_{Ti,k})\Delta t_k F_i(P_{Ti,k})\Delta t_k \\
& + \sum_{k=1}^{t_n} \lambda_k \left( P_{D,k} - \sum_{i=1}^{n} P_{Ti,k} - \sum_{i=1}^{m} P_{Hi,k} \right)\Delta t_k \\
& + \gamma_i \left( \sum_{k=1}^{t_n} \lambda_k W_i(P_{Hi,k})\Delta t_k - C_i \right)
\end{aligned}
\tag{12.17}
$$

To find the optimal solution of the Lagrangian function, all partial derivatives of the Lagrangian function with respect to each of the unknown variables $P_{Ti,k}$, $P_{Hi,k}$, $\lambda_k, \gamma_i$ are set to zero

$$\frac{\partial\mathcal{L}(P_{Ti,k}, P_{Hi,k}, \lambda_k, \gamma_i)}{\partial P_{Ti,k}} = \frac{\partial F_i(P_{Ti,k})}{\partial P_{Ti,k}} - \lambda_k = 0 \quad i = 1, ..., n, k = 1, ..., t_n \tag{12.18}$$

$$\frac{\partial\mathcal{L}(P_{Ti,k}, P_{Hi,k}, \lambda_k, \gamma_i)}{\partial P_{Hi,k}} = \gamma_i \frac{\partial W_i(P_{Hi,k})}{\partial P_{Hi,k}} - \lambda_k = 0 \quad i = 1, ..., m, k = 1, ..., t_n \tag{12.19}$$

$$\sum_{i=1}^{n} P_{Ti,k} + \sum_{i=1}^{m} P_{Hi,k} - P_{D,k} = 0, k = 1, ..., t_n \tag{12.20}$$

$$\sum_{k=1}^{t_n} W_i(P_{Hi,k})\Delta t_k - C_i = 0, i = 1, ..., m \tag{12.21}$$

The last two Eqs. (12.20) and (12.21) are the power balance and water consumption budget equality constraints. Equations (12.18) and (12.19) are combined as:

$$\frac{\partial F_i(P_{Ti,k})}{\partial P_{Ti,k}} = \gamma_j \frac{\partial W_j(P_{Hj,k})}{\partial P_{Hj,k}} = \lambda_k \quad i = 1, ..., n, j = 1, ..., m, k = 1, ..., t_n \tag{12.22}$$

where $\dfrac{\partial F_i(P_{Ti,k})}{\partial P_{Ti,k}}$ is the incremental fuel rate of thermal unit $i$ at time $k$, denoted as $\lambda_{Ti}$. $\dfrac{\partial W_j(P_{Hj,k})}{\partial P_{Hj,k}}$ is the incremental water consumption rate of hydro unit $j$ at time $k$, denoted as $\lambda_{Hj}$. When the thermal unit and hydro unit incremental active power are set as the same, Lagrange multipliers $\gamma_j$ are simplified

$$\gamma_j = \frac{\dfrac{\partial F_i(P_{Ti,k})}{\partial P_{Ti,k}}}{\dfrac{\partial W_j(P_{Hj,k})}{\partial P_{Hj,k}}} = \frac{\lambda_{Ti}}{\lambda_{Hj}} = \frac{\partial F_i(P_{Ti,k})}{\partial W_j(P_{Hj,k})} \tag{12.23}$$

$\gamma_j$ is the equivalent factor which indicates the equivalent incremental fuel cost of the thermal unit to the incremental water consumption of the hydro unit.

Assuming the water consumption is quadratic, the incremental water consumption rate curve is derived as

$$\lambda_{Hj} = \frac{\partial W_j\left(P_{Hj,k}\right)}{\partial P_{Hj,k}} = \frac{\partial\left(2a_{Hj}P_{Hj,k}^2 + b_{Hj}P_{Hj,k} + c_{Hj}\right)}{\partial P_{Hj,k}} = 2a_{Hj}P_{Hj,k} + b_{Hj}\ j = 1, ..., m \quad (12.24)$$

In the partial derivative equation set, the number of equations is $(n \times t_n + m \times t_n + t_n + m)$ which equals the number of variables $(P_{Ti,k}, P_{Hi,k}, \lambda_k, \gamma_i)$. The equations are solvable. When one time period is applied to illustrate the solving process, by defining $P_T = (P_{T1}, ..., P_{Tn})^T$, $P_H = (P_{H1}, ..., P_{Hm})^T$, and $\gamma = (\gamma_1, ..., \gamma_m)^T$, $A_T = diag\,(a_{T1}, ..., a_{Tn})$, $B_T = (b_{T1}, ..., b_{Tn})^T$, $A_H = diag\,(a_{H1}, ..., a_{Hm})$, $B_H = (b_{H1}, ..., b_{Hm})^T$, $C_H = (c_{H1}, ..., c_{Hm})^T$, $C = (C_1, ..., C_m)^T$, in matrix format, the hydrothermal power generation economic dispatch problem is restated as

$$F(P_T, P_H, \lambda, \gamma) = \begin{bmatrix} 2A_TP_T + B_T - \lambda \times [1]_{n \times 1} \\ \gamma \circ (2A_HP_H + B_H) - \lambda \times [1]_{m \times 1} \\ \sum P_T + \sum P_H - P_D \\ \left(A_HP_H^2 + B_H \circ P_H + C_H\right)\Delta t - C \end{bmatrix} = 0 \quad (12.25)$$

Where $\gamma \circ (2A_HP_H + B_H)$ and $B_H \circ P_H$ represent Hadamard product or element-wise multiplication [3]. Hadamard product of two vectors is similar to matrix addition, elements corresponding to the same row and columns of given vectors/matrices are multiplied together to form a new vector/matrix.

The nonlinear Eq. (12.25) can be solved by Newton's method. Newtons method forms a linear model for $F(x, y, s)$ around the current point and obtains the update of the unknown variables $(\Delta P_T, \Delta P_H, \Delta\lambda, \Delta\gamma)$ by solving the following linearized system:

$$J(P_T, P_H, \lambda, \gamma) \begin{bmatrix} \Delta P_T \\ \Delta P_H \\ \Delta\lambda \\ \Delta\gamma \end{bmatrix} = -F(P_T, P_H, \lambda, \gamma) \quad (12.26)$$

where $J(P_T, P_H, \lambda, \gamma)$ is the Jacobian of $F(P_T, P_H, \lambda, \gamma)$. The iteration equation to update the unknown variables is shown as the following linear equation:

$$\begin{bmatrix} 2A_T & & [-1]_{n \times 1} & \\ & 2\gamma^T A_H & [-1]_{m \times 1} & (2A_HP_H + B_H) \\ [1]_{1 \times n} & [1]_{1 \times m} & & \\ & (2A_HP_H + B_H)\Delta t & & \end{bmatrix} \begin{bmatrix} \Delta P_T \\ \Delta P_H \\ \Delta\lambda \\ \Delta\gamma \end{bmatrix}$$

$$= -\begin{bmatrix} 2A_TP_T + B_T - \lambda \times [1]_{n \times 1} \\ \gamma \circ (2A_HP_H + B_H) - \lambda \times [1]_{m \times 1} \\ \sum P_T + \sum P_H - P_D \\ \left(A_HP_H^2 + B_H \circ P_H + C_H\right)\Delta t - C \end{bmatrix} \quad (12.27)$$

The unknown variables are updated for the next iteration as:

$$
\begin{bmatrix} P_T \\ P_H \\ \lambda \\ \gamma \end{bmatrix}_{(k+1)} = \begin{bmatrix} P_T \\ P_H \\ \lambda \\ \gamma \end{bmatrix}_{(k)} + \begin{bmatrix} \Delta P_T \\ \Delta P_H \\ \Delta \lambda \\ \Delta \gamma \end{bmatrix} \tag{12.28}
$$

The iteration is taken until the duality gap is within a given small tolerance $\varepsilon$.

$$
\max(|\Delta P_T|, |\Delta P_H|, |\Delta \lambda|, |\Delta \gamma|) < \varepsilon \tag{12.29}
$$

The pseudocode for the hydrothermal power generation system economic dispatch algorithm is shown as follows.

---

**Algorithm 12.2  Hydrothermal Power System Economic Dispatch**

---

1: **Initialize** $(P_{T0}, P_{H0}, \lambda_0, \gamma_0)$

2: **While** $\max(|\Delta P_T|, |\Delta P_H|, |\Delta \lambda|, |\Delta \gamma|) > \varepsilon$

3: **Solve**
$$
\begin{bmatrix} 2A_T & & [-1]_{n \times 1} & \\ & 2\gamma A_H & [-1]_{m \times 1} & (2A_H P_H + B_H) \\ [1]_{1 \times n} & [1]_{1 \times m} & & \\ & (2A_H P_H + B_H) & & \end{bmatrix} \begin{bmatrix} \Delta P_T \\ \Delta P_H \\ \Delta \lambda \\ \Delta \gamma \end{bmatrix} = \begin{bmatrix} \lambda \times [1]_{n \times 1} - 2A_T P_T + B_T \\ \lambda \times [1]_{m \times 1} - \gamma \circ (2A_H P_H + B_H) \\ P_D - \sum P_T + \sum P_H \\ C - (A_H P_H^2 + B_H \circ P_H + C_H) \Delta t \end{bmatrix}
$$

4:  **Set**
$$
\begin{bmatrix} P_T \\ P_H \\ \lambda \\ \gamma \end{bmatrix}_{(k+1)} = \begin{bmatrix} P_T \\ P_H \\ \lambda \\ \gamma \end{bmatrix}_{(k)} + \begin{bmatrix} \Delta P_T \\ \Delta P_H \\ \Delta \lambda \\ \Delta \gamma \end{bmatrix}
$$

5:  **If** $P_{Gi}^{(k)} > P_{Gi}^{max}$ **then**

6:      $P_{Gi}^{(k)} = P_{Gi}^{max}$, $G \in (T, H)$ **Remove** $P_{Gi}$ from the unknown variable list

7:  **If** $P_{Gi}^{(k)} < P_{Gi}^{min}$ **then**

8:      $P_{Gi}^{(k)} = P_{Gi}^{min}$, $G \in (T, H)$ **Remove** $P_{Gi}$ from the unknown variable list

9:  $k = k + 1$

---

Taking the same economic dispatch problem in Section 12.1.1 as a numerical example, assume G5 is a hydroelectric power unit. The water consumption curve meets $W(P_H) = (0.001P_H^2 + 0.5P_H + 2)$ m$^3$/s. The water hourly consumption budget $C = 8 \times 10^4$ m$^3$. $\Delta t = 3600$ s $= 1$ hour.

By initializing $(P_{T0}, P_{H0}, \lambda_0) = (158.98, 20, 40.32, 20, 20, 9.23)$ as the solution to the economic dispatch problem in Section 12.1.1, $\gamma_0$ is estimated as $\gamma_0 = \dfrac{\lambda_0}{2a_H P_H + b_H} = 17.09$. The iteration Eq. (12.27) is formed numerically as:

$$
\begin{bmatrix} 0.014 & & & & & -1 & \\ & 0.019 & & & & -1 & \\ & & 0.018 & & & -1 & \\ & & & 0.018 & & -1 & \\ & & & & 0.0342 & -1 & 0.54 \\ 1 & 1 & 1 & 1 & 1 & & \\ & & & & 0.54 & & \end{bmatrix} \begin{bmatrix} \Delta P_{T1} \\ \Delta P_{T2} \\ \Delta P_{T3} \\ \Delta P_{T4} \\ \Delta P_{H1} \\ \Delta \lambda \\ \Delta \gamma \end{bmatrix} = \begin{bmatrix} 0.0043 \\ -1.1500 \\ 0.0042 \\ 2.1300 \\ 0.0014 \\ -0.3000 \\ 9.8222 \end{bmatrix} \tag{12.30}
$$

Solving the iteration Eq. (12.30), we get

$$
\begin{bmatrix}
\Delta P_{T1} \\
\Delta P_{T2} \\
\Delta P_{T3} \\
\Delta P_{T4} \\
\Delta P_{H1} \\
\Delta \lambda \\
\Delta \gamma
\end{bmatrix}
=
\begin{bmatrix}
48.85 \\
-24.76 \\
37.99 \\
-80.58 \\
18.19 \\
0.68 \\
0.11
\end{bmatrix}
\tag{12.31}
$$

The unknown variables are updated as

$$
\begin{bmatrix}
P_{T1} \\
P_{T2} \\
P_{T3} \\
P_{T4} \\
P_{H1} \\
\lambda \\
\gamma
\end{bmatrix}_{(k+1)}
=
\begin{bmatrix}
P_{T1} \\
P_{T2} \\
P_{T3} \\
P_{T4} \\
P_{H1} \\
\lambda \\
\gamma
\end{bmatrix}_{(k)}
+
\begin{bmatrix}
\Delta P_{T1} \\
\Delta P_{T2} \\
\Delta P_{T3} \\
\Delta P_{T4} \\
\Delta P_{H1} \\
\Delta \lambda \\
\Delta \gamma
\end{bmatrix}
=
\begin{bmatrix}
207.83 \\
-4.76 \\
78.31 \\
-60.58 \\
38.19 \\
9.91 \\
17.20
\end{bmatrix}
\tag{12.32}
$$

They are limited by the corresponding generation's upper and lower limits

$$
\begin{bmatrix}
P_{T1} \\
P_{T2} \\
P_{T3} \\
P_{T4} \\
P_{H1} \\
\lambda \\
\gamma
\end{bmatrix}
=
\begin{bmatrix}
207.83 \\
20.00 \\
78.31 \\
20.00 \\
38.19 \\
9.91 \\
17.20
\end{bmatrix}
\tag{12.33}
$$

Assuming the tolerance $\varepsilon = 1 \times 10^{-3}$, taking two iterations, the thermal hydroelectric power generation system economic dispatch problem is converged to the optimal solution as shown in Table 12.5.

**Table 12.5** Hydrothermal system unit economic dispatch iteration.

| Iter | $\lambda$ | $\gamma$ | PT1 (MW) | PT2 (MW) | PT3 (MW) | PT4 (MW) | PH1 (MW) | Total generation (MW) | Pd (MW) | Fuel cost ($) | Water consumption (m³) |
|---|---|---|---|---|---|---|---|---|---|---|---|
| 0 | 9.23 | 17.09 | 158.98 | 20 | 40.32 | 20 | 20 | 259.31 | 259 | 2,935 | 44,640 |
| 1 | 9.91 | 17.2 | 207.83 | 20 | 78.31 | 20 | 38.19 | 364.33 | 259 | 3,765 | 81,191 |
| 2 | 9.09 | 15.8 | 148.9 | 20 | 32.48 | 20 | 37.62 | 259 | 259 | 2,770 | 80,001 |

## 12.2 Security-Constrained Economic Dispatch

The classic economic dispatch does not model network security constraints. The optimal solution solved by classic economic dispatch may violate network operation limits. In practice, the economic dispatch model should include network security constraints as follows:

$$\min F = \sum_{i-1}^{n} \sum_{k=1}^{t_n} F_i(P_{Ti,k}) \Delta t_k \tag{12.34}$$

Subject to

$$P_{Ti,k} + P_{Hi,k} - P_{Di,k} + \sum_{j \in i} P_{l(i-j)} = 0 \; i = 1, 2, \cdots, n_b \; k = 1, \ldots, t_n \tag{12.35}$$

$$P_{Ti}^{\min} \leq P_{Ti,k} \leq P_{Ti}^{\max}, \; i = 1, \ldots, n, k = 1, \ldots, t_n \tag{12.36}$$

$$P_{Hi}^{\min} \leq P_{Hi,k} \leq P_{Hi}^{\max}, \; i = 1, \ldots, m, k = 1, \ldots, t_n \tag{12.37}$$

$$\sum_{k=1}^{t_n} W_i(P_{Hi,k}) \Delta t_k = C_i, i = 1, 2 \ldots, m \tag{12.38}$$

$$\left| P_{l(i-j)} \right| = \left| V_i V_j \left( G_{ij} \cos \theta_{ij} + B_{ij} \sin \theta_{ij} \right) \right| \leq P_{l(i-j)}^{\max}, l = 1, \ldots, n_l \tag{12.39}$$

Where $n_b$ is the number of buses, $n_l$ is the number of lines. Equation (12.39) represents the network security constraints. The security-constrained economic dispatch (SCED) model (12.34)–(12.39) is nonlinear. To solve the SCED problem by linear programming, the nonlinear objective function and nonlinear constraints are linearized.

$$F_i(P_{Ti,k}) = F_i\left(P_{Ti,k}^0\right) + \left. \frac{\partial F_i(P_{Ti,k})}{\partial P_{Ti,k}} \right|_{P_{Ti,k}^0} \Delta P_{Ti,k} = F_i\left(P_{Ti,k}^0\right) + 2a_{Ti} P_{Ti,k}^0 \Delta P_{Ti,k} + b_{Ti} \Delta P_{Ti,k} \tag{12.40}$$

Since $F_i\left(P_{Ti,k}^0\right)$ is constant, for a given operating point, the objective function is linearized as

$$\min F = \sum_{i-1}^{n} \sum_{k=1}^{t_n} \left( 2a_{Ti} P_{Ti,k}^0 \Delta P_{Ti,k} + b_{Ti} \Delta P_{Ti,k} \right) \Delta t_k \tag{12.41}$$

Since the load demand $P_{Dk}$ is constant, in DC power flow, the power balance Eq. (12.35) is linearized as follows:

$$\sum_{i=1}^{n} \Delta P_{Ti,k} + \sum_{i=1}^{m} \Delta P_{Hi,k} = 0, k = 1, \ldots, t_n \tag{12.42}$$

By linearizing the branch power flow, the branch incremental power flow is expressed as

$$\Delta P_l = V_i^0 V_j^0 \left( -G_{ij} \sin \theta_{ij}^0 \Delta \theta_{ij} + B_{ij} \cos \theta_{ij}^0 \Delta \theta_{ij} \right) \tag{12.43}$$

Taking the DC power flow assumptions, $V_i = V_j = 1$, $\cos \theta_{ij} = 1$, and $\sin \theta_{ij} = 0$. (12.43) is reduced as

$$\Delta P_l = B_{ij} \Delta \theta_{ij} \tag{12.44}$$

To express $\Delta P_l$ by generator incremental active power $\Delta P_{G,k}$, linearizing the following bus power balance equation

$$P_{Ti,k} + P_{Hi,k} - V_i \sum_{j\in i} V_j \left( G_{ij} cos\, \theta_{ij} + B_{ij}\, sin\, \theta_{ij} \right) = P_{Dk} \quad i = 1, ..., n_b, k = 1, ..., t_n \qquad (12.45)$$

Where $n_b$ is the number of buses. We get

$$\Delta P_{Ti,k} + \Delta P_{Hi,k} = \sum_{j\in i} B_{ij}\Delta \theta_{ij} \qquad (12.46)$$

In matrix format, (12.46) is rewritten as

$$\Delta P_{G,k} = B'\Delta \theta \qquad (12.47)$$

where, $\Delta P_{G,k} = (\Delta P_{T1,k} + \Delta P_{H1,k}, \Delta P_{T2,k} + \Delta P_{H2,k}, ..., \Delta P_{TnB,k} + \Delta P_{Hn_B,k})^T$, $\Delta \theta = (\Delta \theta_1, \Delta \theta_1, ..., \Delta \theta_{nb})^T$, the elements of the susceptance matrix $B'$ are:

$$B'_{ij} = -b_{ij} = -\frac{1}{x_{ij}} \qquad (12.48)$$

$$B'_{ii} = -\sum_{\substack{i=1 \\ j\neq i}}^{n_b} B'_{ij} \qquad (12.49)$$

Then $\Delta \theta$ can be expressed by

$$\Delta \theta = B'^{-1}\Delta P_{G,k} \qquad (12.50)$$

The linearization expressions of the branch active power flow constraints are rewritten as:

$$|\Delta P_l| = \left|B_{ij}\Delta \theta_{ij}\right| = \left|B_{ij}[0, ..., 0, 1, 0, ..., 0, -1, 0, ..., 0]B'^{-1}\Delta P_{G,k}\right| \qquad (12.51)$$
$$\phantom{xxxxxxxxxxxxxxxxxxxxxxxx} i \phantom{xxx} j$$

$$-P_l^{max} \le P_l = P_l^0 + \Delta P_l \le P_l^{max} \qquad (12.52)$$

$$-P_l^{max} - P_l^0 \le \Delta P_l \le P_l^{max} - P_l^0 \qquad (12.53)$$

$$\Delta P_l^{min} \le \Delta P_l \le \Delta P_l^{max} \qquad (12.54)$$

where $P_l^0 = B_{ij}\theta_{ij}^0$. $\Delta P_l^{max} = P_l^{max} - P_l^0$, $\Delta P_l^{min} = -P_l^{max} - P_l^0$.

Equation (12.51) shows that the DC power flow model can be used to calculate the sensitivities of branch flows to changes in generator power injections, called generation shift factors [4]. If $H \in \mathbb{R}^{n_l \times (n+m)}$ is used to denote the generation shift factor matrix, in matrix format, (12.51) together with (12.39) can be rewritten as

$$\Delta \boldsymbol{P_l^{min}} \le H\Delta P_{G,k} \le \Delta \boldsymbol{P_l^{max}} \qquad (12.55)$$

where $\Delta \boldsymbol{P_l^{max}} = \left(\Delta P_1^{max}, \Delta P_2^{max}, ..., \Delta P_{n_l}^{max}\right)^T$, $\Delta \boldsymbol{P_l^{min}} = \left(\Delta P_1^{min}, \Delta P_2^{min}, ..., \Delta P_{n_l}^{min}\right)^T$.

The thermal unit output upper and lower limit constraints (12.36) are linearized as follows:

$$P_{Ti}^{min} \le P_{Ti,k} = P_{Ti,k}^0 + \boldsymbol{\Delta P}_{Ti,k} \le P_{Ti}^{max}, \quad i = 1, ..., n, k = 1, ..., t_n \qquad (12.56)$$

$$P_{Ti}^{min} - P_{Ti,k}^0 \leq \Delta P_{Ti,k} \leq P_{Ti}^{max} - P_{Ti,k}^0, \quad i = 1, ..., n, k = 1, ..., t_n \tag{12.57}$$

$$\Delta P_{Ti}^{min} \leq \Delta P_{Ti,k} \leq \Delta P_{Ti}^{max}, \quad i = 1, ..., n, k = 1, ..., t_n \tag{12.58}$$

where $\Delta P_{Ti}^{min} = P_{Ti}^{min} - P_{Ti,k}^0$, $\Delta P_{Ti}^{max} = P_{Ti}^{max} - P_{Ti,k}^0$.
Similarly

$$\Delta P_{Hi}^{min} \leq \Delta P_{Hi,k} \leq \Delta P_{Hi}^{max}, \quad i = 1, ..., m, k = 1, ..., t_n \tag{12.59}$$

where $\Delta P_{Hi}^{min} = P_{Hi}^{min} - P_{Hi,k}^0$, $\Delta P_{Hi}^{max} = P_{Hi}^{max} - P_{Hi,k}^0$.

Assuming the water consumption curve is quadratic, the water consumption can be linearized as:

$$W_i(P_{Hi,k}) = W_i(P_{Hi,k}^0) + \frac{\partial W_i(P_{Hi,k})}{\partial P_{Hi,k}} \Delta P_{Hi,k} = W_i(P_{Hi,k}^0) + \frac{\partial(2a_{Hi}P_{Hi,k}^2 + b_{Hi}P_{Hi,k} + c_{Hi})}{\partial P_{Hi,k}} \Delta P_{Hi,k}$$

$$= W_i(P_{Hi,k}^0) + 2a_{Hi}P_{Hi,k}^0 \Delta P_{Hi,k} + b_{Hi}\Delta P_{Hi,k} \quad i = 1, ..., m \tag{12.60}$$

$$W_i(\Delta P_{Hi,k}) = W_i(P_{Hi,k}) - W_i(P_{Hi,k}^0) = 2a_{Hi}P_{Hi,k}^0 \Delta P_{Hi,k} + b_{Hi}\Delta P_{Hi,k}$$

$$i = 1, ..., m \tag{12.61}$$

The water consumption constraint (12.38) is then linearized as:

$$\sum_{k=1}^{t_n} (2a_{Hi}P_{Hi,k}^0 \Delta P_{Hi,k} + b_{Hi}\Delta P_{Hi,k})\Delta t_k = \Delta C_i, i = 1, 2..., m \tag{12.62}$$

where $\Delta C_i = C_i - \sum_{k=1}^{t_n} W_i(P_{Hi,k}^0)\Delta t_k$.

In summary, the linearized SCED mathematical model is then rewritten as follows:

$$\min F = \sum_{i-1}^{n}\sum_{k=1}^{t_n} (2a_{Ti}P_{Ti,k}^0 \Delta P_{Ti,k} + b_{Ti}\Delta P_{Ti,k})\Delta t_k \tag{12.63}$$

Subject to

$$\sum_{i=1}^{n} \Delta P_{Ti,k} + \sum_{i=1}^{m} \Delta P_{Hi,k} = 0, k = 1, ..., t_n \tag{12.64}$$

$$\Delta P_{Ti}^{min} \leq \Delta P_{Ti,k} \leq \Delta P_{Ti}^{max}, \quad i = 1, ..., n, k = 1, ..., t_n \tag{12.65}$$

$$\Delta P_{Hi}^{min} \leq \Delta P_{Hi,k} \leq \Delta P_{Hi}^{max}, \quad i = 1, ..., m, k = 1, ..., t_n \tag{12.66}$$

$$\sum_{k=1}^{t_n} (2a_{Hi}P_{Hi,k}^0 \Delta P_{Hi,k} + b_{Hi}\Delta P_{Hi,k})\Delta t_k = \Delta C_i, i = 1, 2..., m \tag{12.67}$$

$$|H\Delta P_{G,k}| \leq \Delta P_l^{max} \tag{12.68}$$

The unknown variables of the linearized SCED optimization problem are $\Delta P_{Ti,k}$ and $\Delta P_{Hi,k}$ which can be solved by linear programming methods, such as the simplex method or the interior-point method.

## 12.2.1 Generation Shift Factor Matrix

The generation shift factor matrix $H = \left[ h_{pq} \right] \in \mathbb{R}^{n_l \times (n + m)}$ stores the sensitivities of the branch flows to the changes in generator power injections. The element $h_{pq}$ of the generation shift factor matrix is calculated by the following process. In the DC power flow, voltage angle changes resulted in the $q$-th unit injection is

$$B'\Delta\theta = e_q \tag{12.69}$$

Where $e_q$ is a unit vector whose $q$-th entry is 1 and 0 elsewhere. $B'$ and its graph are built as addressed in Section 10.4.1. After $B'$ is factorized as addressed in 10.3, $\Delta\theta$ in (12.69) can be solved by forward and backward substitutions. Then the generation shift factor of $p$-th branch flow to change in $q$-th generator injection is calculated

$$h_{pq} = \frac{\Delta\theta_{ij}}{x_p} = \frac{\Delta\theta_i - \Delta\theta_j}{x_p} \tag{12.70}$$

where $x_p$ is the impedance of branch $p$.

To build a generation shift factor graph, the vertices GenND, LineND, and edges Connected_Gen_Line are defined as follows:

CREATE VERTEX GenND (PRIMARY_ID GenID int, bus int)
CREATE VERTEX LineND (PRIMARY_ID LineID int, from_bus int, to_bus int)
CREATE DIRECTED EDGE Connected_Gen_Line (from GenND, to LineND, hij
    double)

With the defined vertices and edge above, the generation shift factor graph schema is then defined as:

CREATE GRAPH GSFGraph(GenND, LineND, Connected_Gen_Line)

In the generation shift factor graph, each generator and line are modeled as a vertex. The edge Connected_Gen_Line connects a generator to a line, and the attribute hij is the sensitivity of the $i$-th branch flow to change in $j$-th generator injection.

The pseudocode of graph-based generation shift factor matrix calculation is shown as follows.

---

**Algorithm 12.3   Graph-Based Generation Shift Factor Matrix Calculation**

---

```
1:   CREATE QUERY calculateGSF() FOR
2:    GRAPH busBranchGraph(V1,E1) and B_pGraph(V2,E3) and GSFGraph(V3,E3)
3:   FACTORIZE B′ on B_pGraph //in hierarchical parallel
4:   SELECT NODE q FROM V1 WHERE q.busType == 2 OR q.busType == 3
5:     SOLVE Δθ by F/B substitutions with the injection of a unit
       vector eq
6:     SELECT NODE p FROM E1
```
7: $\quad$ INSERT INTO EDGE(p, q) in E3 VALUE$\left( \dfrac{\Delta\theta_i - \Delta\theta_j}{x_p} \right)$

---

In the algorithm above, the graph symbolic factorization, elimination tree creation, node partition, and graph numerical factorization are the same as we process power flow calculation by the Newton–Raphson method in 10.3. The only change is to apply these graph-based algorithms to the graph B_pGraph other than the graph JacobianGraph.

The graph-based forward and backward substitutions are the same as addressed in 10.3.6 but with the injection of a unit vector $e_q$.

The graph query to implement the graph-based generation shift factor matrix calculation is presented as follows:

```
CREATE QUERY calculateGSF(
) FOR GRAPH busBranchGraph and B_pGraph and GSFGraph{

T0 = {B_pGraph.B_pND.*}
T1 = {busBranchGraph.TopoND.*}
T2 = {GSFGraph.LineND.*}

int nBus = busBranchGraph.getVertexCount()
FOR GRAPH busBranchGraph and GRAPH B_pGraph
    formFilledGraph() FOR GRAPH B_pGraph
    createEliminationTree() FOR GRAPH B_pGraph and GRAPH eTreeGraph
    int nLevel = nodePartition(1) FOR GRAPH B_pGraph and GRAPH
    eTreeGraph
    factorizeJacobian(int nLevel) FOR GRAPH B_pGraph

T3 = SELECT q FROM T1 WHERE q.busType == 2 OR q.busType == 3
    ACCUM
        T0.bii = 0
        SELECT s FROM T0 WHERE s.B_pID == q.TOPOID
          s.bii = 1
        Substitution (int nLevel, int nBus)
            FOR GRAPH B_pGraph and GRAPH busBranchGraph

    POST-ACCUM
        T4 = SELECT s FROM T1:s-(connected:e)->:t
            ACCUM
                SELECT p FROM T2 WHERE p.from_bus == s.from_bus
                  AND p.to_bus == s.to_bus AND s.from_bus.va!=s.to_bus.va
                INSERT INTO Connected_Gen_Line(q,p)
                        VALUES((s.from_bus.va-s.to_bus.va)/e.X)
}
```

In the query, the $B'$ on the graph B_pGraph is factorized by calling the subqueries defined in Section 10.3.1–10.3.4. In the DC power flow, since $B'$ is constant, the factorized graph is static. It is conducted once without an update nested in ACCUM.

For each generator in the vertex set T3 = SELECT q FROM T1 WHERE q.busType == 2 OR q.busType == 3, a unit vector $e_q$ is set as the right-hand side vector by

```
T0.bii = 0
SELECT s FROM T0 WHERE s.B_pID == q.TOPOID
        s.bii = 1
```

By calling the subquery `Substitution(int nLevel, int. nBus)` defined in Section 10.3.6, the voltage angles are solved.

After the voltage angles are solved, in `POST-ACCUM`, graph traversal `T4 = SELECT s FROM T1: s-(connected:e)->:t` travels all edges from vertex `s` in the vertex set `T1 = {busBranch-Graph.TopoND.*}`. Nested in the `ACCUM`, branch `p` is selected from the vertex set `T2 = {GSFGraph.LineND.*}` by the following `SELECT` statement:

```
SELECT p FROM T2 WHERE p.from_bus == s.from_bus
                   AND p.to_bus == s.to_bus
```

Eventually, the generator shift factor is calculated and inserted to the edge from generator `q` to branch `p` in the graph `GSFGraph`.

## 12.2.2 Graph-Based SCED Modeling

To build the linear program in standard form $\min_x c^T x, Ax = b, x \geq 0$, taking one time period ($k = 1$) as an example to illustrate, the objectives, variables, and constraints of the linearized SCED problem are identified by the following process.

The linearized objective function is

$$\min F = \sum_{i=1}^{n} \left( 2a_{Ti} P_{Ti}^0 \Delta P_{Ti} + b_{Ti} \Delta P_{Ti} \right) \tag{12.71}$$

Subject to the following constraints:

$$\sum_{i=1}^{n} \Delta P_{Ti} + \sum_{i=1}^{m} \Delta P_{Hi} = 0$$

$$\Delta P_{Ti} + x_i = P_{Ti}^{\max} - P_{Ti}^0, \quad i = 1, ..., n$$

$$\Delta P_{Hi} + x_{n+i} = P_{Hi}^{\max} - P_{Hi}^0, i = 1, ..., m$$

$$\Delta P_{Ti} - x_{i+n+m} = P_{Ti}^{\min} - P_{Ti}^0, \quad i = 1, ..., n$$

$$\Delta P_{Hi} - x_{i+2n+m} = P_{Hi}^{\min} - P_{Hi}^0, \quad i = 1, ..., m$$

$$\left( 2a_{Hi} P_{Hi}^0 + b_{Hi} \right) \Delta P_{Hi} = \frac{C_i}{\Delta t} - a_{Hi} \left( P_{Hi}^0 \right)^2 - b_{Hi} P_{Hi}^0 - c_{Hi}, \quad i = 1, ..., m$$

$$H \Delta P_G + X_{2n+2m+1 \sim 2n+2m+n_l} = \Delta P_l^{\max} - P_l^0$$

$$H \Delta P_G - X_{2n+2m+n_l+1 \sim 2n+2m+2n_l} = -\Delta P_l^{\max} - P_l^0$$

where $X_{2n+2m+1 \sim 2n+2m+n_l} = [x_{2n+2m+1}, ..., x_{2n+2m+n_l}]^T$, $X_{2n+2m+n_l+1 \sim 2n+2m+2n_l} = [x_{2n+2m+n_l+1}, ..., x_{2n+2m+2n_l}]^T$. $x_1 \sim x_{2n+2m+2n_l}$ are nonnegative slack variables added to convert inequality constraints to equality constraints.

In matrix format, the above linear program problem is restated as

$$\min F = \sum_{i=1}^{n} \left( 2a_{Ti} P_{Ti}^0 + b_{Ti} \right) \Delta P_{Ti} \tag{12.72}$$

Subject to

$$
\begin{bmatrix}
[1]_{1 \times (n+m)} & & & \\
[I]_{(n+m) \times (n+m)} & & & \\
[I]_{(n+m) \times (n+m)} & [I]_{(n+m) \times (n+m)} & & \\
diag\left[2a_{Hi}P_{Hi}^0 + b_{Hi}\right]_{m \times m} & & -[I]_{(n+m) \times (n+m)} & \\
H_{n_l} \times (n+m) & & & [I]_{n_l \times n_l} \\
H_{n_l} \times (n+m) & & & -[I]_{n_l \times n_l}
\end{bmatrix}
$$

$$
\begin{bmatrix}
\Delta P_T \\
\Delta P_H \\
x_{1-n+m} \\
x_{n+m+1-2n+2m} \\
x_{2n+2m+1-2n+2m+n_l} \\
x_{2n+2m+n_l+1 \sim 2n+2m+2n_l}
\end{bmatrix}
=
\begin{bmatrix}
0 \\
\begin{bmatrix} P_T^{\max} - P_T^0 \\ P_H^{\max} - P_H^0 \end{bmatrix}_{(n+m) \times 1} \\
\begin{bmatrix} P_T^{\min} - P_T^0 \\ P_H^{\min} - P_H^0 \end{bmatrix}_{(n+m) \times 1} \\
\left[\dfrac{C_i}{\Delta t} - a_{Hi}\left(P_{Hi}^0\right)^2 - b_{Hi}P_{Hi}^0 - C_{Hi}\right]_{m \times 1} \\
\left[P_l^{\max} - P_l^0\right]_{n_l \times 1} \\
\left[-P_l^{\max} - P_l^0\right]_{n_l \times 1}
\end{bmatrix}
$$

(12.73)

In the linear program problem, $\Delta P_T$ and $\Delta P_H$ are not nonnegative variables. They are free to take on either positive, negative values, or zero. Two nonnegative variables $u$ and $v$ are introduced to represent $\Delta P_G = (\Delta P_T, \Delta P_H) = u - v$.

Then linear program problem in standard form is written as

$$
\min F = \sum_{i=1}^{n}\left(\left(2a_{Ti}P_{Ti}^0 + b_{Ti}\right)\left(u_{Ti} - v_{Ti}\right)\right)
$$

(12.74)

Subject to

$$
\begin{bmatrix}
[1]_{1 \times (n+m)} & -[1]_{1 \times (n+m)} & & \\
[I]_{(n+m) \times (n+m)} & -[I]_{(n+m) \times (n+m)} & [I]_{(n+m) \times (n+m)} & \\
[I]_{(n+m) \times (n+m)} & -[I]_{(n+m) \times (n+m)} & & -[I]_{(n+m) \times (n+m)} \\
diag\left[2a_{Hi}P_{Hi}^0 + b_{Hi}\right]_{m \times m} & diag\left[2a_{Hi}P_{Hi}^0 + b_{Hi}\right]_{m \times m} & & \\
H_{n_l} \times (n+m) & -H_{n_l} \times (n+m) & & [I]_{n_l \times n_l} \\
H_{n_l} \times (n+m) & -H_{n_l} \times (n+m) & & -[I]_{n_l \times n_l}
\end{bmatrix}
$$

$$
\begin{bmatrix}
u_{1-n+m} \\
v_{1-n+m} \\
x_{1-n+m} \\
x_{n+m+1-2n+2m} \\
x_{2n+2m+1-2n+2m+n_l} \\
x_{2n+2m+n_l+1-2n+2m+2n_l}
\end{bmatrix}
=
\begin{bmatrix}
0 \\
\begin{bmatrix} P_T^{\max} - P_T^0 \\ P_H^{\max} - P_H^0 \end{bmatrix}_{(n+m) \times 1} \\
\begin{bmatrix} P_T^{\min} - P_T^0 \\ P_H^{\min} - P_H^0 \end{bmatrix}_{(n+m) \times 1} \\
\left[\dfrac{C_i}{\Delta t} - a_{Hi}\left(P_{Hi}^0\right)^2 - b_{Hi}P_{Hi}^0 - C_{Hi}\right]_{m \times 1} \\
\left[P_l^{\max} - P_l^0\right]_{n_l \times 1} \\
\left[-P_l^{\max} - P_l^0\right]_{n_l \times 1}
\end{bmatrix}
$$

(12.75)

### 12.2.3 Graph-Based SCED

The pseudocode to solve the SCED problem by linear programming is then shown as follows.

---

**Algorithm 12.4  SCED Linear Programming**

---

1:   **Set** iteration number $k$ = 0
2:   **Initialize** $\left(P_T^k, P_H^k, P_l^k\right)$ by feasible power flow solution
3:   **Linearize** objective function and constraints at the power flow solution
4:   **Calculate** generation shift factor matrix
5:   **While** $max(|\Delta P_T|, |\Delta P_H|) > \varepsilon$
6:       **Solve** linearized SCED optimization problem (12.63)–(12.68) by simplex
7:       **Update** $\begin{bmatrix} P_T \\ P_H \end{bmatrix}_{(k+1)} = \begin{bmatrix} P_T \\ P_H \end{bmatrix}_{(k)} + \begin{bmatrix} \Delta P_T \\ \Delta P_H \end{bmatrix}$
8:       **Solve** branch power flow $P_l^{k+1}$ with the generation of $P_T^{k+1}, P_H^{k+1}$
9:       $k = k + 1$

---

To implement the graph-based SCED linear programming algorithm, the simplex graph schema defined in 6.5 is applied as copied as follows:

```
CREATE VERTEX V (primary_id id string)
CREATE DIRECTED EDGE E (FROM V, TO V, double a_ij)
CREAT GRAPH SimplexGraph (V, E)
```

To support SCED for a hydrothermal system, the bus-branch graph is further extended as follows.

```
CREATE VERTEX TopoND (PRIMARY_ID topoid int, TOPOID int, island int,
    bus_name string, area string, loss_zone uint, busType int, Vm
    double, Va double, M_Vm double, M_Va double, base_kV double,
    desired_volts double, control_bus_number int, up_V double, lo_V
    double, GenP double, GenQ double, M_Gen_P double, M_Gen_Q double,
    qUp double, qLower double, LdP double, LdQ double, P double,
    Q double, M_Load_P double, M_Load_Q double, G double, B double, Sub
    string, OV int, UV int, sumB double, sumG double, sumBi double,
    volt double, double, M_C_P double, SE_Vm double, SE_Va double,
    Ri_vP double, Ri_vQ double, Ri_eP double, Ri_eQ double, SE_Inj
    tuple, GainP tuple, GainQ tuple, H_r_P double, H_r_Q double, Pmax
    double, Pmin double, a double, b double, c double, unitType string,
    waterConsum tuple)
```

Two new attributes unitType and waterConsum are highlighted in black bold in the vertex TopoND definition. unitType defines the type of thermal or hydro unit. The water consumption limits waterConsum are defined in the data structure tuple to constrain the water consumption in a defined time period.

```
typedef tuple< double Ck, double Tk > waterConsum;
```

The table header of the unit information file unitInfo.csv is extended in Table 12.6.

**Table 12.6** Unit information table header.

| Bus ID | Unit name | Unit type | Pmax | Pmin | a | b | c | Ck | Tk |
|--------|-----------|-----------|------|------|---|---|---|----|----|

When the unit information has been read to the bus-branch graph by the query `readUnitInfo`, the graph-based SCED optimization is performed in two steps iteratively: build up a simplex graph model and solve the optimization problem by the simplex graph.

### 12.2.3.1 Buildup Simplex Graph

Assuming the unit information has been read to the bus-branch graph by the query `readUnitInfo`, the graph query to build up the simplex graph is shown as follow:

```
CREATE QUERY buildSimplexGraph ()
FOR GRAPH busBranchGraph and GSFGraph and SimplexGraph {

T1 = {GSFGraph.GenND.*}
T2 = {busBranchGraph.TopoND.*}
E2 = {busBranchGraph.TopoConnect.*}

int nUnit = T1.getCount()
int nBranch = E2.getCount()

// Build up simplex table
// Insert vertices
INSERT INTO SimplexGraph.V("V1") //Vertex for objective function
INSERT INTO SimplexGraph.V("S")  //Vertex for power balance equation
FOREACH s IN T1 DO
  INSERT INTO SimplexGraph.V("u"+str(s.GenID)) //Vertex ui
  INSERT INTO SimplexGraph.V("v"+str(s.GenID)) //Vertex vi

FOREACH i IN RANGE[1, 2*(nUnit+nBranch)] DO
  ACCUM
    INSERT INTO SimplexGraph.V("x"+str(i)) //Vertex xi

SELECT s1 FROM T1
  SELECT s2 FROM T2 WHERE s1.bus == s2.TOPOID AND s2.unitType ==
  "Hydro"
    ACCUM
        //Vertex for water consumption equation
        INSERT INTO SimplexGraph.V("H"+str(s.GenID))

INSERT INTO SimplexGraph.V("b") //Vertex b
```

```
// Insert edges
SELECT s1 FROM T1
  ACCUM
    SELECT s2 FROM T2 WHERE s1.bus == s2.TOPOID
      //Objective function
      INSERT INTO SimplexGraph.E("V1","u"+to_str(s1.GenID),
                              2*s2.a*s2.GenP+s2.b)
      INSERT INTO SimplexGraph.E("V1","v"+to_str(s1.GenID),
                              -(2*s2.a*s2.GenP+s2.b))
      INSERT INTO SimplexGraph.E("V1","b",0)

      //Power balance constraint
      INSERT INTO SimplexGraph.E("S","u"+to_str(s1.GenID),1)
      INSERT INTO SimplexGraph.E("S","v"+to_str(s1.GenID),-1)
      INSERT INTO SimplexGraph.E("S","b",0)

      //PGmax limit constraints
      INSERT INTO SimplexGraph.E("x"+to_str(s1.GenID),
                              "u"+to_str(s1.GenID),1)
      INSERT INTO SimplexGraph.E("x"+to_str(s1.GenID),
                              "v"+to_str(s1.GenID),-1)
      INSERT INTO SimplexGraph.E("x"+to_str(s1.GenID),
                              "x"+to_str(s1.GenID),1)
      INSERT INTO SimplexGraph.E("x"+to_str(s1.GenID),
                              "b",s2.Pmax-s2.GenP)


      //PGmin limit constraints
      INSERT INTO SimplexGraph.E("x"+to_str(s1.GenID+nUnit),
                              "u"+to_str(s1.GenID),1)
      INSERT INTO SimplexGraph.E("x"+to_str(s1.GenID+nUnit),
                              "v"+to_str(s1.GenID),-1)
      INSERT INTO SimplexGraph.E("x"+to_str(s1.GenID+nUnit),
                              "x"+to_str(s1.GenID),-1)
      INSERT INTO SimplexGraph.E("x"+to_str(s1.GenID+nUnit),
                              "b",s2.Pmin-s2.GenP)

SELECT s FROM T1:s-(connected:e)->:t
  ACCUM
    //Branch flow upper limit
    INSERT INTO SimplexGraph.E("x"+to_str(2*nUnit+e.LineND.LineID),
                            "u"+to_str(e.GenND.GenID),e.hij)
    INSERT INTO SimplexGraph.E("x"+to_str(2*nUnit+e.LineND.LineID),
                            "v"+to_str(e.GenND.GenID),-e.hij)
    INSERT INTO SimplexGraph.E("x"+to_str(2*nUnit+e.LineND.LineID),
                            "x"+to_str(2*nUnit+e.LineND.LineID),1)
    //Branch flow lower limit
```

```
     INSERT INTO SimplexGraph.E("x"+ to_str
             (2*nUnit+nBranch+e.LineND.LineID),
             "u"+to_str(e.GenND.GenID),e.hij)
     INSERT INTO SimplexGraph.E("x"+to_str
             (2*nUnit+nBranch+e.LineND.LineID),
             "v"+to_str(e.GenND.GenID),-e.hij)
     INSERT INTO SimplexGraph.E("x"+to_str
             (2*nUnit+nBranch+e.LineND.LineID),
             "x"+to_str(2*nUnit+nBranch+e.LineND.LineID),-1)


   POST-ACCUM
     SELECT s2 FROM E2
             WHERE e.from_bus == s2.from_bus and e.to_bus == s2.to_bus
       INSERT INTO SimplexGraph.E("x"+to_str(2*nUnit+e.LineND.LineID),
                                  "b",s2.line_Q1_list-s2.P_BR)
       INSERT INTO SimplexGraph.E("x"+to_str
          (2*nUnit+nBranch+e.LineND.LineID),
          "b",-s2.line_Q1_list-s2.P_BR)

//Water Consumption Equation
SELECT s1 FROM T1
  ACCUM
    SELECT s2 FROM T2 WHERE s1.bus == s2.TOPOID AND s2.unitType ==
    "Hydro"
      INSERT INTO SimplexGraph.E("H"+str(s1.GenID),
                            "u"+to_str(s1.GenID),2*s2.a*s2.GenP+s2.b)
      INSERT INTO SimplexGraph.E("H"+str(s1.GenID),
                        "v"+to_str(s1.GenID),-(2*s2.a*s2.GenP+s2.b))
      INSERT INTO SimplexGraph.E("H"+str(s.GenID),
                          "b",s1.waterConsum.Ck/s1.waterConsum.Tk
                          -(s2.a*s2.a*s2.GenP+s2.b*s2.GenP+s2.c))
}
```

In the query `buildSimplexGraph`, all vertices are virtual. Vertex `V1` specifies a vertex for the objective function. Vertices `ui` and `vi` represent the artificial nonnegative variables for generation output `PGi = ui-vi`. `xi` represents the slack variables for inequality constraints. Vertices `S` and `Hi` refer to the power balance equation and the water consumption equations.

After vertices are created, the statement `INSERT INTO SimplexGraph.E(FROM Vertex, TO Vertex, Value)` insert edges to link objective function or constraints (`FROM Vertex`) to variables (`TO Vertex`).

For example, the following two `INSERT` statements link objective function vertex `V1` to generation output vertices.

```
     INSERT INTO SimplexGraph.E("V1","u"+to_str(s1.GenID),
                           2*s2.a*s2.GenP+s2.b)
     INSERT INTO SimplexGraph.E("V1","v"+to_str(s1.GenID),
                           -(2*s2.a*s2.GenP+s2.b))
```

Where the build-in function `to_str` converts a number to a string.

### 12.2.3.2  Graph-Based Simplex Method

As addressed in Section 6.5.1, the key steps in the simplex method are (i) selecting the entering variable, (ii) selecting the leaving variable, and (iii) pivot operation. The graph query for the simplex method is implemented as follows:

```
CREATE QUERY simplexMethodGraph()
FOR GRAPH SimplexGraph {

T = {SimplexGraph.V.*}
E = {SimplexGraph.E.*}

NegCostCoeff = True
WHILE (NegCostCoeff) DO
  SELECT s FROM T:s-(connected:e)->:t ORDER by e.a_ij ASC LIMIT 1
    IF (e.a_ij>0)
      NegCostCoeff = False
    ELSE
      // select entering variable
      SELECT s1 FROM T:s1-(connected:e)->:t WHERE s1.id="V1"
        SELECT q FROM e ORDER by e.a_ij ASC LIMIT 1
      // select leaving variable
      SELECT s2 FROM E WHERE s2.TO="b"
        SELECT eq FROM E WHERE eq.TO=q.TO
          SELECT p FROM eq ORDER by s2.a_ij/eq.a_ij
                               WHERE eq.a_ij>0 ASC LIMIT 1
      pivotOperation(p.FROM, q.TO)
}
```

In the query `simplexMethodGraph`, the boolean variable `NegCostCoeff` is defined to check the negative cost coefficient. When the lowest cost coefficient is nonnegative, the optimal solution is found. The process is terminated.

To select the entering variable (vertex), the edges adjacent to the cost vertex V1 are selected by `SELECT s1 FROM T:s1-(connected:e)->:t WHERE s1.id="V1"` and their coefficient attribute a_ij is ordered in ascending order. In the query, the `LIMIT` clause ensures one entering variable is returned when there are tied lowest cost coefficients.

The leaving variable (vertex) is then selected by ratio test $p = \min\left(\dfrac{b_i}{a_{iq}}\right)$, where $b_i$ is right-hand side in (12.75) represented by attribute a_ij of edges connecting to vertex b, and $a_{iq}$ is the attribute a_ij of edges connecting to entering vertex q. The `LIMIT` clause ensures one leaving variable is returned when there are tied ratios. After the entering vertex and leaving vertex are selected, the pivot operation is performed by calling the query `pivotOperation(p.FROM, q.TO)`.

### 12.2.3.3  Update Power Flow

When the linearized SCED problem is optimized, the generation output and branch power flow are updated by

$$\begin{bmatrix} P_T \\ P_H \end{bmatrix}_{(k+1)} = \begin{bmatrix} P_T \\ P_H \end{bmatrix}_{(k)} + \begin{bmatrix} \Delta P_T \\ \Delta P_H \end{bmatrix} \tag{12.76}$$

$$[P_l]_{(k+1)} = [P_l]_{(k)} + [\Delta P_l] = [P_l]_{(k)} + H \begin{bmatrix} \Delta P_T \\ \Delta P_H \end{bmatrix} \tag{12.77}$$

The following pseudocode in graph query is developed to update power flow in parallel.

```
CREATE QUERY updatePowerFlow()
FOR GRAPH busBranchGraph and GSFGraph and SimplexGraph {

SumAccum<double> @dPg = 0
SumAccum<double> @dPl = 0

T0 = {SimplexGraph.V.*}
T1 = {GSFGraph.GenND.*}
T2 = {busBranchGraph.TopoND.*}

SELECT s0 FROM T0:s0-(connected:eu)->:t WHERE s0.TO.getChars() == "u"
  ACCUM
    SELECT s1 FROM T1 WHERE s1.GenID == s0.TO.getNums()
     SELECT s2 FROM T2 WHERE s2.TOPOID == s1.bus
      SELECT s FROM s0:s-(connected:eb)->:t WHERE s.TO.getChars()== "b"
            s1.@dPg = eu.a_ij*eb.a_ij
            s2.@dPg = eu.a_ij*eb.a_ij
            s2.GenP = s2.GenP+s2.@dPg

   POST_ACCUM
    SELECT s1 FROM T1:s1-(connected:e)->:t
      ACCUM
       t.@dPl = t.@dPl+s1.@dPg*e.hij
      POST-ACCUM
         SELECT s2 FROM T2:s2-(connected:el)->:tl
            WHERE s2.from_bus == t.from_bus AND s2.to_bus == t.to_bus
         el.P_BR = el.P_BR+t.@dPl
}
```

In the query `updatePowerFlow()`, vertex-attached accumulator variables `SumAccum @dPg`, `@dPl` are defined with a single add-sign @ to store the generation and branch power incremental changes.

The attributes connecting to the vertex "u" are either 1 or −1 in the final optimal simplex graph and the attribute of the paired vertex "v" is either −1 or 1. The incremental generation output `@dPg` can be calculated by `eu.a_ij` times corresponding attribute `eb.a_ij` of the edge connecting to the "b" vertex (the final right-hand side).

When the incremental generation `@dPg` is known, incremental branch flows `@dPl` are calculated by the following statements:

```
SELECT s1 FROM T1:s1-(connected:e)->:t
    ACCUM
       t.@dPl = t.@dPl+s1.@dPg*e.hij
```

Where the statement `SELECT s1 FROM T1:s1-(connected:e)->:t` travels all edges from the vertices in `T1 = {GSFGraph.GenND.*}` which are all generator nodes in `GSFGraph` to branch `LineND` in `t`. The branch flow is updated in parallel nested in `ACCUM` by `t.@dPl = t.@dPl + s1.@dPg*e.hij`. Since `GSFGraph` contains nonzero shift factors as edges, the sparsity of computing is preserved.

#### 12.2.3.4 Graph-Based SCED Implementation

The query and subquery to build the simplex graph, solve the incremental linear SCED optimization problem, and update power flow have been developed. They are called to implement the graph-based SCED as shown in the following pseudocode.

```
CREATE QUERY securityConstrainedEconomicDispatch(int MaxIteration,
double Tolerance) FOR GRAPH busBranchGraph and GSFGraph and
SimplexGraph{

MaxAccum<double> @@Pmismatch = -INF
SumAccum<double> @dPg = 0

T1 = {GSFGraph.GenND.*}

WHILE (abs(@@Pmismatch)>Tolerance) LIMIT MaxIteration DO
  buildSimplexGraph() FOR GRAPH busBranchGraph and GSFGraph
                        and SimplexGraph
  simplexMethodGraph() FOR GRAPH SimplexGraph
  updatePowerFlow() FOR GRAPH busBranchGraph and GSFGraph and
  SimplexGraph

  SELECT s1 FROM T1
    ACCUM
      @@dPgMax = abs(s1.@dPg)
}
```

In the query `securityConstrainedEconomicDispatch`, maximum iteration number `MaxIteration` and active power mismatch tolerance `Tolerance` are passed through. Global variable `@@Pmismatch` and vertex-attached variable `@dPg` are defined to store updated active power mismatch and incremental generation.

The `WHILE` loop statement checks the active power mismatch absolute value against the `Tolerance` with the `LIMIT` clause which counts the iterations in the `WHILE` loop against the maximum iteration number `MaxIteration`.

```
WHILE (abs(@@Pmismatch)>Tolerance) LIMIT MaxIteration DO
```

By calling the queries `buildSimplexGraph`, `simplexMethodGraph`, and `updatePowerFlow`, the simplex graph is built and solved and power flow is updated in the iteration till the biggest incremental generation is less than the tolerance or the maximum iterations is reached.

## 12.3 Security-Constrained Unit Commitment

Security-constrained unit commitment (SCUC) is a vital tool for power generation schedules and electricity market settlement. Widely adopted two-settlement electricity markets involve day-ahead market and real-time market which require solving SCUC and SCED problems to determine the optimal commitment and dispatch decision [5–8].

Many simulation platforms have been developed for modeling the market-clearance process. AMES Wholesale Power Market Test Bed allows users to vary the generation mix in the wholesale markets through a stochastic SCUC formulation [9]. Commercial electricity market simulation platforms [10–12] are also adopted to support market participants. In general, SCUC models embedded in the electricity market simulation framework prefer to use the DC power flow model [13] with modifications, such as a fictitious nodal demand model [14], to offset the effect of active power losses.

With the continuous expansion of the power system scale, the SCUC calculation algorithm requires intensive computation. To meet the day-ahead and real-time market requirements, high-performance computing technologies are crucial for solving the large-scale SCUC problem [15–19].

A fast SCUC approach was presented in [15] for large-scale power systems using the Mixed Integer Programming (MIP) framework. Midcontinent Independent System Operator first introduced alternative optimization methods [16] and advanced mathematical formulations [17] into the MIP framework to accelerate the day-ahead SCUC calculation. In [20], a parallel graph power flow is applied to facilitate the electricity market clearing process. A graph computing-based approach is developed for solving the large-scale hydrothermal SCUC problem [21].

### 12.3.1 SCUC Model

The objective of SCUC is to determine a day-ahead unit commitment for minimizing the cost of supplying energy and ancillary services based on generating bids while meeting the prevailing constraints.

Considering both the system reserve and energy requirement, the objective function of a SCUC model can be formulated, with $F_{c,i}(P_{i,t,j})$ indicating stepwise bidding cost function; $SU_{i,t}(SD_{i,t})$for start-up (shut-down) cost of generator $i$ at time $t$ with a binary indicator $W_{i,t}(Y_{i,t})$ and $F_{.,i}(R_{i,t})$ for various types of reserve of generator $i$ at time $t$.

The SCUC is explicitly described by the following mathematical optimization model:

$$\min \sum_{i \in \mathcal{G}} \sum_{t=1}^{NT} \left[ F_{c,i}(P_{i,t,j}) + SU_{i,t}W_{i,t} + SD_{i,t}Y_{i,t} + F_{r,i}(R_{i,t}^r) + F_{sp,i}(R_{i,t}^{sp}) + F_{n1,i}(R_{i,t}^{n1}) + F_{n3,i}(R_{i,t}^{n3}) \right]$$

(12.78)

Subject to

$$\sum_{i \in \mathcal{G}} \sum_{j=1}^{NL} P_{i,t,j} = D_t, \quad t = 1, 2, ..., N$$

(12.79)

$$I_{i,t} - I_{i,t-1} = W_{i,t} - Y_{i,t}, \quad t = 1, ..., NT, \ \forall i \in \mathcal{G}$$

(12.80)

$$W_{i,t} + Y_{i,t} \leq 1, \quad t = 1, ..., NT, \ \forall i \in \mathcal{G}$$

(12.81)

$$\min[NT - t, T_i^{up} - 1] W_{i,t} \leq \sum_{s=t}^{t+\min(NT-t,T_i^{up}-1)} I_{i,s}, \quad t = 1, ...NT - 1, \forall i \in \mathcal{G}$$

(12.82)

$$\min\left[NT - t, T_i^{dn} - 1\right]Y_{i,t} + \sum_{s=t}^{t+\min\left(NT - t, T_i^{dn} - 1\right)} I_{i,s} \leq \min\left(NT - t, T_i^{dn} - 1\right), t = 1, \dots NT - 1, \forall i \in \mathcal{G}$$
(12.83)

$$\sum_{t=1}^{NT} W_{i,t} \leq NS_i, \forall i \in \mathcal{G}$$
(12.84)

$$I_{i,0} = I_{i,initial}, \ \forall i \in \mathcal{G}$$
(12.85)

$$-ramp_i \leq \sum_{j=1}^{NL} P_{i,t,j} - \sum_{j=1}^{NL} P_{i,t-1,j} \leq ramp_i, \ t = 2, \dots, NT, \forall i \in \mathcal{G}$$
(12.86)

$$0 \leq P_{i,t,j} \leq P_{i,t,j}^{UB} I_{i,t}, \ t = 1, \dots, NT, j = 1, \dots, NL, \forall i \in \mathcal{G}$$
(12.87)

$$P_l^{(k-1)} + \sum_i \left[GSF_{i,l} \times \sum_{j=1}^{NL} \left(P_{i,t,j} - P_{i,t,j}^{(k-1)}\right)\right] \leq P_l^{UB}, \ t = 1, \dots, NT, \forall l \in \mathcal{L}$$
(12.88)

$$0 \leq R_{i,t}^r \leq R_{max,i}^r I_{i,t}, \ 0 \leq R_{i,t}^{sp} \leq R_{max,i}^{sp} I_{i,t}, \ t = 1, \dots, NT, \forall i \in \mathcal{G}$$
(12.89)

$$0 \leq R_{i,t}^{n1} \leq R_{max,i}^{n1}, \ 0 \leq R_{i,t}^{n3} \leq R_{max,i}^{n3}, \ t = 1, \dots, NT, \forall i \in \mathcal{G}$$
(12.90)

$$\sum_i R_{i,t}^r \geq R_t^r, \ t = 1, \dots, NT$$
(12.91)

$$\sum_i R_{i,t}^{sp} \geq R_t^{sp}, \ t = 1, \dots, NT$$
(12.92)

$$\sum_i \left(R_{i,t}^{sp} + R_{i,t}^{n1}\right) \geq R_t^{sp} + R_t^{n1}, \ t = 1, \dots, NT$$
(12.93)

$$\sum_i \left(R_{i,t}^{sp} + R_{i,t}^{n1} + R_{i,t}^{n3}\right) \geq R_t^{sp} + R_t^{n1} + R_t^{n3}, \ t = 1, \dots, NT$$
(12.94)

$$R_{i,t}^r + R_{i,t}^r + R_{i,t}^r + R_{i,t}^r + \sum_{j=1}^{NL} P_{i,t,j} \leq P_i^{max}, \ t = 1, \dots, NT, \forall i \in \mathcal{G}$$
(12.95)

$$I_{i,t}, W_{i,t}, Y_{i,t} \in \{0,1\}, \ t = 1, \dots, NT, \forall i \in \mathcal{G}$$
(12.96)

Constraint (12.79) ensures system balance between supply and demand, and (12.80)–(12.85) guarantees that start-up and shunt-down characteristics, including minimum up (12.82) and down (12.83) time limits, are satisfied with generator turn-on indicator, $I_{i,t}$. The consideration for ramp rate limit and output lower bound are given by (12.86) and (12.87). Equation (12.88) provides transmission line capacity limit constraint with generation shift factor, $GSF_{i,l}$, of generator $i$ for transmission line $l$. The subscript $k - 1$ and $k$ indicates the $(k - 1)$th and $k$th iteration respectively. The reserves requirement constraints are given in (12.89)–(12.95). Equation (12.96) explicitly indicates all the binary variable constraints.

## 12.3.2 Graph-Based SCUC

SCUC modeled in (12.78)–(12.96) is a mixed integer linear programming problem. Optimization solver CPLEX can be used to solve the MIP problem. Then the spot electricity prices are calculated in the real-time market by the SCED model addressed in Section 12.2. In the SCUC model, the number of security constraints is large and the number of binding security constraints is small. To reduce the complexity and size of the optimization problem, a simultaneous feasibility test is performed on the given SCUC solution to check network security [22]. The violated security constraints are added to the SCUC MIP model to solve, and the iterative process is repeated until no new violations are found or the maximum iteration is met. The security check based on power flow is leveraging the graph-based power flow parallel computing to improve the computation efficiency. The graph computing-based framework for power market simulation is illustrated in Figure 12.1.

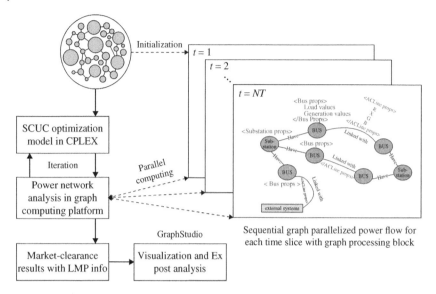

Sequential graph parallelized power flow for each time slice with graph processing block

**Figure 12.1** Graph computing based power market simulation framework [20].

## 12.4 Numerical Case Study

Taking the IEEE-14 bus system economic dispatch problem in Section 12.1.2 as a numerical example, the system bus and branch information are given in Tables 12.3 and 12.7.

All values in Tables 12.3 and 12.7 are in per unit and the bus voltage angle is in degrees. Buses in type 2 or 3 are generator buses. Keeping the same assumption in Section 12.1.2, generators at buses 1,2,3,6 are thermal units, and the generator at bus 8 is a hydro unit. We assume the loading job statement loads the data into the bus-branch graph. Line Q1, Line Q2, and Line Q3 are the branch's normal, emergency, and load-shedding limits. They are set as equal in the study case.

### 12.4.1 Graph-Based SCED Modeling

The IEEE-14 bus system has 4 thermal units, 1 hydro unit, and 20 branches. To build the linear program in standard form $\min_{x} c^T x$, $Ax = b$, $x \geq 0$, following the same assumption as in Section 12.1.2, such that one time period ($k = 1$, $\Delta t_k = 3600$ s $= 1$ hour) is illustrated, the objectives, variables, and constraints for the IEEE-14 bus system linearized SCED problem are identified by the following process.

The linearized objective function is

$$\min F = \left(2a_{T1}P_{T1}^0 + b_{T1}\right)\Delta P_{T1} + \left(2a_{T2}P_{T2}^0 + b_{T2}\right)\Delta P_{T2} \\ + \left(2a_{T3}P_{T3}^0 + b_{T3}\right)\Delta P_{T3} + \left(2a_{T4}P_{T4}^0 + b_{T4}\right)\Delta P_{T4} \tag{12.97}$$

Subject to the following constraints:

$$\Delta P_{T1} + x_1 = P_{T1}^{\max} - P_{T1}^0$$
$$\Delta P_{T2} + x_2 = P_{T2}^{\max} - P_{T2}^0$$
$$\Delta P_{T3} + x_3 = P_{T3}^{\max} - P_{T3}^0$$

**Table 12.7** IEEE-14 bus system branch table.

| From bus | To bus | Area | Zone | Circuit | R | X | B | Line Q1 | Line Q2 | Line Q3 | Turns ratio |
|---|---|---|---|---|---|---|---|---|---|---|---|
| 1 | 2 | 1 | 1 | 1 | 0.01938 | 0.05917 | 0.0528 | 120 | 120 | 120 | 0 |
| 1 | 5 | 1 | 1 | 1 | 0.05403 | 0.22304 | 0.0492 | 65 | 65 | 65 | 0 |
| 2 | 3 | 1 | 1 | 1 | 0.04699 | 0.19797 | 0.0438 | 36 | 36 | 36 | 0 |
| 2 | 4 | 1 | 1 | 1 | 0.05811 | 0.17632 | 0.034 | 65 | 65 | 65 | 0 |
| 2 | 5 | 1 | 1 | 1 | 0.05695 | 0.17388 | 0.0346 | 50 | 50 | 50 | 0 |
| 3 | 4 | 1 | 1 | 1 | 0.06701 | 0.17103 | 0.0128 | 65 | 65 | 65 | 0 |
| 4 | 5 | 1 | 1 | 1 | 0.01335 | 0.04211 | 0 | 45 | 45 | 45 | 0 |
| 4 | 7 | 1 | 1 | 1 | 0 | 0.20912 | 0 | 55 | 55 | 55 | 0.978 |
| 4 | 9 | 1 | 1 | 1 | 0 | 0.55618 | 0 | 32 | 32 | 32 | 0.969 |
| 5 | 6 | 1 | 1 | 1 | 0 | 0.25202 | 0 | 45 | 45 | 45 | 0.932 |
| 6 | 11 | 1 | 1 | 1 | 0.09498 | 0.19890 | 0 | 18 | 18 | 18 | 0 |
| 6 | 12 | 1 | 1 | 1 | 0.12291 | 0.25581 | 0 | 32 | 32 | 32 | 0 |
| 6 | 13 | 1 | 1 | 1 | 0.06615 | 0.13027 | 0 | 32 | 32 | 32 | 0 |
| 7 | 8 | 1 | 1 | 1 | 0 | 0.17615 | 0 | 42 | 42 | 42 | 0 |
| 7 | 9 | 1 | 1 | 1 | 0 | 0.11001 | 0 | 32 | 32 | 32 | 0 |
| 9 | 10 | 1 | 1 | 1 | 0.03181 | 0.08450 | 0 | 32 | 32 | 32 | 0 |
| 9 | 14 | 1 | 1 | 1 | 0.12711 | 0.27038 | 0 | 32 | 32 | 32 | 0 |
| 10 | 11 | 1 | 1 | 1 | 0.08205 | 0.19207 | 0 | 12 | 12 | 12 | 0 |
| 12 | 13 | 1 | 1 | 1 | 0.22092 | 0.19988 | 0 | 12 | 12 | 12 | 0 |
| 13 | 14 | 1 | 1 | 1 | 0.17093 | 0.34802 | 0 | 12 | 12 | 12 | 0 |

$$\Delta P_{T4} + x_4 = P_{T4}^{\max} - P_{T4}^0$$

$$\Delta P_{H1} + x_5 = P_{H1}^{\max} - P_{H1}^0$$

$$\Delta P_{T1} - x_6 = P_{T1}^{\min} - P_{T1}^0$$

$$\Delta P_{T2} - x_7 = P_{T2}^{\min} - P_{T2}^0$$

$$\Delta P_{T3} - x_8 = P_{T3}^{\min} - P_{T3}^0$$

$$\Delta P_{T4} - x_9 = P_{T4}^{\min} - P_{T4}^0$$

$$\Delta P_{H1} - x_{10} = P_{H1}^{\min} - P_{H1}^0$$

$$\left(2a_{H1}P_{H1}^0 + b_{H1}\right)\Delta P_{H1} = \frac{C_1 - a_{H1}\left(P_{H1}^0\right)^2 + b_{H1}P_{H1}^0 - c_{H1}}{\Delta t}$$

$$h_{11}\Delta P_{T1} + h_{12}\Delta P_{T2} + h_{13}\Delta P_{T3} + h_{14}\Delta P_{T4} + h_{15}\Delta P_{H1} + x_{11} = P_{l1}^{\max} - P_{l1}^0$$

$$\vdots$$

**Table 12.8** Generation shift factor matrix.

|  | G1 | G2 | G3 | G4 | G5 |
|---|---|---|---|---|---|
| L1 | 0 | −0.8380 | −0.7465 | −0.6291 | −0.6573 |
| L2 | 0 | −0.1620 | −0.2535 | −0.3709 | −0.3428 |
| L3 | 0 | 0.0274 | −0.5320 | −0.1188 | −0.1427 |
| L4 | 0 | 0.0573 | −0.1434 | −0.2487 | −0.2986 |
| L5 | 0 | 0.0774 | −0.0711 | −0.2616 | −0.2160 |
| L6 | 0 | 0.0274 | 0.4680 | −0.1188 | −0.1427 |
| L7 | 0 | 0.0799 | 0.3067 | −0.0389 | 0.3584 |
| L8 | 0 | 0.0030 | 0.0113 | −0.2075 | −0.6338 |
| L9 | 0 | 0.0017 | 0.0066 | −0.1211 | −0.1658 |
| L10 | 0 | −0.0046 | −0.0179 | −0.6714 | −0.2004 |
| L11 | 0 | −0.0028 | −0.0108 | 0.1979 | −0.1207 |
| L12 | 0 | −0.0004 | −0.0016 | 0.0291 | −0.0177 |
| L13 | 0 | −0.0015 | −0.0056 | 0.1017 | −0.0620 |
| L14 | 0 | 0 | 0 | 0 | −1 |
| L15 | 0 | 0.0030 | 0.0113 | −0.2075 | 0.3662 |
| L16 | 0 | 0.0028 | 0.0108 | −0.1979 | 0.1207 |
| L17 | 0 | 0.00186 | 0.0071 | −0.1307 | 0.0797 |
| L18 | 0 | 0.00282 | 0.0108 | −0.1979 | 0.1207 |
| L19 | 0 | −0.00041 | −0.0016 | 0.0291 | −0.0177 |
| L20 | 0 | −0.00186 | −0.0071 | 0.1307 | −0.0797 |

$$h_{20,1}\Delta P_{T1} + h_{20,2}\Delta P_{T2} + h_{20,3}\Delta P_{T3} + h_{20,4}\Delta P_{T4} + h_{20,5}\Delta P_{H1} + x_{30} = P_{l20}^{\max} - P_{l20}^{0}$$

$$h_{11}\Delta P_{T1} + h_{12}\Delta P_{T2} + h_{13}\Delta P_{T3} + h_{14}\Delta P_{T4} + h_{15}\Delta P_{H1} - x_{31} = -P_{l1}^{\max} - P_{l1}^{0}$$

$$\vdots$$

$$h_{20,1}\Delta P_{T1} + h_{20,2}\Delta P_{T2} + h_{20,3}\Delta P_{T3} + h_{20,4}\Delta P_{T4} + h_{20,5}\Delta P_{H1} - x_{50} = -P_{l20}^{\max} - P_{l20}^{0}$$

Where $x_1 - x_{50}$ are nonnegative slack variables added to convert the inequality constraints to equality constraints.

And assuming generator 1 to be a slack bus, the generation shift factor matrix $H$ is calculated as shown in Table 12.8.

In matrix format, the above linear program problem is restated as

$$\min F = \sum_{i=1}^{4}\left(\left(2a_{Ti}P_{Ti}^{0} + b_{Ti}\right)\Delta P_{Ti}\right) \tag{12.98}$$

Subject to

$$
\begin{bmatrix}
[1]_{1 \times 5} & & \\
[I]_{5 \times 5} & [I]_{5 \times 5} & \\
[I]_{5 \times 5} & & -[I]_{5 \times 5} \\
2a_{H1}P^0_{H1} + b_{H1} & & \\
H_{20 \times 5} & & [I]_{20 \times 20} \\
H_{20 \times 5} & & -[I]_{20 \times 20}
\end{bmatrix}
\begin{bmatrix}
\Delta P_T \\
\Delta P_H \\
x_{1-5} \\
x_{6-10} \\
x_{11-30} \\
x_{31-50}
\end{bmatrix}
=
$$

$$
\begin{bmatrix}
0 \\
\begin{bmatrix} P^{max}_T - P^0_T \\ P^{max}_H - P^0_H \end{bmatrix}_{5 \times 1} \\
\begin{bmatrix} P^{min}_T - P^0_T \\ P^{min}_H - P^0_H \end{bmatrix}_{5 \times 1} \\
\dfrac{C_1 - a_{H1}\left(P^0_{H1}\right)^2 - b_{H1}P^0_{H1} - C_{H1}}{\Delta t} \\
\left[ P^{max}_l - P^0_l \right]_{20 \times 1} \\
\left[ -P^{max}_l - P^0_l \right]_{20 \times 1}
\end{bmatrix}
\tag{12.99}
$$

In the linear program, $\Delta P_T$ and $\Delta P_H$ are not nonnegative variables. They are free to take on either positive, negative values, or zero. Two nonnegative variables $u = (u_T, u_H) = (u_{T1}, u_{T2}, u_{T3}, u_{T4}, u_H)$ and $v = (v_T, v_H) = (v_{T1}, v_{T2}, v_{T3}, v_{T4}, v_H)$ are introduced to represent $\Delta P_T$ and $\Delta P_H$ as $\Delta P_T = u_T - v_T$ and $\Delta P_H = u_H - v_H$.

Then linear program problem in the standard form is written as

$$
\min \ F = \sum_{i=1}^{4} \left( \left( 2a_{Ti}P^0_{Ti} + b_{Ti} \right)\left( u_{Ti} - v_{Ti} \right) \right)
\tag{12.100}
$$

Subject to

$$
\begin{bmatrix}
[1]_{1 \times 5} & -[1]_{1 \times 5} & & \\
[I]_{5 \times 5} & -[I]_{5 \times 5} & [I]_{5 \times 5} & \\
[I]_{5 \times 5} & -[I]_{5 \times 5} & & -[I]_{5 \times 5} \\
2a_{H1}P^0_{H1} + b_{H1} & -2a_{H1}P^0_{H1} - b_{H1} & & \\
H_{20 \times 5} & -H_{20 \times 5} & & [I]_{20 \times 20} \\
H_{20 \times 5} & -H_{20 \times 5} & & -[I]_{20 \times 20}
\end{bmatrix}
\begin{bmatrix}
u_T \\
u_H \\
v_T \\
x_{1-5} \\
x_{6-10} \\
x_{11-30} \\
x_{31-50}
\end{bmatrix}
$$

$$
=
\begin{bmatrix}
0 \\
\begin{bmatrix} P_T^{\max} - P_T^0 \\ P_H^{\max} - P_H^0 \end{bmatrix}_{5 \times 1} \\
\begin{bmatrix} P_T^{\min} - P_T^0 \\ P_H^{\min} - P_H^0 \end{bmatrix}_{5 \times 1} \\
\dfrac{C_1 - a_{H1}\left(P_{H1}^0\right)^2 - b_{H1}P_{H1}^0 - C_{H1}}{\Delta t} \\
\begin{bmatrix} P_l^{\max} - P_l^0 \end{bmatrix}_{20 \times 1} \\
\begin{bmatrix} -P_l^{\max} - P_l^0 \end{bmatrix}_{20 \times 1}
\end{bmatrix}
\tag{12.101}
$$

Or

$$
\min_{x} c^T x, \quad Ax = b, \quad x \ge 0 \tag{12.102}
$$

Where

$$
c^T = [2a_{T1}P_{T1}^0 + b_{T1}, 2a_{T2}P_{T2}^0 + b_{T2}, 2a_{T3}P_{T3}^0 + b_{T3}, 2a_{T4}
$$
$$
P_{T4}^0 + b_{T4}, \ 0, \ -2a_{T1}P_{T1}^0 - b_{T1}, \ -2a_{T2}P_{T2}^0 - b_{T2}, \ -2a_{T3}P_{T3}^0 - b_{T3}, \ -2a_{T4}P_{T4}^0 - b_{T4}, 0, ..., 0]
$$

$$
A =
\begin{bmatrix}
[1]_{1 \times 5} & -[1]_{1 \times 5} & & & \\
[I]_{5 \times 5} & -[I]_{5 \times 5} & [I]_{5 \times 5} & & \\
[I]_{5 \times 5} & -[I]_{5 \times 5} & -[I]_{5 \times 5} & & \\
2a_{H1}P_{H1}^0 + b_{H1} & -2a_{H1}P_{H1}^0 - b_{H1} & & & \\
H_{20 \times 5} & -H_{20 \times 5} & & [I]_{20 \times 20} & \\
H_{20 \times 5} & -H_{20 \times 5} & & & -[I]_{20 \times 20}
\end{bmatrix}
$$

$$
b =
\begin{bmatrix}
0 \\
\begin{bmatrix} P_T^{\max} - P_T^0 \\ P_H^{\max} - P_H^0 \end{bmatrix}_{5 \times 1} \\
\begin{bmatrix} P_T^{\min} - P_T^0 \\ P_H^{\min} - P_H^0 \end{bmatrix}_{5 \times 1} \\
\dfrac{C_1 - a_{H1}\left(P_{H1}^0\right)^2 - b_{H1}P_{H1}^0 - C_{H1}}{\Delta t} \\
\begin{bmatrix} P_l^{\max} - P_l^0 \end{bmatrix}_{20 \times 1} \\
\begin{bmatrix} -P_l^{\max} - P_l^0 \end{bmatrix}_{20 \times 1}
\end{bmatrix}
$$

### 12.4.2 Basic Feasible Solution

By initializing $\left(P_{T0}^T, P_{H0}^T\right)^T = (148.9, 20, 32.48, 20, 37.62)^T$ as the optimal solution of the thermal and hydroelectric power generation system economic dispatch problem in Section 12.1.2, the branch power flow results in the power flow solution are $P_l^0 = (103.0,\ 45.9,\ \mathbf{44.4},\ 33.1,\ 23.8,$

$-17.3$, $-40.7$, $0.7$, $8.1$, $21.3$, $5.9$, $7.5$, $16.8$, $-37.6$, **38.3**, $6.7$, $10.2$, $-2.4$, $1.4$, $4.7)^T$ in MW. Where $P_{l3}^0 = 44.4$ MW and $P_{l15}^0 = 38.4$ MW highlighted in black bold violate the branch flow limit.

To set up a feasible solution for the linearized SCED problem (12.102), $Ax = b$ are regrouped as

$$[A_N \quad A_B]\begin{bmatrix} x_N \\ x_B \end{bmatrix} = b \tag{12.103}$$

Where

$$A_N = \begin{bmatrix} [1]_{1 \times 5} & -[1]_{1 \times 5} \\ [I]_{5 \times 5} & -[I]_{5 \times 5} \\ [I]_{5 \times 5} & -[I]_{5 \times 5} \\ 2a_{H1}P_{H1}^0 + b_{H1} & -2a_{H1}P_{H1}^0 - b_{H1} \\ H_{20 \times 5} & -H_{20 \times 5} \\ H_{20 \times 5} & -H_{20 \times 5} \end{bmatrix} \tag{12.104}$$

$$A_B = \begin{bmatrix} [I]_{5 \times 5} & & & \\ & -[I]_{5 \times 5} & & \\ & & [I]_{20 \times 20} & \\ & & & -[I]_{20 \times 20} \end{bmatrix} \tag{12.105}$$

$$x_N = \begin{bmatrix} u_T \\ u_H \\ v_T \\ v_H \end{bmatrix} \tag{12.106}$$

$$x_B = \begin{bmatrix} x_{1-5} \\ x_{6-10} \\ x_{11-30} \\ x_{31-50} \end{bmatrix} \tag{12.107}$$

The right-hand side vector $b$ is computed as
$b = (0.0, 351.1, 180.0, 267.5, 130.0, 162.4, -98.9, 0.0, -12.5, 0.0, -17.6, 0.0, 17.0, 19.1, -8.4, 31.9, 26.2, 47.7, 4.3, 54.3, 23.9, 23.7, 12.2, 24.5, 15.2, 4.4, -6.3, 25.4, 21.8, 9.7, 10.6, 7.3, -223.0, -110.9, -80.4, -98.1, -73.8, -82.3, -85.7, -55.7, -40.1, -66.3, -23.9, -39.5, -48.8, -79.6, -70.3, -38.7, -42.2, -14.4, -13.4, -16.7)^T$,

By assigning the nonbasic variables $x_N$ to be equal to 0, (12.103) is reduced to

$$A_B x_B = b \tag{12.108}$$

By solving (12.108), the values of the basic variables $x_B$ are calculated as
$x_B = \hat{b} = (351.1, 180.0, 267.5, 130.0, 162.4, 98.9, 0.0, 12.5, 0.0, 17.6, 17.0, 19.1, \mathbf{-8.4}, 31.9, 26.2, 47.7, 4.3, 54.3, 23.9, 23.7, 12.2, 24.5, 15.2, 4.4, \mathbf{-6.3}, 25.4, 21.8, 9.7, 10.6, 7.3, 223.0, 110.9, 80.4, 98.1, 73.8, 82.3, 85.7, 55.7, 40.1, 66.3, 23.9, 39.5, 48.8, 79.6, 70.3, 38.7, 42.2, 14.4, 13.4, 16.7)^T$

Because of the two branch power flow violations ($P_{l3}^0$ and $P_{l15}^0$) in the base case, the basic solution to the economic dispatch problem is infeasible since $x_B$ has two negative elements ($x_{13} = -8.4$ and $x_{25} = -6.3$) highlighted in black bold which do not meet the requirement that $x$ is nonnegative. To mitigate the conflict, as addressed in Section 6.2.1.1, two nonnegative variables $y_k \geq 0$ and $z_k \geq 0$ are

**Table 12.9** SCED iteration.

| Iter | $\Delta P_{G1}$ (MW) | $\Delta P_{G2}$ (MW) | $\Delta P_{G3}$ (MW) | $\Delta P_{G4}$ (MW) | $\Delta P_{G5}$ (MW) | $P_{G1}$ (MW) | $P_{G2}$ (MW) | $P_{G3}$ (MW) | $P_{G4}$ (MW) | $P_{G5}$ (MW) | $\Sigma P_G$ (MW) | $\Sigma P_L$ (MW) | Total Cost ($) | Violations |
|---|---|---|---|---|---|---|---|---|---|---|---|---|---|---|
| 0 | 0.00 | 0.00 | 0.00 | 0.00 | 0.00 | 148.90 | 20.00 | 32.48 | 20.00 | 37.62 | 259.00 | 259.00 | 2766.87 | L3,L15 |
| 1 | −39.83 | 0.00 | 8.98 | 30.85 | 0.00 | 109.07 | 20.00 | 41.46 | 50.85 | 37.62 | 259.00 | 259.00 | 2837.79 | None |
| 2 | 0.00 | 0.00 | 0.00 | 0.00 | 0.00 | 109.07 | 20.00 | 41.46 | 50.85 | 37.62 | 259.00 | 259.00 | 2837.79 | None |

introduced to represent the corresponding basic variables $x_k$ as $x_k = y_k - z_k$. ($x_k = x_{13}$ or $x_{25}$ in the example case). And the basic solution for $y_k$ and $z_k$ are 0 and $-\hat{b}_k \left( -\hat{b}_k > 0 \right)$ respectively to ensure the basic solution is feasible.

### 12.4.3 Economic Dispatch Optimal Solution

Starting with the basic feasible solution, in the first iteration, the linearized SCED optimizes the incremental generation of active power at $\Delta P_G^{(1)} = (-39.83, 0, 8.98, 30.85, 0)^T$ in MW. The generation active power outputs are updated as $P_G^{(1)} = (109.07, 20, 41.46, 50.85, 37.62)^T$. With the updated generation power injection, the DC power flow is recalculated. The branch power flow is updated as $P_l^{(1)} = (76.93, 32.14, 36, 24.18, 15.05, -16.74, -39.13, -5.62, 4.4, 0.46, 11.86, 8.36, 19.89, -37.62, 32, 0.64, 6.25, -8.36, 2.26, 8.65)^T$ in MW with no violation.

In the second iteration, the right-hand side vector $b$ is updated as
$b = (0.0, 390.9, 180.0, 258.5, 99.2, 162.4, -59.1, 0.0, -21.5, -30.9, -17.6, 0.0, 43.1, 32.9, 0.0, 40.8, 35.0, 48.3, 5.9, 49.4, 27.6, 44.5, 6.1, 23.6, 12.1, 4.4, 0.0, 31.4, 25.8, 3.6, 9.7, 3.4, -196.9, -97.1, -72.0, -89.2, -65.1, -81.7, -84.1, -60.6, -36.4, -45.5, -29.9, -40.4, -51.9, -79.6, -64.0, -32.6, -38.3, -20.4, -14.3, -20.7)^T$

Since $P_{H1}^2 = P_{H1}^1 = P_{H1}^0$, element $2a_{H1}P_{H1}^2 + b_{H1}$ in matrix $A$ does not change. $x_B$ is calculated as
$x_B = (390.9, 180.0, 258.5, 99.2, 162.4, 59.1, 0.0, 21.5, 30.9, 17.6, 0.0, 43.1, 32.9, 0.0, 40.8, 35.0, 48.3, 5.9, 49.4, 27.6, 44.5, 6.1, 23.6, 12.1, 4.4, 0.0, 31.4, 25.8, 3.6, 9.7, 3.4, 196.9, 97.1, 72.0, 89.2, 65.1, 81.7, 84.1, 60.6, 36.4, 45.5, 29.9, 40.4, 51.9, 79.6, 64.0, 32.6, 38.3, 20.4, 14.3, 20.7)^T$

All elements in the basic solution are nonnegative. With the basic feasible solution, the new linearized SCED optimization problem in the third iteration is solved at $\Delta P_G^{(3)} = (0, 0, 0, 0, 0)^T$ in MW. The iteration is converged as summarized in Table 12.9.

## References

1 Kirchmayer, L.K. (1958). *Economic Operation of Power Systems*. New York: Wiley.
2 Souag, S. and Benhamida, F. (2014). Secured economic dispatch algorithm using GSDF matrix. *Leonardo Journal of Sciences* 13: 1–14.
3 Ando, T. (1995). Majorization relations for hadamard products. *Linear Algebra and its Applications* 223–224: 57–64.
4 Zhang, B. and Chen, S. (1998). *Advanced Power Grid Analysis*. Science Press.

**5** Fu, Y., Shahidehpour, M., and Li, Z. (2005). Security-constrained unit commitment with ac constraints. *IEEE transactions on power systems* 20 (3): 1538–1550.

**6** Zheng, T. and Litvinov, E. (2011). On ex post pricing in the real-time electricity market. *IEEE Transactions on Power Systems* 26 (1): 153–164.

**7** Li, Z. and Shahidehpour, M. (2005). Security-constrained unit commitment for simultaneous clearing of energy and ancillary services markets. *IEEE transactions on power systems* 20 (2): 1079–1088.

**8** Wu, L. (2016). Accelerating ncuc via binary variable-based locally ideal formulation and dynamic global cuts. *IEEE Transactions on Power Systems* 31 (5): 4097–4107.

**9** Krishnamurthy, D., Li, W., and Tesfatsion, L. (2016). An 8-zone test system based on ISO New England data: development and application. *IEEE Transactions on Power Systems* 31 (1): 234–246.

**10** Wang, Q., McCalley, J.D., Zheng, T., and Litvinov, E. (2013). A computational strategy to solve preventive risk-based security-constrained opf. *IEEE Transactions on Power Systems* 28 (2): 1666–1675.

**11** Exemplar, E. (2011). Plexos for power systems-power market simulation and analysis software [software]. https://www.energyexemplar.com/plexos

**12** Veselka, T., Boyd, G., Conzelmann, G. et al. (2002). Simulating the behavior of electricity markets with an agent-based methodology: the electric market complex adaptive systems (EMCAS) model. *2002 USAEE Conference*, Vancouver, Canada. https://ceeesa.es.anl.gov/pubs/43943.pdf.

**13** Hu, Z., Cheng, H., Yan, Z., and Li, F. (2010). An iterative LMP calculation method considering loss distributions. *IEEE Transactions on Power Systems* 25 (3): 1469–1477.

**14** Li, F. and Bo, R. (2007). DCOPF-based LMP simulation: algorithm, comparison with ACOPF, and sensitivity. *IEEE Transactions on Power Systems* 22 (4): 1475–1485.

**15** Fu, Y. and Shahidehpour, M. (2007). Fast SCUC for large-scale power systems. *IEEE Transactions on Power Systems* 22 (4): 2144–2151.

**16** Chen, Y., Casto, A., Wang, F. et al. (2016). Improving large scale day-ahead security constrained unit commitment performance. *IEEE Transactions on Power Systems* 31 (6): 4732–4743.

**17** Chen, Y. and Wang, F. (2017). MIP formulation improvement for large scale security constrained unit commitment with configuration based combined cycle modeling. *Electric Power Systems Research* 148: 147–154.

**18** Wang, C. and Fu, Y. (2016). Fully parallel stochastic security-constrained unit commitment. *IEEE Transactions on Power Systems* 31 (5): 3561–3571.

**19** Papavasiliou, A. and Oren, S.S. (2013). A comparative study of stochastic unit commitment and security-constrained unit commitment using high performance computing. *2013 European Control Conference (ECC)*, Zurich, Switzerland, pp. 2507–2512. https://doi.org/10.23919/ECC.2013.6669244.

**20** Chen, T., Yuan, C., Liu, G., and Dai, R. (2018). Graph based platform for electricity market study, education and training. *2018 IEEE Power & Energy Society General Meeting (PESGM)*, Portland, OR, USA, pp. 1–5. https://doi.org/10.1109/PESGM.2018.8586243.

**21** Wei, L., Tang, Y., Zhang, X. et al. (2021). Graph computing based ADMM approach for security constrained unit commitment in hydrothermal power systems. *PES & PELS 2021 6th IEEE Workshop on the Electronic Grid (eGrid 2021)*, New Orleans, Louisiana (8–10 November 2021).

**22** Zhu, J. and Cheung, K. (2010). Flexible simultaneous feasibility test in energy market. *IEEE PES General Meeting*, Minneapolis, MN, pp. 1–7. https://doi.org/10.1109/PES.2010.5589447.

# 13

## Automatic Generation Control

In power system operation, the system frequency is controlled to remain nearly constant. The system frequency is dependent on the active power balance. An active power demand change results in a change in system frequency. To maintain the system frequency, generators in the system are controlled to share the power demand change. A speed governor on each generator responds to the primary speed (frequency) control. With the primary speed control, system load change will result in a steady-state frequency bias [1]. A supplementary control is required to restore the system frequency to the nominal value. The supplementary control is a central control adjusting load reference set-point. The function to adjust load reference set-points on generating units automatically responding to system load changes is automatic generation control (AGC) [2].

Typically, a power system is an interconnected system with multiple controlled areas. In addition to control of frequency, generators will be controlled to maintain the scheduled interchange power on flowgates between controlled areas.

Besides balancing active power to maintain the system frequency and regulating flowgate power flow at scheduled value, one of the objectives of AGC is to minimize the power generation cost.

The classic AGC does not model network security constraints [3, 4]. Therefore, the power system operation point controlled by the classic AGC may violate network operation limits. To secure power system operation, security-constrained AGC is proposed and addressed in this Chapter.

## 13.1 Classic Automatic Generation Control

A power system's frequency depends on the balance of active power. Since the static frequency in an interconnected AC system is a common factor and observable at any point of the system, a change in active power load at any point in the system will be reflected by a frequency change in the system. A speed governor will respond to the frequency deviation primarily with a steady-state frequency bias, and AGC will regulate active power to control system frequency back to the nominal value supplementarily.

### 13.1.1 Speed Governor Control

The basic principle of speed governing involves prime mover speed measurement and adjusting the energy supplied to the prime mover to maintain desired speed [5]. The turbine-governor model connecting to a synchronous generator diagram is shown in Figure 13.1.

*Graph Database and Graph Computing for Power System Analysis*, First Edition. Renchang Dai and Guangyi Liu.
© 2024 The Institute of Electrical and Electronics Engineers, Inc. Published 2024 by John Wiley & Sons, Inc.

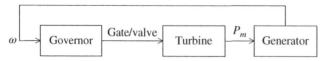

**Figure 13.1** Turbine-governor diagram.

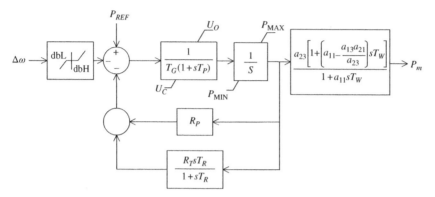

**Figure 13.2** IEEE type 3 speed-governor model IEEEG3 transfer function.

There are many types of governor and turbine models in power system analysis. To illustrate the basic principle of speed governing, the IEEE standard governor model type 3 (IEEEG3) [6] is depicted in Figure 13.2.

Where $dbL$ and $dbH$ are the deadband widths that are only in the IEEEG3D model. $T_G$ and $T_P$ are the gate servo time constant and the pilot servo valve time constant. $U_o$ and $U_c$ are the maximum and minimum gate opening velocities. $P_{MAX}$ and $P_{MIN}$ are the maximum and minimum gate openings. $R_P$ and $R_T$ are the permanent and temporary droop, $T_R$ is the dashpot time constant, $T_w$ is the water inertia time constant, $a_{11}$, $a_{13}$, $a_{21}$, and $a_{23}$ are the turbine coefficients.

When the system load varies, the active power imbalance results in generator speed fluctuation. If the speed difference is over the deadband width, the gate is regulated within its upper and lower limits. Since hydro turbine has water inertia effect, a change in the gate produces an initial power change opposite to the attempt. Temporary transient droop and permanent droop feedback provide compensation to stabilize the gate position control.

The mechanical power $P_m$ of the output of the turbine will create a mechanical torque onto the generator. When there is an active power load change, the generator's electrical power $P_e$ will change instantaneously to balance the load change. The mismatch between the mechanical power $P_m$ and the electrical power $P_e$ results in speed variations determined by the following rotor swing equation:

$$M\frac{d\omega}{dt} = P_m - P_e - D\omega \qquad (13.1)$$

where $M$ is the inertia, $D$ is the damping factor, and $D\omega$ represents frequency-dependent load. Equation (13.1) is linearized and expressed in terms of $\Delta P_m$, $\Delta P_e$, and $\Delta\omega$ as shown in Figure 13.3 as follows.

Combining the turbine, governor, and generator, the generating unit transfer function diagram is illustrated as shown in Figure 13.4 as follows.

When the system is subjected to an increase in load, the positive $\Delta P_e$ causes the rotor speed to drop. As the speed drops, the steam valve of the thermal unit or the water gate of the hydro unit will increase to provide more mechanical power to balance the increased load. The control diagram works when a single generation unit is regulated to balance the load. It cannot be used to coordinate two or more generation units connected to the same system since they are trying to control system frequency to its own setting.

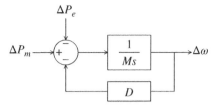

**Figure 13.3** Speed and power transfer function.

## 13.1.2 Speed Droop Function [7]

To share load changes, a speed droop function is designed by adding a feedback loop, as shown in Figure 13.5.

To illustrate how two or more generators are sharing the load in parallel, the frequency drop functions of two generators are shown in Figure 13.6.

The ratio of frequency deviation $\Delta f = \Delta \omega$ to the change in power output ($\Delta P_1$ and $\Delta P_2$ as shown in Figure 13.6) is $R$. With the frequency droop function, two or more generators will respond to a load change and share the change. When a load increase causes the two units to slow down, the

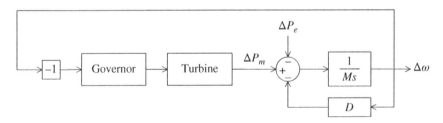

**Figure 13.4** Generating unit control block diagram.

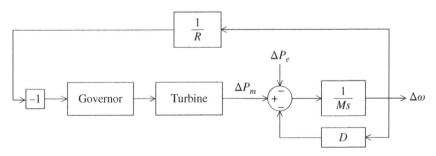

**Figure 13.5** Generating unit control block diagram with droop function.

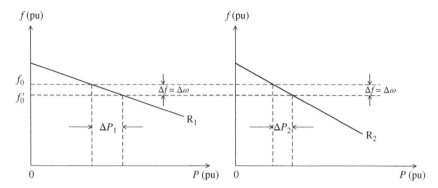

**Figure 13.6** Frequency droop function.

governors regulate the valve/gate to increase the power output to reach a new frequency. The amount of load shared by each unit depends on the droop ratio $R$ as follows:

$$\Delta P_1 = -\frac{\Delta f}{R_1} \tag{13.2}$$

$$\Delta P_2 = -\frac{\Delta f}{R_2} \tag{13.3}$$

With primary speed control, system load change will result in a steady-state frequency deviation from the nominal frequency. The frequency response on a load change $\Delta P_L$ depends on the droop function of the generator and the frequency-damping characteristics of the load. The steady-state frequency deviation is given by

$$\Delta f = \frac{-\Delta P_L}{\frac{1}{R} + D} \tag{13.4}$$

The steady-state frequency deviation resulting from the governor speed droop and load frequency dependence is illustrated in Figure 13.7.

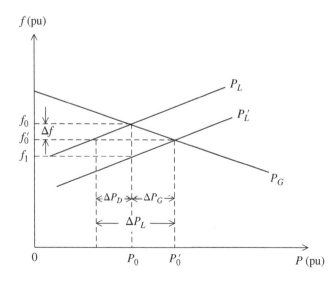

**Figure 13.7** Governor droop function and load frequency dependence.

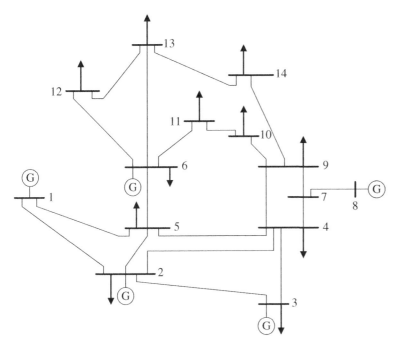

**Figure 13.8** Single line diagram of the IEEE 14-bus system.

The system initially operates at the nominal frequency $f_0$. When the load increases, represented by moving the load frequency dependency line from $P_L$ to $P'_L$, the system frequency drops down to $f_1$. Governor regulates the valve or gate to increase the generator power output by $\Delta P_G$. Meanwhile, the load is reduced by $\Delta P_D$ due to its frequency dependency. The system will be stable at a new frequency $f'_0$, resulting in a steady-state frequency deviation $\Delta f = f_0 - f'_0$.

The IEEE-14 bus system shown in Figure 13.8 is taken as an example to study generator primary frequency control.

In the base case, the IEEE 14-bus system consists of 5 synchronous machines, 20 branches, and 14 buses with 11 loads. The total load is 259 MW. Taking the security-constrained economic dispatch solution based on the DC power flow in 12.4 as a base case solution, the generation power output is $P_G^{(0)} = (109.07, 20, 41.46, 50.85, 37.62)^T$ in MW. Assume that the system has two areas in which buses 1, 2, 3, 4, and 5 are in area 1, while other buses are in area 2 as highlighted as bold in Table 13.1.

The two areas of the system are connected by three tie lines, which power flows in the base case are $P_{l4-7}^{(0)} = -5.62$ MW, $P_{l4-9}^{(0)} = 4.4$ MW, $P_{l5-6}^{(0)} = 0.46$ MW. By defining the flowgate power flow as $P_{FG} = P_{l4-7} + P_{l4-9} + P_{l5-6}$, the system nodal model is simplified as a zonal model in Figure 13.9 where the flowgate power flow is scheduled as $P_{FG}^{(s)} = P_{l4-7}^{(0)} + P_{l4-9}^{(0)} + P_{l5-6}^{(0)} = -0.76$ MW.

In the zonal model, the droop ratio $R$ and the damping factor $D$ are in per unit. With the 100 MW and 60 Hz base, the real value of the droop ratio $R$ and the damping factor $D$ of each area are calculated as

$$\frac{1}{R_1} = \frac{1}{0.05} \times \frac{100\text{ MW}}{60\text{ Hz}} = 33.33\text{ MW/Hz} \tag{13.5}$$

$$\frac{1}{R_2} = \frac{1}{0.1} \times \frac{100\text{ MW}}{60\text{ Hz}} = 16.67\text{ MW/Hz} \tag{13.6}$$

**Table 13.1** Modified IEEE-14 bus system bus table.

| Bus ID | Area | Loss zone | Type | Voltage | Angle | Load_P | Load_Q | Unit_P | Unit_Q | Desired voltage | MAX Q | MIN Q | G | B |
|---|---|---|---|---|---|---|---|---|---|---|---|---|---|---|
| 1 | 1 | 1 | 3 | 1.060 | 0.00 | 0.000 | 0.000 | 1.0907 | −0.169 | 1.06 | 0 | 0 | 0 | 0 |
| 2 | 1 | 1 | 2 | 1.045 | −4.98 | 0.217 | 0.127 | 0.2000 | 0.424 | 1.045 | 0.50 | −0.4 | 0 | 0 |
| 3 | 1 | 1 | 2 | 1.010 | −12.72 | 0.942 | 0.190 | 0.4146 | 0.234 | 1.01 | 0.40 | 0 | 0 | 0 |
| 4 | 1 | 1 | 0 | 1.019 | −10.33 | 0.478 | −0.039 | 0 | 0 | 0 | 0 | 0 | 0 | 0 |
| 5 | 1 | 1 | 0 | 1.020 | −8.78 | 0.076 | 0.016 | 0 | 0 | 0 | 0 | 0 | 0 | 0 |
| 6 | 2 | 1 | 2 | 1.070 | −14.22 | 0.112 | 0.075 | 0.5084 | 0.122 | 1.07 | 0.24 | −0.6 | 0 | 0 |
| 7 | 2 | 1 | 0 | 1.062 | −13.37 | 0 | 0 | 0 | 0 | 0 | 0 | 0 | 0 | 0 |
| 8 | 2 | 1 | 2 | 1.090 | −13.36 | 0 | 0 | 0.3762 | 0.174 | 1.09 | 0.24 | −0.6 | 0 | 0 |
| 9 | 2 | 1 | 0 | 1.056 | −14.94 | 0.295 | 0.166 | 0 | 0 | 0 | 0 | 0 | 0 | 0.19 |
| 10 | 2 | 1 | 0 | 1.051 | −15.10 | 0.090 | 0.058 | 0 | 0 | 0 | 0 | 0 | 0 | 0 |
| 11 | 2 | 1 | 0 | 1.057 | −14.79 | 0.035 | 0.018 | 0 | 0 | 0 | 0 | 0 | 0 | 0 |
| 12 | 2 | 1 | 0 | 1.055 | −15.07 | 0.061 | 0.016 | 0 | 0 | 0 | 0 | 0 | 0 | 0 |
| 13 | 2 | 1 | 0 | 1.050 | −15.16 | 0.135 | 0.058 | 0 | 0 | 0 | 0 | 0 | 0 | 0 |
| 14 | 2 | 1 | 0 | 1.036 | −16.04 | 0.149 | 0.050 | 0 | 0 | 0 | 0 | 0 | 0 | 0 |

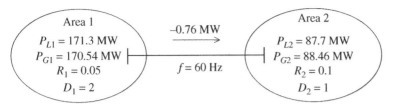

**Figure 13.9** The IEEE 14-bus system zonal model.

$$\frac{1}{R} = \frac{1}{R_1} + \frac{1}{R_2} = 50 \, \text{MW/Hz} \tag{13.7}$$

$$D_1 = 2 \times \frac{100 \, \text{MW}}{60 \, \text{Hz}} = 3.33 \, \text{MW/Hz} \tag{13.8}$$

$$D_2 = 1 \times \frac{100 \, \text{MW}}{60 \, \text{Hz}} = 1.67 \, \text{MW/Hz} \tag{13.9}$$

$$D = D_1 + D_2 = 5 \, \text{MW/Hz} \tag{13.10}$$

When the system in both areas loses 5% load, with the primary frequency control, the system steady-state frequency deviation is

$$\Delta f = \frac{-\Delta P_L}{\left(\dfrac{1}{R} + D\right)} = \frac{-(-P_L \times 5\%)}{50 + 5} = \frac{-(-259 \times 5\%)}{55} = 0.2355 (\text{Hz}) \tag{13.11}$$

The new system frequency, area load, and area generation are

$$f' = f + \Delta f = 60.2355 \, \text{Hz} \tag{13.12}$$

$$\begin{aligned} P'_{L1} &= P_{L1} \times 0.95 + \Delta P_{L1} = P_{L1} \times 0.95 + D_1 \Delta f = 162.735 + 3.33 \times 0.2355 \\ &= 163.52 (\text{MW}) \end{aligned} \tag{13.13}$$

$$\begin{aligned} P'_{L2} &= P_{L2} \times 0.95 + \Delta P_{L2} = P_{L2} \times 0.95 + D_2 \Delta f = 83.315 + 1.67 \times 0.2355 \\ &= 83.71 (\text{MW}) \end{aligned} \tag{13.14}$$

$$P'_{G1} = P_{G1} + \Delta P_{G1} = P_{G1} - \frac{1}{R_1} \Delta f = 170.54 - 33.33 \times 0.2355 = 162.69 (\text{MW}) \tag{13.15}$$

$$P'_{G2} = P_{G2} + \Delta P_{G2} = P_{G2} - \frac{1}{R_2} \Delta f = 88.46 - 16.67 \times 0.2355 = 84.54 (\text{MW}) \tag{13.16}$$

$$P'_{G1} - P'_{L1} = -\left(P'_{G2} - P'_{L2}\right) = -0.83 \text{MW} \tag{13.17}$$

The primary frequency response and system frequency after the load reduction are shown in Figure 13.10.

Assume the unit has the same droop ratio and the load has the same damping factor. With respect to the unit generation upper and lower limits, the generation output and load are updated and highlighted as bold in Table 13.2.

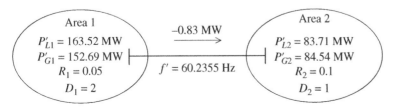

**Figure 13.10** Primary frequency response with 5% load reduction.

**Table 13.2** IEEE-14 bus system bus table.

| Bus ID | Area | Loss zone | Type | Voltage | Angle | Load_P | Load_Q | Unit_P | Unit_Q | Desired voltage | MAX Q | MIN Q | G | B |
|---|---|---|---|---|---|---|---|---|---|---|---|---|---|---|
| 1 | 1 | 1 | 3 | 1.060 | 0 | 0 | 0 | 1.05145 | −0.169 | 1.060 | 0 | 0 | 0 | 0 |
| 2 | 1 | 1 | 2 | 1.045 | −4.98 | 0.2081 | 0.127 | 0.20000 | 0.424 | 1.045 | 0.50 | −0.4 | 0 | 0 |
| 3 | 1 | 1 | 2 | 1.010 | −12.72 | 0.8969 | 0.190 | 0.37535 | 0.234 | 1.010 | 0.40 | 0 | 0 | 0 |
| 4 | 1 | 1 | 0 | 1.019 | −10.33 | 0.4561 | −0.039 | 0 | 0 | 0 | 0 | 0 | 0 | 0 |
| 5 | 1 | 1 | 0 | 1.020 | −8.78 | 0.0742 | 0.016 | 0 | 0 | 0 | 0 | 0 | 0 | 0 |
| 6 | 1 | 1 | 2 | 1.070 | −14.22 | 0.1070 | 0.075 | 0.48890 | 0.122 | 1.070 | 0.24 | −0.6 | 0 | 0 |
| 7 | 1 | 1 | 0 | 1.062 | −13.37 | 0 | 0 | 0 | 0 | 0 | 0 | 0 | 0 | 0 |
| 8 | 1 | 1 | 2 | 1.090 | −13.36 | 0 | 0 | 0.35660 | 0.174 | 1.090 | 0.24 | −0.6 | 0 | 0 |
| 9 | 1 | 1 | 0 | 1.056 | −14.94 | 0.2808 | 0.166 | 0 | 0 | 0 | 0 | 0 | 0 | 0.19 |
| 10 | 1 | 1 | 0 | 1.051 | −15.10 | 0.0861 | 0.058 | 0 | 0 | 0 | 0 | 0 | 0 | 0 |
| 11 | 1 | 1 | 0 | 1.057 | −14.79 | 0.0338 | 0.018 | 0 | 0 | 0 | 0 | 0 | 0 | 0 |
| 12 | 1 | 1 | 0 | 1.055 | −15.07 | 0.0585 | 0.016 | 0 | 0 | 0 | 0 | 0 | 0 | 0 |
| 13 | 1 | 1 | 0 | 1.050 | −15.16 | 0.1288 | 0.058 | 0 | 0 | 0 | 0 | 0 | 0 | 0 |
| 14 | 1 | 1 | 0 | 1.036 | −16.04 | 0.1421 | 0.050 | 0 | 0 | 0 | 0 | 0 | 0 | 0 |

With the updated generation power output and load, the branch power flow is updated as $P'_l = (74.14, 31.02, 35.36, 23.41, 14.56, -16.80, -37.87, -5.31, 4.18, 0.29, 11.44, 8.01, 19.03, -35.66, 30.35, 0.55, 5.90, -8.06, 2.16, 8.31)^T$ in MW with no violation.

When the system load in both areas increases by 5%, with the primary frequency control, the system steady-state frequency deviation is

$$\Delta f = \frac{-\Delta P_L}{\left(\dfrac{1}{R} + D\right)} = \frac{-(P_L \times 5\%)}{50 + 5} = \frac{-(259 \times 5\%)}{55} = -0.2355 (\text{Hz}) \tag{13.18}$$

The system's new frequency, area load, and area generation are

$$f'' = f - \Delta f = 59.7645 \text{ Hz} \tag{13.19}$$

$$\begin{aligned} P''_{L1} &= P_{L1} \times 1.05 + \Delta P_{L1} = P_{L1} \times 1.05 + D_1 \Delta f = 179.865 + 3.33 \times (-0.2355) \\ &= 179.08 (\text{MW}) \end{aligned} \tag{13.20}$$

$$\begin{aligned} P''_{L2} &= P_{L2} \times 1.05 + \Delta P_{L2} = P_{L2} \times 1.05 + D_2 \Delta f = 92.085 + 1.67 \times (-0.2355) \\ &= 91.69 (\text{MW}) \end{aligned} \tag{13.21}$$

$$P''_{G1} = P_{G1} + \Delta P_{G1} = P_{G1} - \frac{1}{R_1} \Delta f = 170.54 - 33.33 \times (-0.2355) = 178.39 (\text{MW}) \tag{13.22}$$

$$P''_{G2} = P_{G2} + \Delta P_{G2} = P_{G2} - \frac{1}{R_2} \Delta f = 88.46 - 16.67 \times (-0.2355) = 92.38 (\text{MW}) \tag{13.23}$$

$$P''_{G1} - P''_{L1} = -\left(P''_{G2} - P''_{L2}\right) = -0.69 \text{ MW} \tag{13.24}$$

The primary frequency response and system frequency after the load increase are shown in Figure 13.11.

Assume each unit has the same droop ratio and each load has the same damping factor. With respect to the unit generation upper and lower limits, the generation output and load are updated and highlighted as bold in Table 13.3.

With the updated generation power output and load, the branch power flow is updated as $P''_l = (78.51, 33.17, \mathbf{37.41}, 25.30, 15.83, -17.22, -40.58, -5.94, 4.61, 0.63, 12.28, 8.72, 20.74, -39.58, \mathbf{33.64}, 0.73, 6.60, -8.66, 2.37, 8.99)^T$ in MW. Branch power flows $P''_{l3}$ and $P''_{l15}$ violate their branch power flow limits.

### 13.1.3 Frequency Supplementary Control

Primary speed control will result in a steady-state frequency bias. Therefore, regulating system frequency to the nominal value requires a supplementary control that adds additional control set-point to move the valve or gate through a speed-changer motor till the steady-state frequency deviation is zero. The control block diagram is shown in Figure 13.12.

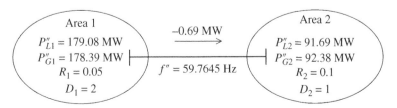

**Figure 13.11** Primary frequency response with 5% load increase.

**Table 13.3** IEEE-14 bus system bus table.

| Bus ID | Area | Loss zone | Type | Voltage | Angle | Load_P | Load_Q | Unit_P | Unit_Q | Desired voltage | MAX Q | MIN Q | G | B |
|---|---|---|---|---|---|---|---|---|---|---|---|---|---|---|
| 1 | 1 | 1 | 3 | 1.060 | 0 | 0 | 0 | 1.1169 | −0.169 | 1.060 | 0 | 0 | 0 | 0 |
| 2 | 1 | 1 | 2 | 1.045 | −4.98 | 0.2259 | 0.127 | 0.2262 | 0.424 | 1.045 | 0.50 | −0.4 | 0 | 0 |
| 3 | 1 | 1 | 2 | 1.010 | −12.72 | 0.9871 | 0.190 | 0.4408 | 0.234 | 1.010 | 0.40 | 0 | 0 | 0 |
| 4 | 1 | 1 | 0 | 1.019 | −10.33 | 0.4999 | −0.039 | 0 | 0 | 0 | 0 | 0 | 0 | 0 |
| 5 | 1 | 1 | 0 | 1.020 | −8.78 | 0.0778 | 0.016 | 0 | 0 | 0 | 0 | 0 | 0 | 0 |
| 6 | 1 | 1 | 2 | 1.070 | −14.22 | 0.1170 | 0.075 | 0.5281 | 0.122 | 1.070 | 0.24 | −0.6 | 0 | 0 |
| 7 | 1 | 1 | 0 | 1.062 | −13.37 | 0 | 0 | 0 | 0 | 0 | 0 | 0 | 0 | 0 |
| 8 | 1 | 1 | 2 | 1.090 | −13.36 | 0 | 0 | 0.3958 | 0.174 | 1.090 | 0.24 | −0.6 | 0 | 0 |
| 9 | 1 | 1 | 0 | 1.056 | −14.94 | 0.3092 | 0.166 | 0 | 0 | 0 | 0 | 0 | 0 | 0.19 |
| 10 | 1 | 1 | 0 | 1.051 | −15.10 | 0.0939 | 0.058 | 0 | 0 | 0 | 0 | 0 | 0 | 0 |
| 11 | 1 | 1 | 0 | 1.057 | −14.79 | 0.0362 | 0.018 | 0 | 0 | 0 | 0 | 0 | 0 | 0 |
| 12 | 1 | 1 | 0 | 1.055 | −15.07 | 0.0635 | 0.016 | 0 | 0 | 0 | 0 | 0 | 0 | 0 |
| 13 | 1 | 1 | 0 | 1.050 | −15.16 | 0.1412 | 0.058 | 0 | 0 | 0 | 0 | 0 | 0 | 0 |
| 14 | 1 | 1 | 0 | 1.036 | −16.04 | 0.1559 | 0.050 | 0 | 0 | 0 | 0 | 0 | 0 | 0 |

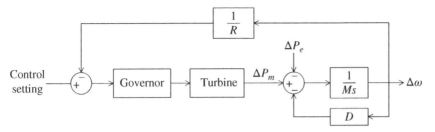

**Figure 13.12** Generating unit control block diagram with supplementary control.

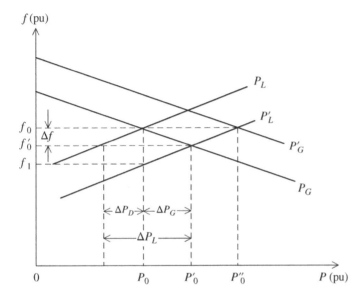

**Figure 13.13** Governor supplementary control.

The supplementary control function is illustrated by moving the governor droop function line from $P_G$ to $P'_G$ as shown in Figure 13.13.

The supplementary control by means of changing the control set-point to regulate the prime mover to increase the generator output from $P'_0$ to $P''_0$ making the system frequency back to the nominal frequency $f_0$. The set-point needs to be changed automatically to follow the load changes. This automatic function is called AGC.

### 13.1.4 Fundamentals of Automatic Generation Control

A practical power system has more than one interconnected control area. AGC is designed to be fully functional in interconnected power systems [8]. Taking an interconnected power system with two control areas as an example shown in Figure 13.14, the AGC function is illustrated.

In the interconnected system, the two areas are connected by a tie line with reactance. $X_t$. The electrical equivalent of the system is shown in Figure 13.15.

In the equivalent electrical system, each area is represented by an equivalent voltage source behind an equivalent reactance. The tie-line power flow from area 1 to area 2 is

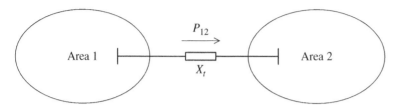

**Figure 13.14** Two-area interconnected system.

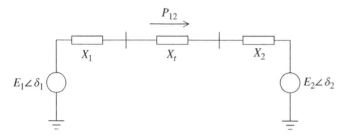

**Figure 13.15** Electrical equivalent system.

$$P_{12} = \frac{E_1 E_2}{X_T} sin(\delta_1 - \delta_2)$$ (13.25)

where $X_T = X_1 + X_t + X_2$. Linearizing the above equation at an initial point $E_1 \angle \delta_{10}$ and $E_2 \angle \delta_{20}$, we have

$$\Delta P_{12} = \frac{E_1 E_2}{X_T} cos(\delta_{10} - \delta_{20})(\Delta\delta_1 - \Delta\delta_2)$$ (13.26)

The block diagram representing the two-area system is shown in Figure 13.16.

Without the supplementary control, the steady-state frequency deviation $\Delta f$ under a generator electric power change of $\Delta P_e$ to balance a load change of $\Delta P_L$ in the system is calculated by the follows:

$$\Delta f = \Delta\omega_1 = \Delta\omega_2 = \frac{-\Delta P_L}{\left(\frac{1}{R_1} + D_1\right) + \left(\frac{1}{R_2} + D_2\right)} = \frac{-\Delta P_L}{\beta_1 + \beta_2}$$ (13.27)

where $\beta_1 = \left(\frac{1}{R_1} + D_1\right)$ and $\beta_2 = \left(\frac{1}{R_2} + D_2\right)$ are defined as the composite frequency response characteristic of the two areas.

Besides the objective of balancing active power to maintain the system frequency, AGC regulates tie-line power flow at scheduled value as well. A control signal made of the tie line flow deviation $\Delta P_{12}$ added to the frequency deviation weighted by the frequency response characteristic $\beta_1$ or $\beta_2$ to accomplish the tie-line flow control objective. The control signal is known as area control error (ACE) and is defined as follows:

$$ACE_1 = \Delta P_{12} + \left(\frac{1}{R_1} + D_1\right)\Delta f = \Delta P_{12} + \beta_1 \Delta f = \Delta P_{12} + B_1 \Delta f$$ (13.28)

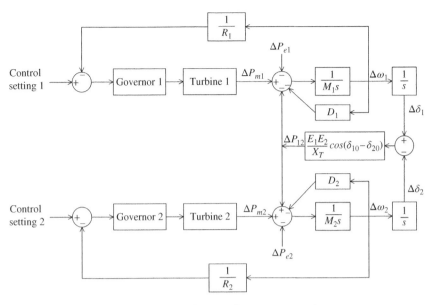

**Figure 13.16** Two-area interconnected system frequency control.

$$ACE_2 = -\Delta P_{12} + \left(\frac{1}{R_2} + D_2\right)\Delta f = -\Delta P_{12} + \beta_2\Delta f = -\Delta P_{12} + B_2\Delta f \qquad (13.29)$$

where $B_1 = \beta_1 = \left(\dfrac{1}{R_1} + D_1\right)$ and $B_2 = \beta_2 = \left(\dfrac{1}{R_2} + D_2\right)$ are the frequency deviation bias factors.

And ACE represents the required changes in each area generation to correct the tie-line power deviation and the frequency deviation. The control block diagram shown in Figure 13.17 illustrates the AGC function for a two-area interconnected system. The concept can be extended for systems with more than two areas.

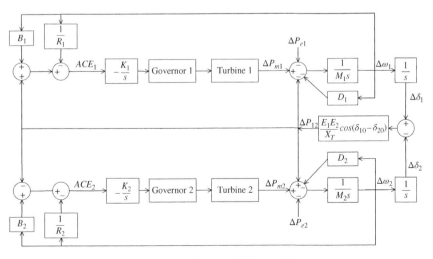

**Figure 13.17** Two-area interconnected system AGC control.

## 13.2 Network Security-Constrained Automatic Generation Control

The classic AGC base point is determined by economic dispatch. And AGC regulates the system frequency deviation to be zero and controls the flowgate power flow to the scheduled value.

Economic dispatch optimizes generation under network security constraints. There are two commonly used methods for active economic dispatch in power systems: (i) Offline Economic Dispatch [9] and (ii) Online Economic Dispatch [10].

Offline economic dispatch calculates unit active power output for the next day or the next few days in a time interval of hours based on the generation capacity, grid network constraints, as well as the forecasted load.

Since offline economic dispatch is based on load forecasting, the generation dispatch may not accurately meet the actual load. The operating conditions of the power system are changing, and the active power output of the generators may deviate from the scheduled power generation set-point. Therefore, online economic dispatch adjusts the generation output which is set by offline economic dispatch continuously to satisfy the power system's actual operating point in a short time interval, typically 5–15 minutes.

AGC provides automatic control based on online economic dispatch. AGC is used to control tie-line power flow as well. Large-scale power systems usually is divided into areas. The areas are connected by tie lines. Each inter-area tie-line power flow is controlled to be a desired value within its transmission capacity limit to ensure operation security. AGC is functionally designed to control the inter-area tie-line power flows. In modern power system, advanced high-voltage direct current (HVDC) [11, 12], energy storage system [13, 14], and demand side management [15, 16] also participate in tie-line power flow control and frequency regulation in the same fashion of synchronous machine AGC function.

For a large-scale power system, economic dispatch optimization with network security constraints takes minutes. Due to the short cycle of AGC (typically five seconds) conflicting with the extensive computation efforts of the security-constrained optimization, network security constraints are not modeled in AGC in real-time. When power system operation changes significantly in the economic dispatch cycle, AGC commands without network security constraints cannot guarantee that the power system is secure. The present state of the art assumes the power system operation point does not significantly change between two economic dispatch optimization executions. Under the assumption, the AGC base point determined by economic dispatch will not lead the system to violate network security constraints. However, this assumption is not always true in power system operations.

In the case of intermittent renewable energy highly penetrated in power systems and fast-response power electronics-based generation integrated into transmission and distribution power network, the system power flow has a great chance of shifting away from the base point which is optimized in the cycle of economic dispatch. Classic AGC does not optimize power flow within network security constraints. The power flow may result in network constraint violations with the AGC command. Thus, the AGC without network security constraints presents risks in power system operation.

As addressed in this Chapter, AGC regulates generation unit active power to control inter-area flowgate power flow at the desired value. When one time period is applied to illustrate, the network security-constrained AGC base point is the solution to the following optimization problem:

$$\min F = \sum_{i=1}^{n} f_i(P_{Ti}) \tag{13.30}$$

Subject to

$$P_{Ti} + P_{Hi} - P_{Di} + \sum_{j \in i} P_{l(i-j)} = 0, i = 1, 2, ..., n_b \tag{13.31}$$

$$P_{FG} = P_{FG}^{(s)} \tag{13.32}$$

$$P_{Ti}^{\min} \le P_{Ti} \le P_{Ti}^{\max}, \quad i = 1, ..., n \tag{13.33}$$

$$P_{Hi}^{\min} \le P_{Hi} \le P_{Hi}^{\max}, \quad i = 1, ..., m \tag{13.34}$$

$$W_i(P_{Hi})\Delta t = C_i, i = 1, 2..., m \tag{13.35}$$

$$\left| P_{l(i-j)} \right| = \left| V_i V_j \left( G_{ij} \cos \theta_{ij} + B_{ij} \sin \theta_{ij} \right) \right| \le P_{l(i-j)}^{\max}, l = 1, ..., n_l \tag{13.36}$$

In the optimization model, the representation of the symbols is listed as follows:

$f_i$ – power generation cost function of the $i$-th unit;
$P_{Ti}$ – active power output of the $i$-th thermal unit;
$P_{Hi}$ – active power output of the $i$-th hydro unit;
$n$ – number of the thermal units;
$P_{Di}$ – active power of load at bus $i$;
$P_{l(i-j)}$ – active power of branch $l$ from bus $i$ to bus $j$;
$n_b$ – number of the buses;
$P_{FG}$ – active power of flowgate;
$P_{FG}^{(s)}$ – active power schedule of flowgate;
$m$ – number of the hydro units;
$P_{Ti}^{\min}$ – active power lower limit of the $i$-th thermal unit;
$P_{Ti}^{\max}$ – active power upper limit of the $i$-th thermal unit;
$P_{Hi}^{\min}$ – active power lower limit of the $i$-th hydro unit;
$P_{Hi}^{\max}$ – active power upper limit of the $i$-th hydro unit;
$W_i(P_{Hi})$ – water consumption function of the $i$-th hydro unit;
$C_i$ – water consumption budget;
$P_{l(i-j)}^{\max}$ – active power limit of branch $l$;
$n_l$ – number of the branches;

Flowgate flow constraint is modeled by (13.32). The security-constrained AGC model is nonlinear. To solve the security-constrained generation control model by linear programming, the linearized model is developed based on the DC power flow as follows, with the same assumption that the fuel cost curve and the water consumption curve are quadratic, as in Chapter 12

$$\min F = \sum_{i=1}^{n} \left( 2a_{Ti} P_{Ti}^0 \Delta P_{Ti} + b_{Ti} \Delta P_{Ti} \right) \tag{13.37}$$

Subject to:

$$\sum_{i=1}^{n} \Delta P_{Ti} + \sum_{i=1}^{m} \Delta P_{Hi} = 0 \tag{13.38}$$

$$H_{FG} \Delta P_G = P_{FG}^{(s)} - P_{FG}^0 \tag{13.39}$$

$$P_{Ti}^{\min} - P_{Ti}^0 \le \Delta P_{Ti} \le P_{Ti}^{\max} - P_{Ti}^0, \quad i = 1, ..., n \tag{13.40}$$

$$P_{Hi}^{\min} - P_{Hi}^0 \le \Delta P_{Hi} \le P_{Hi}^{\max} - P_{Hi}^0, \quad i = 1, ..., m \tag{13.41}$$

$$(2a_{Hi}P_{Hi}^0 + b_{Hi})\Delta P_{Hi} = \frac{C_i}{\Delta t} - a_{Hi}\left(P_{Hi}^0\right)^2 - b_{Hi}P_{Hi}^0 - c_{Hi}, i = 1, 2..., m \tag{13.42}$$

$$-\Delta P_l^{max} - P_l^0 \le H\Delta P_G \le \Delta P_l^{max} - P_l^0 \tag{13.43}$$

Where $H \in \mathbb{R}^{n_l \times (n + m)}$ is the generation shift factor matrix and $H_{FG}$ is the submatrix of $H$ for flowgate branches.

To build the linear program in standard form $\min_x c^T x, Ax = b, x \ge 0$, the objectives, variables, and constraints of the linearized security-constrained AGC are identified by the following process.

$$\min F = \sum_{i=1}^n \left(2a_{Ti}P_{Ti}^0\Delta P_{Ti} + b_{Ti}\Delta P_{Ti}\right)$$

Subject to the following constraints:

$$\sum_{i=1}^n \Delta P_{Ti} + \sum_{i=1}^m \Delta P_{Hi} = 0$$

$$H_{FG}\Delta P_G = P_{FG}^{(s)} - P_{FG}$$

$$\Delta P_{Ti} + x_i = P_{Ti}^{max} - P_{Ti}^0, \quad i = 1, ..., n$$

$$\Delta P_{Hi} + x_{n+i} = P_{Hi}^{max} - P_{Hi}^0, i = 1, ..., m$$

$$\Delta P_{Ti} - x_{i+n+m} = P_{Ti}^{min} - P_{Ti}^0, \quad i = 1, ..., n$$

$$\Delta P_{Hi} - x_{i+2n+m} = P_{Hi}^{min} - P_{Hi}^0, \quad i = 1, ..., m$$

$$(2a_{Hi}P_{Hi}^0 + b_{Hi})\Delta P_{Hi} = \frac{C_i}{\Delta t} - a_{Hi}\left(P_{Hi}^0\right)^2 - b_{Hi}P_{Hi}^0 - c_{Hi}, \quad i = 1, ..., m$$

$$H\Delta P_G + X_{2n+2m+1 - 2n+2m+n_l} = \Delta P_l^{max} - P_l^0$$

$$H\Delta P_G - X_{2n+2m+n_l+1 - 2n+2m+2n_l} = -\Delta P_l^{max} - P_l^0$$

where $x_1 - x_{2n+2m+2n_l}$ are nonnegative slack variables added to convert the inequality constraints to equality constraints.

In the linear program, $\Delta P_T$ and $\Delta P_H$ are not nonnegative variables. They are free to take on either positive, negative values, or zero. Two nonnegative variables $u = (u_T, u_H) = (u_{T1}, u_{T2}, u_{T3}, u_{T4}, u_H)$ and $v = (v_T, v_H) = (v_{T1}, v_{T2}, v_{T3}, v_{T4}, v_H)$ are introduced to represent $\Delta P_T$ and $\Delta P_H$ as $\Delta P_T = u_T - v_T$ and $\Delta P_H = u_H - v_H$.

When branch power flow violation occurs in the base case, the basic solution of the economic dispatch problem as the initial solution makes the basic variable $x_k$ referring to the slack variable which corresponds to violated branch flow constraints infeasible in the simplex method. To resolve the issue, as addressed in 6.2.1.1, two nonnegative variables $y_k \ge 0$ and $z_k \ge 0$ are introduced to represent the corresponding basic variables $x_k$ as $x_k = y_k - z_k$ to ensure the basic solution is feasible.

The IEEE-14 bus system is taken as an example to study the security-constrained AGC. When the system loses 5% load, to make an initial feasible solution, assume initially the slack bus balances the load reduction. The initial generation output is $P_G^{(0)} = (96.12, 20, 41.46, 50.85, 37.62)^T$ in MW, and the branch power flow is $P_l^{(0)} = (67.86, 28.26, 32.55, 21.54, 13.15, -15.48, -35.88, -7.04, 3.57, -1.69, 11.52, 7.98, 19.02, -37.62, 30.58, 0.36, 5.77, -8.19, 2.18, 8.38)^T$ in MW. The flowgate power flow is $-5.16$ MW which is away from the scheduled flowgate power flow of $-0.76$ MW.

In the first iteration, the linearized security-constrained AGC problem optimizes the incremental generation of active power at $\Delta P_G^{(1)} = (9.8903, 0, -5.4953, -4.3950, 0)^T$ in MW. The generation active power outputs are updated as $P_G^{(1)} = (106.0103, 20, 35.9647, 46.455, 37.62)^T$. The branch power flows are updated as $P_l^{(1)} = (74.73, 31.28, 36.00, 23.42, 14.69, -17.53, -37.40, -6.19, 4.07, 1.36, 10.71, 7.86, 18.61, -37.62, 31.43, 1.17, 6.31, -7.38, 2.07, 7.85)^T$ in MW with no violation. The flowgate power flow is controlled to its scheduled value at $-0.76$ MW.

In the second iteration, the linearized security-constrained AGC problem is solved at $\Delta P_G^{(2)} = (0, 0, 0, 0, 0, 0)^T$ in MW. The iteration is converged, as summarized in Table 13.4.

When the system load increases by 5%, to make an initial feasible solution, assume the slack bus balances the load increase. The initial generation output is $P_G^{(0)} = (122.02, 20, 41.46, 50.85, 37.62)^T$ in MW, and the branch power flow is $P_l^{(0)} = (86.00, 36.02, \mathbf{39.45}, 26.83, 16.94, -18.00, -42.38, -4.21, 5.22, 2.61, 12.20, 8.74, 20.75, -37.62, \mathbf{33.41}, 0.92, 6.73, -8.53, 2.34, 8.91)^T$ in MW with two branch flow violations ($P_{l3}^0$ and $P_{l15}^0$). The flowgate power flow is 3.62 MW which is away from the scheduled flowgate power flow at $-0.76$ MW.

In the first iteration, the linearized security-constrained AGC problem optimizes the incremental generation active power at $\Delta P_G^{(1)} = (-9.8764, 0, 5.5014, 4.3750, 0)^T$ in MW. The generation active power outputs are updated as $P_G^{(1)} = (112.1436, 20, 46.9614, 55.225, 37.62)^T$. The branch power flow is updated as $P_l^{(1)} = (79.14, 33.00, 36.00, 24.95, 15.40, -15.95, -40.86, -5.06, 4.73, -0.43, 13.01, 8.86, 21.17, -37.62, 32.56, 0.12, 6.20, -9.33, 2.46, 9.45)^T$ in MW with no violation. The flowgate power flow is controlled to its scheduled value at $-0.76$ MW.

In the second iteration, the linearized security-constrained AGC problem is solved at $\Delta P_G^{(2)} = (0, 0, 0, 0, 0, 0)^T$ in MW. The iteration is converged as summarized in Table 13.5.

## 13.3 Security-Constrained AGC Graph Computing

The linearized optimization problem (13.37)–(13.43) is standardized as

$$\min_x c^T x, \quad Ax = b, \quad x \geq 0 \tag{13.44}$$

Where $A$ is the extended linearized constraint coefficient matrix. The linear program can be solved by the graph-based simplex method addressed in 12.2. By using the graph-based simplex method, the high-level pseudo-code to solve security-constrained AGC in parallel is shown as follows:

---

**Algorithm 13.1   Graph-Based Security-Constrained AGC**

---

```
1: Set iteration number k = 0
2: Initialize (P_G^k, P_l^k) by feasible power flow solution
3: Linearize objective function and constraints at the power flow
   solution
4: Calculate generation shift factor matrix
5: While max(|ΔP_G|) > ε
6:    Solve linearized optimization problem (13.44) by simplex
7:    Update P_G^{k+1} = P_G^k + ΔP_G
8:    Solve branch power flow P_l^{k+1} with the generation of P_G^{k+1}
9:    k = k + 1
```

---

**Table 13.4** Security-constrained automatic generation control iteration.

| Iter | $\Delta P_{G1}$ (MW) | $\Delta P_{G2}$ (MW) | $\Delta P_{G3}$ (MW) | $\Delta P_{G4}$ (MW) | $\Delta P_{G5}$ (MW) | $P_{G1}$ (MW) | $P_{G2}$ (MW) | $P_{G3}$ (MW) | $P_{G4}$ (MW) | $P_{G5}$ (MW) | Total cost ($) | Violations | $\Delta P_{FG}$ (MW) |
|---|---|---|---|---|---|---|---|---|---|---|---|---|---|
| 0 | 0.00 | 0.00 | 0.00 | 0.00 | 0.00 | 96.12 | 20.00 | 41.46 | 50.85 | 37.62 | 2751.82 | None | 4.4 |
| 1 | 9.89 | 0.00 | −5.50 | −4.40 | 0.00 | 106.01 | 20.00 | 35.96 | 46.46 | 37.62 | 2732.31 | None | 0 |
| 2 | 0.00 | 0.00 | 0.00 | 0.00 | 0.00 | 106.01 | 20.00 | 35.96 | 46.46 | 37.62 | 2732.31 | None | 0 |

**Table 13.5** Security-constrained automatic generation control iteration.

| Iter | $\Delta P_{G1}$ (MW) | $\Delta P_{G2}$ (MW) | $\Delta P_{G3}$ (MW) | $\Delta P_{G4}$ (MW) | $\Delta P_{G5}$ (MW) | $P_{G1}$ (MW) | $P_{G2}$ (MW) | $P_{G3}$ (MW) | $P_{G4}$ (MW) | $P_{G5}$ (MW) | Total cost ($) | Violations | $\Delta P_{FG}$ (MW) |
|---|---|---|---|---|---|---|---|---|---|---|---|---|---|
| 0 | 0.00 | 0.00 | 0.00 | 0.00 | 0.00 | 122.02 | 20.00 | 41.46 | 50.85 | 37.62 | 2972.66 | L3,L15 | −4.38 |
| 1 | −9.88 | 0.00 | 5.50 | 4.38 | 0.00 | 112.14 | 20.00 | 46.96 | 55.23 | 37.62 | 2990.78 | None | 0 |
| 2 | 0.00 | 0.00 | 0.00 | 0.00 | 0.00 | 112.14 | 20.00 | 46.96 | 55.23 | 37.62 | 2990.78 | None | 0 |

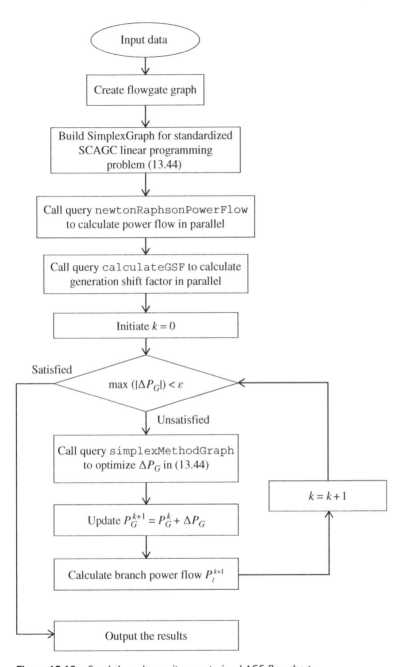

**Figure 13.18** Graph-based security-constrained AGC flowchart.

The flowchart to illustrate the graph-based security-constrained AGC with more granular is shown in Figure 13.18.

Except for the steps to create a flowgate graph and build a simplex graph for standardized Security Constrained Automatic Generation Control (SCAGC) linear programming problem (13.44), queries `newtonRaphsonPowerFlow`, `calculateGSF`, and `simplexMethodGraph` are created in 10.3.7, 10.3.6, and 12.2.3 respectively. The pseudo-code to build flowgate graph and simplex graph for standardized SCAGC linear programming problem is provided in Appendix.

## References

1 Wood, A.J., Wollenberg, B.F., Sheblé, G.B., and Generation, P. (2014). *Operation, and Control*, 3e. Hoboken: Wiley.

2 Jaleeli, N., VanSlyck, L.S., Ewart, D.N. et al. (1992). Understanding automatic generation control. *IEEE Transactions on Power Systems* 7 (3): 1106–1122.

3 Mills, R.J. and B'Rells, W.F. (1973). Automatic generation control part I-process modeling. *IEEE Transactions on Power Apparatus and Systems* PAS-92 (2): 710–715.

4 Taylor, C.W. and Cresap, R.L. (1976). Real-time power system simulation for automatic generation control. *IEEE Transactions on Power Apparatus and Systems* 95 (1): 375–384.

5 Hagihara, S., Yokota, H., Goda, K., and Isobe, K. (1979). Stability of a hydraulic turbine generating unit controlled by P.I.D. governor. *IEEE Transactions on Power Apparatus and Systems* PAS-98 (6): 2294–2298.

6 IEEE (2011). IEEE guide for the application of turbine governing systems for hydroelectric generating units - redline. *IEEE Std 1207–2011 (Revision to IEEE Std 1207–2004) - Redline*, pp. 1–139, 20 June 2011.

7 Kundur, P.S. and Malik, O.P. (2022). *Power System Stability and Control*, 2e. McGraw Hill.

8 Kennedy, T., Hoyt, S.M., and Abell, C.F. (1988). Variable, nonlinear tie-line frequency bias for interconnected systems control. *IEEE Transactions on Power Systems* 3 (3): 1244–1253.

9 Grainger, J.J., William, J., and Stevenson, D. (1994). *Power System Analysis, Ser. Electrical and Computer Engineering* (ed. A.B. Akay and E. Castellano). McGraw Hill International Editions ISBN: 0-07- 061493-5.

10 Demartini, G., Granelli, G.P., Marannino, P. et al. (1996). Co-ordinated economic and advance dispatch procedures. *IEEE Transactions on Power Systems* 11 (4): 1785–1791.

11 Pathak, N., Verma, A., Bhatti, T.S., and Nasiruddin, I. (2019). Modeling of HVDC tie links and their utilization in AGC/LFC operations of multiarea power systems. *IEEE Transactions on Industrial Electronics* 66 (3): 2185–2197.

12 McNamara, P. and Milano, F. (2018). Model predictive control-based AGC for multi-terminal HVDC-connected AC GRIDS. *IEEE Transactions on Power Systems* 33 (1): 1036–1048.

13 Donadee, J. and Wang, J. (2014). AGC signal modeling for energy storage operations. *IEEE Transactions on Power Systems* 29 (5): 2567–2568.

14 Tripathy, S.C., Balasubramanian, R., and Chandramohanan Nair, P.S. (1992). Adaptive automatic generation control with superconducting magnetic energy storage in power systems. *IEEE Transactions on Energy Conversion* 7 (3): 434–441.

15 Shiltz, D.J., Baros, S., Cvetković, M., and Annaswamy, A.M. (2019). Integration of automatic generation control and demand response via a dynamic regulation market mechanism. *IEEE Transactions on Control Systems Technology* 27 (2): 631–646.

16 Pourmousavi, S.A. and Nehrir, M.H. (2012). Real-time central demand response for primary frequency regulation in microgrids. *IEEE Transactions on Smart Grid* 3 (4): 1988–1996.

# 14

# Small-Signal Stability

In practical operations, power systems are subject to large disturbances, such as short circuit faults, switching of circuit elements, and small disturbances like load changes. The system must be able to keep its synchronism under large disturbances and suppress oscillations caused by small disturbances. The ability of a power system to maintain its synchronism when subject to a small disturbance is small-signal stability [1, 2]. Frequency-domain (eigenvalue) analysis is the most common analytical method to study power system small-signal stability [3–5]. And novel methods have been developed over the past decade to analyze large-scale power system's small-signal stability [6, 7].

With the high penetration of power electronics interfaced with renewable energy resources [8–13], high voltage direct current transmission systems [14], and flexible AC transmission systems (FACT) [15], the concerns of small-signal stability have encouraged research and study to address the potential issues with reasonable assumptions and effective approaches [16–19].

## 14.1 Small-Signal Stability of a Dynamic System

The power system is a multi-dimensional high-order nonlinear dynamic system. The behavior of a dynamic system is described by a set of $n$-order nonlinear ordinary differential equations and $m$-dimensional algebraic equations:

$$\dot{X} = f(X, U) \quad X \in \mathbb{R}^n \tag{14.1}$$

$$Y = g(X, U) \quad Y \in \mathbb{R}^m \tag{14.2}$$

The equilibrium points are these points where all the derivatives are simultaneously zero. The equilibrium or singular point must satisfy the equation

$$0 = f(X_0) \tag{14.3}$$

where $X_0$ is the state vector $X$ at the equilibrium point. For a nonlinear system, there may be more than one equilibrium point. Each equilibrium point may be in the category of local, finite, or global stability. Most equilibrium points in a practical large-scale power system are locally stable or finite stable. Investigating linearized nonlinear systems with zero input truly uncovers the characteristic of the behavior of the dynamic system's local stability and finite stability.

## 14.2  System Linearization

Let $X_0$ be the initial state vector and $U_0$ the input vector corresponding to the equilibrium point at which the local stability performance is investigated

$$\dot{X}_0 = f(X_0, U_0) \tag{14.4}$$

$$\Delta \dot{x}_i = \sum_{i=1}^{i=n} \frac{\partial f_i}{\partial x_i} \Delta x_i + \sum_{j=1}^{j=m} \frac{\partial f_i}{\partial u_j} \Delta u_j \tag{14.5}$$

$$\Delta y_k = \sum_{i=1}^{i=n} \frac{\partial g_k}{\partial x_i} \Delta x_i + \sum_{j=1}^{j=m} \frac{\partial g_k}{\partial u_j} \Delta u_j \tag{14.6}$$

where $i = 1, 2, ..., n$ and $k = 1, 2, ..., m$, therefore, the linearized forms of Eqs. (14.5) and (14.6) are

$$\Delta \dot{x} = A\Delta x + B\Delta u \tag{14.7}$$

$$\Delta y = C\Delta x + D\Delta u \tag{14.8}$$

By taking the Laplace transform of the above equations, the state equations in the frequency domain are:

$$s\Delta x(s) - \Delta x(0) = A\Delta x(s) + B\Delta u(s) \tag{14.9}$$

$$\Delta y(s) = C\Delta x(s) + D\Delta u(s) \tag{14.10}$$

The poles of $\Delta x$ (s) and $\Delta y$ (s) are the roots of the equation

$$det(sI - A) = 0 \tag{14.11}$$

The time-dependent characteristic of a mode corresponding to an eigenvalue $\lambda_i$ of matrix $A$ is $e^{\lambda_i t}$. Therefore, the stability of the system is determined by the sign of the real component of the eigenvalue. The real component of the eigenvalues gives the damping. The negative real part represents a damped oscillation, whereas the positive real part represents an oscillation of increasing amplitude.

If eigenvalues are complex conjugates as $\lambda = -\zeta\omega_n \pm j\omega_n\sqrt{1-\zeta^2} = \sigma \pm j\omega$, the damped frequency is $\omega$ in rad/s and $\zeta = \dfrac{|\sigma|}{\sqrt{\sigma^2 + \omega^2}}$ is the damping ratio [1].

## 14.3 Small-Signal Stability Mode

In large power systems, small-signal stability problems may be either local or global in nature.

Local oscillations involve a small part of the overall system. In this mode, a single generator or a single power plant usually oscillates against the rest of the power system. Such oscillations are called local mode oscillations. The stability problems associated with this oscillation are similar to those of a single-machine infinite bus system. The most common small-signal stability problems fall into this category.

Local mode oscillation may also be associated with oscillations between the rotors of a few generators close to each other. Such oscillations are called inter-machine or inter-plant mode oscillations.

Global small-signal stability problems are usually excited by interactions among large groups of generators. They involve oscillations of a group of generators in one area swinging against a group of generators in another area. Such oscillations are called inter-area mode oscillations.

## 14.4 Single-Machine Infinite Bus System

Single-machine infinite bus system is a starting point used to study the local problems of the small-signal performance of a system. The equivalent circuit of a single-machine infinite bus system is shown in Figure 14.1. In the single-machine infinite bus system, to focus on a single generator or a single plant oscillation against the rest of the power system, the rest of the power system is reduced to an infinite bus and the adjacent transmission using Thevenin's equivalent where the infinite bus voltage $U \angle 0°$ remains constant and the voltage angle is set to be 0 as a reference.

The synchronous generator can be represented by models of varying degrees of detail. The synchronous generator models in different level details are discussed in Appendix.

### 14.4.1 Classical Generator Model

Begin with the classical model, the generator is represented by $E' \angle \delta$ behind $x'_d$, as shown in Figure 14.2.

**Figure 14.1** Single-machine infinite bus system equivalent circuit.

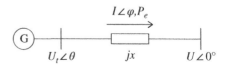

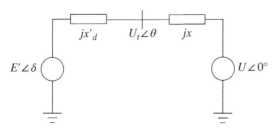

**Figure 14.2** Single-machine infinite bus system with classical generator model.

In the classical model, the magnitude of $E'$ is assumed to be constant

$$I = \frac{E'(cos\delta + jsin\delta) - U}{j(x + x'_d)} \tag{14.12}$$

$$P_e = \frac{E'Usin\delta}{x + x'_d} \tag{14.13}$$

$$2H\frac{d\omega}{dt} = P_m - P_e - D(\omega - 1) \tag{14.14}$$

$$\frac{d\delta}{dt} = (\omega - 1) \tag{14.15}$$

Linearizing the Eqs. (14.13)–(14.15) at an initial operating condition by $\delta = \delta_0$, the small-signal stability model with the classical generator system is

$$\Delta P_e = \frac{E'Ucos\delta_0}{x + x'_d}\Delta\delta \tag{14.16}$$

$$s\Delta\omega = \frac{1}{2H}(-\Delta P_e - D\Delta\omega) = \frac{1}{2H}\left(-\frac{E'Ucos\delta_0}{x + x'_d}\Delta\delta - D\Delta\omega\right) \tag{14.17}$$

$$s\Delta\delta = \omega_0\Delta\omega \tag{14.18}$$

Writing Eqs. (14.17) and (14.18) in the matrix form, we have

$$\begin{bmatrix}\Delta\dot{\omega}\\\Delta\dot{\delta}\end{bmatrix} = \begin{bmatrix}-\dfrac{D}{2H} & -\dfrac{\frac{E'U\cos\delta_0}{x + x'_d}}{2H}\\\omega_0 & 0\end{bmatrix}\begin{bmatrix}\Delta\omega\\\Delta\delta\end{bmatrix} = A\begin{bmatrix}\Delta\omega\\\Delta\delta\end{bmatrix} \tag{14.19}$$

To illustrate the small-signal stability of a single-machine infinite bus system with a classical generator model, assume the system parameters are listed as follows (Table 14.1).

**Table 14.1** System parameters.

| U | P | Q | x | $x_d'$ | H(s) | D |
|-----|-----|-----|-----|-------|-------|---|
| 1.0 | 0.8 | 0.2 | 0.2 | 0.256 | 4.130 | 5 |

$P + jQ$ is the load at the infinite bus. Then the generator stator current is given by

$$I = \frac{(P + jQ)^*}{U^*} = \frac{0.8 - j0.2}{1.0} = 0.8 - j0.2 \tag{14.20}$$

The voltage of the generator Thevenin's equivalent behind the equivalent transient reactance is

$$E' = U + j(x + x_d')I = 1.0 + j(0.2 + 0.256)(0.8 - j0.2) = 1.0912 + j\,0.3648$$
$$= 1.1505\angle 18.48° \tag{14.21}$$

$E'$ is leading $U$ by $\delta_0 = 18.48°$. Then the system linearized Eq. (14.19) is

$$\begin{bmatrix} \Delta\dot{\omega} \\ \Delta\dot{\delta} \end{bmatrix} = \begin{bmatrix} -\dfrac{D}{2H} & -\dfrac{E'U\cos\delta_0}{x + x_d'}\dfrac{1}{2H} \\ \omega_0 & 0 \end{bmatrix} \begin{bmatrix} \Delta\omega \\ \Delta\delta \end{bmatrix} = A \begin{bmatrix} \Delta\omega \\ \Delta\delta \end{bmatrix} = \begin{bmatrix} -0.6053 & -0.2897 \\ 377 & 0 \end{bmatrix} \begin{bmatrix} \Delta\omega \\ \Delta\delta \end{bmatrix}$$
$$\tag{14.22}$$

The eigenvalues of the system state matrix $A$ are solved by

$$\begin{bmatrix} -0.6053 - \lambda & -0.2897 \\ 377 & -\lambda \end{bmatrix} = 0 \tag{14.23}$$

The eigenvalues are $\lambda_1, \lambda_2 = -0.3026 \pm j10.4463$. The damped frequency is 10.4463 rad/s = 1.6626 Hz. The damping ratio is $\dfrac{0.3026}{\sqrt{(-0.3026)^2 + 10.4463^2}} = 0.0290$. The system is stable.

## 14.4.2 Third-Order Generator Model

In the classical model, the magnitude of $E'$ remains constant during disturbance is a strong assumption that ignores the excitation system dynamics. In the third-order generator model, the exciter field voltage dynamic is modeled. And the generator dynamic equations are given by

$$T_{d0}' \frac{dE_q'}{dt} = E_f - (x_d - x_d')I_d - E_q' \tag{14.24}$$

(a)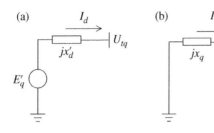

(b)

**Figure 14.3** Generator equivalent circuits in *d–q* reference frame. (a) *q*-axis; (b) *d*-axis.

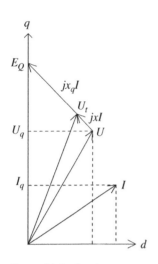

**Figure 14.4** Synchronous machine third-order model vector diagram.

$$2H\frac{d\omega}{dt} = P_m - P_e - D(\omega - 1) \tag{14.25}$$

$$\frac{d\delta}{dt} = (\omega - 1) \tag{14.26}$$

The generator equivalent circuits are shown in Figure 14.3. The generator terminal voltage equations in the *d–q* reference frame are given by

$$U_{td} = jx_q I_q \tag{14.27}$$

$$U_{tq} = E_q' - x_d' I_d \tag{14.28}$$

The voltage equations can be represented by the synchronous machine vector diagram in Figure 14.4.

The equivalent circuit representing a single-machine infinite bus system with a third-order generator model is shown in Figure 14.5.

When the components in the *d–q* reference frame interface to the network in the *x–y* reference frame, the network equations are given by

$$U_d = U_{td} + xI_q = \left(x + x_q\right)I_q \tag{14.29}$$

$$U_q = U_{tq} + xI_d = E_q' - \left(x + x_d'\right)I_d \tag{14.30}$$

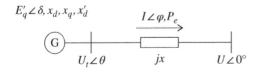

**Figure 14.5** Single-machine infinite bus system with third-order generator model.

$$
\begin{bmatrix} I_d \\ I_q \end{bmatrix} = \begin{bmatrix} 0 & -(x+x_q) \\ (x+x_d') & 0 \end{bmatrix}^{-1} \begin{bmatrix} -U\sin\delta \\ E_q' - U\cos\delta \end{bmatrix}
$$

$$
= \begin{bmatrix} 0 & \dfrac{1}{(x+x_d')} \\ \dfrac{-1}{(x+x_q)} & 0 \end{bmatrix} \begin{bmatrix} -U\sin\delta \\ E_q' - U\cos\delta \end{bmatrix} \tag{14.31}
$$

$$
\begin{bmatrix} U_{td} \\ U_{tq} \end{bmatrix} = \frac{1}{(x+x_q)(x+x_d')} \left\{ \begin{bmatrix} -x_q(x+x_d') & 0 \\ 0 & -x_d'(x+x_q) \end{bmatrix} \begin{bmatrix} -U\sin\delta \\ -U\cos\delta \end{bmatrix} + \begin{bmatrix} 0 \\ x(x+x_q) \end{bmatrix} E_q' \right\} \tag{14.32}
$$

Linearizing the Eqs. (14.27)–(14.32), the small-signal stability model of the classical generator system is

$$
\begin{bmatrix} \Delta I_d \\ \Delta I_q \end{bmatrix} = \begin{bmatrix} \dfrac{U}{x+x_d'}\sin\delta_0 \\ \dfrac{U}{x+x_q}\cos\delta_0 \end{bmatrix} \Delta\delta + \begin{bmatrix} \dfrac{1}{x+x_d'} \\ 0 \end{bmatrix} \Delta E_q' \tag{14.33}
$$

$$
\begin{bmatrix} \Delta U_d \\ \Delta U_q \end{bmatrix} = \begin{bmatrix} \dfrac{x_q}{x+x_q}U\cos\delta_0 \\ \dfrac{x_d'}{x+x_d'}U\sin\delta_0 \end{bmatrix} \Delta\delta + \begin{bmatrix} 0 \\ \dfrac{x}{x+x_d'} \end{bmatrix} \Delta E_q' \tag{14.34}
$$

$$
T_{d0}'s\Delta E_q' = \Delta E_{fd} - \Delta E_q' \tag{14.35}
$$

$$
s\Delta\omega = \frac{1}{2H}(-\Delta P_e - D\Delta\omega) \tag{14.36}
$$

$$
s\Delta\delta = \omega_0\Delta\omega \tag{14.37}
$$

The basic dynamic characteristic of the excitation system is represented in Figure 14.6.

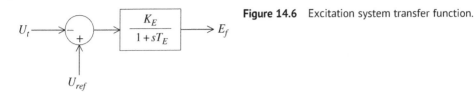

**Figure 14.6** Excitation system transfer function.

Where $K_E$ is the excitation system gain and $T_E$ is the excitation system time constant. $U_t$ is the generator terminal voltage, $U_{ref}$ is the controlled voltage reference. The transfer function of the excitation system is

$$\frac{\Delta E_f}{-\Delta U_t} = \frac{K_E}{1 + sT_E} \tag{14.38}$$

The system linearization state equations are derived as

$$\begin{bmatrix} \Delta \dot{\omega} \\ \Delta \dot{\delta} \\ \Delta \dot{E}'_q \\ \Delta \dot{E}_f \end{bmatrix} = \begin{bmatrix} -\dfrac{D}{2H} & -\dfrac{\dfrac{\partial P_e}{\partial \delta}}{2H} & -\dfrac{\dfrac{\partial P_e}{\partial E'_q}}{2H} & 0 \\ \omega_0 & 0 & 0 & 0 \\ 0 & -\dfrac{\dfrac{\partial E_q}{\partial \delta}}{T'_{d0}} & -\dfrac{\dfrac{\partial E_q}{\partial E'_q}}{T'_{d0}} & \dfrac{1}{T'_{d0}} \\ 0 & \dfrac{-K_E \dfrac{\partial U_t}{\partial \delta}}{T_E} & \dfrac{-K_E \dfrac{\partial U_t}{\partial E'_q}}{T_E} & -\dfrac{1}{T_E} \end{bmatrix} \begin{bmatrix} \Delta \omega \\ \Delta \delta \\ \Delta E'_q \\ \Delta E_f \end{bmatrix} = A \begin{bmatrix} \Delta \omega \\ \Delta \delta \\ \Delta E'_q \\ \Delta E_f \end{bmatrix} \tag{14.39}$$

where

$$\frac{\partial P_e}{\partial \delta} = \left( \frac{U}{x + x'_d} \sin \delta_0 \right) (x_q - x'_d) I_{q0} + \left( \frac{U}{x + x_q} \cos \delta_0 \right) \left[ E'_{q0} + (x_q - x'_d) I_{d0} \right] \tag{14.40}$$

$$\frac{\partial P_e}{\partial E'_q} = I_{q0} + \left( \frac{1}{x + x'_d} \right) (x_q - x'_d) I_{q0} \tag{14.41}$$

$$\frac{\partial E_q}{\partial \delta} = (x_d - x'_d) \left( \frac{U}{x + x'_d} \sin \delta_0 \right) \tag{14.42}$$

$$\frac{\partial E_q}{\partial E'_q} = 1 + \left( \frac{1}{x + x'_d} \right) (x_d - x'_d) \tag{14.43}$$

$$\frac{\partial U_t}{\partial \delta} = -\left(\frac{U}{x+x_d'}\sin\delta_0\right)x_d'\frac{U_{q0}}{U_{t0}} + \left(\frac{U}{x+x_q}\cos\delta_0\right)x_q\frac{U_{d0}}{U_{t0}} \tag{14.44}$$

$$\frac{\partial U_t}{\partial E_q'} = \frac{U_{q0}}{U_{t0}} - \left(\frac{1}{x+x_d'}\right)x_d'\frac{U_{q0}}{U_{t0}} \tag{14.45}$$

The single-machine infinite bus system small-signal stability is determined by the eigenvalues of the state Eq. (14.39) coefficient matrix $A$ as follows

$$det(sI - A) = 0 \tag{14.46}$$

The stability of the system is determined by the sign of the real component of the eigenvalue. The real component of the eigenvalues gives the damping. The negative real part represents a damped oscillation, whereas the positive real part represents an oscillation of increasing amplitude. Note that the elements in the matrix $A$, $D$, $M$, $T_{d0}'$, $T_E$ are parameters of the machine or controller which are positive and independent of the operation point. $\frac{\partial E_q}{\partial E_q'} = 1 + \left(\frac{1}{x+x_d'}\right)(x_d - x_d')$ is also positive and independent of the operation point. $\frac{\partial P_e}{\partial \delta}$, $\frac{\partial P_e}{\partial E_q'}$, $\frac{\partial E_q}{\partial \delta}$, $\frac{\partial U_t}{\partial \delta}$, $\frac{\partial U_t}{\partial E_q'}$ are varied at operation points, in which $\frac{\partial P_e}{\partial \delta}$, $\frac{\partial P_e}{\partial E_q'}$, $\frac{\partial E_q}{\partial \delta}$, $\frac{\partial U_t}{\partial E_q'}$ are positive but $\frac{\partial U_t}{\partial \delta} = -\left(\frac{U}{x+x_d'}\sin\delta_0\right)x_d'\frac{U_{q0}}{U_{t0}} + \left(\frac{U}{x+x_q}\cos\delta_0\right)x_q\frac{U_{d0}}{U_{t0}}$ may be negative at heavy loading conditions when $\tan\delta_0 > \frac{x+x_d'}{x+x_q}\frac{U_{d0}x_q}{U_{q0}x_d'}$. When $\frac{\partial U_t}{\partial \delta}$ changes the sign from positive to negative, the system may be excited at small-signal instable risk.

### 14.4.3 Numerical Case Study

#### 14.4.3.1 Stable Case
Taking the single-machine infinite bus system shown in Figure 14.1 as an example, the generator parameters are given in Table 14.2. The value of reactance in the table is given in per unit.

The excitation system gain $K_E$ is 30, time constant $T_E$ is 0.5 seconds.

In the single-machine infinite bus system, the generator stator current is given by

$$I = \frac{(P+jQ)^*}{U^*} = \frac{0.8-j0.2}{1.0} = 0.8 - j0.2 = 0.8246\angle - 14.04° \tag{14.47}$$

**Table 14.2** System parameters.

| $U$ | $P$ | $Q$ | $x$ | $x_d$ | $x_q$ | $x_d'$ | $T_{d0}'$(s) | $H$(s) | $D$ |
| --- | --- | --- | --- | --- | --- | --- | --- | --- | --- |
| 1.0 | 0.8 | 0.2 | 0.2 | 1.7 | 1.62 | 0.256 | 4.8 | 4.130 | 5 |

The generator terminal voltage and the voltage of the generator on the q-axis are calculated by

$$U_t = U + jxI = 1.0 + j0.2(0.8 - j0.2) = 1.04 + j0.16 = 1.0522\angle 8.7462° \quad (14.48)$$

$$E_Q = U_t + jx_q I = (1.04 + j0.16) + j1.62(0.8 - j0.2) = 1.364 + j1.456 = 1.9951\angle 46.87° \quad (14.49)$$

$E_Q$ is leading $U$ by $\delta_0 = 46.87°$.

$$I_{d0} + jI_{q0} = 0.8246 \times (\sin 60.91° + j\cos 60.91°) = 0.7206 + j0.4010 \quad (14.50)$$

$$U_{d0} + jU_{q0} = 1.0 \times (\sin 46.87° + j\cos 46.87°) = 0.7298 + j0.6837 \quad (14.51)$$

$$E'_{q0} = U_{q0} + j(x + x'_d)I_{d0} = 0.6837 + 0.456 \times 0.7206 = 1.0123 \quad (14.52)$$

$$\frac{\partial P_e}{\partial \delta} = \left(\frac{U}{x + x'_d}\sin\delta_0\right)(x_q - x'_d)I_{q0} + \left(\frac{U}{x + x_q}\cos\delta_0\right)\left[E'_{q0} + (x_q - x'_d)I_{d0}\right]$$

$$= \left(\frac{1}{0.456}\sin 46.87°\right) \times 1.364 \times 0.4010 + \left(\frac{1}{1.82}\cos 46.87°\right)[1.0123 + 1.364 \times 0.7206]$$

$$= 1.6248$$

$$(14.53)$$

$$\frac{\partial P_e}{\partial E'_q} = I_{q0} + \left(\frac{1}{x + x'_d}\right)(x_q - x'_d)I_{q0} = 0.4010 + \frac{1}{0.456} \times 1.364 \times 0.4010 = 1.6004 \quad (14.54)$$

$$\frac{\partial E_q}{\partial \delta} = (x_d - x'_d)\left(\frac{U}{x + x'_d}\sin\delta_0\right) = 1.444 \times \frac{1}{0.456}\sin 46.87° = 2.311 \quad (14.55)$$

$$\frac{\partial E_q}{\partial E'_q} = 1 + \left(\frac{1}{x + x'_d}\right)(x_d - x'_d) = 1 + \frac{1}{0.456} \times 1.444 = 4.1667 \quad (14.56)$$

$$\frac{\partial U_t}{\partial \delta} = -\left(\frac{U}{x + x'_d}\sin\delta_0\right)x'_d\frac{U_{q0}}{U_{t0}} + \left(\frac{U}{x + x_q}\cos\delta_0\right)x_q\frac{U_{d0}}{U_{t0}}$$

$$= -\frac{1}{0.456}\sin 46.87° \times 0.256 \times \frac{0.6837}{1.0522} + \frac{1}{1.82}\cos 46.87° \times 1.62 \times \frac{0.7298}{1.0522} = 0.1559$$

$$(14.57)$$

$$\frac{\partial U_t}{\partial E'_q} = \frac{U_{q0}}{U_{t0}} - \left(\frac{1}{x + x'_d}\right)x'_d\frac{U_{q0}}{U_{t0}} = \frac{0.6837}{1.0522} - \frac{0.256}{0.456} \times \frac{0.6837}{1.0522} = 0.2850 \tag{14.58}$$

Then the system linearized Eq. (14.39) is formulated as

$$
\begin{bmatrix} \Delta\dot\omega \\ \Delta\dot\delta \\ \Delta\dot E'_q \\ \Delta\dot E_f \end{bmatrix} =
\begin{bmatrix}
-\dfrac{D}{2H} & -\dfrac{\dfrac{\partial P_e}{\partial\delta}}{2H} & -\dfrac{\dfrac{\partial P_e}{\partial E'_q}}{2H} & 0 \\[2ex]
\omega_0 & 0 & 0 & 0 \\[2ex]
0 & -\dfrac{\dfrac{\partial E_q}{\partial\delta}}{T'_{d0}} & -\dfrac{\dfrac{\partial E_q}{\partial E'_q}}{T'_{d0}} & \dfrac{1}{T'_{d0}} \\[2ex]
0 & \dfrac{-K_E\dfrac{\partial U_t}{\partial\delta}}{T_E} & \dfrac{-K_E\dfrac{\partial U_t}{\partial E'_q}}{T_E} & -\dfrac{1}{T_E}
\end{bmatrix}
\begin{bmatrix} \Delta\omega \\ \Delta\delta \\ \Delta E'_q \\ \Delta E_f \end{bmatrix}
$$

$$
=
\begin{bmatrix}
-0.6053 & -0.1967 & -0.1938 & 0 \\[1ex]
377 & 0 & 0 & 0 \\[1ex]
0 & -0.4815 & -0.8681 & 0.2083 \\[1ex]
0 & -9.3517 & -17.0983 & -2.0
\end{bmatrix}
\begin{bmatrix} \Delta\omega \\ \Delta\delta \\ \Delta E'_q \\ \Delta E_f \end{bmatrix}
\tag{14.59}
$$

The eigenvalues of the system state matrix $A$ are solved by

$$
\begin{bmatrix}
-0.6053 - \lambda & -0.2041 & -0.2085 & 0 \\[1ex]
377 & -\lambda & 0 & 0 \\[1ex]
0 & -0.4617 & -0.8681 - \lambda & 0.2083 \\[1ex]
0 & 9.468 & -18.066 & -2.0 - \lambda
\end{bmatrix} = 0
\tag{14.60}
$$

The eigenvalues, damped frequencies, and damping ratios are listed in Table 14.3. The system is stable under small disturbances.

**Table 14.3** Small-signal stability analysis result.

| Eigenvalue | Damped frequency (Hz) | Damping ratio |
| --- | --- | --- |
| $-0.5681 \pm j8.7090$ | 1.3861 | 0.0651 |
| $-1.1687 \pm j1.0003$ | 0.1592 | 0.7597 |

**14.4.3.2 Instable Case**

Taking the same single-machine infinite bus system shown in Figure 14.1 as an example and keeping the generator parameters the same as given in Table 14.2, but the system increases the reactive load as leading as $jQ = -j0.6$ and increases the time constant $T_E$ to 2 seconds to introduce phase lag in between the terminal voltage $U_t$ and excitation $E_f$.

In the single-machine infinite bus system, the generator stator current is given by

$$I = \frac{(P + jQ)^*}{U^*} = \frac{0.8 + j0.6}{1.0} = 0.8 + j0.6 = 1.0\angle 36.87° \tag{14.61}$$

The generator terminal voltage and the voltage of the generator on the q-axis are calculated by

$$U_t = U + jxI = 1.0 + j0.2(0.8 + j0.6) = 0.88 + j0.16 = 0.8944\angle 10.30° \tag{14.62}$$

$$E_Q = U_t + jx_q I = (0.88 + j0.16) + j1.62(0.8 + j0.6) = -0.092 + j1.456 = 1.4589\angle 93.62° \tag{14.63}$$

$E_Q$ is leading $U$ by $\delta_0 = 93.62°$.

$$I_{d0} + jI_{q0} = 1.0 \times (\sin 56.75° + j\cos 56.75°) = 0.8362 + j0.5484 \tag{14.64}$$

$$U_{d0} + jU_{q0} = 1.0 \times (\sin 93.62° + j\cos 93.62°) = 0.9980 - j0.0631 \tag{14.65}$$

$$E'_{q0} = U_{q0} + j(x + x'_d)I_{d0} = -0.0631 + 0.456 \times 0.8362 = 0.3183 \tag{14.66}$$

$$\frac{\partial P_e}{\partial \delta} = \left(\frac{U}{x + x'_d} \sin \delta_0\right)(x_q - x'_d)I_{q0} + \left(\frac{U}{x + x_q} \cos \delta_0\right)\left[E'_{q0} + (x_q - x'_d)I_{d0}\right]$$

$$= \left(\frac{1}{0.456} \sin 93.62°\right) \times 1.364 \times 0.5484 + \left(\frac{1}{1.82} \cos 93.62°\right)[0.3183 + 1.364 \times 0.8362]$$

$$= 1.5864 \tag{14.67}$$

$$\frac{\partial P_e}{\partial E'_q} = I_{q0} + \left(\frac{1}{x + x'_d}\right)(x_q - x'_d)I_{q0} = 0.5484 + \frac{1}{0.456} \times 1.364 \times 0.5484 = 2.1886 \tag{14.68}$$

$$\frac{\partial E_q}{\partial \delta} = (x_d - x'_d)\left(\frac{U}{x + x'_d} \sin \delta_0\right) = 1.444 \times \frac{1}{0.456} \sin 93.62° = 3.1604 \tag{14.69}$$

$$\frac{\partial E_q}{\partial E'_q} = 1 + \left(\frac{1}{x + x'_d}\right)(x_d - x'_d) = 1 + \frac{1}{0.456} \times 1.444 = 4.1667 \tag{14.70}$$

$$\frac{\partial U_t}{\partial \delta} = -\left(\frac{U}{x + x'_d} \sin \delta_0\right) x'_d \frac{U_{q0}}{U_{t0}} + \left(\frac{U}{x + x_q} \cos \delta_0\right) x_q \frac{U_{d0}}{U_{t0}}$$

$$= -\frac{1}{0.456} \sin 93.62° \times 0.256 \times \frac{-0.0631}{0.8944} + \frac{1}{1.82} \cos 93.62° \times 1.62 \times \frac{0.9980}{0.8944}$$

$$= -0.0231$$

$$(14.71)$$

$$\frac{\partial U_t}{\partial E'_q} = \frac{U_{q0}}{U_{t0}} - \left(\frac{1}{x + x'_d}\right) x'_d \frac{U_{q0}}{U_{t0}} = \frac{-0.0631}{0.8944} - \frac{0.256}{0.456} \times \frac{-0.0631}{0.8944} = 0.0309 \qquad (14.72)$$

Then the system linearized Eq. (14.39) is formulated as

$$
\begin{bmatrix} \Delta\dot{\omega} \\ \Delta\dot{\delta} \\ \Delta\dot{E}'_q \\ \Delta\dot{E}_f \end{bmatrix} =
\begin{bmatrix}
-\dfrac{D}{2H} & -\dfrac{\partial P_e}{\partial \delta}{2H} & -\dfrac{\partial P_e}{\partial E'_q}{2H} & 0 \\[2ex]
\omega_0 & 0 & 0 & 0 \\[2ex]
0 & -\dfrac{\partial E_q}{\partial \delta}{T'_{d0}} & -\dfrac{\partial E_q}{\partial E'_q}{T'_{d0}} & \dfrac{1}{T'_{d0}} \\[2ex]
0 & \dfrac{-K_E \dfrac{\partial U_t}{\partial \delta}}{T_E} & \dfrac{-K_E \dfrac{\partial U_t}{\partial E'_q}}{T_E} & -\dfrac{1}{T_E}
\end{bmatrix}
\begin{bmatrix} \Delta\omega \\ \Delta\delta \\ \Delta E'_q \\ \Delta E_f \end{bmatrix}
$$

$$
=
\begin{bmatrix}
-0.6053 & -0.1921 & -0.2650 & 0 \\
377 & 0 & 0 & 0 \\
0 & -0.6584 & -0.8681 & 0.2083 \\
0 & 0.3469 & 0.4638 & -0.5
\end{bmatrix}
\begin{bmatrix} \Delta\omega \\ \Delta\delta \\ \Delta E'_q \\ \Delta E_f \end{bmatrix}
$$

$$(14.73)$$

The eigenvalues of the system state matrix $A$ are solved by

$$
\begin{vmatrix}
-0.6053 - \lambda & -0.1921 & -0.2650 & 0 \\
377 & -\lambda & 0 & 0 \\
0 & -0.6584 & -0.8681 - \lambda & 0.2083 \\
0 & 0.3469 & 0.4638 & -0.5 - \lambda
\end{vmatrix} = 0 \qquad (14.74)
$$

The eigenvalues, damped frequencies, and damping ratios are listed in Table 14.4. The system is instable under small disturbances.

**Table 14.4** Small-signal stability analysis result.

| Eigenvalue | Damped frequency (Hz) | Damping ratio |
| --- | --- | --- |
| $-0.7568 \pm j8.5042$ | 1.3535 | 0.0866 |
| $-0.4941$ | 0 | 1.0 |
| 0.0343 | 0 | $-1.0$ |

## 14.5 Small-Signal Oscillation Stabilization

High damping torque coefficient $D$ contributes to stabilizing the oscillation. The damping torque coefficient is driven by two major factors, the physical parameter of the machine and system loading. The physical parameter does not change but operation loading changes. The higher the loading, the higher the damping torque coefficient. Virtual damping control could provide extra damping torque at the cost of dynamic performance degradation.

Optimizing excitation system control parameters has a great impact on small-signal stability. To avoid the eigenvalue associated with $E_f$ moving to the positive on the real axis, the time constant $T_E$ of the terminal voltage transducer should be enlarged, and the gain $K_E$ of the exciter should be depressed. However, the control parameters changing in such a way will degrade the exciter response on voltage regulation.

The large time constant $T_E$ introduces phase lag in between the terminal voltage $U_t$ and excitation $E_f$. To balance the dynamic and steady-state performance of the excitation system, extra signal $V_s$ regulated by a power system stabilizer is designed to depress the generator oscillation driven by the phase lag of an excitation system as follows (Figure 14.7).

The power system stabilizer regulates an additional voltage control signal to stabilize generator rotor oscillations. The power system stabilizer detects the oscillation of generator speed using phase leading-lagging blocks to shift the phase of the controlled signal to compensate for the oscillation source from the system. The transfer function is shown as follows (Figure 14.8).

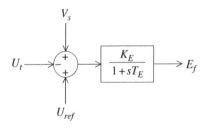

**Figure 14.7** Excitation system transfer function with power system stabilizer signal.

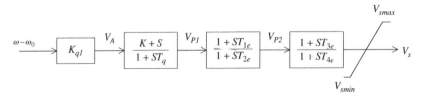

**Figure 14.8** Power system stabilizer transfer function.

Because of the nature of the control system time delay, the exciter input and the generator electrical torque have phase lag. The lead–lag compensator designed in the power system stabilizer provides the appropriate phase compensation to the phase lag. The overall system stability should be ensured in applying the power system stabilizer, not just the small-signal stability. To enhance system stability, its parameters should be carefully tuned.

## 14.6 Eigenvalue Calculation

As generally introduced in Sections 14.1 and 14.2, in the power system small-signal stability analysis, system state Eqs. (14.1) and (14.2) are linearized at the operating point. The analytical technique applied in the small-signal stability study is essentially a calculation of the eigenvalue $\lambda$ of the linearized system state matrix $A$ by solving the determinant equations. $|A - \lambda I| = 0$ and associated damped frequency and damping ratio of each oscillation mode are calculated to determine the small-signal stability of the system.

For a simple $2 \times 2$ matrix $A = \begin{bmatrix} a_{11} & a_{12} \\ a_{21} & a_{22} \end{bmatrix}$, the determinant equation $|A - \lambda I| = 0$ is rewritten as the following quadratic equation

$$\lambda^2 - (a_{11} + a_{22})\lambda + a_{12}a_{21} = 0 \tag{14.75}$$

Then the eigenvalues can be analytically solved as $\lambda_{1,2} = \dfrac{(a_{11} + a_{22}) \pm \sqrt{(a_{11} + a_{22})^2 - 4a_{12}a_{21}}}{2}$.

For a large-scale power system, state matrix $A$ is high-dimensional. There is no analytical formula to calculate eigenvalues. In numerical analysis, one of the most efficient and stable algorithms for finding the eigenvalues of a matrix is the QR algorithm, in which state matrix $A$ is transformed to be a similar upper triangular matrix or a block-diagonal upper triangular matrix. To develop the QR algorithm, the following theorems are first introduced.

**Theorem 14.1**  Suppose a general state matrix $A \in \mathbb{R}^{n \times n}$ is a square matrix, and $P \in \mathbb{R}^{n \times n}$ is nonsingular. Then the matrix $B = P^{-1}AP$ is said to be similar to matrix $A$. The eigenvalues of a matrix are preserved under similarity transformation.

Suppose

$$B = P^{-1}AP \tag{14.76}$$

We have

$$|B - \lambda I| = |P^{-1}AP - \lambda I| |P^{-1}(A - \lambda I)P| = \frac{1}{|P|}(A - \lambda I)|P|(A - \lambda I) \tag{14.77}$$

Thus, the eigenvalues of a matrix are preserved under similarity transformation.

**Theorem 14.2**   The eigenvalues of a diagonal matrix $\Lambda = \text{diag}\,(a_{11}, a_{22}, ..., a_{nn})$ are the diagonal elements.

The eigenvalues of a diagonal matrix are solved by the following equation

$$|\Lambda - \lambda I| = \prod_{i=1}^{n}(a_{ii} - \lambda_i) = 0 \tag{14.78}$$

The solution of Eq. (14.78) is $\lambda = (a_{11}, a_{22}, ..., a_{nn})$.

**Theorem 14.3**   The eigenvalues of an upper triangular matrix $U$ are the diagonal elements.

The eigenvalues of an upper triangular matrix $U$ are solved by the following equation

$$|U - \lambda I| = \begin{vmatrix} u_{11} - \lambda & u_{12} & \cdots & u_{1n} \\ 0 & u_{22} - \lambda & \cdots & u_{2n} \\ \vdots & \vdots & \ddots & \vdots \\ 0 & \cdots & 0 & u_{nn} - \lambda \end{vmatrix} = \prod_{i=1}^{n}(u_{ii} - \lambda_i) = 0 \tag{14.79}$$

The solution of Eq. (14.79) is $\lambda = (u_{11}, u_{22}, ..., u_{nn})$.

**Theorem 14.4**   The eigenvalues of block-diagonal upper triangular matrix $A$ are composed of the eigenvalues of each block-diagonal matrix.

Assume $A$ has $m$ blocks which can be rewritten as two blocks, as shown in (14.80).

$$A = \begin{bmatrix} A_{11} & A_{12} & \cdots & A_{1m} \\ 0 & A_{22} & \cdots & A_{2m} \\ \vdots & \vdots & \ddots & \vdots \\ 0 & \cdots & 0 & A_{mm} \end{bmatrix} = B = \begin{bmatrix} B_{11} & B_{12} \\ & B_{22} \end{bmatrix} \tag{14.80}$$

where $A \epsilon \mathbb{R}^{n \times n}$, $B_{11} = A_{11} \epsilon \mathbb{R}^{p \times p}$, $B_{12} = [A_{12} \; ... \; A_{1m}] \epsilon \mathbb{R}^{p \times q}$, $B_{22} = \begin{bmatrix} A_{22} & \cdots & A_{2m} \\ \vdots & \ddots & \vdots \\ 0 & 0 & A_{mm} \end{bmatrix} \epsilon \mathbb{R}^{q \times q}$,

$p + q = n$.

Assume $P = \begin{bmatrix} P_1^T & P_2^T \end{bmatrix}^T$ is an eigenvector of $B$, where $P_1 \epsilon \mathbb{R}^p$ and $P_2 \epsilon \mathbb{R}^q$. For $\lambda \epsilon \lambda(B)$, we have

$$\begin{bmatrix} A_{11} & A_{12} & \cdots & A_{1m} \\ 0 & A_{22} & \cdots & A_{2m} \\ \vdots & \vdots & \ddots & \vdots \\ 0 & \cdots & 0 & A_{mm} \end{bmatrix} \begin{bmatrix} P_1 \\ P_2 \end{bmatrix} = \begin{bmatrix} B_{11} & B_{12} \\ & B_{22} \end{bmatrix} \begin{bmatrix} P_1 \\ P_2 \end{bmatrix} = \lambda \begin{bmatrix} P_1 \\ P_2 \end{bmatrix} \tag{14.81}$$

If $P_2 \neq 0$, then $B_{22}P_2 = \lambda P_2$, and $\lambda \epsilon \lambda(B_{22})$. If $P_2 = 0$, then $B_{11}P_1 = \lambda P_1$, and $\lambda \epsilon \lambda(B_{11})$. It follows that $\lambda(A) = \lambda(B)\epsilon\lambda(B_{11}) \cup \lambda(B_{22}) = \lambda(A_{11}) \cup \lambda(B_{22})$.

Recursively, the block-diagonal upper-triangular matrix $B_{22} = \begin{bmatrix} A_{22} & \cdots & A_{2m} \\ \vdots & \ddots & \vdots \\ 0 & 0 & A_{mm} \end{bmatrix}$ can be

rewritten as

$$B_{22} = \begin{bmatrix} A_{22} & \cdots & A_{2m} \\ \vdots & \ddots & \vdots \\ 0 & 0 & A_{mm} \end{bmatrix} = C = \begin{bmatrix} C_{11} & C_{12} \\ 0 & C_{22} \end{bmatrix} \tag{14.82}$$

Where $C_{11} = A_{22}$, $C_{12} = [A_{23} \quad \cdots \quad A_{2m}]$, $C_{22} = \begin{bmatrix} A_{33} & \cdots & A_{3m} \\ \vdots & \ddots & \vdots \\ 0 & 0 & A_{mm} \end{bmatrix}$.

We can approve $\lambda(B_{22}) = \lambda(C_{11}) \cup \lambda(C_{22}) = \lambda(A_{22}) \cup \lambda(C_{22})$. Then $\lambda(A) = \lambda(B)\epsilon\lambda$ $(B_{11}) \cup \lambda(B_{22}) = \lambda(A_{11}) \cup \lambda(B_{22}) = \lambda(A_{11}) \cup \lambda(A_{22}) \cup \lambda(C_{22})$. The process is continued till the last block $A_{mm}$ eigenvalues are calculated as $\lambda(A_{mm})$. Then we conclude that

$$\lambda(A) = \lambda(A_{11}) \cup \lambda(A_{22}) \cup \ldots \cup \lambda(A_{mm}) = \bigcup_{i=1}^{m} \lambda(A_{ii}) \tag{14.83}$$

QR algorithm is a procedure to calculate the eigenvalues of a matrix [3, 4]. By a QR decomposition, a matrix is written as a product of an orthogonal matrix $Q$ (i.e. $Q^T = Q^{-1}$) and an upper triangular matrix $R$

$$A_k = Q_k R_k \tag{14.84}$$

By multiplying the two matrices $Q$ and $R$ reversely, a matrix $A_{k+1}$ similar to $A_k$ is updated as

$$A_{k+1} = R_k Q_k = Q_k^{-1} Q_k R_k Q_k = Q_k^{-1} A_k Q_k = Q_k^T A_k Q_k \tag{14.85}$$

The procedure is continued till lower triangle elements are approaching 0 by the convergence test

$$|a_{i,i-1}| < \varepsilon(|a_{i-1,i-1}| + |a_{i,i}|)\forall i = 2, 3, \ldots, n \tag{14.86}$$

where $\varepsilon$ is the tolerance for a converged solution.

In the procedure, all the $A_k$ are similar, and their eigenvalues are preserved.

The QR decomposition algorithm is shown as follows,

---

**Algorithm 14.1   QR Decomposition QR(A)**

---

1: $A = [a_1 | a_2 | \ldots | a_n]$
2: $q_1 = \dfrac{a_1}{\|a_1\|}$ //Normalization
3: **For** $j = 2 : n$
4:     $q_j = a_j$
5:     **For** $i = 1 : j - 1$
6:         $q_j = q_j - \left(a_j' q_i\right) q_i$ // Orthogonalization
7:     $q_j = \dfrac{q_j}{\|q_j\|}$ // Normalization
8: $Q = [q_1 | q_2 | \ldots | q_n]$
9: $R = Q'A$

---

In the algorithm, $\|V\|$ returns the Euclidean norm, or 2-norm of the vector $V$. And the algorithm of QR iteration of finding eigenvalues is shown as follows:

---

**Algorithm 14.2   Eigenvalue Calculation by QR Iteration**

---

1: $k = 0$, $A_k = A$
2: **While** $(|a_{i,i-1}| > \varepsilon(|a_{i-1,i-1}| + |a_{i,i}|))$ $\forall i = 2, 3, \ldots, n$
3:     $[Q_k \; R_k] = QR(A_k)$
4:     $A_{k+1} = R_k Q_k$
5:     $k = k + 1$
6: **Calculate** the eigenvalues of each block-diagonal matrix

---

### 14.6.1   Graph-Based Small-Signal Stability Analysis

To implement the graph-based small-signal stability analysis based on eigenvalues of the system state matrix $A$, the small-signal stability graph schema is defined as follows:

```
CREATE VERTEX V (primary_id id int, eigenvalue tuple)
CREATE DIRECTED EDGE E (FROM V, TO V, a_ij double, q_ij double, r_ij double)
CREAT GRAPH SmallSignalStabilityGraph (V, E)
```

The vertex attribute `id` represents the state variable number in order. The attribute `eigenvalue` is defined in the data structure `tuple` to constrain the real component and imaginary component of the complex eigenvalue of an associated state variable.

```
typedef tuple< double a, double b > eigenvalue;
```

The edge attributes `a_ij`, `q_ij`, `r_ij` represent the element of the original system state matrix $A$ and transformed matrix during the iteration, matrix $Q$ and $R$. A self-loop edge that connects a vertex to itself is defined as an edge attribute other than a vertex attribute to facilitate graph computing in small-signal stability analysis.

## 14.6.2  Buildup Small-Signal Stability Graph

Assume the system state matrix $A$ is built and saved in a raw data file `stateMatrix.csv`. And the table header of `stateMatrix.csv` is shown in Table 14.5.

Where `row` and `col` are the row and col number of the element $a_{ij}$ in the system state matrix, the system state matrix table is read to the small-signal stability graph `SmallSignalStability-Graph` by the query `readSystemStateMatrix`.

```
CREATE TABLE systemStateMatrixTable Table (row int, col int, aij double)
CREATE QUERY readSystemStateMatrix (int nState)
      FOR GRAPH SmallSignalStabilityGraph {
BULK INSERT systemStateMatrixTable
FROM '$sys.data_root/stateMatrix.csv'
FOREACH i in range(1:n)
  ACCUM
   INSERT INTO SmallSignalStabilityGraph.V(i)
WITH(FIRSTROW = 2,FIELDTERMINATOR = ',',ROWTERMINATOR = '\n',TABLOCK)
   SELECT A FROM systemStateMatrixTable
     ACCUM
         INSERT INTO SmallSignalStabilityGraph.E(A.row, A.col, A.aij)
}
```

Statement BULK INSERT reads data from `stateMatrix.csv`. The control parameters defined by a WITH clause WITH(FIRSTROW = 2, FIELDTERMINATOR = ',', ROWTERMINATOR = '\n', TABLOCK) tells the first row is a table header, the field delimiter is a comma, the row terminator is '\n', and the table is locked when it is read.

In the query, the total number of state variables `nState` is passed through as an argument. When the CSV file is read into the `systemStateMatrixTable`, the data are inserted into `Small-SignalStabilityGraph` attributes.

After the system state matrix has been read to the small-signal stability graph `SmallSignalStability-Graph`, the pseudo-code of the graph query to perform QR decomposition and eigenvalue calculation is implemented and provided in Appendix.

## 14.6.3  Numerical Example

Taking matrix $A$ in the instable case in Section 14.4.3.2 as a numerical example, matrix $A$ is copied as follows:

$$A_0 = \begin{bmatrix} -0.6053 & -0.1921 & -0.2650 & 0 \\ 377 & 0 & 0 & 0 \\ 0 & -0.6584 & -0.8681 & 0.2083 \\ 0 & 0.3469 & 0.4638 & -0.5 \end{bmatrix} = [a_1|a_2|a_3|a_4]$$

**Table 14.5**  State matrix table header.

| Row | Col | $a_{ij}$ |
| --- | --- | --- |

In the first QR decomposition iteration, we have

$$
q_1 = \frac{a_1}{\|a_1\|} = \frac{1}{377.0005}
\begin{bmatrix} -0.6053 \\ 377 \\ 0 \\ 0 \end{bmatrix}
=
\begin{bmatrix} -0.0016 \\ 1 \\ 0 \\ 0 \end{bmatrix}
$$

$$
q_2 = a_2 - (a_2' q_1) q_1
$$

$$
=
\begin{bmatrix} -0.1921 \\ 0 \\ -0.6584 \\ 0.3469 \end{bmatrix}
-
\left(
\begin{bmatrix} -0.1921 & 0 & -0.6584 & 0.3469 \end{bmatrix}
\begin{bmatrix} -0.0016 \\ 1 \\ 0 \\ 0 \end{bmatrix}
\right)
\begin{bmatrix} -0.0016 \\ 1 \\ 0 \\ 0 \end{bmatrix}
$$

$$
=
\begin{bmatrix} -0.1921 \\ 0 \\ -0.6584 \\ 0.3469 \end{bmatrix}
- 3.0838e^{-4}
\begin{bmatrix} -0.0016 \\ 1 \\ 0 \\ 0 \end{bmatrix}
=
\begin{bmatrix} -0.1921 \\ -0.0003 \\ -0.6584 \\ 0.3469 \end{bmatrix}
$$

$$
q_2 = \frac{q_2}{\|q_2\|} = \frac{1}{0.7686}
\begin{bmatrix} -0.1921 \\ -0.0003 \\ -0.6584 \\ 0.3469 \end{bmatrix}
=
\begin{bmatrix} -0.2499 \\ -0.0004 \\ -0.8566 \\ 0.4514 \end{bmatrix}
$$

$$
q_3 = a_3 - (a_3' q_1) q_1 - (a_3' q_2) q_2
$$

$$
=
\begin{bmatrix} -0.2650 \\ 0 \\ -0.8681 \\ 0.4638 \end{bmatrix}
-
\left(
\begin{bmatrix} -0.2650 & 0 & -0.8681 & 0.4638 \end{bmatrix}
\begin{bmatrix} -0.0016 \\ 1 \\ 0 \\ 0 \end{bmatrix}
\right)
\begin{bmatrix} -0.0016 \\ 1 \\ 0 \\ 0 \end{bmatrix}
$$

$$
-
\left(
\begin{bmatrix} -0.2650 & 0 & -0.8681 & 0.4638 \end{bmatrix}
\begin{bmatrix} -0.2499 \\ -0.0004 \\ -0.8566 \\ 0.4514 \end{bmatrix}
\right)
\begin{bmatrix} -0.2499 \\ -0.0004 \\ -0.8566 \\ 0.4514 \end{bmatrix}
$$

$$
=
\begin{bmatrix} -0.2650 \\ 0 \\ -0.8681 \\ 0.4638 \end{bmatrix}
- 4.24e^{-4}
\begin{bmatrix} -0.0016 \\ 1 \\ 0 \\ 0 \end{bmatrix}
- 1.0192
\begin{bmatrix} -0.2499 \\ -0.0004 \\ -0.8566 \\ 0.4514 \end{bmatrix}
=
\begin{bmatrix} -0.0103 \\ -0.0000 \\ -0.0050 \\ 0.0038 \end{bmatrix}
$$

$$q_3 = \frac{q_3}{\|q_3\|} = \frac{1}{0.0121} \begin{bmatrix} -0.0103 \\ -0.0000 \\ -0.0050 \\ 0.0038 \end{bmatrix} = \begin{bmatrix} -0.8534 \\ -0.0014 \\ -0.4151 \\ 0.3154 \end{bmatrix}$$

$$q_4 = a_4 - (a_4'q_1)q_1 - (a_4'q_2)q_2 - (a_4'q_3)q_3$$

$$= \begin{bmatrix} 0 \\ 0 \\ 0.2083 \\ -0.5 \end{bmatrix} - \left( \begin{bmatrix} 0 & 0 & 0.2083 & -0.5 \end{bmatrix} \begin{bmatrix} -0.0016 \\ 1 \\ 0 \\ 0 \end{bmatrix} \right) \begin{bmatrix} -0.0016 \\ 1 \\ 0 \\ 0 \end{bmatrix}$$

$$- \left( \begin{bmatrix} 0 & 0 & 0.2083 & -0.5 \end{bmatrix} \begin{bmatrix} -0.2499 \\ -0.0004 \\ -0.8566 \\ 0.4514 \end{bmatrix} \right) \begin{bmatrix} -0.2499 \\ -0.0004 \\ -0.8566 \\ 0.4514 \end{bmatrix}$$

$$- \left( \begin{bmatrix} 0 & 0 & 0.2083 & -0.5 \end{bmatrix} \begin{bmatrix} -0.8534 \\ -0.0014 \\ -0.4151 \\ 0.3154 \end{bmatrix} \right) \begin{bmatrix} -0.8534 \\ -0.0014 \\ -0.4151 \\ 0.3154 \end{bmatrix} = \begin{bmatrix} -0.1618 \\ -0.0003 \\ -0.1083 \\ -0.2951 \end{bmatrix}$$

$$q_4 = \frac{q_4}{\|q_4\|} = \frac{1}{0.3535} \begin{bmatrix} -0.1618 \\ -0.0003 \\ -0.1083 \\ -0.2951 \end{bmatrix} = \begin{bmatrix} -0.4545 \\ -0.0007 \\ -0.3064 \\ -0.8347 \end{bmatrix}$$

$$Q_0 = \begin{bmatrix} -0.0016 & -0.2499 & -0.8534 & -0.4545 \\ 1 & -0.0004 & -0.0014 & -0.0007 \\ 0 & -0.8566 & -0.4151 & -0.3064 \\ 0 & 0.4514 & 0.3154 & -0.8347 \end{bmatrix}$$

$$R_0 = Q_0'A_0 \begin{bmatrix} -0.0016 & -0.2499 & -0.8534 & -0.4545 \\ 1 & -0.0004 & -0.0014 & -0.0007 \\ 0 & -0.8566 & -0.4151 & -0.3064 \\ 0 & 0.4514 & 0.3154 & -0.8347 \end{bmatrix}' \begin{bmatrix} -0.6053 & -0.1921 & -0.2650 & 0 \\ 377 & 0 & 0 & 0 \\ 0 & -0.6584 & -0.8681 & 0.2083 \\ 0 & 0.3469 & 0.4638 & -0.5 \end{bmatrix}$$

$$= \begin{bmatrix} 377.0005 & 0.0003 & 0.0004 & 0 \\ 0 & 0.7686 & 1.0192 & -0.4042 \\ 0 & 0 & 0.0121 & -0.0712 \\ 0 & 0 & 0 & 0.3535 \end{bmatrix}$$

State matrix is transformed into a similar matrix by

$$
A_1 = R_0 Q_0 \begin{bmatrix} 377.0005 & 0.0003 & 0.0004 & 0 \\ 0 & 0.7686 & 1.0192 & -0.4042 \\ 0 & 0 & 0.0121 & -0.0712 \\ 0 & 0 & 0 & 0.3535 \end{bmatrix} \begin{bmatrix} -0.0016 & -0.2499 & -0.8534 & -0.4545 \\ 1 & -0.0004 & -0.0014 & -0.0007 \\ 0 & -0.8566 & -0.4151 & -0.3064 \\ 0 & 0.4514 & 0.3154 & -0.8347 \end{bmatrix}
$$

$$
= \begin{bmatrix} -0.6065 & -94.2071 & -321.7134 & -172.4950 \\ 0.7686 & -1.0558 & 0.2946 & 0.0245 \\ 0 & -0.0425 & -0.0175 & 0.0557 \\ 0 & 0.1596 & 0.1115 & -0.2951 \end{bmatrix}
$$

In the second iteration, $A_2$ is decomposed as

$$
A_2 = Q_2 R_2
$$

$$
= \begin{bmatrix} -0.6185 & -0.7858 & -0.0017 & 0 \\ 0.7858 & -0.6185 & -0.0014 & 0 \\ 0 & -0.0006 & 0.2826 & 0.9592 \\ 0 & 0.0021 & 0.9592 & -0.2826 \end{bmatrix} \begin{bmatrix} 0.9782 & 57.4399 & 199.2194 & 106.7119 \\ 0 & 74.6779 & 252.6084 & 135.5243 \\ 0 & 0 & 0.4465 & 0.5984 \\ 0 & 0 & 0 & 0.0379 \end{bmatrix}
$$

Matrix $A_1$ is transformed into a similar matrix $A_2$ by

$$
A_2 = R_1 Q_1
$$

$$
= \begin{bmatrix} 0.9782 & 57.4399 & 199.2194 & 106.7119 \\ 0 & 74.6779 & 252.6084 & 135.5243 \\ 0 & 0 & 0.4465 & 0.5984 \\ 0 & 0 & 0 & 0.0379 \end{bmatrix} \begin{bmatrix} -0.6185 & -0.7858 & -0.0017 & 0 \\ 0.7858 & -0.6185 & -0.0014 & 0 \\ 0 & -0.0006 & 0.2826 & 0.9592 \\ 0 & 0.0021 & 0.9592 & -0.2826 \end{bmatrix}
$$

$$
= \begin{bmatrix} 44.5292 & -36.1818 & -46.1327 & -221.2597 \\ 58.6792 & -46.0441 & -58.7021 & -280.6161 \\ 0 & 0.0010 & -0.4478 & -0.5974 \\ 0 & 0.0001 & -0.0363 & -0.0107 \end{bmatrix}
$$

In the third iteration, $A_2$ is decomposed as

$$A_2 = Q_2 R_2$$

$$= \begin{bmatrix} 0.6045 & -0.7966 & -0.0008 & 0 \\ 0.7966 & 0.6045 & 0.0006 & 0 \\ 0 & 0.0010 & -0.9967 & -0.0808 \\ 0 & 0.0001 & -0.0808 & 0.9967 \end{bmatrix} \begin{bmatrix} 73.6621 & -58.5509 & -74.6496 & -357.2918 \\ 0 & 0.9885 & 1.2630 & 6.6206 \\ 0 & 0 & 0.4506 & 0.6032 \\ 0 & 0 & 0 & 0.0376 \end{bmatrix}$$

Matrix $A_2$ is then transformed into a similar matrix $A_3$ by

$$A_3 = R_2 Q_2$$

$$= \begin{bmatrix} 73.6621 & -58.5509 & -74.6496 & -357.2918 \\ 0 & 0.9885 & 1.2630 & 6.6206 \\ 0 & 0 & 0.4506 & 0.6032 \\ 0 & 0 & 0 & 0.0376 \end{bmatrix} \begin{bmatrix} 0.6045 & -0.7966 & -0.0008 & 0 \\ 0.7966 & 0.6045 & 0.0006 & 0 \\ 0 & 0.0010 & -0.9967 & -0.0808 \\ 0 & 0.0001 & -0.0808 & 0.9967 \end{bmatrix}$$

$$= \begin{bmatrix} -2.1124 & -94.1802 & 103.1863 & -350.0893 \\ 0.7874 & 0.5994 & -1.7934 & 6.4968 \\ 0 & 0.0005 & -0.4979 & 0.5648 \\ 0 & 0 & -0.0030 & 0.0375 \end{bmatrix}$$

In the fourth iteration, $A_3$ is decomposed as

$$A_3 = Q_3 R_3$$

$$= \begin{bmatrix} -0.9370 & -0.3493 & 0 & 0 \\ 0.3493 & -0.9370 & 0 & 0 \\ 0 & 0 & -1 & -0.0061 \\ 0 & 0 & -0.0061 & 1 \end{bmatrix} \begin{bmatrix} 2.2544 & 88.4579 & -97.3138 & 330.3091 \\ 0 & 32.3338 & -34.3607 & 116.1921 \\ 0 & 0 & 0.4973 & -0.5632 \\ 0 & 0 & 0 & 0.0341 \end{bmatrix}$$

Matrix $A_3$ is then transformed into a similar matrix $A_4$ by

$$A_4 = R_3 Q_3$$

$$= \begin{bmatrix} 2.2544 & 88.4579 & -97.3138 & 330.3091 \\ 0 & 32.3338 & -34.3607 & 116.1921 \\ 0 & 0 & 0.4973 & -0.5632 \\ 0 & 0 & 0 & 0.0341 \end{bmatrix} \begin{bmatrix} -0.9370 & -0.3493 & 0 & 0 \\ 0.3493 & -0.9370 & 0 & 0 \\ 0 & 0 & -1 & -0.0061 \\ 0 & 0 & -0.0061 & 1 \end{bmatrix}$$

$$= \begin{bmatrix} 28.7843 & -83.6756 & 95.2924 & 330.8976 \\ 11.2936 & -30.2979 & 33.6496 & 116.3999 \\ 0 & 0 & -0.4939 & -0.5662 \\ 0 & 0 & -0.0002 & 0.0341 \end{bmatrix}$$

The matrix $A_4$ is converged to a block-diagonal upper triangular matrix. Since $A_4$ is similar to the system state matrix $A$, they share the same eigenvalues, which are the roots of the following equation

$$\begin{vmatrix} 28.7843 - \lambda & -83.6756 & 95.2924 & 330.8976 \\ 11.2936 & -30.2979 - \lambda & 33.6496 & 116.3999 \\ 0 & 0 & -0.4939 - \lambda & -0.5662 \\ 0 & 0 & -0.0002 & 0.0371 - \lambda \end{vmatrix} = 0 \tag{14.87}$$

The above equation can be reduced to

$$\begin{vmatrix} 28.7843 - \lambda & -83.6756 \\ 11.2936 & -30.2979 - \lambda \end{vmatrix} \begin{vmatrix} -0.4939 - \lambda & -0.5662 \\ -0.0002 & 0.0371 - \lambda \end{vmatrix} = 0 \tag{14.88}$$

The solutions of the Eq. (14.88) are equivalent to the solutions of the following two quadratic equations

$$\begin{vmatrix} 28.7843 - \lambda & -83.6756 \\ 11.2936 & -30.2979 - \lambda \end{vmatrix} = 0 \tag{14.89}$$

$$\begin{vmatrix} -0.4939 - \lambda & -0.5662 \\ -0.0002 & 0.0371 - \lambda \end{vmatrix} = 0 \tag{14.90}$$

Or

$$\lambda^2 + 1.5136\lambda + 72.8948 = 0 \tag{14.91}$$

$$\lambda^2 + 0.4568\lambda - 0.0169 = 0 \tag{14.92}$$

The solutions of Eq. (14.91) are $\lambda_{1,2} = -0.7568 \pm j8.5042$. The solutions of the Eq. (14.92) are $\lambda_3 = -0.4941$ and $\lambda_4 = 0.0343$. The eigenvalue $\lambda_4 = 0.0343$ tells us the system is instable under small disturbances.

## References

1 Kundur, P.S. (1993). *Power System Stability and Control*. McGraw-Hill.
2 Vittal, V., McCalley, J.D., Anderson, P.M., and Fouad, A.A. (2019). *Power System Control and Stability*, 3e. Wiley-IEEE Press.
3 Francis, J.G.F. The QR transformation, I. *The Computer Journal* 4 (3): 265–271.
4 Francis, J.G.F. The QR transformation, II. *The Computer Journal* 4 (4): 332–345.
5 Lin, K.-H., Lin, C.-H., Chang, R.C.-H. et al. (2009). Iterative QR decomposition architecture using the modified Gram-Schmidt algorithm. *2009 IEEE International Symposium on Circuits and Systems*, Taipei, Taiwan, pp. 1409–1412. https://doi.org/10.1109/ISCAS.2009.5118029.

**6** Li, C. and Du, Z. (2013). A novel method for computing small-signal stability boundaries of large-scale power systems. *IEEE Transactions on Power Systems* 28 (2): 877–883.

**7** Rommes, J. and Martins, N. (2008). Computing large-scale system eigenvalues most sensitive to parameter changes, with applications to power system small-signal stability. *IEEE Transactions on Power Systems* 23 (2): 434–442.

**8** Qin, C. and Yu, Y. (2013). Small signal stability region of power systems with DFIG in injection space. *Journal of Modern Power Systems and Clean Energy* 1 (2): 127–133.

**9** Du, W., Bi, J., Wang, T., and Wang, H. (2015). Impact of grid connection of large-scale wind farms on power system small-signal angular stability. *CSEE Journal of Power and Energy Systems* 1 (2): 83–89.

**10** Dong, W., Xin, H., Wu, D., and Huang, L. (2019). Small signal stability analysis of multi-infeed power electronic systems based on grid strength assessment. *IEEE Transactions on Power Systems* 34 (2): 1393–1403.

**11** Liu, S., Liu, P.X., and Wang, X. (2016). Stochastic small-signal stability analysis of grid-connected photovoltaic systems. *IEEE Transactions on Industrial Electronics* 63 (2): 1027–1038.

**12** Ma, J., Qiu, Y., Li, Y. et al. (2017). Research on the impact of DFIG virtual inertia control on power system small-signal stability considering the phase-locked loop. *IEEE Transactions on Power Systems* 32 (3): 2094–2105.

**13** Rueda, J.L., Guaman, W.H., Cepeda, J.C. et al. (2013). Hybrid approach for power system operational planning with smart grid and small-signal stability enhancement considerations. *IEEE Transactions on Smart Grid* 4 (1): 530–539.

**14** Du, W., Fu, Q., and Wang, H. (2018). Small-signal stability of an AC/MTDC power system as affected by open-loop modal coupling between the VSCs. *IEEE Transactions on Power Systems* 33 (3): 3143–3152.

**15** Acha, E., Fuerte-Esquivel, C.R., Ambriz-Pérez, H., and Angeles-Camacho, C. (2004). *FACTS: Modelling and Simulation in Power Networks*. Wiley-IEEE Press.

**16** De Marco, F.J., Apolinário, J.A., Pellanda, P.C., and Martins, N. (2013). Efficient online estimation of electromechanical modes in large power systems. *2013 IEEE 4th Latin American Symposium on Circuits and Systems (LASCAS)*, Cusco, Peru, pp. 1–4. https://doi.org/10.1109/LASCAS.2013.6518979.

**17** Santos, M., Calvaittis Santana, G., de Campos, M. et al. (2021). Performance of controller designs in small-disturbance angle stability of power systems with parametric uncertainties. *IEEE Latin America Transactions* 19 (12): 2054–2061.

**18** Men, Y., Du, Y., and Lu, X. (2021). Distributed control framework and scalable small-signal stability analysis for dynamic microgrids. *Chinese Journal of Electrical Engineering* 7 (4): 49–59.

**19** Wen, B., Boroyevich, D., Burgos, R. et al. (2015). Small-signal stability analysis of three-phase AC systems in the presence of constant power loads based on measured $d$-$q$ frame impedances. *IEEE Transactions on Power Electronics* 30 (10): 5952–5963.

# 15

# Transient Stability

Fundamentally, the ability of a power system to maintain its synchronism when subject to a large disturbance is referred to as transient stability. In general, power system transient stability is studied by time domain simulation. The energy function method is also adopted to analyze power system transient stability.

## 15.1  Transient Stability Theory

The power system is a multi-dimensional high-order nonlinear dynamic system. The behavior of a dynamic system is described by a set of high-order nonlinear ordinary differential equations and high dimensional algebraic equations.

Transient stability theory addresses the stability of the solution of the differential algebraic equations and the trajectories of the dynamic system under large disturbances.

### 15.1.1  Stability Region and Boundary

A nonlinear power system is designed to be operated at an equilibrium point. Stability requires robustness of the equilibrium point to disturbances, i.e. the system could return to the equilibrium point or a new equilibrium point after disturbances. For a real power system, being globally stable is not practical. The equilibrium state can be restored under a limited amount of disturbances and the equilibrium state is within a certain stability region. A stability region is defined as a set of initial conditions of state variables whose trajectories tend to an equilibrium point when the time goes to infinity. The stability boundary refers to the boundary of the stability region.

### 15.1.2  Energy Function Method

Estimation of stability regions is challenging. The energy function method has been used to estimate stability regions and stability boundaries. Different from the time domain simulation, the energy function method is called the direct method. Aleksandr Mikhailovich Lyapunov derived sufficient conditions for stability based on energy function, called Lyapunov energy function [1, 2].

To avoid time domain simulation, the direct method uses Lyapunov functions to screen out stable trajectories. Only a small number of trajectories will be subjected to time domain simulation. The region of attraction of the postfault stable equilibrium point is characterized by the property that all trajectories inside this region will converge to the postfault stable equilibrium point [3].

*Graph Database and Graph Computing for Power System Analysis*, First Edition. Renchang Dai and Guangyi Liu.
© 2024 The Institute of Electrical and Electronics Engineers, Inc. Published 2024 by John Wiley & Sons, Inc.

The ideal Lyapunov function would estimate the entire region of attraction. However, this is difficult to achieve in practice for formulating an ideal Lyapunov function. Lyapunov functions tend to be conservative, in that they cover part of the region of attraction. Hence, a trajectory not deemed stable by a Lyapunov function does not necessarily imply instability. The efficiency of the online transient stability screening tool depends strongly on the choice of Lyapunov functions.

Although the Lyapunov energy function method opens a window to estimate physical system stability without simulating transient trajectories, it has at least two major limitations. One limitation is there is no generic approach to formulate a Lyapunov energy function to represent a generic physic system. The other limitation is Lyapunov energy function derives sufficient conditions for stability. The derived stability boundary and region are typically conservative.

With the knowledge of power system stability characteristics, researchers and engineers developed several effective methods to more accurately estimate power system stability regions, such as the Controlling Unstable Equilibrium Point method (controlling UEP) [4], the Boundary of Stability-Region-based Controlling Unstable Equilibrium Point method (BCU) [5], and the Extended Equal Area Criterion (EEAC) method [6, 7], etc. These methods have been applied to large-scale power system stability analysis.

### 15.1.2.1 Controlling UEP Method

Several methods are available for determining critical energy values for direct stability analysis. When used for transient stability analysis of power systems, the classical method of determining the closest UEP has been found to produce excessive results. The potential energy boundary surface (PEBS) method [8] gives fairly fast stability assessments, but it can be inaccurate (i.e. giving overestimated and unduly underestimated stability assessments). A desirable method for determining the critical energy value can provide the most accurate approximation of the relevant stability boundary toward which the fault-on trajectory is heading. This is the spirit of the controlling UEP method.

It is now well recognized that among several methods for determining the critical energy value, the controlling UEP method is the most viable for direct stability analysis of practical power systems. As shown in Figure 15.1, the controlling UEP method uses the constant energy surface passing through the controlling UEP to approximate the relevant part of the stability boundary toward which the fault-on trajectory is heading. If, when the fault is cleared, the system state lies inside the energy surface passing through the controlling UEP, then the postfault system is stable [9].

Where EP is the exit point; MGP is the minimum gradient point; $X_{CO}$ is the controlling UEP; $R(X_{CO})$ is the region of convergence of the controlling UEP; $X_s$ is the postfault stable equilibrium point; $X_s^{pre}$ is the prefault stable equilibrium point; $A(X_s)$ is the region of attraction of postfault stable equilibrium point; $\partial A(X_s)$ is the boundary of the region of attraction of postfault stable equilibrium point; and $W^s(X_{CO})$ is the stable manifold of the controlling UEP.

If the fault is cleared before the fault-on trajectory reaches the exit point, then the fault-clearing point must be inside the stability region of the postfault stable

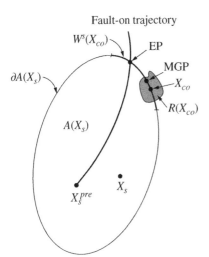

**Figure 15.1** Fault-on trajectory.

equilibrium point. The controlling UEP method approximates the relevant stability boundary, the stable manifold of the controlling UEP, by the constant energy surface, passing through the controlling UEP.

### 15.1.2.2 Stability-Region-Based Controlling UEP Method

The BCU method computes the reduced-state controlling UEP and explores the dynamic relationship between the controlling UEP of the original system and the reduced-state controlling UEP of the reduced-state system to achieve its key objective: computing the controlling UEP of the original system [5]. Hence, the dynamic relationship and computation of the reduced-state controlling UEP of the reduced-state system play a key role in the success of the BCU method. In other words, the successful computation of the controlling UEP by the BCU method depends on the computation of the reduced-state controlling UEP and the correspondence between the reduced-state controlling UEP and the controlling UEP.

The procedure of the BCU method to find the controlling UEP relative to a fault-on trajectory is shown as follows:

Step 1: From the fault-on trajectory $(\delta(t), \omega(t))$, detect the exit point $\delta^*$ which is the point projected trajectory $\delta(t)$ exits the stability boundary of the reduced system.
Step 2: Use the point $\delta^*$ as the initial condition and integrate the postfault reduced system to find the first local minimum $\delta_0^*$;
Step 3: Use the point $\delta_0^*$ as the initial guess to solve $\delta_{co}^*$
Step 4: Assign the controlling UEP with respect to the fault-on trajectory to be $(\delta_0^*, 0)$.

Once the controlling UEP is found, the BCU method uses the same procedure as that of the conceptual controlling UEP method presented to perform stability assessment.

## 15.2    Transient Simulation Model

The power system transient simulation is essentially solving the following large-scale differential algebraic equations (DAE) which model the physics of power system components, such as generators, exciters, governors, power system stabilizer (PSS), power network, and loads.

$$\dot{X} = f(X, U) \ \ X \in \mathbb{R}^n \tag{15.1}$$
$$Y = g(X, U) \ \ Y \in \mathbb{R}^m \tag{15.2}$$

The ordinary differential equations (ODE) (15.1) represent the dynamics of synchronous machines and the associated control systems. The algebraic Eq. (15.2) describes the characteristics of a power network. The DAE can be represented in a system block diagram (Figure 15.2).

A variety of excitation systems, governors, and stabilizers are running in practical power systems. This section is not intended to cover all models comprehensively but introduce representatives of each subsystem to highlight the characteristics of graph computing on transient stability.

### 15.2.1    Generator Rotor Model

The generator rotor model can be simplified by the swing equations.

$$T_J \frac{d\omega}{dt} = P_T - P_e - D(\omega - 1) \tag{15.3}$$

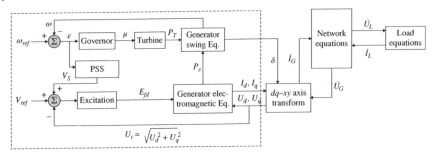

**Figure 15.2** System block diagram.

$$\frac{d\delta}{dt} = \omega - 1 \tag{15.4}$$

### 15.2.2 Generator Electro-Magnetic Model

For different study purposes, a synchronous generator is modeled in varying degrees of detail. For example, the Type-6 model is represented by the following differential equations:

$$T'_{d0}\frac{dE'_q}{dt} = E_{fd} - (K_G - 1)E'_q - \left(x_d - x'_d\right)I_d - E'_q \tag{15.5}$$

$$T''_{d0}\frac{dE''_q}{dt} = -E''_q - \left(x'_d - x''_d\right)I_d + E'_q + T''_{d0}\frac{dE'_q}{dt} \tag{15.6}$$

$$T'_{q0}\frac{dE'_d}{dt} = -E'_d + \left(x_q - x'_q\right)I_q \tag{15.7}$$

$$T''_{q0}\frac{dE''_d}{dt} = -E''_d - \left(x'_q - x''_q\right)I_q + E'_d + T''_{q0}\frac{dE'_d}{dt} \tag{15.8}$$

Other types of synchronous generator models in different levels of detail are included in Appendix.

### 15.2.3 Excitation System Model

An excitation system is functioning to provide direct current to the synchronous generator field winding. In addition, the control function of an excitation system regulates field voltage and field current to control the generator terminal voltage. To support power system stability analysis, the excitation system model has been developed in different forms. They are typically classified into three categories based on the excitation power source used [10]:

- DC excitation systems
- AC excitation systems
- Static excitation systems

The transfer function blocks to the excitation system response are covered in great detail in the IEEE report [11]. DC excitation systems utilize DC generators as a source of excitation power and prove current to the synchronous machine rotor through slip rings or an inverter. The DC excitation systems are either self-excited or separately excited [10]. In this section, the two types of DC

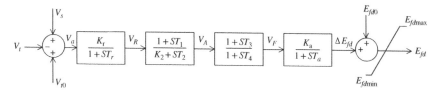

**Figure 15.3** Self-excited AVR transfer function block diagram.

excitation systems are taken as examples to illustrate the excitation system models suitable for use in large-scale system stability analysis.

The self-excited excitation system is the simplest generator automatic voltage regulator (AVR). It is considered the shunt method because a shunt winding is arranged to provide a portion of the output armature current to the voltage regulator. A typical AVR is composed of a measurement block, a regulator, and an exciter. The transfer function block diagram of a self-excited excitation system is given in Figure 15.3.

The input of this model is the terminal voltage measurement, the voltage signal from the PSS, and the voltage target. The voltage target is subtracted from the terminal voltage and the PSS output signal is added to produce an error voltage. The noise of the error voltage is filtered out as the input of the regulator. $K_2$ is a switcher which changes the block $\dfrac{1 + ST_1}{K_2 + + ST_2}$ to be a proportional-integral controller (when $K_2 = 0$) or a lag–lead compensator (when $K_2 = 1$). The lag–lead element $\dfrac{1 + ST_3}{1 + + ST_4}$ in the voltage regulator provides phase compensation. The compensated signal is amplified in the regulator to output the field voltage $E_{fd}$.

The self-excited AVR controller parameters are given in Table 15.1.

The separately excited AVR transfer function block diagram is shown as follows (Figure 15.4):

The separately excited AVR parameters are given in Table 15.2.

The other model structures with excitation limiters and supplementary controls are available in [12].

**Table 15.1** Self-excited AVR parameters.

| Parameter | Annotation | Units |
| --- | --- | --- |
| $K_r$ | Measurement gain | p.u. |
| $T_r$ | Measurement time constant | Seconds |
| $K_2$ | Block switcher | p.u. |
| $T_1, T_3$ | Lead time constant | Seconds |
| $T_2, T_4$ | Lag time constant | Seconds |
| $K_a$ | AVR gain | p.u. |
| $T_a$ | AVR time constant | Seconds |
| $E_{fdmax}$ | Field voltage upper limit | p.u. |
| $E_{fdmin}$ | Field voltage lower limit | p.u. |

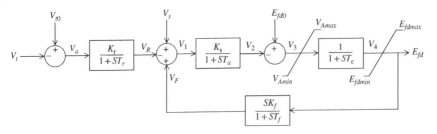

**Figure 15.4** Separately excited AVR transfer function block diagram.

**Table 15.2** Separately excited AVR parameters.

| Parameter | Annotation | Units |
|---|---|---|
| $K_r$ | Measurement gain | p.u. |
| $T_r$ | Transducer time constant | Seconds |
| $K_s$ | Main AVR gain | p.u. |
| $T_a$ | Main AVR time constant | Seconds |
| $K_f$ | Feedback gain | p.u. |
| $T_f$ | Feedback time constant | Seconds |
| $T_e$ | Exciter time constant | Seconds |
| $V_{Amax}$ | AVR input upper limit | p.u. |
| $V_{Amin}$ | AVR input lower limit | p.u. |
| $E_{fdmax}$ | Field voltage upper limit | p.u. |
| $E_{fdmin}$ | Field voltage lower limit | p.u. |

### 15.2.4 Governor Model

As illustrated in the system block diagram Figure 15.2, the mechanical power $P_T$ is an input to the generator rotor equation. The value of $P_T$ is controlled by a governor system. Typically, a governor system is composed of a prime mover, a governor controller, and a turbine. To build a generic model for both steam turbine and hydraulic turbine, the IEEE standard governor model type 3 (IEEEG3) [13] is modified and the transfer function block diagram is shown in Figure 15.5.

The parameters in the diagram are listed in Table 15.3.

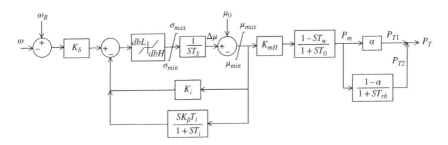

**Figure 15.5** Generic speed-governor transfer function.

**Table 15.3** Governor parameters.

| Parameter | Annotation | Units |
|---|---|---|
| $K_\delta$ | Measurement gain | p.u. |
| $dbL$ | Dead band | p.u. |
| $dbH$ | Dead band | p.u. |
| $\sigma_{max}$ | Maximum valve position | p.u. |
| $\sigma_{min}$ | Minimum valve position | p.u. |
| $T_s$ | Servo motor time constant | Seconds |
| $T_p$ | Pilot lag time constant | Seconds |
| $\mu_{max}$ | Gate position at maximum power | p.u. |
| $\mu_{min}$ | Gate position at minimum power | p.u. |
| $K_i$ | Permanent droop coefficient | p.u. |
| $K_\beta$ | Temporary droop coefficient | p.u. |
| $T_i$ | Temporary droop time constant | Seconds |
| $K_{mH}$ | Capacity coefficient | p.u. |
| $T_W$ | Water inertia time constant | Seconds |
| $T_0$ | Steam inertia time constant | Seconds |
| $\alpha$ | Reheat | p.u. |
| $T_{rh}$ | Reheat time constant | Seconds |

### 15.2.5 PSS Model

An auxiliary stabilizing signal is used to regulate the excitation of a generator and dampen its rotor oscillations, which is the primary function of a PSS. The PSS detects the oscillation of generator speed using phase lead–lag blocks to shift the phase of the controlled signal to compensate for the oscillation source from the expiration system. The transfer function of a conventional PSS is shown as follows (Figure 15.6).

The PSS is composed of a gain block, a washout block, and two lead–lag blocks for phase compensation. The parameters in the diagram are listed in Table 15.4.

## 15.3 Transient Simulation Approach

The power system transient simulation solves large-scale differential algebraic Eqs. (15.1) and (15.2) by time domain simulation. Equations (15.1) and (15.2) can be solved simultaneously by an integrated method or alternatively by a sequential method.

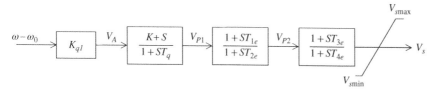

**Figure 15.6** Type I PSS function block diagram.

**Table 15.4** PSS parameters.

| Parameter | Annotation | Units |
|---|---|---|
| $K_{q1}$ | Transducer gain | p.u. |
| $K$ | Block switcher | p.u. |
| $T_q$ | Washout time constant | Seconds |
| $T_{1e}, T_{2e}, T_{3e}, T_{4e}$ | Lead–lag time constant | Seconds |
| $V_{smax}$ | Output upper limit | p.u. |
| $V_{smin}$ | Output lower limit | p.u. |

### 15.3.1 Transient Simulation Algorithm

When a sequential method is adopted, differential Eq. (15.1) are solved after the bus voltage in the algebraic Eq. (15.2) is calculated. The DAE structure is shown in Figure 15.2.

By the sequential method, in each iteration, the network equations are solved to update the network bus voltage including the generator terminal voltage. The differentiated equations use the generator terminal bus voltage as boundary conditions to solve the dynamic states of the generator, exciter, governor, PSS, and current injections from generators to the network. The updated current injections are applied to solve the network equations in the next iteration until the converged solution is achieved. The flowchart of the sequential method is shown in Figure 15.7.

In the transient simulation, power flow initialization (step ①), $Y$ matrix formation and factorization (step ③), and bus voltage calculation (step ④) have been addressed in Chapter 8 and Chapter 9. This chapter focuses on steady-state equilibrium condition calculation (step ②), solving differential equations (step ⑤), generator injection current calculation (step ⑥), and graph-based transient simulation implementation.

### 15.3.2 Steady-State Equilibrium Condition

Power system state variables represent the dynamic states of generators, excitation systems, governors, PSSs, and so on. When the generator terminal voltages are solved by Eq. (15.2), their equilibrium points can be derived.

Synchronous machine voltage equations can be represented by the following phasor diagram (Figure 15.8).

When generator terminal voltages are solved by the initial power flow equations, the generator steady-state equilibrium operating point is solved as follows:

$$\dot{E}_{Q(0)} = \dot{U}_{(0)} + \left(R_a + jx_q\right)\dot{I}_{t(0)} \tag{15.9}$$

$$\delta_{(0)} = arctg\left(\frac{E_{Qy(0)}}{E_{Qx(0)}}\right) \tag{15.10}$$

$$E_{q(0)} = U_{q(0)} + R_a I_{q(0)} + x_d I_{d(0)} \tag{15.11}$$

$$E'_{q(0)} = U_{q(0)} + R_a I_{q(0)} + x'_d I_{d(0)} \tag{15.12}$$

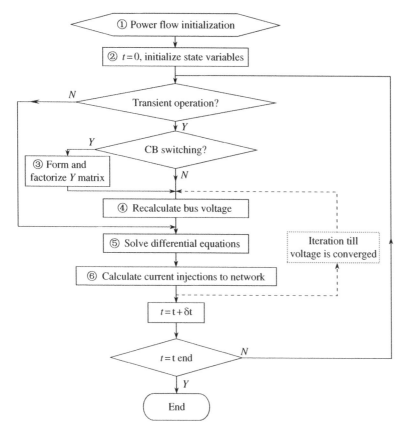

**Figure 15.7** Flowchart of sequential method to transient simulation.

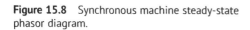

**Figure 15.8** Synchronous machine steady-state phasor diagram.

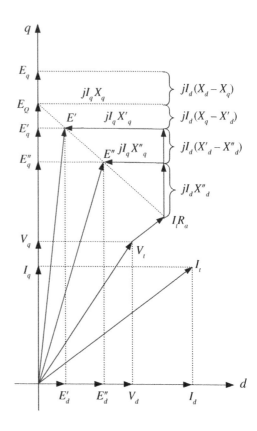

$$E'_{d(0)} = U_{d(0)} + R_a I_{d(0)} - x'_q I_{q(0)} \tag{15.13}$$

$$E''_{q(0)} = U_{q(0)} + R_a I_{q(0)} + x''_d I_{d(0)} \tag{15.14}$$

$$E''_{d(0)} = U_{d(0)} + R_a I_{q(0)} + x''_q I_{q(0)} \tag{15.15}$$

$$P_{e(0)} = Re(\dot{V}\dot{I}_t) \tag{15.16}$$

When the resistance $R_a$ is ignored ($R_a = 0$), the generator steady-state equilibrium operating point is simplified as

$$E''_{d(0)} = I_{q(0)} \cdot \left(x_q - x''_q\right) = \left(x_q - x''_q\right)\left(I_{tx}\cos\delta + I_{ty}\sin\delta\right) \tag{15.17}$$

$$I''_{t(0)x} = \frac{1}{x''_d}\left(E''_q \sin\delta - E''_d \cos\delta\right) \tag{15.18}$$

$$I''_{t(0)y} = \frac{1}{x''_d}\left(-E''_q \cos\delta - E''_d \sin\delta\right) \tag{15.19}$$

$$I''_{t(0)x}\cos\delta + I''_{t(0)y}\sin\delta$$

$$= \frac{1}{x''_d}\left(E''_q \sin\delta\cos\delta - E''_d \cos^2\delta\right) + \frac{1}{x''_d}\left(-E''_q \sin\delta\cos\delta - E''_d \sin^2\delta\right) \tag{15.20}$$

$$I''_{t(0)x}\cos\delta + I''_{t(0)y}\sin\delta = -\frac{1}{x''_d}E''_d = -\frac{1}{x''_d}\left(x_q - x''_q\right)\left(I_{tx}\cos\delta + I_{ty}\sin\delta\right) \tag{15.21}$$

Since,

$$\cos\delta\left(I''_{t(0)x}x''_d + \left(x_q - x''_d\right)I_{tx}\right) = \sin\delta\left(I_{ty}\left(x_q - x''_q\right) - I''_{t(0)y}x''_d\right) \tag{15.22}$$

We have,

$$tg\delta = \frac{I''_{t(0)(x)}x''_d + \left(x_q - x''_q\right)I_{tx}}{I_{ty}\left(x''_q - x_q\right) - I''_{t(0)y}x''_d} \tag{15.23}$$

The excitation system, governor, and PSS equilibrium points can be easily derived by setting the input error as zero and the control variables are not saturated.

### 15.3.3  Generator Injection Current

In each iteration, when the terminal voltages are solved by (15.2) and state variables are solved by (15.1), the generator injection current to the network in the $x - y$ reference frame can be calculated by

$$\begin{bmatrix} I_x \\ I_y \end{bmatrix} = \begin{bmatrix} g_x & b_x \\ b_y & g_y \end{bmatrix}\begin{bmatrix} E''_q \\ E''_d \end{bmatrix} - \begin{bmatrix} G_x & B_x \\ B_y & G_y \end{bmatrix}\begin{bmatrix} U_x \\ U_y \end{bmatrix} \tag{15.24}$$

where

$$G_x = \frac{R_a - \left(x''_d - x''_q\right)\sin\delta\cos\delta}{R_a^2 + x''_d x''_q} = \frac{R_a - \dfrac{\left(x''_d - x''_q\right)}{2}\sin 2\delta}{R_a^2 + x''_d x''_q}$$

$$B_x = \frac{x_d'' \cos^2\delta + x_q'' \sin^2\delta}{R_a^2 + x_d''x_q''} = \frac{\left(x_d'' + x_q''\right) + \left(x_d'' - x_q''\right)\cos 2\delta}{2\left(R_a^2 + x_d''x_q''\right)}$$

$$B_y = \frac{-x_d'' \sin^2\delta - x_q'' \cos^2\delta}{R_a^2 + x_d''x_q''} = \frac{-\left(x_d'' + x_q''\right) + \left(x_d'' - x_q''\right)\cos 2\delta}{2\left(R_a^2 + x_d''x_q''\right)}$$

$$G_y = \frac{R_a + \left(x_d'' - x_q''\right)\sin\delta\cos\delta}{R_a^2 + x_d''x_q''} = \frac{R_a + \frac{\left(x_d'' - x_q''\right)}{2}\sin 2\delta}{R_a^2 + x_d''x_q''}$$

$$g_x = \frac{R_a\cos\delta + x_q''\sin\delta}{R_a^2 + x_d''x_q''}$$

$$b_x = \frac{R_a\sin\delta - x_d''\cos\delta}{R_a^2 + x_d''x_q''}$$

$$b_y = \frac{R_a\sin\delta - x_q''\cos\delta}{R_a^2 + x_d''x_q''}$$

$$g_y = \frac{-R_a\cos\delta - x_d''\sin\delta}{R_a^2 + x_d''x_q''}$$

## 15.4 Transient Simulation by Graph Parallel Computing

Power system transient simulation involves solving complicated DAEs. To improve transient simulation efficiency for large-scale networks, graph parallel computing is developed.

In previous chapters, we investigated the feasibility of adopting graph parallel computing on topology processing [14, 15], state estimation [16–18], power flow analysis [19–27], contingency analysis [28–32], and power system optimization [33–37]. In this section, the graph-based transient simulation will be discussed.

### 15.4.1 Transient Simulation Graph

In power system transient simulation, a bus-branch model is used to describe a power system network. As addressed in Chapter 8, the vertex `TopoND` and the edge `TopoConnect` in the base bus-branch graph model are defined to contain the parameters and connections of elements. The bus-branch graph model is extendable to support state estimation, contingency analysis, economic dispatch, etc. by adding attributes of the vertex and/or edge. To model a generator and its control system dynamics, the vertex `TopoND` is further extended as follows:

```
CREATE VERTEX TopoND (PRIMARY_ID topoid int, TOPOID int, island int,
    bus_name string, area string, loss_zone uint, busType int, Vm
    double, Va double, M_Vm double, M_Va double, base_kV double,
    desired_volts double, control_bus_number int, up_V double, lo_V
    double, GenP double, GenQ double, M_Gen_P double, M_Gen_Q double,
    qUp double, qLower double, LdP double, LdQ double, P double,
```

```
Q double, M_Load_P double, M_Load_Q double, G double, B double, Sub
string, OV int, UV int, sumB double, sumG double, sumBi double,
volt double, double, M_C_P double, SE_Vm double, SE_Va double,
Ri_vP double, Ri_vQ double, Ri_eP double, Ri_eQ double, SE_Inj
tuple, GainP tuple, GainQ tuple, H_r_P double, H_r_Q double, Pmax
double, Pmin double, a double, b double, c double, Gen_Model int,
Gen_Par tuple, AVR_Model int, AVR_Par List<tuple>, GOV_Model int,
GOV_Par List<tuple>, PSS_Model int, PSS_Par List<tuple>, Gen_Var
tuple, AVR_Var List<tuple>, GOV_Var List<tuple>, PSS_Var
List<tuple>)
```

There is no update in the edge definition. In the vertex definition above, the new attributes **Gen_Par, AVR_Model, AVR_Par, GOV_Model, GOV_Par, PSS_Model, PSS_Par** are defined in the type of tuple to store generator and control system parameters. **Gen_Var, AVR_Var, GOV_Var, PSS_Var** are defined in the type of tuple to store generator and control system variables.

```
typedef tuple<double Xd, double Xdp, double Xdpp, double Xq, double
    Xqp, double Xqpp, double X2, double Ra, double Td0p, double Td0pp,
    double Tq0p, double Tq0pp, double Tj, double a, double b, double n,
    double D, double omega_ref> Gen_Par
```

In transient simulation modeling, generators are modeled in different types from the classical 2-order model to the detailed 6-order model. Tuple Gen_Par contains parameters covering different generator models. For different generator models from the classical 2-order model to the 6-order model, its parameters are a subset of Gen_Par (for Type-1 to Type-5 models) or the whole Gen_Par set (for Type-6 6-order model).

AVR, governor, and PSS have multiple types as well. However, the control structure and parameters are different from one type of model to another. They cannot be consolidated in one parameter set as the generator does. AVR_Model, GOV_Model, and PSS_Model are the indexes of the controller type. The control parameters of the referred controller type are defined in the list of tuples AVR_Par, GOV_Par, and PSS_Par as follows:

```
typedef tuple<double Kr, double Tr, double K2, double T1, double T2,
    double T3, double T4, double Ka, double Ta, double Efdmax, double
    Efdmin> AVR_Par[1]
typedef tuple<double Kr, double Tr, double Ks, double Ta, double Kf,
    double Tf, double Te, double Efdmax, double Efdmin> AVR_Par[2]
typedef tuple<double Kdelt, double dbL, double dbH, double sigmaMax,
    double sigmaMin, double Ts, double Tp, double muMax, double muMin,
    double Ki, double Kbeta, double Ti, double KmH, double Tw, double
    alpha, double Trh, double omega_ref> GOV_Par[1]
typedef tuple<double Kq1, double K, double Tq, double T1e, double T2e,
    double T3e, double T4e, double Vsmax, double Vsmin> PSS_Par[1]
```

Two types of AVR parameter sets are defined by the list of tuples AVR_Par[1] for self-excited AVR and AVR_Par[2] for separately excited AVR. The types of AVR are differentiated by the index 1 or 2 assigned in the vertex attribute AVR_Model. The list of the tuple is extendable to support more types of AVR. Governor and PSS parameter sets are defined in the same way. To simplify the illustration, one type of governor and one type of PSS are given without loss of generality.

Similarly, `Gen_Var`, `AVR_Var`, `GOV_Var`, `PSS_Var` are defined as follows:

```
typedef tuple<double Vx, double Vy, double Ix, double Iy, double EQx,
    double EQy, double omega_ref,
    double Vd, double Vq, double Id, double Iq, double omega, double
    delta, double Pe, double Edp, double Eqp, double Edpp, double Eqpp,
    double Vd0, double Vq0, double Id0, double Iq0, double omega0,
    double delta0, double Pe0, double Edp0, double Eqp0, double Edpp0,
    double Eqpp0> Gen_Var;
typedef tuple<double Va, double VR, double VA, double VF, double
    Vt_ref, double Efd_ref, double Efd,
    double Va0, double VR0, double VA0, double VF0, double dEfd0,
    double Efd0> AVR_Var[1];
typedef tuple<double Va, double V1, double V2, double V3, double V4,
    double Vt_ref, double Efd_ref, double Efd,
    double Va0, double V10, double V20, double V30, double V40, double
    Efd0> AVR_Var[2];
typedef tuple<double omega_ref, double mu_ref, double delta, double
    dmu, double mu, double F1, double F2, double Pm, double PT1, double
    PT2, double PT,
    double delta0, double dmu0, double mu0, double F20, double Pm0,
    double PT20, double PT0> GOV_Var[1];
typedef tuple<double dOmega, double VA, double VP1, double VP2, double
    Vs, double VA0, double VP10, double VP20, double Vs0> PSS_Var[1];
```

With the extended vertex, the `busBranchGraph` graph schema is updated by the following statement.

```
CREATE GRAPH busBranchGraph(TopoND, TopoConnect)
```

The bus-branch graph is not necessarily reformed since the additional attributes do not change the network topology and parameters.

### 15.4.2 Loading Data into Graph

Once a graph schema has been established, the next step is to input data into the graph. Using the loading job command offered by the graph database management system, the generator parameters and controller parameters are loaded into the bus-branch graph.

A variety of excitation systems, governors, and stabilizers are running in practical power systems. They are of different types and associated with different parameters. Assuming the parameters are given in the files `genPar.csv`, `avr1Par.csv`, `avr2Par.csv`, `gov1Par.csv`, and `pss1Par.csv`. The table header of each file is shown in Tables 15.5–15.9 as follows:

**Table 15.5**  Generator model parameter table header.

| Bus | $x_d$ | $x'_d$ | $x''_d$ | $x_q$ | $x'_q$ | $x''_q$ | $x_2$ | $R_a$ | $T'_{d0}$ | $T''_{d0}$ | $T'_{q0}$ | $T''_{q0}$ | $T_J$ | $a$ | $b$ | $n$ | D |
|---|---|---|---|---|---|---|---|---|---|---|---|---|---|---|---|---|---|

**Table 15.6** Self-excited AVR model parameter table header.

| Bus | $K_r$ | $T_r$ | $K_2$ | $T_1$ | $T_2$ | $T_3$ | $T_4$ | $K_a$ | $T_a$ | $E_{fdmax}$ | $E_{fdmin}$ |
|---|---|---|---|---|---|---|---|---|---|---|---|

**Table 15.7** Separately excited AVR parameter table header.

| Bus | $K_r$ | $T_r$ | $K_s$ | $T_a$ | $K_f$ | $T_f$ | $T_e$ | $E_{fdmax}$ | $E_{fdmin}$ |
|---|---|---|---|---|---|---|---|---|---|

**Table 15.8** Generic governor model parameter table header.

| Bus | $db$ | $T_G$ | $T_P$ | $R_P$ | $R_T$ | $T_R$ | $T_W$ | $U_C$ | $U_o$ | $P_{MIN}$ | $P_{MAX}$ | $a_{11}$ | $a_{13}$ | $a_{21}$ | $a_{23}$ |
|---|---|---|---|---|---|---|---|---|---|---|---|---|---|---|---|

**Table 15.9** PSS model parameter table header.

| Bus | $K_{q1}$ | $K$ | $T_q$ | $T_{1e}$ | $T_{2e}$ | $T_{3e}$ | $T_{4e}$ | $V_{smax}$ | $V_{smin}$ |
|---|---|---|---|---|---|---|---|---|---|

   With the updated graph vertex attributes, the data files are read to the new `busBranchGraph` by the following query `readParInfo`.

```
CREATE TABLE genParTable (double Xd, double Xdp, double Xdpp, double
Xq, double Xqp, double Xqpp, double X2, double Ra, double Td0p, double
Td0pp, double Tq0p, double Tq0pp, double Tj, double a, double b,
double n, double D)

CREATE TABLE avr1ParTable (double Kr, double Tr, double K2, double T1,
double T2, double T3, double T4, double Ka, double Ta, double Efdmax,
double Efdmin)

CREATE TABLE avr2ParTable (double Kr, double Tr, double Ks, double Ta,
double Kf, double Tf, double Te, double Efdmax, double Efdmin)

CREATE TABLE gov1ParTable (double dbL, double dbH, double TG, double
TP, double Rp, double TR, double TW, double Uc, double Uo, double Pmax,
double Pmin, double a11, double a13, double a21, double a23)

CREATE TABLE pss1ParTable (double Kq1, double K, double Tq, double
T1e, double T2e, double T3e, double T4e, double Vsmax, double Vsmi)

CREATE QUERY readParInfo() FOR GRAPH busBranchGraph {

//Read generator parameters
BULK INSERT genParTable
FROM '$sys.data_root/genPar.csv'
WITH(FIRSTROW = 2,FIELDTERMINATOR = ',',ROWTERMINATOR = '\n',TABLOCK)
T0= {TopoND.*}
T1 = SELECT s FROM T0
```

```
  ACCUM
    SELECT gen FROM genParTable WHERE gen.bus == s.TOPOID
      ACCUM
        INSERT INTO s.Gen_Par(gen.Xd, gen.Xdp, gen.Xdpp, gen.Xq, gen.Xqp,
          gen.Xqpp, gen.X2, gen.Ra, gen.Td0p, gen.Td0pp, gen.Tq0p, gen.
          Tq0pp, gen.Tj, gen.a, gen.b, gen.n, gen.D)
//Read self-excited AVR parameters
BULK INSERT avr1ParTable
FROM '$sys.data_root/avr1Par.csv'
WITH(FIRSTROW = 2,FIELDTERMINATOR = ',',ROWTERMINATOR = '\n',TABLOCK)
ACCUM
  SELECT avr FROM avr1ParTable WHERE avr.bus == s.TOPOID
    ACCUM
      INSERT INTO s.AVR_Par[1](avr.Kr, avr.Tr, avr.K2, avr.T1, avr.T2,
          avr.T3, avr.T4, avr.Ka, avr.Ta, avr.Efdmax, avr.Efdmin)
//Read separately excited AVR parameters
BULK INSERT avr2ParTable
FROM '$sys.data_root/avr2Par.csv'
WITH(FIRSTROW = 2,FIELDTERMINATOR = ',',ROWTERMINATOR = '\n',TABLOCK)
ACCUM
  SELECT avr FROM avr2ParTable WHERE avr.bus == s.TOPOID
    ACCUM
      INSERT INTO s.AVR_Par[2](avr.Kr, avr.Tr, avr.Ks, avr.Ta, avr.Kf,
          avr.Tf, avr.Te, avr.Efdmax, avr.Efdmin)
//Read governor parameters
BULK INSERT gov1ParTable
FROM '$sys.data_root/gov1Par.csv'
WITH(FIRSTROW = 2,FIELDTERMINATOR = ',',ROWTERMINATOR = '\n',TABLOCK)
ACCUM
 SELECT gov FROM gov1ParTable WHERE gov.bus == s.TOPOID
  ACCUM
    INSERT INTO s.GOV_Par[1](gov.dbL, gov.dbH, gov.TG, gov.TP, gov.Rp,
      gov.TR, gov.TW, gov.Uc, gov.Uo, gov.Pmax, gov.Pmin, gov.a11,
      gov.a13, gov.a21, gov.a23)
//Read power system stabilizer parameters
BULK INSERT pss1ParTable
FROM '$sys.data_root/pss1Par.csv'
WITH(FIRSTROW = 2,FIELDTERMINATOR = ',',ROWTERMINATOR = '\n',TABLOCK)
ACCUM
  SELECT pss FROM gov1ParTable WHERE pss.bus == s.TOPOID
    ACCUM
      INSERT INTO s.PSS_Par[1](pss.Kq1, pss.K, pss.Tq, pss.T1e, pss.
        T2e, pss.T3e, pss.T4e, pss.Vsmax, pss.Vsmin)
}
```

### 15.4.3 Graph-Based Transient Simulation Implementation

As shown in Figure 15.7, the power system transient simulation involves power flow initialization (step ①), steady-state equilibrium condition calculation (step ②), $Y$ matrix formation and factorization (step ③), bus voltage calculation (step ④), solving differential equations (step ⑤), and generator injection current calculation (step ⑥) iteratively.

Among these steps, power flow initialization (step ①), $Y$ matrix formation and factorization (step ③), and bus voltage calculation (step ④) have been addressed in Chapters 8 and 9. The implementation of steady-state equilibrium condition calculation (step ②), solving differential equations (step ⑤), and generator injection current calculation (step ⑥) are included in Appendix.

## 15.5 Numerical Example

To illustrate the transient simulation by graph parallel computing, the IEEE 14-bus modified test system is used. The single-line diagram of the IEEE-14 bus system is shown in Figure 15.9.

The IEEE 14-bus modified test system consists of 5 synchronous machines with IEEE type-1 exciters, 3 of which are synchronous compensators used only for reactive power support, 20 branches, and 14 buses with 11 loads. The total load demand is 259 MW and 73.5 MVAr.

### 15.5.1 Power Flow Data

The IEEE 14-bus modified test system branch data and bus data are listed in Tables 15.10 and 15.11.

### 15.5.2 Dynamic Data

The IEEE 14-bus modified test system machine data, exciter data, and governor data are listed in the following Tables 15.12–15.14 [38].

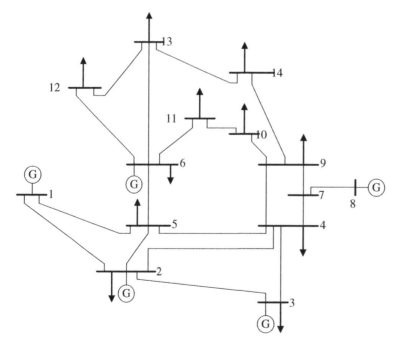

**Figure 15.9** Single-line diagram of the IEEE 14-bus system.

**Table 15.10** IEEE 14-bus test system branch data.

| Line number | From bus | To bus | Line impedance (p.u.) | | Line charging susceptance (p.u.) | Transformer turns ratio |
|---|---|---|---|---|---|---|
| | | | Resistance | Reactance | | |
| 1 | 1 | 2 | 0.0194 | 0.0592 | 0.0528 | 0.0000 |
| 2 | 1 | 5 | 0.0540 | 0.2230 | 0.0492 | 0.0000 |
| 3 | 2 | 3 | 0.0470 | 0.1980 | 0.0438 | 0.0000 |
| 4 | 2 | 4 | 0.0581 | 0.1763 | 0.0340 | 0.0000 |
| 5 | 2 | 5 | 0.0570 | 0.1739 | 0.0346 | 0.0000 |
| 6 | 3 | 4 | 0.0670 | 0.1710 | 0.0128 | 0.0000 |
| 7 | 4 | 5 | 0.0134 | 0.0421 | 0.0000 | 0.0000 |
| 8 | 4 | 7 | 0.0000 | 0.2091 | 0.0000 | 0.9780 |
| 9 | 4 | 9 | 0.0000 | 0.5562 | 0.0000 | 0.9690 |
| 10 | 5 | 6 | 0.0000 | 0.2520 | 0.0000 | 0.9320 |
| 11 | 6 | 11 | 0.0950 | 0.1989 | 0.0000 | 0.0000 |
| 12 | 6 | 12 | 0.1229 | 0.2558 | 0.0000 | 0.0000 |
| 13 | 6 | 13 | 0.0662 | 0.1303 | 0.0000 | 0.0000 |
| 14 | 7 | 8 | 0.0000 | 0.1762 | 0.0000 | 0.0000 |
| 15 | 7 | 9 | 0.0000 | 0.1100 | 0.0000 | 0.0000 |
| 16 | 9 | 10 | 0.0318 | 0.0845 | 0.0000 | 0.0000 |
| 17 | 9 | 14 | 0.1271 | 0.2704 | 0.0000 | 0.0000 |
| 18 | 10 | 11 | 0.0821 | 0.1921 | 0.0000 | 0.0000 |
| 19 | 12 | 13 | 0.2209 | 0.1999 | 0.0000 | 0.0000 |
| 20 | 13 | 14 | 0.1709 | 0.3480 | 0.0000 | 0.0000 |

**Table 15.11** IEEE 14-bus test system bus data.

| Bus number | Load_P (p.u.) | Load_Q (p.u.) | Unit_P (p.u.) | Unit_Q (p.u.) | Shunt_B (p.u.) |
|---|---|---|---|---|---|
| 1 | 0.0000 | 0.0000 | 2.3240 | −0.1690 | 0.0000 |
| 2 | 0.2170 | 0.1270 | 0.4000 | 0.4240 | 0.0000 |
| 3 | 0.9420 | 0.1900 | 0.0000 | 0.2340 | 0.0000 |
| 4 | 0.4780 | −0.0390 | 0.0000 | 0.0000 | 0.0000 |
| 5 | 0.0760 | 0.0160 | 0.0000 | 0.0000 | 0.0000 |
| 6 | 0.1120 | 0.0750 | 0.0000 | 0.1220 | 0.0000 |
| 7 | 0.0000 | 0.0000 | 0.0000 | 0.0000 | 0.0000 |
| 8 | 0.0000 | 0.0000 | 0.0000 | 0.1740 | 0.0000 |
| 9 | 0.2950 | 0.1660 | 0.0000 | 0.0000 | 0.1909 |
| 10 | 0.0900 | 0.0580 | 0.0000 | 0.0000 | 0.0000 |
| 11 | 0.0350 | 0.0180 | 0.0000 | 0.0000 | 0.0000 |
| 12 | 0.0610 | 0.0160 | 0.0000 | 0.0000 | 0.0000 |
| 13 | 0.1350 | 0.0580 | 0.0000 | 0.0000 | 0.0000 |
| 14 | 0.1490 | 0.0500 | 0.0000 | 0.0000 | 0.0000 |

**Table 15.12** IEEE 14-bus modified test system machine data.

| Type<br>Operation | GENROU<br>Sync. gen. | GENROU<br>Sync. gen. | GENROU<br>Condenser | GENROU<br>Condenser |
|---|---|---|---|---|
| Unit bus no. | 1 | 2 | 3 | 6, 8 |
| Rated power (MVA) | 448 | 100 | 40 | 25 |
| Rated voltage (kV) | 22 | 13.8 | 13.8 | 13.8 |
| Rated pf | 0.85 | 0.8 | 0.0 | 0.0 |
| $H$ (s) | 2.656 | 4.985 | 1.520 | 1.200 |
| $D$ | 2.000 | 2.000 | 0.000 | 0.000 |
| $R_a$ (p.u) | 0.0043 | 0.0035 | 0.000 | 0.0025 |
| $x_d$ (p.u) | 1.670 | 1.180 | 2.373 | 1.769 |
| $x_q$ (p.u) | 1.600 | 1.050 | 1.172 | 0.855 |
| $x'_d$ (p.u) | 0.265 | 0.220 | 0.343 | 0.304 |
| $x'_q$ (p.u) | 0.460 | 0.380 | 1.172 | 0.5795 |
| $x''_d$ (p.u) | 0.205 | 0.145 | 0.231 | 0.2035 |
| $x''_q$ (p.u) | 0.205 | 0.145 | 0.231 | 0.2035 |
| $x_l$ or $x_p$ (p.u) | 0.150 | 0.075 | 0.132 | 0.1045 |
| $T'_{d0}$ (s) | 0.5871 | 1.100 | 11.600 | 8.000 |
| $T'_{q0}$ (s) | 0.1351 | 0.1086 | 0.159 | 0.008 |
| $T''_{d0}$ (s) | 0.0248 | 0.0277 | 0.058 | 0.0525 |
| $T''_{q0}$ (s) | 0.0267 | 0.0351 | 0.201 | 0.0151 |
| $S(1.0)$ | 0.091 | 0.0933 | 0.295 | 0.304 |
| $S(1.2)$ | 0.400 | 0.4044 | 0.776 | 0.666 |

**Table 15.13** IEEE 14-bus modified test system exciter data.

| Type | IEEET1 | IEEET1 | IEEET1 | IEEET1 |
|---|---|---|---|---|
| Unit bus no. | 1 | 2 | 3 | 6, 8 |
| Rated power (MVA) | 448 | 100 | 40 | 25 |
| Rated voltage (kV) | 22 | 13.8 | 13.8 | 13.8 |
| $T_r$ (s) | 0.000 | 0.060 | 0.000 | 0.000 |
| $K_a$ (p.u) | 50 | 25 | 400 | 400 |
| $T_a$ (s) | 0.060 | 0.200 | 0.050 | 0.050 |
| $V_{Rmax}$ (p.u) | 1.000 | 1.000 | 6.630 | 4.407 |
| $V_{Rmin}$ (p.u) | −1.000 | −1.000 | −6.630 | −4.407 |
| $K_e$ (p.u) | −0.0465 | −0.0582 | −0.170 | −0.170 |
| $T_e$ (s) | 0.520 | 0.6544 | 0.950 | 0.950 |
| $K_f$ (p.u) | 0.0832 | 0.105 | 0.040 | 0.040 |
| $T_f$ (s) | 1.000 | 0.350 | 1.000 | 1.000 |
| $E_1$ (p.u) | 3.240 | 2.5785 | 6.375 | 4.2375 |
| $S_E (E_1)$ | 0.072 | 0.0889 | 0.2174 | 0.2174 |
| $E_2$ (p.u) | 4.320 | 3.438 | 8.500 | 5.650 |
| $S_E (E_2)$ | 0.2821 | 0.3468 | 0.9388 | 0.9386 |

**Table 15.14** IEEE 14-bus modified test system governor data.

| Type | BPA_GG | BPA_GG |
|---|---|---|
| Unit bus no. | 1 | 2 |
| Rated power (MVA) | 448 | 100 |
| Rated voltage (kV) | 22 | 13.8 |
| $P_{max}$ (p.u) | 0.870 | 1.050 |
| $R$ (p.u) | 0.011 | 0.050 |
| $T_1$ (s) | 0.100 | 0.090 |
| $T_2$ (s) | 0.000 | 0.000 |
| $T_3$ (s) | 0.300 | 0.200 |
| $T_4$ (s) | 0.050 | 0.300 |
| $T_5$ (s) | 10.000 | 0.000 |
| $F$ | 0.250 | 1.000 |

### 15.5.3 Power Flow Results

The first step of simulating the transient of the IEEE-14 bus system is to solve the network power flow. With the given network topology, parameters, and load assumptions, the power flow results are shown in Table 15.15.

**Table 15.15** IEEE 14-bus modified test system power flow solution.

| Bus number | Bus_Vm (p.u.) | Bus_Va (degree) | Unit_P (p.u.) | Unit_Q (p.u.) |
|---|---|---|---|---|
| 1 | 1.0600 | 0.0000 | 2.3415 | −0.1777 |
| 2 | 1.0450 | −5.2969 | 0.4000 | 0.1657 |
| 3 | 1.0021 | −7.3642 | 0.0000 | 0.4000 |
| 4 | 1.0159 | −8.8895 | 0.0000 | 0.0000 |
| 5 | 1.0190 | −7.8294 | 0.0000 | 0.0000 |
| 6 | 1.0700 | −13.1560 | 0.0000 | 0.1417 |
| 7 | 1.0599 | −12.0156 | 0.0000 | 0.0000 |
| 8 | 1.0900 | −12.0156 | 0.0000 | 0.1861 |
| 9 | 1.0536 | −13.6364 | 0.0000 | 0.0000 |
| 10 | 1.0491 | −13.8363 | 0.0000 | 0.0000 |
| 11 | 1.0560 | −13.6248 | 0.0000 | 0.0000 |
| 12 | 1.0550 | −13.9961 | 0.0000 | 0.0000 |
| 13 | 1.0501 | −14.0576 | 0.0000 | 0.0000 |
| 14 | 1.0341 | −14.8223 | 0.0000 | 0.0000 |

### 15.5.4  Steady-State Equilibrium Point

To solve the ordinary differential equations (ODE) (15.1), the power system state variable equilibrium points are calculated with the generator terminal voltages solved by power flow equations.

For the unit at Bus 1, the generator power output is converted to its rated power base as follows:

$$S_{(0)} = \left(P_{(0)} + jQ_{(0)}\right) \times \frac{S_B}{Rated\_MVA} = (2.3415 - j0.1777) \times \frac{100}{448} = 0.5227 - j0.0397$$

Then the current and power angle are calculated by (15.25) and (15.27)

$$\dot{I}_{t(0)} = I_{x(0)} + jI_{y(0)} \left(\frac{P_{(0)} + jQ_{(0)}}{\dot{U}_{(0)}}\right)^* = \frac{0.5227 + j0.0397}{1.06} = 0.4931 + j0.0375 \tag{15.25}$$

$$\dot{E}_{Q(0)} = \dot{U}_{(0)} + \left(R_a + jx_q\right)\dot{I}_{t(0)} = 1.06 + (0.0043 + j1.6)(0.4931 + j0.0375) = 1.0022 + j0.7891 \tag{15.26}$$

$$\delta_{(0)} = arctg\left(\frac{E_{Qy(0)}}{E_{Qx(0)}}\right) = 38.22° \tag{15.27}$$

To initialize $E'_{q(0)}$, $E'_{d(0)}$, $E''_{q(0)}$, and $E''_{d(0)}$, current and voltage in $x - y$ coordinates are converted to be in $d - q$ coordinates as follows:

$$\begin{bmatrix} I_{d(0)} \\ I_{q(0)} \end{bmatrix} = \begin{bmatrix} sin(\delta_{(0)}) & -cos(\delta_{(0)}) \\ cos(\delta_{(0)}) & sin(\delta_{(0)}) \end{bmatrix} \begin{bmatrix} I_{x(0)} \\ I_{y(0)} \end{bmatrix} = \begin{bmatrix} sin(38.22°) & -cos(38.22°) \\ cos(38.22°) & sin(38.22°) \end{bmatrix} \begin{bmatrix} 0.4931 \\ 0.0375 \end{bmatrix} = \begin{bmatrix} 0.2756 \\ 0.4106 \end{bmatrix} \tag{15.28}$$

$$\begin{bmatrix} U_{d(0)} \\ U_{q(0)} \end{bmatrix} = \begin{bmatrix} sin(\delta_{(0)}) & -cos(\delta_{(0)}) \\ cos(\delta_{(0)}) & sin(\delta_{(0)}) \end{bmatrix} \begin{bmatrix} U_{x(0)} \\ U_{y(0)} \end{bmatrix} = \begin{bmatrix} sin(38.22°) & -cos(38.22°) \\ cos(38.22°) & sin(38.22°) \end{bmatrix} \begin{bmatrix} 1.06 \\ 0 \end{bmatrix} = \begin{bmatrix} 0.6557 \\ 0.8326 \end{bmatrix} \tag{15.29}$$

Then $E'_{q(0)}$, $E'_{d(0)}$, $E''_{q(0)}$, and $E''_{d(0)}$ are calculated by (15.30)–(15.33).

$$E'_{q(0)} = U_{q(0)} + R_a I_{q(0)} + x'_d I_{d(0)} = 0.8326 + 0.0043 \times 0.4106 + 0.265 \times 0.2756 = 0.9074 \tag{15.30}$$

$$E'_{d(0)} = U_{d(0)} + R_a I_{d(0)} - x'_q I_{q(0)} = 0.6557 + 0.0043 \times 0.2756 - 0.460 \times 0.4106 = 0.4680 \tag{15.31}$$

$$E''_{q(0)} = U_{q(0)} + R_a I_{q(0)} + x''_d I_{d(0)} = 0.8326 + 0.0043 \times 0.4106 + 0.205 \times 0.2756 = 0.8909 \tag{15.32}$$

$$E''_{d(0)} = U_{d(0)} + R_a I_{d(0)} - x''_q I_{q(0)} = 0.6557 + 0.0043 \times 0.2756 - 0.205 \times 0.4106 = 0.5727$$

$$(15.33)$$

For the unit at Bus 2, the generator power output is converted to its rated power base as follows:

$$S_{(0)} = \left(P_{(0)} + jQ_{(0)}\right) \times \frac{S_B}{Rated\_MVA} = (0.4 + j0.1657) \times \frac{100}{100} = 0.4 + j0.1657$$

Then the current and power angle are calculated by (15.34) and (15.36)

$$\dot{I}_{t(0)} = I_{x(0)} + jI_{y(0)} \left(\frac{P_{(0)} + jQ_{(0)}}{\dot{U}_{(0)}}\right)^* = \frac{0.4 - j0.1657}{1.045(cos(-5.30°) - sin(-5.30°))}$$

$$= \frac{0.4 - j0.1657}{1.0405 + j0.0965} = 0.3665 - j0.1932$$

$$(15.34)$$

$$\dot{E}_{Q(0)} = \dot{U}_{(0)} + (R_a + jx_q)\dot{I}_{t(0)} = 1.0405 - j0.0965 + (0.0035 + j1.05)(0.3665 - j0.1932)$$

$$= 1.2447 + j0.2876$$

$$(15.35)$$

$$\delta_{(0)} = arctg\left(\frac{E_{Qy(0)}}{E_{Qx(0)}}\right) = 13.01° \qquad (15.36)$$

To initialize $E'_{q(0)}$, $E'_{d(0)}$, $E''_{q(0)}$, and $E''_{d(0)}$, current and voltage in $x - y$ coordinates are converted to be in $d - q$ coordinates as follows:

$$\begin{bmatrix} I_{d(0)} \\ I_{q(0)} \end{bmatrix} = \begin{bmatrix} sin(\delta_{(0)}) & -cos(\delta_{(0)}) \\ cos(\delta_{(0)}) & sin(\delta_{(0)}) \end{bmatrix} \begin{bmatrix} I_{x(0)} \\ I_{y(0)} \end{bmatrix} = \begin{bmatrix} sin(13.01°) & -cos(13.01°) \\ cos(13.01°) & sin(13.01°) \end{bmatrix} \begin{bmatrix} 0.3665 \\ -0.1932 \end{bmatrix}$$

$$= \begin{bmatrix} 0.2707 \\ 0.3136 \end{bmatrix}$$

$$(15.37)$$

$$\begin{bmatrix} U_{d(0)} \\ U_{q(0)} \end{bmatrix} = \begin{bmatrix} sin(\delta_{(0)}) & -cos(\delta_{(0)}) \\ cos(\delta_{(0)}) & sin(\delta_{(0)}) \end{bmatrix} \begin{bmatrix} U_{x(0)} \\ U_{y(0)} \end{bmatrix} = \begin{bmatrix} sin(13.01°) & -cos(13.01°) \\ cos(13.01°) & sin(13.01°) \end{bmatrix} \begin{bmatrix} 1.0405 \\ -0.0965 \end{bmatrix}$$

$$= \begin{bmatrix} 0.3283 \\ 0.9921 \end{bmatrix}$$

$$(15.38)$$

Then $E'_{q(0)}$, $E'_{d(0)}$, $E''_{q(0)}$, and $E''_{d(0)}$ are calculated by (15.39)–(15.42).

$$E'_{q(0)} = U_{q(0)} + R_a I_{q(0)} + x'_d I_{d(0)} = 0.9921 + 0.0035 \times 0.3136 + 0.220 \times 0.2707 = 1.0528$$

$$(15.39)$$

$$E'_{d(0)} = U_{d(0)} + R_a I_{d(0)} - x'_q I_{q(0)} = 0.3283 + 0.0035 \times 0.2707 - 0.380 \times 0.3136 = 0.2101$$

$$(15.40)$$

$$E''_{q(0)} = U_{q(0)} + R_a I_{q(0)} + x''_d I_{d(0)} = 0.9921 + 0.0035 \times 0.3136 + 0.145 \times 0.2707 = 1.0324$$

$$(15.41)$$

$$E''_{d(0)} = U_{d(0)} + R_a I_{d(0)} - x''_q I_{q(0)} = 0.3283 + 0.0035 \times 0.2707 - 0.145 \times 0.3136 = 0.2838$$

$$(15.42)$$

For the condenser at Bus 3, the condenser output is converted to its rated power base as follows:

$$S_{(0)} = \left(P_{(0)} + jQ_{(0)}\right) \times \frac{S_B}{Rated\_MVA} = (0 + j0.4) \times \frac{100}{40} = 0 + j1$$

Then the current and power angle are calculated by (15.43) and (15.45)

$$\dot{I}_{t(0)} = I_{x(0)} + jI_{y(0)} \left(\frac{P_{(0)} + jQ_{(0)}}{\dot{U}_{(0)}}\right)^* = \frac{0 - j1}{1.0021\left(cos(-7.36°) - sin(-7.36°)\right)}$$

$$(15.43)$$

$$= \frac{0 - j1}{0.9938 + j0.1284} = -0.1278 - j0.9897$$

$$\dot{E}_{Q(0)} = \dot{U}_{(0)} + \left(R_a + jx_q\right)\dot{I}_{t(0)} = 0.9938 - j0.1284 + (0 + j1.172)(-0.1278 - j0.9897)$$

$$= 2.1538 - j0.2782$$

$$(15.44)$$

$$\delta_{(0)} = arctg\left(\frac{E_{Qy(0)}}{E_{Qx(0)}}\right) = -7.36°$$

$$(15.45)$$

To initialize $E'_{q(0)}$, $E'_{d(0)}$, $E''_{q(0)}$, and $E''_{d(0)}$, current and voltage in $x - y$ coordinates are converted to be in $d - q$ coordinates as follows:

$$\begin{bmatrix} I_{d(0)} \\ I_{q(0)} \end{bmatrix} = \begin{bmatrix} sin(\delta_{(0)}) & -cos(\delta_{(0)}) \\ cos(\delta_{(0)}) & sin(\delta_{(0)}) \end{bmatrix} \begin{bmatrix} I_{x(0)} \\ I_{y(0)} \end{bmatrix} = \begin{bmatrix} sin(-7.36°) & -cos(-7.36°) \\ cos(-7.36°) & sin(-7.36°) \end{bmatrix} \begin{bmatrix} -0.1278 \\ -0.9897 \end{bmatrix}$$

$$= \begin{bmatrix} 0.9979 \\ 0 \end{bmatrix}$$

$$(15.46)$$

$$\begin{bmatrix} U_{d(0)} \\ U_{q(0)} \end{bmatrix} = \begin{bmatrix} sin(\delta_{(0)}) & -cos(\delta_{(n)}) \\ cos(\delta_{(0)}) & sin(\delta_{(0)}) \end{bmatrix} \begin{bmatrix} U_{x(0)} \\ U_{y(0)} \end{bmatrix} = \begin{bmatrix} sin(-7.36°) & -cos(-7.36°) \\ cos(-7.36°) & sin(-7.36°) \end{bmatrix} \begin{bmatrix} 0.9938 \\ -0.1284 \end{bmatrix}$$

$$= \begin{bmatrix} 0 \\ 1.0021 \end{bmatrix}$$

$$(15.47)$$

Then $E'_{q(0)}$, $E'_{d(0)}$, $E''_{q(0)}$, and $E''_{d(0)}$ are calculated by (15.48)–(15.51)

$$E'_{q(0)} = U_{q(0)} + R_a I_{q(0)} + x'_d I_{d(0)} = 1.0021 + 0.343 \times 0.9979 = 1.3443 \tag{15.48}$$

$$E'_{d(0)} = U_{d(0)} + R_a I_{d(0)} - x'_q I_{q(0)} = 0 - 1.172 \times 0 = 0 \tag{15.49}$$

$$E''_{q(0)} = U_{q(0)} + R_a I_{q(0)} + x''_d I_{d(0)} = 1.0021 + 0.231 \times 0.9979 = 1.2326 \tag{15.50}$$

$$E''_{d(0)} = U_{d(0)} + R_a I_{d(0)} - x''_q I_{q(0)} = 0 - 0.231 \times 0 = 0 \tag{15.51}$$

For the condenser at Bus 6, the condenser output is converted to its rated power base as follows:

$$S_{(0)} = \left(P_{(0)} + jQ_{(0)}\right) \times \frac{S_B}{Rated\_MVA} = (0 + j0.1417) \times \frac{100}{25} = 0 + j0.5668$$

Then the current and power angle are calculated by (15.52) and (15.54)

$$\dot{I}_{t(0)} = I_{x(0)} + jI_{y(0)} \left(\frac{P_{(0)} + jQ_{(0)}}{\dot{U}_{(0)}}\right)^* = \frac{0 - j0.5668}{1.07(cos(-13.16°) - sin(-13.16°))} = \frac{0 - j0.5668}{1.0419 + j0.2436}$$

$$= -0.1206 - j0.5158$$

$$\tag{15.52}$$

$$\dot{E}_{Q(0)} = \dot{U}_{(0)} + \left(R_a + jx_q\right)\dot{I}_{t(0)} = 1.0419 - j0.2436 + (0.0025 + j0.855)(-0.1206 - j0.5158)$$

$$= 1.4826 - j0.3480$$

$$\tag{15.53}$$

$$\delta_{(0)} = arctg\left(\frac{E_{Qy(0)}}{E_{Qx(0)}}\right) = -13.21° \tag{15.54}$$

To initialize $E'_{q(0)}$, $E'_{d(0)}$, $E''_{q(0)}$, and $E''_{d(0)}$ current and voltage in $x - y$ coordinates is converted to be in $d - q$ coordinates as follows:

$$\begin{bmatrix} I_{d(0)} \\ I_{q(0)} \end{bmatrix} = \begin{bmatrix} sin(\delta_{(0)}) & -cos(\delta_{(0)}) \\ cos(\delta_{(0)}) & sin(\delta_{(0)}) \end{bmatrix} \begin{bmatrix} I_{x(0)} \\ I_{y(0)} \end{bmatrix}$$

$$= \begin{bmatrix} sin(-13.21°) & -cos(-13.21°) \\ cos(-13.21°) & sin(-13.21°) \end{bmatrix} \begin{bmatrix} -0.1206 \\ -0.5158 \end{bmatrix} = \begin{bmatrix} 0.5297 \\ 0.0004 \end{bmatrix}$$

$$\tag{15.55}$$

$$\begin{bmatrix} U_{d(0)} \\ U_{q(0)} \end{bmatrix} = \begin{bmatrix} sin(\delta_{(0)}) & -cos(\delta_{(0)}) \\ cos(\delta_{(0)}) & sin(\delta_{(0)}) \end{bmatrix} \begin{bmatrix} U_{x(0)} \\ U_{y(0)} \end{bmatrix}$$

$$= \begin{bmatrix} sin(-13.21°) & -cos(-13.21°) \\ cos(-13.21°) & sin(-13.21°) \end{bmatrix} \begin{bmatrix} 1.0419 \\ -0.2436 \end{bmatrix} = \begin{bmatrix} -0.0009 \\ 1.0700 \end{bmatrix}$$

$$\tag{15.56}$$

Then $E'_{q(0)}$, $E'_{d(0)}$, $E''_{q(0)}$, and $E''_{d(0)}$ are calculated by (15.57)–(15.60)

$$E'_{q(0)} = U_{q(0)} + R_a I_{q(0)} + x'_d I_{d(0)} = 1.0700 + 0.0025 \times 0.0004 + 0.304 \times 0.5297 = 1.2310$$

$$(15.57)$$

$$E'_{d(0)} = U_{d(0)} + R_a I_{d(0)} - x'_q I_{q(0)} = -0.0009 + 0.0025 \times 0.5297 - 0.5795 \times 0.0004 = 0.0002$$

$$(15.58)$$

$$E''_{q(0)} = U_{q(0)} + R_a I_{q(0)} + x''_d I_{d(0)} = 1.0700 + 0.0025 \times 0.0004 + 0.2035 \times 0.5297 = 1.1778$$

$$(15.59)$$

$$E''_{d(0)} = U_{d(0)} + R_a I_{d(0)} - x''_q I_{q(0)} = -0.0009 + 0.0025 \times 0.5297 - 0.2035 \times 0.0004 = 0.0003$$

$$(15.60)$$

For the condenser at Bus 8, the condenser output is converted to its rated power base as follows:

$$S_{(0)} = \left(P_{(0)} + jQ_{(0)}\right) \times \frac{S_B}{Rated\_MVA} = (0 + j0.1861) \times \frac{100}{25} = 0 + j0.7444$$

Then the current and power angle are calculated by (15.52) and (15.54)

$$\dot{I}_{t(0)} = I_{x(0)} + jI_{y(0)} \left(\frac{P_{(0)} + jQ_{(0)}}{\dot{U}_{(0)}}\right)^* = \frac{0 - j0.7444}{1.09(cos(-12.02°) - sin(-12.02°))}$$

$$= \frac{0 - j0.7444}{1.0661 + j0.2270} = -0.1422 - j0.6680$$

$$(15.61)$$

$$\dot{E}_{Q(0)} = \dot{U}_{(0)} + \left(R_a + jx_q\right)\dot{I}_{t(0)} = 1.0661 - j0.227 + (0.0025 + j0.855)(-0.1422 - j0.6680)$$

$$= 1.6367 - j0.3503$$

$$(15.62)$$

$$\delta_{(0)} = arctg\left(\frac{E_{Qy(0)}}{E_{Qx(0)}}\right) = -12.09°$$

$$(15.63)$$

To initialize $E'_{q(0)}$, $E'_{d(0)}$, $E''_{q(0)}$, and $E''_{d(0)}$ current and voltage in $x - y$ coordinates are converted to be in $d - q$ coordinates as follows:

$$\begin{bmatrix} I_{d(0)} \\ I_{q(0)} \end{bmatrix} = \begin{bmatrix} sin(\delta_{(0)}) & -cos(\delta_{(0)}) \\ cos(\delta_{(0)}) & sin(\delta_{(0)}) \end{bmatrix} \begin{bmatrix} I_{x(0)} \\ I_{y(0)} \end{bmatrix}$$

$$= \begin{bmatrix} sin(-12.09°) & -cos(-12.09°) \\ cos(-12.09°°) & sin(-12.09°) \end{bmatrix} \begin{bmatrix} -0.1422 \\ -0.6680 \end{bmatrix} = \begin{bmatrix} 0.6829 \\ 0.0007 \end{bmatrix}$$

$$(15.64)$$

$$\begin{bmatrix} U_{d(0)} \\ U_{q(0)} \end{bmatrix} = \begin{bmatrix} sin(\delta_{(0)}) & -cos(\delta_{(0)}) \\ cos(\delta_{(0)}) & sin(\delta_{(0)}) \end{bmatrix} \begin{bmatrix} U_{x(0)} \\ U_{y(0)} \end{bmatrix}$$

$$= \begin{bmatrix} sin(-12.09°) & -cos(-12.09°) \\ cos(-12.09°°) & sin(-12.09°) \end{bmatrix} \begin{bmatrix} 1.0661 \\ -0.2270 \end{bmatrix} = \begin{bmatrix} -0.0011 \\ 1.0900 \end{bmatrix}$$

$$(15.65)$$

Then $E'_{q(0)}$, $E'_{d(0)}$, $E''_{q(0)}$, and $E''_{d(0)}$ are calculated by (15.66)–(15.69).

$$E'_{q(0)} = U_{q(0)} + R_a I_{q(0)} + x'_d I_{d(0)} = 1.0900 + 0.0025 \times 0.0007 + 0.304 \times 0.6829 = 1.2976 \tag{15.66}$$

$$E'_{d(0)} = U_{d(0)} + R_a I_{d(0)} - x'_q I_{q(0)} = -0.0011 + 0.0025 \times 0.6829 - 0.5795 \times 0.0007 = 0.0002 \tag{15.67}$$

$$E''_{q(0)} = U_{q(0)} + R_a I_{q(0)} + x''_d I_{d(0)} = 1.0900 + 0.0025 \times 0.0007 + 0.2035 \times 0.6829 = 1.2290 \tag{15.68}$$

$$E''_{d(0)} = U_{d(0)} + R_a I_{d(0)} - x''_q I_{q(0)} = -0.0011 + 0.0025 \times 0.6829 - 0.2035 \times 0.0007 = 0.0005 \tag{15.69}$$

The steady-state equilibrium points of all units are summarized in Table 15.16 as an initial condition to solve generator differential equations.

### 15.5.5 Generator Injection Current Calculation

The generator injection current is a function of the generator terminal voltage and generator state variables. In each iteration, when the terminal voltages and state variables are solved, the generator injection current to the network in the $x - y$ reference frame can be calculated.

In the case of the IEEE 14-Bus system, in the first iteration, for the unit at Bus 1, the injection current in the $d - q$ reference frame by (15.70)

$$\begin{bmatrix} U_d \\ U_q \end{bmatrix} = \begin{bmatrix} E''_d \\ E''_q \end{bmatrix} - \begin{bmatrix} R_a & -x''_q \\ x''_d & R_a \end{bmatrix} \begin{bmatrix} I_d \\ I_q \end{bmatrix} \tag{15.70}$$

Then, we have

$$\begin{bmatrix} U_d \\ U_q \end{bmatrix} = \begin{bmatrix} E''_d \\ E''_q \end{bmatrix} - \begin{bmatrix} R_a & -x''_q \\ x''_d & R_a \end{bmatrix} \begin{bmatrix} I_d \\ I_q \end{bmatrix} \tag{15.71}$$

**Table 15.16** Unit steady-state equilibrium points.

| Unit bus no. | $\delta_{(0)}$ | $E''_{d(0)}$ | $E''_{q(0)}$ | $E'_{d(0)}$ | $E'_{q(0)}$ | $\omega_{(0)}$ |
|---|---|---|---|---|---|---|
| 1 | 0.6670 | 0.5727 | 0.8909 | 0.4680 | 0.9074 | 1.0000 |
| 2 | 0.2271 | 0.2838 | 1.0324 | 0.2101 | 1.0528 | 1.0000 |
| 3 | −0.1285 | 0.0000 | 1.2326 | 0.0000 | 1.3443 | 1.0000 |
| 6 | −0.2306 | 0.0003 | 1.1778 | 0.0002 | 1.2310 | 1.0000 |
| 8 | −0.2110 | 0.0005 | 1.2290 | 0.0002 | 1.2976 | 1.0000 |

**Table 15.17** IEEE 14-bus modified test system generator injection current.

| Unit bus no. | $I_x$ | $I_y$ |
|---|---|---|
| 1 | 2.2149 | 0.0748 |
| 2 | 0.3565 | −0.2103 |
| 3 | −0.0511 | −0.3958 |
| 6 | −0.0333 | −0.1282 |
| 8 | −0.0396 | −0.1661 |

$$\begin{bmatrix} 0.6557 \\ 0.8326 \end{bmatrix} = \begin{bmatrix} 0.5727 \\ 0.8909 \end{bmatrix} - \begin{bmatrix} 0.0043 & 0.205 \\ -0.205 & 0.0043 \end{bmatrix} \begin{bmatrix} I_d \\ I_q \end{bmatrix} \tag{15.72}$$

By solving (15.72), we get

$$\begin{bmatrix} I_d \\ I_q \end{bmatrix} = - \begin{bmatrix} 0.2928 \\ 0.3987 \end{bmatrix} \tag{15.73}$$

Converting (15.73) to the system base in the $x - y$ reference frame, the generator injection current is

$$\begin{bmatrix} I_x \\ I_y \end{bmatrix} = \begin{bmatrix} sin(\delta_{(0)}) & - cos(\delta_{(0)}) \\ cos(\delta_{(0)}) & sin(\delta_{(0)}) \end{bmatrix} \begin{bmatrix} I_d \\ I_q \end{bmatrix} \times \frac{Rated\_MVA}{S_B}$$

$$= \begin{bmatrix} sin(38.22°) & cos(38.22°) \\ - cos(38.22°) & sin(38.22°) \end{bmatrix} \begin{bmatrix} 0.2928 \\ 0.3987 \end{bmatrix} \times \frac{448}{100} = \begin{bmatrix} 2.2149 \\ 0.0748 \end{bmatrix}$$

Similarly, the injection currents of other units are calculated as summarized in Table 15.17.

### 15.5.6 Calculate Bus Voltage

With the calculated generator injection current, when no fault occurs and topology does not change, the admittance matrix does not need to reform and factorize. The bus voltage in the algebraic Eq. (15.2) is updated as shown in Table 15.18.

The differences in the bus voltage in the two iterations are within tolerance, the simulation is moved to the next time step.

### 15.5.7 Simulation Results

To simulate the transient under a fault condition, a case study is constructed. Initially, the system is operated at the steady-state condition. a three-phase fault at bus 5 at 2.0 seconds and clean the fault at 2.2 seconds. The terminal bus voltage magnitude of Generators right before, during, and right after the fault are listed in Table 15.19.

**Table 15.18** Bus voltage updates.

| Bus number | Bus_$Vm$ (p.u.) | Bus_$Va$ (degree) |
|---|---|---|
| 1 | 1.0600 | −0.0724 |
| 2 | 1.0450 | −5.3654 |
| 3 | 1.0021 | −7.4325 |
| 4 | 1.0158 | −8.9595 |
| 5 | 1.0190 | −7.8995 |
| 6 | 1.0700 | −13.2276 |
| 7 | 1.0599 | −12.0882 |
| 8 | 1.0900 | −12.0912 |
| 9 | 1.0536 | −13.7085 |
| 10 | 1.0491 | −13.9084 |
| 11 | 1.0560 | −13.6966 |
| 12 | 1.0550 | −14.0677 |
| 13 | 1.0500 | −14.1292 |
| 14 | 1.0340 | −14.8942 |

**Table 15.19** Generator bus and faulted bus voltage.

| Time (s) | $U_x$ | | | | | | $U_y$ | | | | | |
|---|---|---|---|---|---|---|---|---|---|---|---|---|
| | Bus 1 | Bus 2 | Bus 3 | Bus 5 | Bus 6 | Bus 8 | Bus 1 | Bus 2 | Bus 3 | Bus 5 | Bus 6 | Bus 8 |
| 0.00 | 1.06 | 1.04 | 0.99 | 1.01 | 1.04 | 1.07 | 0.00 | −0.10 | −0.13 | −0.14 | −0.24 | −0.23 |
| ... | ... | ... | ... | ... | ... | ... | ... | ... | ... | ... | ... | ... |
| 1.98 | 1.06 | 1.03 | 0.98 | 0.99 | 1.01 | 1.04 | −0.10 | −0.19 | −0.22 | −0.23 | −0.34 | −0.32 |
| 2.00 | 1.06 | 1.03 | 0.98 | 0.99 | 1.01 | 1.04 | −0.10 | −0.19 | −0.22 | −0.23 | −0.34 | −0.33 |
| 2.02 | 0.71 | 0.53 | 0.47 | 0.00 | 0.24 | 0.43 | −0.05 | −0.10 | −0.09 | 0.00 | −0.10 | −0.15 |
| 2.04 | 0.66 | 0.49 | 0.43 | 0.00 | 0.22 | 0.41 | −0.02 | −0.08 | −0.08 | 0.00 | −0.09 | −0.14 |
| 2.06 | 0.63 | 0.47 | 0.41 | 0.00 | 0.21 | 0.39 | 0.00 | −0.06 | −0.06 | 0.00 | −0.09 | −0.13 |
| 2.08 | 0.61 | 0.46 | 0.40 | 0.00 | 0.20 | 0.38 | 0.03 | −0.04 | −0.05 | 0.00 | −0.09 | −0.13 |
| 2.10 | 0.60 | 0.45 | 0.40 | 0.00 | 0.20 | 0.37 | 0.05 | −0.03 | −0.04 | 0.00 | −0.09 | −0.13 |
| 2.12 | 0.60 | 0.45 | 0.40 | 0.00 | 0.19 | 0.37 | 0.08 | −0.01 | −0.02 | 0.00 | −0.09 | −0.13 |
| 2.14 | 0.61 | 0.46 | 0.40 | 0.00 | 0.19 | 0.37 | 0.10 | 0.00 | −0.01 | 0.00 | −0.09 | −0.12 |
| 2.16 | 0.62 | 0.47 | 0.41 | 0.00 | 0.19 | 0.37 | 0.14 | 0.02 | 0.01 | 0.00 | −0.09 | −0.12 |
| 2.18 | 0.65 | 0.49 | 0.42 | 0.00 | 0.19 | 0.37 | 0.17 | 0.05 | 0.03 | 0.00 | −0.09 | −0.12 |
| 2.20 | 0.67 | 0.50 | 0.44 | 0.00 | 0.19 | 0.38 | 0.21 | 0.07 | 0.06 | 0.00 | −0.09 | −0.11 |
| 2.22 | 1.00 | 0.98 | 0.94 | 0.96 | 0.99 | 1.00 | 0.35 | 0.21 | 0.17 | 0.16 | 0.03 | 0.02 |

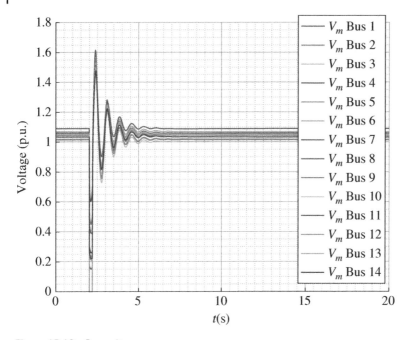

**Figure 15.10** Bus voltage.

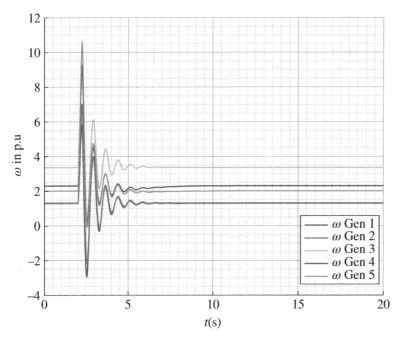

**Figure 15.11** Generator rotor speed.

The bus voltage magnitude, generator rotor speed $\omega$, filed voltage $E_{fd}$, transient voltage $E'_d$, $E'_q$, and subtransient voltage $E''_d$, $E''_q$ are shown in Figures 15.10–15.16.

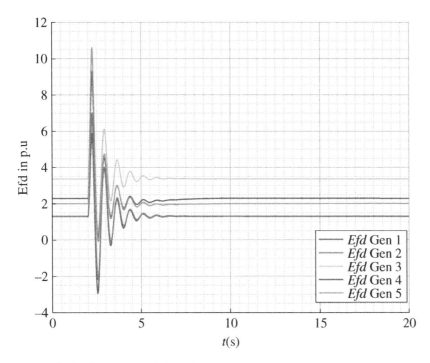

**Figure 15.12** Generator field voltage.

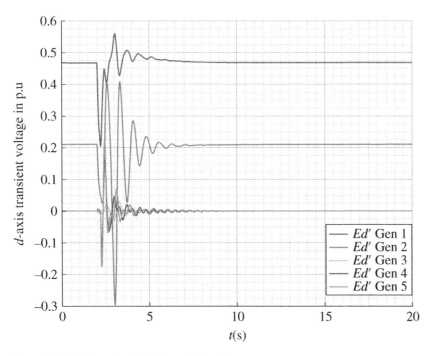

**Figure 15.13** Generator d-axis transient voltage.

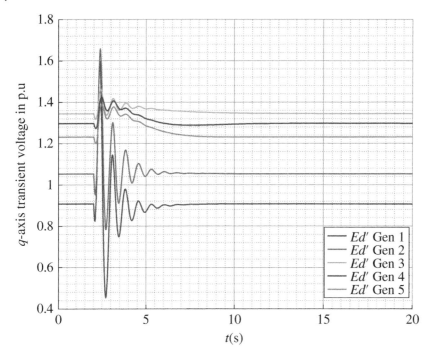

**Figure 15.14** Generator *q*-axis transient voltage.

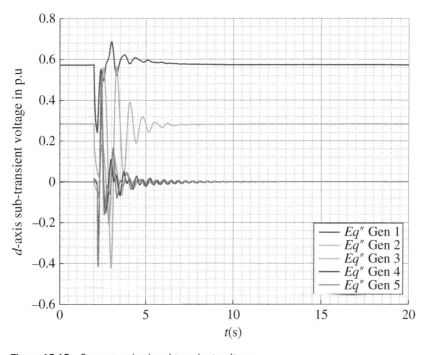

**Figure 15.15** Generator *d*-axis subtransient voltage.

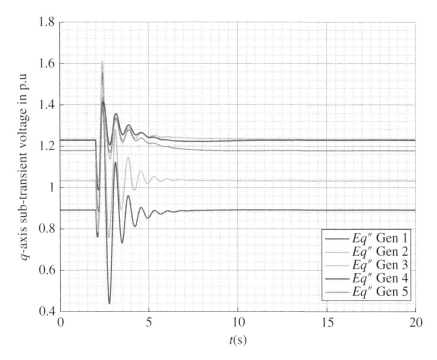

**Figure 15.16** Generator $q$-axis subtransient voltage.

# References

**1** Lyapunov, A.M. The general problem of the stability of motion. *International Journal of Control* 55 (3): 531–534.

**2** Lyapunov, A.M. (1967). *Stability of Motion*, English translation. Academic Press.

**3** Chang, H.-D., Chu, C.-C., and Cauley, G. (1995). Direct stability analysis of electric power systems using energy functions: theory, applications, and perspective. *Proceedings of the IEEE* 83 (11): 1497–1529.

**4** Alberto, L.F.C. and Chiang, H.D. (2008). Controlling unstable equilibrium point theory for stability assessment of two-time scale power system models. *2008 IEEE Power and Energy Society General Meeting - Conversion and Delivery of Electrical Energy in the 21st Century*, Pittsburgh, PA, USA, pp. 1–9. https://doi.org/10.1109/PES.2008.4596222.

**5** Chiang, H.D., Wu, F.F., and Varaiya, P. (1994). A BCU for direct analysis of power system transient stability. *IEEE Transactions on Power System* 9 (3): 1194–1208.

**6** Xue, Y., Van Cutsem, T., and Ribbens-Pavella, M. (1988). A simple direct method for fast transient stability assessment of large power systems. *IEEE Transactions on Power Systems* 3 (2): 400–412.

**7** Xue, Y., Van Custem, T., and Ribbens-Pavella, M. (1989). Extended equal area criterion justifications, generalizations, applications. *IEEE Transactions on Power Systems* 4 (1): 44–52.

**8** Chiang, H.D., Wu, F.F., and Varaiya, P.P. (1988). Foundations of the potential energy boundary surface method for power system transient stability analysis. *IEEE Transactions on Circuits and Systems* 35 (6): 712–728.

**9** Mitra, J. and Benidris, M. (2014). Real-time remedial action screening using direct stability analysis methods. *2014 IEEE PES General Meeting | Conference & Exposition*, National Harbor, MD, USA, pp. 1–1. https://doi.org/10.1109/PESGM.2014.6939093.

**10** Kundur, P. (1994). *Power System Stability and Control*. New York: McGraw- Hill.

**11** Report, I.C. (1973). Excitation system dynamic characteristics. *IEEE Transactions on Power Apparatus and Systems* PAS-92 (1): 64–75.

**12** IEEE (2016). IEEE recommended practice for excitation system models for power system stability studies. *IEEE Std 421.5-2016 (Revision of IEEE Std 421.5-2005)* (26 August 2016), pp. 1–207. https://doi.org/10.1109/IEEESTD.2016.7553421.

**13** IEEE (2011). IEEE guide for the application of turbine governing systems for hydroelectric generating units. *IEEE Std 1207-2011 (Revision to IEEE Std 1207-2004)* (20 June 2011), pp. 1–131. https://doi.org/10.1109/IEEESTD.2011.5936081.

**14** Dai, J., Yao, Z., Zhang, G. et al. (2019). Graph computing-based real-time network topology analysis for power system. *2019 IEEE Power & Energy Society General Meeting (PESGM)*, Atlanta, GA, USA, pp. 1–5. https://doi.org/10.1109/PESGM40551.2019.8973614.

**15** Zhou, Z., Yuan, C., Yao, Z. et al. (2018). CIM/E oriented graph database model architecture and parallel network topology processing. *2018 IEEE Power & Energy Society General Meeting (PESGM)*, Portland, OR, USA, pp. 1–5. https://doi.org/10.1109/PESGM.2018.8586367.

**16** Yuan, C., Zhou, Y., Zhang, G. et al. (2018). Exploration of graph computing in power system state estimation. *2018 IEEE Power & Energy Society General Meeting (PESGM)*, Portland, OR, USA, pp. 1–5. https://doi.org/10.1109/PESGM.2018.8586535.

**17** Yuan, C., Zhou, Y., Liu, G. et al. (2020). Graph computing-based WLS fast decoupled state estimation. *IEEE Transactions on Smart Grid* 11 (3): 2440–2451.

**18** Lu, Y., Yuan, C., Zhang, X. et al. (2020). Graph computing based distributed state estimation with PMUs. *2020 IEEE Power & Energy Society General Meeting (PESGM)*, Montreal, QC, Canada, pp. 1–5. https://doi.org/10.1109/PESGM41954.2020.9281976.

**19** Shi, J., Liu, G., Dai, R. et al. (2018). Graph based power flow calculation for energy management system. *2018 IEEE Power & Energy Society General Meeting (PESGM)*, Portland, OR, USA, pp. 1–5. https://doi.org/10.1109/PESGM.2018.8586233.

**20** Feng, W., Yuan, C., Shi, Q. et al. (2021). Graph computing based distributed parallel power flow for AC/DC systems with improved initial estimate. *Journal of Modern Power Systems and Clean Energy* 9 (2): 253–263.

**21** Yuan, C., Lu, Y., Liu, K. et al. (2018). Exploration of Bi-level pagerank algorithm for power flow analysis using graph database. *2018 IEEE International Congress on Big Data (BigData Congress)*, San Francisco, CA, USA, pp. 143–149. https://doi.org/10.1109/BigDataCongress.2018.00026.

**22** Feng, W., Yuan, C., Dai, R. et al. (2018). Graph computation based power flow for large-scale AC/DC system. *2019 IEEE Power & Energy Society General Meeting (PESGM)*, Atlanta, GA, USA, pp. 1–5. https://doi.org/10.1109/PESGM40551.2019.8973563.

**23** Yuan, C., Liu, G., Dai, R. et al. (2018). Power flow analysis using graph based combination of iterative methods and vertex contraction approach. *2018 International Conference on Power System Technology (POWERCON)*, Guangzhou, China, pp. 4480–4487. https://doi.org/10.1109/POWERCON.2018.8601938.

**24** Shi, Q., Yuan, C., Feng, W. et al. (2019). Enabling model-based LTI for large-scale power system security monitoring and enhancement with graph-computing-based power flow calculation. *IEEE Access* 7: 167010–167018.

**25** Feng, W., Wu, J., Yuan, C. et al. (2019). A graph computation based sequential power flow calculation for large-scale AC/DC systems. *2019 IEEE Power & Energy Society General Meeting (PESGM)*, Atlanta, GA, USA, pp. 1–5. https://doi.org/10.1109/PESGM40551.2019.8973563.

**26** Tan, J., Lu, Y., Liu, K. et al. (2019). Graph computing based parallel power flow algorithm and graph visualization for power distribution networks. *2019 IEEE Power & Energy Society General Meeting (PESGM)*, Atlanta, GA, USA, 2019, pp. 1–5. https://doi.org/10.1109/PESGM40551.2019.8973902.

**27** Feng, W., Yuan, C., Shi, Q. et al. (2021). Graph computing based distributed parallel power flow for AC/DC systems with improved initial estimate. *Journal of Modern Power Systems and Clean Energy* 9 (2): 253–263.

**28** Zhao, Y., Yuan, C., Liu, G., and Grinberg, I. (2018). Graph-based preconditioning conjugate gradient algorithm for "N-1" contingency analysis. *2018 IEEE Power & Energy Society General Meeting (PESGM)*, Portland, OR, USA, pp. 1–5. https://doi.org/10.1109/PESGM.2018.8586214.

**29** Mitchell-Colgan, E., Wu, J., Zhao, Y. et al. (2018). Symbolic factorization re-utilization for contingency analysis. *2018 IEEE Power & Energy Society General Meeting (PESGM)*, Portland, OR, USA, pp. 1–5. https://doi.org/10.1109/PESGM.2018.8586437.

**30** Zhao, Y., Yuan, C., Li, S. et al. (2019). Graph computing based fast screening in contingency analysis. *2019 IEEE Innovative Smart Grid Technologies - Asia (ISGT Asia)*, Chengdu, China, pp. 667–672. https://doi.org/10.1109/ISGT-Asia.2019.8881406.

**31** Zhu, Y., Shi, L., Dai, R., and Liu, G. (2019). Fast grid splitting detection for N-1 contingency analysis by graph computing. *2019 IEEE Innovative Smart Grid Technologies - Asia (ISGT Asia)*, Chengdu, China, pp. 673–677. https://doi.org/10.1109/ISGT-Asia.2019.8880879.

**32** Zhu, Y., Dai, R., and Liu, G. (2020). Parallel betweenness computation in graph database for contingency selection. *2020 IEEE Power & Energy Society General Meeting (PESGM)*, Montreal, QC, Canada, pp. 1–5. https://doi.org/10.1109/PESGM41954.2020.9281492.

**33** Wei, L., Liu, G., Tan, J. et al. (2020). GraphVPP: enabling optimal bidding strategy of virtual power plants in graph computing frameworks. *2020 IEEE Power & Energy Society General Meeting (PESGM)*, Montreal, QC, Canada, pp. 1–5. https://doi.org/10.1109/PESGM41954.2020.9281401.

**34** Tan, J., He, M., Zhang, G. et al. (2020). Volt/Var optimization for active power distribution systems on a graph computing platform: an paralleled PSO approach. *2020 IEEE Power & Energy Society General Meeting (PESGM)*, Montreal, QC, Canada, pp. 1–5. https://doi.org/10.1109/PESGM41954.2020.9281603.

**35** Zhu, Y., Liu, C., Dai, R. et al. (2019). Optimal battery energy storage placement for transient voltage stability enhancement. *2019 IEEE Power & Energy Society General Meeting (PESGM)*, Atlanta, GA, USA, pp. 1–5. https://doi.org/10.1109/PESGM40551.2019.8973610.

**36** Chen, T., Yuan, C., Liu, G., and Dai, R. (2018). Graph based platform for electricity market study, education and training. *2018 IEEE Power & Energy Society General Meeting (PESGM)*, Portland, OR, USA, pp. 1–5. https://doi.org/10.1109/PESGM.2018.8586243.

**37** Wei, L., Tang, Y., Zhang, X. et al. (2021). Graph computing based ADMM approach for security constrained unit commitment in hydro-thermal power systems. *PES & PELS 2021 6th IEEE Workshop on the Electronic Grid (eGrid 2021)*, 8–10 November 2021.

**38** Demetriou, P., Asprou, M., Quiros-Tortos, J., and Kyriakides, E. (2017). Dynamic IEEE test Systems for Transient Analysis. *IEEE Systems Journal* 11 (4): 2108–2117.

# 16

# Graph-Based Deep Reinforcement Learning on Overload Control

## 16.1 Introduction

Power systems are designed and operated with the main objective of having a stable, secure, and continuous supply of electrical power to customers. However, an unpredictable disturbance may result in branch overload which may trigger a sequence of one or more dependent component outages and lead to a server interruption of the energy supply across a major territory of the system.

Overloaded branches are reported to be one of the main contributors to cascading outages and system blackouts [1]. Depending on the level of overload and the branch protection settings, the overloaded branch(s) might trip out. This, in turn, might cause additional branches to exceed their loading limits and trip out.

To prevent such cascading outages, the overload on branches must be managed and relieved. Generation redispatch is one of the effective control approaches to mitigate branch power flow overload violations.

In general, the problem of generation redispatch is typically formulated as an optimal power flow (OPF) problem [2–4]. Power systems are typically large and highly nonlinear and complex systems which make the OPF problem one of a high computational burden. The generation redispatch ideally takes steps to reach the optimal goal in which the objective function is optimal and the constraints are satisfied. However, the OPF solution does not inherently offer guidance on the correct sequence of steps to achieve the desired dispatch [5].

Within this context, the method of deep reinforcement learning enables control agents to learn optimal control policies from interaction with a system. Then the trained control agents take control actions to redispatch generation till the optimal solution is reached [6–8].

Deep learning is combined with reinforcement learning, resulting in the Deep Q Network algorithm [9, 10] which can solve problems with low-dimensional observation spaces. The Deep Q Network algorithm handles discrete action spaces by defining an action-value lookup table. It cannot be straightforwardly applied to continuous domains. A simple approach to adapting the Deep Q Network algorithm is to discretize the continuous action space. However, this is infeasible to define and inefficient to use a large lookup table for power system OPF problem which involves high dimensional and continuous actions.

To conquer the challenge, the deep deterministic policy gradient (DDPG) algorithm [9] using deep neural network approximators is introduced to learn control policies in high dimensional and continuous power system OPF action spaces. The work is based on the deterministic policy gradient algorithm [11].

## 16.2 DDPG Algorithm

The DDPG algorithm is a model-free, actor-critic algorithm. As illustrated in Figure 16.1, in the DDPG algorithm, the agent interacts with the environment in discrete time steps. At each time step $t$, the agent observes the system states $s$ from the environment, decides on an action $a$ and receives a reward $r$. The environment in the generation redispatch application case is a power system. The system state $s$ is the power flow solution. The action $a$ is the generation MW output adjustment.

DDPG learning is a reinforcement learning in which the learning agent learns policies progressively as it accumulates more and more experience until it learns optimal control policies. In this section, the DDPG learning terminologies are first defined. The policy gradient and the DDPG algorithms are then discussed.

### 16.2.1 Terminology

The key terminologies, concepts, and mathematical relations of the DDPG are defined as follows:

1) Policy: A policy $\pi$, is a function that defines the behavior of the agent which maps states to either specific actions or a probability distribution of actions. According to the difference, a policy $\pi$ is categorized to be deterministic or stochastic. A deterministic policy, $\pi : S \rightarrow A$, maps state to specific actions. On the other hand, a stochastic policy outputs a probability distribution over a predefined set of actions $\pi : S \rightarrow \mathcal{P}(A)$.

2) Return: Return is defined as the sum of discounted future rewards

$$R_t = \sum_{i=t}^{T} \gamma^{i-t} r_i(s_i, a_i) \tag{16.1}$$

where $\gamma$ is a discount factor, $\gamma \in [0, 1]$. The goal of deep reinforcement learning is to learn a policy that maximizes the expected return from the initial distribution as follows:

$$J = \mathbb{E}_{r_i, s_i \sim E, a_i \sim \pi}[R_1] \tag{16.2}$$

3) Value Function: Value functions represent the expected value of a given state. There are two types of value functions: (i) state-value function, $V(s_t)$, which is defined as the expected return starting from the state $s_t$ and following policy $\pi$. (ii) action-value function, $Q(s_t, a_t)$, which is defined as the expected return after the agent takes action $a_t$ in state $s_t$ and following policy $\pi$ thereafter, which is expressed in (16.3) as follows:

$$Q^\pi(s_t, a_t) = \mathbb{E}_{r_i \geq t, s_i > t \sim E, a_i > t \sim \pi}[R_t | s_t, a_t] \tag{16.3}$$

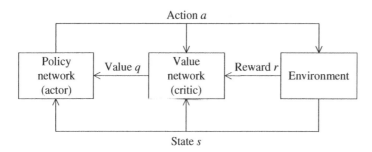

**Figure 16.1** Actor-critic algorithm structure.

4) Bellman's Equation: In reinforcement learning, the goal is to learn a policy that maximizes the expected return in every state, i.e. its value function. The majority of learning algorithms utilize value functions to infer or learn the optimal policy. Those algorithms make use of recursive relationships to calculate action value based on the following Bellman's equation [12].

$$Q^{\pi}(s_t, a_t) = \mathbb{E}_{r_t, s_{t+1} \sim E}[r(s_t, a_t) + \gamma \mathbb{E}_{a_{t+1} \sim \pi}[Q^{\pi}(s_{t+1}, a_{t+1})]] \tag{16.4}$$

The above equation allows iterative approaches to solving the optimal action-value function $Q^{\pi}(s, a)$.

## 16.2.2 Q Function

In reinforcement learning, an agent tries to learn the optimal policy from its history of interaction with the environment. A history of an agent is a sequence of state-action-rewards:

$$\langle s_0, a_0, r_1, s_1, a_1, r_2, s_2, a_2, r_3, s_3, a_3, r_4, s_4... \rangle \tag{16.5}$$

which means that the agent was in the state $s_0$ and did action $a_0$, which resulted in it receiving a reward $r_1$ and being in the state $s_1$; then it did action $a_1$, received reward $r_2$, and ended up in the state $s_2$; then it did action $a_2$, received reward $r_3$, and ended up in the state $s_3$; and so on.

We treat this history of interaction as a sequence of experiences, where an experience is a tuple:

$$\langle s, a, r, s' \rangle \tag{16.6}$$

which means that the agent was in state $s$, it did action $a$, it received reward $r$, and it went into state $s'$. These experiences will be the data from which the agent can learn. As in decision-theoretic planning, the aim of the agent is to maximize its value, which is usually the discounted reward.

Specifically, $Q^*(s, a)$, where $s$ is a state and $a$ is an action, is the expected value of doing an action $a$ in state $s$ and then following the optimal policy.

Q-learning uses temporal differences to estimate the value of $Q^*(s, a)$. In Q-learning, the agent maintains a table of $Q(s, a)$, where $S$ is the set of states and $A$ is the set of actions. $Q(s, a)$ represents its current estimate of $Q^*(s, a)$.

An experience $\langle s, a, r, s' \rangle$ provides one data point for the value of $Q(s, a)$. The agent observes the future value of $r + \gamma V(s')$, where $V(s') = \max_{a'} Q(s', a')$ which is the actual current reward plus the discounted estimated future value. This new data point is called a return. The agent can use the temporal difference equation to update its estimate for $Q(s, a)$:

$$Q(s, a) \leftarrow Q(s, a) + \alpha[r + \gamma \max_{a'} Q(s', a') - Q(s, a)] \tag{16.7}$$

Or, equivalently,

$$Q(s, a) \leftarrow (1 - \alpha)Q(s, a) + \alpha[r + \gamma \max_{a'} Q(s', a')] \tag{16.8}$$

## 16.2.3 Q Value Approximation

$Q(s, a)$ estimation is challenging. Theoretically, we can memorize $Q^*(\cdot)$ for all state-action pairs in Q-learning, in a gigantic table. However, Q-learning becomes impractical when the state and action space are large. Thus, a machine learning model is adopted to approximate $Q$ values and this is called $Q$ value approximation. For example, if we use a function with parameter $\theta$ to calculate $Q$ values, we can label $Q$ value function as $Q(s, a; \theta)$.

To approximate the $Q$ function, the greedy policy is used to optimize the critic parameters $\theta^Q$ by minimizing the mean-squared Bellman error as follows [12]:

$$L(\theta^Q) = \mathbb{E}\big[(Q(s_t, a_t \mid \theta) - y_t)^2\big] \tag{16.9}$$

Where $y_t$ is known as the target.

$$y_t = r(s_t, a_t) + \gamma Q\big(s_{t+1}, \pi(s_{t+1}) \mid \theta^Q\big) \tag{16.10}$$

This error indicates how close the value of $Q$ comes to satisfy the optimal condition. The optimal critic parameters $\theta^Q$ are learned by using a critic neural network.

When using a neural network model, $Q$-learning may suffer from instability and divergence. To address the issues, two technologies: replay buffer and periodically updated target are used:

- Replay Buffer: All the episode steps $e_t = (s_t, a_t, r_t, s_{t+1})$ are stored in one replay memory $D_t = \{e_1, ..., e_t\}$. $D_t$ has experience tuples over many episodes. Applying replay buffer, samples are drawn randomly from the replay memory to a sample buffer during $Q$-learning updates which remove correlations in the samples.
- Periodically Updated Target: $Q$ function is optimized toward target values that are only periodically updated. The $Q$ network is cloned and kept frozen as the optimization target periodically. This modification improves the stability of training.

### 16.2.4 Policy Gradient [11]

While the $Q$-learning is approximating the $Q$ function, parameterized actor function $\pi(s \mid \theta^\pi)$ is updated by policy learning. The optimal policy satisfies the following:

$$\theta^{\pi*} = \arg \max_{\theta^\pi} (J = \mathbb{E}[Q(s, \pi(s \mid \theta^\pi))]) \tag{16.11}$$

where, $\theta^\pi$ are the actor parameters. Since the action space is continuous and the $Q$ function is differentiable to an action, gradient ascent is used to optimize the actor parameters $\theta^\pi$ as follows:

$$\nabla_{\theta^\pi} J = \frac{1}{N} \sum_i \nabla_a Q(s, a \mid \theta^Q)\big|_{s = s_i, a = \pi(s_i)} \nabla_{\theta^\pi} \pi(s \mid \theta^\pi)\big|_{s = s_i} \tag{16.12}$$

Eq. (16.12) has no simple analytical solution. In the policy gradient, a neural network is used to model the action probabilities. Each time the agent interacts with the environment, the parameters $\theta^\pi$ of the neural network are tuned so that valued actions will be sampled more likely in the future. This process is repeated until the policy network converges to the optimal policy $\pi^*$.

As shown in Figure 16.1, the DDPG algorithm uses the actor-critic framework to learn the actor-network for policy distribution and the critic network for the $Q$ function concurrently.

A major challenge of DDPG learning is explored in high-dimensional continuous action spaces. Domain knowledge will help to improve exploration efficiency. The domain knowledge is represented by adding the additional term $\mathcal{N}$ in the exploration policy $\mu'$:

$$\mu'(s) = \mu_\theta(s) + \mathcal{N} \tag{16.13}$$

To smooth the parameter updates for both actor and critic network, the DDPG takes soft updates by:

$$\theta^{Q'} \leftarrow \tau\theta^Q + (1 - \tau)\theta^{Q'} \tag{16.14}$$

$$\theta^{\pi'} \leftarrow \tau\theta^\pi + (1 - \tau)\theta^{\pi'} \tag{16.15}$$

By selecting $\tau \ll 1$, The target network values are constrained to change slowly.

In summary, the pseudo code of the DDPG algorithm is shown as follows:

---

**Algorithm 16.1   DDPG Algorithm**

---

1:   **Initialize** critic network $Q(s_0, a_0 | \theta^Q)$ and actor-network $\mu(s_0 | \theta^\pi)$
2:   **Initialize** target networks $Q'$ and $\mu'$ with weights $\theta^{Q'}$ and $\theta^{\pi'}$
3:   **Initialize** replay buffer $R$
4:   **For** episode = 1, M **Do**
5:      **Initialize** a random process $\mathcal{N}$ for action exploration
6:      **Receive** initial observation state $s_1$ from the power flow solution
7:      **For** t = 1, T **Do**
8:         **Select** action $a_t = \mu(s_t | \theta^\pi) + \mathcal{N}$
9:         **Take** action $a_t$ to the power system and run power flow
10:        **Calculate** reward $r_t$ and observe new state $s_{t+1}$
11:        **Store** transition $\langle s_t, a_t, r_t, s_{t+1}, \rangle$ in replay buffer $R$
12:        **Sample** a random batch $\langle s_i, a_i, r_i, s_{i+1}, \rangle$ from replay buffer $R$
13:        **Set** $y_i = r_i + \gamma Q'(s_{i+1}, \mu'(s_{i+1} | \theta^{\pi'}) | \theta^{Q'})$
14:        **Update** critic network by minimizing the loss

$$L = \frac{1}{N} \sum_i \left[ (Q(s_i, a_i | \theta^Q) - y_i)^2 \right]$$

15:        **Update** the actor policy by

$$\nabla_{\theta^\pi} J = \frac{1}{N} \sum_i \nabla_a Q(s, a | \theta^Q)|_{s = s_i, a = \pi(s_i)} \nabla_{\theta^\pi} \pi(s | \theta^\pi)|_{s_i}$$

16:        **Update** the target critic and actor-networks by

17:
$$\theta^{Q'} \leftarrow \tau \theta^Q + (1 - \tau) \theta^{Q'}$$
18:
$$\theta^{\pi'} \leftarrow \tau \theta^\pi + (1 - \tau) \theta^{\pi'}$$
19:  **End For**
20: **End For**

---

## 16.3   Branch Overload Control

To mitigate branch overload, by using the deep reinforcement learning algorithm addressed above, generation active power outputs are optimally adjusted under a variety of power system operating conditions and contingencies. The output of the learning process is to select the optimal active power adjustment among the generators to relieve the congestion of branches.

To apply the deep reinforcement learning algorithm to control generation active power outputs, state, action, and reward are firstly defined and formulated.

### 16.3.1   States

Given an $N$-bus system where $\mathcal{G}$ is the set of generators, $\mathcal{B}$ is the set of branches. The system states are defined as:

$$S = \{ s | s_t = \{ S_{i,t} \forall i \in \mathcal{B} \} \cup \{ P_{gj,t} \forall j \in \mathcal{G} \} \} \tag{16.16}$$

where $S_{i,t}$ is the magnitude of the apparent power flow on the $i$-th branch, $P_{gj,t}$ is the active power output of the $j$-th generator at time step $t$.

### 16.3.2 Actions

The power adjustment action is defined as:

$$A = \left\{ a \,|\, a_{j,t} = \Delta P_{g_j,t} \forall j \in \mathcal{G}, \Delta P_{gj} \in \left[ P_{gi}^{\min} - P_{g_j,t-1}, P_{gi}^{\max} - P_{g_j,t-1} \right] \right\} \tag{16.17}$$

where $\Delta P_{gj,t}$ is the amount of the power output change of the $j$-th generator at time step $t$. The action $\Delta P_{gj,t}$ is continuous in the range of $\left[ P_{gi}^{\min} - P_{g_j,t-1}, P_{gi}^{\max} - P_{g_j,t-1} \right]$ where $P_{g_j,t-1}$ is the active power output of the $j$-th generator at time step $t-1$. $P_{gi}^{\min}$ and $P_{gi}^{\max}$ are active power output lower and upper limits of the $j$-th generator.

$\Delta P_{gj,t}$ is controlled by the agent in each step of the iteration based on the current state and the action selection policy.

### 16.3.3 Rewards

The success of a deep reinforcement learning algorithm relies on the design of the reward function. Several works have proposed the use of the distance-to-goal as a measure of success for the actions taken by the agent [13–15]. For the branch overload control problem, a reward function is composed of two terms:

$$r_t = r_{IM,t} + r_{TML} \tag{16.18}$$

Where

$$r_{IM,t} = -\sum_{i \in B} \frac{|S_{i,t}|}{S_i^{\max}} \tag{16.19}$$

And

$$r_{TML} = \begin{cases} +50 & \text{if all branch flows are within limits} \\ -50 & \text{if power flow has diverged} \end{cases} \tag{16.20}$$

The reward function makes the agent try to reach states where it gets a positive reward. The intermediate reward $r_{IM,t}$ penalizes the agent action at heavily loaded branch condition at each time step $t$. The terminal reward $r_{TML}$ awards the agent when all branch flows are within limits and penalizes the agent if the power flow diverges.

## 16.4 Graph-Based Deep Reinforcement Learning Implementation

As shown in Figure 16.1, the deep reinforcement learning structure consists of three major components: environment, actor-network, and critic network. To minimize correlations between samples and improve learning stability, a replay buffer is used to store and restore samples randomly. The three components and the replay buffer are implemented in Python code.

The class structure of the replay buffer and the actor-critic network are listed as follows:

```
class replayBuffer(object):
   def __init__(self, buffer_size, random_seed):
   ...
   def add_sample(self, s, a, r, t, s2):
   ...
   def sample_batch(self, batch_size):
   ...


class Actor-Critic(object):
   def __init__(self, state_dim, action_dim, learning_rate):
   ...
   def get_actor(self):
   ...
   def get_critic(self):
   ...
   def learn(self):
   ...


class powerSystem():
   def __init__(self):
   ...
   def step(self, action):
   ...
```

The pseudo codes of the above functions are included in Appendix. By calling the functions above, the DDPG algorithm main function is implemented as the following pseudo code:

```
def deepReinforcementLearning():
  env = powerSystem()
  state_dim = env.s_dim
  action_dim = env.s_dim
  agent = Actor-Critic(action_dim, state_dim)
  buffer = replayBuffer()

  if args.train:
    for episode in range(TRAIN_EPISODES):
      state = env.states
      for iSetp in range(MAX_STEPS):
        if RENDER:
          env.render()
        action = agent.get_action(state)
        state_, reward = env.step(action)
        buffer.add_sample(state, action, reward, state_)
        if buffer.count > buffer_size:
          batch = buffer.sample_batch(buffer_size)
          agent.learn(batch)
        state = state_
    agent.save()
```

```
if args.test:
  agent.load()
  for episode in range(TEST_EPISODES):
    for iStep in range(MAX_STEPS):
      state, reward, done, info = env.step(agent.get_actor(state))
```

In order to implement graph-based deep reinforcement learning to optimize power unit active power generation by graph query interfacing with Python functions, the edge `TopoConnect` of the bus-branch graph model `busBranchGraph` is extended as follows:

```
CREATE DIRECTED EDGE TopoConnect (from TopoND, to TopoND, typename
    string, edge_name List<string>, area List<string>, zone
    List<string>, from_bus int, to_bus int, flag uint, R List<double>,
    X List<double>, hB_list List<double>, line_Q1_list List<double>,
    line_Q2_list List<double>, line_Q3_list List<double>, control_bus
    int, K List<double>, shifting_transformer_angle List<double>,
    min_tap double, max_tap double, step_size double, min_volt
    List<double>, max_volt List<double>, M_P_BR List<double>, M_Q_BR
    List<double>, G List<double>, B List<double>, BIJ List<double>,
    circuit uint, P_BR double, Q_BR double, S_BR double, R_BR double,
    from_open List<int>, to_open List<int>, bridge boolean)
```

There is no update in the vertex definition. In the edge definition above, the new attributes **S_BR** and **R_BR** are defined to store the apparent power flow and reward of a branch.

When power flow is calculated, the reward associated with each branch is calculated by the following query `calculateReward`.

```
CREATE QUERY calculateReward() FOR GRAPH busBranchGraph {
  T0 = {TopoND.*}
  T1 = SELECT s FROM T0:s-(connected:e)->:t
    ACCUM:
        s.S_BR = sqrt(s.P_BR * s.P_BR + s.Q_BR * s.Q_BR)
        s.R_BR = - abs(s.S_BR)/s.line_Q1
}
```

Using REST API provided by the graph analytic platform, the implemented Python functions are integrated by graph query.

As addressed in Chapter 11, in graph query, user-defined functions are supplemented in the language of C/C++ and are called in queries to interface Python functions. The inline function to integrate Python code is demonstrated as follows:

```
#include <Python.h>
inline int overloadControl (){
  PyArg_ParseTuple(args, "deepReinforcementLearning");
}
```

The `PyArg_ParseTuple()` function is declared to pass the Python function "deepReinforcementLearning" from Python to the C inline function to be called in the C inline function. And the inline function `overloadControl` can be callable in graph query.

# References

**1** Fliscounakis, S., Panciatici, P., Capitanescu, F., and Wehenkel, L. (2013). Contingency ranking with respect to overloads in very large power systems taking into account uncertainty, preventive and corrective actions. *IEEE Transactions on Power Systems* 28 (4): 4909–4917.

**2** Lavei, J., Rantzer, A., and Low, S. (2011). Power flow optimization using positive quadratic programming. *IFAC Proceedings Volumes* 44 (1): 10481–10486.

**3** Bai, X., Wei, H., Fujisawa, K., and Wang, Y. (2008). Semidefinite programming for optimal power flow problems. *International Journal of Electrical Power & Energy Systems* 30 (6-7): 383–392.

**4** Kocuk, B., Dey, S.S., and Sun, X.A. (2016). Strong SOCP relaxations for the optimal power flow problem. *Operations Research* 64 (6): 1177–1196.

**5** Kamel, M., Dai, R., Wang, Y. et al. (2021). Data-driven and model-based hybrid reinforcement learning to reduce stress on power systems branches. *CSEE Journal of Power and Energy Systems* 7 (3): 433–442.

**6** Kamel, M., Wang, Y., Yuan, C. et al. (2020). A reinforcement learning approach for branch overload relief in power systems. *IFAC Proceedings Volumes* 44 (1): 10481–10486.

**7** Glavic, M. (2019). (deep) reinforcement learning for electric power system control and related problems: a short review and perspectives. *Annual Reviews in Control* 48: 22–35.

**8** Zhang, Z., Zhang, D., and Qiu, R.C. (2019). Deep reinforcement learning for power system applications: an overview. *CSEE Journal of Power and Energy Systems* 6 (1): 213–225.

**9** Mnih, V., Kavukcuoglu, K., Silver, D. et al. (2015). Human-level control through deep reinforcement learning. *Nature* 518: 529–533.

**10** Lillicrap, T.P., Hunt, J.J., Pritzel, A. et al. (2015). Continuous control with deep reinforcement learning, pp. 1–14. https://arxiv.org/abs/1509.02971.

**11** Silver, D., Lever, G., Heess, N. et al. (2014). Deterministic policy gradient algorithms. *International Conference on Machine Learning,* Beijing, China.

**12** Bellman, R. (1954). The theory of dynamic programming. *Bulletin of the American Mathematical Society* 60 (6): 503–515.

**13** Nazari, M., Oroojlooy, A., Snyder, L., and Takác, M. (2018). Reinforcement learning for solving the vehicle routing problem. *Advances in Neural Information Processing Systems* 31: 9839–9849.

**14** Nair, A.V., Pong, V., Dalal, M. et al. (2018). Visual reinforcement learning with imagined goals. *Advances in Neural Information Processing Systems* 31: 9191–9200.

**15** Trott, A., Zheng, S., Xiong, C., and Socher, R. (2019). Keeping your distance: solving sparse reward tasks using self-balancing shaped rewards. *Advances in Neural Information Processing Systems* 32: 10376–10386.

# 17

# Conclusions

The electrical power system is one of the largest systems. In more than a century, the electric power system is evolving into a highly interconnected, large, and complex network that significantly impacts the daily life of human beings. The goal of maintaining a power system to be secure, reliable, cost-effective, and environmentally friendly is crucial. To achieve the challenging goals, power system analysis with detailed system representation is essential.

Power system analysis is typically performed by a model-based, data-driven, or hybrid approach. The model-based algorithms fundamentally are running into solving linear or nonlinear algebraic equations, optimization problems, differential equations, and their combinations. The data-driven algorithm, such as machine learning, creates a mapping from one hyperspace to another. Both model-based and data-driven algorithms require intensive effort in modeling and computing for a large-scale power system analysis. Graph computing provides an effective way to model power systems and conduct power system analysis in parallel.

Graph computing technology has been widely used in many fields such as the Internet, social media, network security, traffic control, e-commerce, and so on. The concept, structure, and management of the graph database are different from the relational database. A relational database organizes data into tables. The database stores structured records and their attributes in tables. Ideally, data relationships of arbitrary complexity can be presented by a relational database. However, the limitations of a relational database on power system applications are obvious when unstructured data are managed and data relationships are convoluted.

The relationships between tables are logically connected by separated tables or by using a join operation to search common attributes in different tables to find the relationships. Maintaining a large dataset in a relational database is challenging. When data reside in multiple tables, they are linked to each other through shared key values to represent the data relationships. The database structure creates unnecessary complexity to analyze larger-scale power systems with high fidelity. To deal with complex data structures, queries require sophisticated join operations inviting more computation time.

To accommodate online calculation for large-scale electric power systems, a graph database is adapted to fulfill the power system calculation requirements on complex database store, traversal, concurrent access, and flexible expansion and reduction. Contrary to the relational database, a graph database uses graph structures for semantic queries with nodes and edges to store data. Unstructured attributes of node or edge are stored in the node or edge. The key concept and merit of a graph database management system is the edge directly defines the data relationship. The relationships allow stored data to be linked together directly and be retrieved with one operation other

*Graph Database and Graph Computing for Power System Analysis*, First Edition. Renchang Dai and Guangyi Liu.
© 2024 The Institute of Electrical and Electronics Engineers, Inc. Published 2024 by John Wiley & Sons, Inc.

than join operations. A graph database, by design, allows simple and fast retrieval of complex hierarchical structures.

Graph computing models power systems as a graph in light of that power system physically is a graph – buses are connected by branches as a graph. The graph data structure tells the topology of the power network and the relations of power system components naturally. The graph computing mechanism by using queries on nodes and graph partitions benefits parallel computing for power system applications from steady-state, optimizations, to dynamics.

This book summarizes cutting-edge technologies in graph data management and graph parallel computing, particularly in power system analysis as well as industry experience in designing and implementing graph computing-based power system applications. The knowledge and experience would be used to change the way we manage power system data and solve power system problems. The book covers practical subjects that are useful for readers in the power industry as guidance to design and develop power system planning tools and real-time operating systems in a different way.

The book introduces a graph database architecture to power systems that supports fine-grained parallel computing to improve power system computation efficiency. The traditional relational database is replaced by a graph database to model power systems and implement applications. Using the graph database, the programs to solve large-scale algebraic equations, high dimensional differential equations, optimization problems, and interfacing to machine learning programs are reconfigured and redesigned to accommodate graph parallel computing in this book unprecedentedly. The graph-based nodal and hierarchical parallelism are explored to achieve high computation efficiency. The implementations of graph computing-based topology analysis, state estimation, power flow calculation, contingency analysis, security-constrained economic dispatch, security-constrained unit commitment, automatic generation control, small-signal stability, transient stability, and deep reinforcement learning are demonstrated in detail in the book. The pseudo-code of algorithms and application implementations are listed and explained in depth.

Graph computing is new to the power industry. We hope this book provides a map and guidance for researchers and engineers on their exploration of this challenging and rewarding technology that benefits power system planning and operation.

# Appendix

## A CIM/E Tables

**Table A.1** Base value table in CIM/E model.

| Attributes | Annotation |
|---|---|
| id | Identification |
| name | Base value name |
| value | Base value |
| unit | Unit |

**Table A.2** Substation table in CIM/E model.

| Attributes | Annotation |
|---|---|
| name | Substation name |
| volt | Voltage level |
| config | Substation configuration |
| nodes | Number of nodes |
| islands | Number of islands |
| island | Island name |
| dvname | Division name |

*Graph Database and Graph Computing for Power System Analysis*, First Edition. Renchang Dai and Guangyi Liu.
© 2024 The Institute of Electrical and Electronics Engineers, Inc. Published 2024 by John Wiley & Sons, Inc.

**Table A.3**  Bus table in CIM/E model.

| Attributes | Annotation |
| --- | --- |
| id | Identification |
| name | Bus name |
| volt | Rated voltage |
| node | Topology node name |
| V | Bus voltage magnitude |
| Ang | Bus voltage angle |
| off | Offline flag |
| V_meas | Voltage magnitude measure |
| Ang_meas | Voltage angle measure |
| nd | Physical node number |
| bs | Bus number |
| island | Island number |
| v_max | Voltage upper limit |
| v_min | Voltage lower limit |

**Table A.4**  ACline table in CIM/E model.

| Attributes | Annotation |
| --- | --- |
| id | Identification |
| name | AC line name |
| volt | Rated voltage |
| Eq | Equivalent line flag |
| R | Resistance |
| X | Reactance |
| B | Charging susceptance |
| I_node | Topology node at I end |
| J_node | Topology node at J end |
| I_P | Active power at I end |
| I_Q | Reactive power at I end |
| J_P | Active power at J End |
| J_Q | Reactive power at J end |
| I_off | Line open flag at I end |
| J_off | Line open flag at J end |
| Ih | Maximum current |
| Pi_meas | Active power measure at I end |
| Qi_meas | Reactive power measure at I end |
| Pj_meas | Active power measure at J end |
| Qj_meas | Reactive power measure at J end |
| I_nd | Physical node number at I end |

**Table A.4**   (Continued)

| Attributes | Annotation |
|---|---|
| J_nd | Physical node number at J end |
| I_bs | Bus number at I end |
| J_bs | Bus number at J end |
| I_island | Island number at I end |
| J_island | Island number at J end |
| R* | Resistance (per unit) |
| X* | Reactance (per unit) |
| B* | Susceptance (per unit) |

**Table A.5**   Unit table in CIM/E model.

| Attributes | Annotation |
|---|---|
| id | Identification |
| name | Generator name |
| Eq | Equivalent generator flag |
| position | Equivalent generator position |
| V_rate | Rated voltage |
| P_rate | Rated power |
| volt_n | Volage level |
| node | Topology node name |
| P | Active power |
| Q | Reactive power |
| Ue | Terminal voltage magnitude |
| Ang | Terminal voltage angle |
| off | Offline flag |
| P_meas | Active power measurement |
| Q_meas | Reactive power measurement |
| Ue_meas | Terminal voltage magnitude measurement |
| Ang_meas | Terminal voltage angle measurement |
| P_max | Active power upper limit |
| P_min | Active power lower limit |
| Q_max | Reactive power upper limit |
| Q_min | Reactive power lower limit |
| pf | Power factor |
| nd | Physical node number |
| bs | Bus number |
| island | Island number |

**Table A.6** Transformer table in CIM/E model.

| Attributes | Annotation |
|---|---|
| id | Identification |
| name | Transformer name |
| Type | Transformer type |
| I_Vol | Voltage level at high voltage side |
| K_Vol | Voltage level at medium voltage side |
| J_Vol | Voltage level at low voltage side |
| I_S | Rated capacity at high voltage side |
| K_S | Rated capacity at medium voltage side |
| J_S | Rated capacity at low voltage side |
| Itap_H | Maximum tap position at high voltage side |
| Itap_L | Minimum tap position at high voltage side |
| Itap_E | Rated tap position at high voltage side |
| Itap_C | Tap step size at high voltage side |
| Itap_V | Rated voltage at high voltage side |
| Ktap_H | Maximum tap position at medium voltage side |
| Ktap_L | Minimum tap position at medium voltage side |
| Ktap_E | Rated tap position at medium voltage side |
| Ktap_C | Tap step size at medium voltage side |
| Ktap_V | Rated voltage at medium voltage side |
| Jtap_V | Rated voltage at low voltage side |
| Ri | Resistance at high voltage side |
| Xi | Reactance at high voltage side |
| Rk | Resistance at medium voltage side |
| Xk | Reactance at medium voltage side |
| Rj | Resistance at low voltage side |
| Xj | Reactance at low voltage side |
| I_node | Topology node name at high voltage side |
| K_node | Topology node name at medium voltage side |
| J_node | Topology node name at low voltage side |
| I_P | Active power at high voltage side |
| I_Q | Reactive power at high voltage side |
| K_P | Active power at medium voltage side |
| K_Q | Reactive power at medium voltage side |
| J_P | Active power at low voltage side |
| J_Q | Reactive power at low voltage side |
| I_tap | Tap position at high voltage side |

**Table A.6** (Continued)

| Attributes | Annotation |
| --- | --- |
| K_tap | Tap position at medium voltage side |
| I_off | High voltage side offline flag |
| K_off | Medium voltage side offline flag |
| J_off | Low voltage side offline flag |
| Pi_meas | Active power measurement at high voltage side |
| Qi_meas | Reactive power measurement at high voltage side |
| Pk_meas | Active power measurement at medium voltage side |
| Qk_meas | Reactive power measurement at medium voltage side |
| Pj_meas | Active power measurement at low voltage side |
| Qj_meas | Reactive power measurement at low voltage side |
| Ti_meas | Tap position measurement at high voltage side |
| Tk_meas | Tap position measurement at medium voltage side |
| G | Conductance |
| B | Susceptance |
| I_nd | Physical node number at high voltage side |
| K_nd | Physical node number at medium voltage side |
| J_nd | Physical node number at low voltage side |
| I_bs | Bus number at high voltage side |
| K_bs | Bus number at medium voltage side |
| J_bs | Bus number at low voltage side |
| I_island | Island number at high voltage side |
| K_island | Island number at medium voltage side |
| J_island | Island number at low voltage side |
| Ri* | High voltage side resistance per unit |
| Xi* | High voltage side reactance per unit |
| Rk* | Medium voltage side resistance per unit |
| Xk* | Medium voltage side reactance per unit |
| Rj* | Low voltage side resistance per unit |
| Xj* | Low voltage side reactance per unit |
| I_t | High voltage side turns ratio |
| K_t | Medium voltage side turns ratio |
| J_t | Low voltage side turns ratio |
| Ibase_V | High voltage side base voltage |
| Kbase_V | Medium voltage side base voltage |
| Jbase_V | Low voltage side base voltage |

**Table A.7**  Load table in CIM/E model.

| Attributes | Annotation |
|---|---|
| id | Identification |
| name | Load name |
| volt | Rated voltage |
| Eq | Equivalent load flag |
| position | Equivalent load position |
| node | Topology node name |
| P | Active power |
| Q | Reactive power |
| off | Offline flag |
| P_meas | Active power measure |
| Q_meas | Reactive power measure |
| nd | Physical node number |
| bs | Bus number |
| island | Island number |

**Table A.8**  Shunt compensator table in CIM/E model.

| Attributes | Annotation |
|---|---|
| id | Identification |
| name | Shunt compensator name |
| volt | Voltage level |
| Q_rate | Rated capacity |
| V_rate | Rated voltage |
| position | Connecting position |
| node | Topology node name |
| P | Active power |
| Q | Reactive power |
| off | Offline flag |
| Q_meas | Reactive power measure |
| Nd | Physical node number |
| bs | Bus number |
| island | Island number |

**Table A.9** Series compensator table in CIM/E model.

| Attributes | Annotation |
| --- | --- |
| id | Identification |
| name | Series compensator name |
| volt | Voltage level |
| Q_rate | Rated capacity |
| V_rate | Rated voltage |
| I_node | Topology node name at I end |
| J_node | Topology node name at J end |
| Zk | Impedance |
| off | Offline flag |
| Pi_meas | Active power measure at I end |
| Qi_meas | Reactive power measure at I end |
| Pj_meas | Active power measure at I end |
| Qj_meas | Reactive power measure at I end |
| Ih | Rated current |
| Zk_max | Impedance upper limit |
| Zk_min | Impedance lower limit |
| Line | Compensated line |
| I_nd | Physical node number at I end |
| J_nd | Physical node number at J end |
| I_bs | Bus number at I end |
| J_bs | Bus number at J end |
| I_island | Island number at I end |
| J_island | Island number at J end |
| Zk* | Actual impedance per unit |

**Table A.10** Converter table in CIM/E model.

| Attributes | Annotation |
| --- | --- |
| id | Identification |
| name | Converter name |
| N | Bridge number |
| I_DCName | Topology node name at DC node I |
| J_DCName | Topology node name at DC node J |
| N_ACName | Topology node name at AC side |
| Vdc | DC voltage |
| Idc | DC current |
| Wdc | DC power |
| Pac | AC active power |
| Qac | AC reactive power |

(*Continued*)

**Table A.10** (Continued)

| Attributes | Annotation |
| --- | --- |
| Mode | Control mode |
| state | Converter state |
| off | Offline flag |
| Ang | Firing angle |
| Vdc_meas | DC voltage measurement |
| Idc_meas | DC current measurement |
| Wdc_meas | DC power measurement |
| Pac_meas | AC active power measurement |
| Qac_meas | AC reactive power measurement |
| Ang_meas | Firing angle measurement |
| Wdc_rate | DC rated power |
| Ang_max | Firing angle upper limit |
| Ang_min | Firing angle lower limit |
| Wdc_Rate | DC rated power |
| Kt | Converter transformer turns ratio |
| I_dcnd | Physical node at DC node I |
| J_dcnd | Physical node at DC node J |
| N_acnd | Physical node at AC side |
| I_dcbs | Bus at DC node I |
| J_dcbs | Bus at DC node J |
| N_acbs | Bus at AC side |

**Table A.11** DC line table in CIM/E model.

| Attributes | Annotation |
| --- | --- |
| id | Identification |
| name | DC Line name |
| R | Bridge number |
| I_node | Topology node I |
| J_node | Topology node J |
| Conduct | Conduct flag |
| off | Offline flag |
| Ih | Maximum current |
| I_nd | Physical node number at I end |
| J_nd | Physical node number at J end |
| I_bs | Bus number at I end |
| J_bs | Bus number at J end |
| I_island | Island number at I end |
| J_island | Island number at J end |

**Table A.12**  Island table in CIM/E model.

| Attributes | Annotation |
| --- | --- |
| id | Identification |
| name | Island name |
| Ref_bus | Reference bus name |
| Off | Dead island flag |
| F_sys | System frequency measurement |

**Table A.13**  Topological node table in CIM/E model.

| Attributes | Annotation |
| --- | --- |
| id | Identification |
| name | Topology node name |
| Island | Island name |
| V | Voltage magnitude |
| Ang | Voltage angle |
| Vbase | Base voltage |

**Table A.14**  Breaker table in CIM/E model.

| Attributes | Annotation |
| --- | --- |
| id | Identification |
| name | Breaker name |
| volt | Rated voltage |
| point | Telemetry point |
| I_nd | Physical node number at I end |
| J_nd | Physical node number at J end |
| I_node | Topology node name at I end |
| J_node | Topology node name at J End |

**Table A.15**  Disconnector table in CIM/E model.

| Attributes | Annotation |
| --- | --- |
| id | Identification |
| name | Disconnector name |
| volt | Rated voltage |
| Point | Telemetry point |
| I_nd | Node number at I end |
| J_nd | Node number at J end |
| I_node | Topology node name at I end |
| J_node | Topology node name at J end |

## A.1 Vertices of Bus-Branch Model

**Table A.16** Vertex attributes in bus-branch model.

| Attributes | Annotation |
|---|---|
| TOPOID | Topo node identification |
| island | Island number |
| bus_name | Bus name |
| area | Area name |
| loss_zone | Loss zone |
| busType | Bus type |
| Vm | Voltage magnitude |
| Va | Voltage angle |
| M_Vm | Voltage magnitude measurement |
| M_Va | Voltage angle measurement |
| base_kV | Base voltage |
| desired_volts | Desired control voltage |
| control_bus_number | Voltage control bus number |
| up_V | Voltage upper limit |
| lo_V | Voltage lower limit |
| GenP | Unit active power |
| GenQ | Unit reactive power |
| M_Gen_P | Unit active power measurement |
| M_Gen_Q | Unit reactive power measurement |
| qUp | Reactive power upper limit |
| qLower | Reactive power lower limit |
| LdP | Load active power |
| LdQ | Load reactive power |
| P | Active power |
| Q | Reactive power |
| M_Load_P | Load active power measurement |
| M_Load_Q | Load reactive power measurement |
| G | Shunt conductance |
| B | Shunt susceptance |
| Sub | Substation name |
| OV | Upper voltage violation flag |
| UV | Lower voltage violation flag |
| SE_Vm | Voltage magnitude estimate |
| SE_Va | Voltage angle estimate |
| M_C_P | Shunt capacitor/reactor measurement |

## A.2 Edge of Bus-Branch Model

**Table A.17** Edge attributes in bus-branch model.

| Attributes | Annotation |
|---|---|
| From_TopoND | Topology node name at from end |
| to_TopoND | Topology node name at to end |
| edge_name | Edge name |
| area | Area name |
| zone | Loss zone |
| from_bus | From bus |
| to_bus | To bus |
| flag | Offline flag |
| R | Resistance |
| X | Reactance |
| hB | Half of charging capacitance |
| line_Q1 | Normal limit |
| line_Q2 | Emergency limit |
| line_Q3 | Load shedding limit |
| control_bus | Controlled bus |
| K | Transformer turns ratio |
| shifting_transformer angle | Shifting transformer angle |
| min_tap | Minimum tap position |
| max_tap | Maximum tap position |
| step_size | Tap step size |
| min_volt | Minimum voltage |
| max_volt | Maximum voltage |
| M_P_BR | Active power measurement |
| M_Q_BR | Reactive power measurement |
| G | Conductance |
| B | Susceptance |
| BIJ | Charging susceptance |
| circuit | Circuit number |
| P_BR | Active power |
| Q_BR | Reactive power |
| from_open | From end open flag |
| to_open | To end open flag |

## B  Synchronous Machine Model

For a power system stability analysis, the mathematical model of a synchronous machine is developed as Park's equations in the $d$–$q$–$0$ reference frame. The Park's equations describe the detailed electrical dynamics of a synchronous machine by flux linkage equations and voltage equations. Solving Park's equations is computationally intensive. In practical power system stability analysis, simplifications with different level assumptions are adopted.

Typically, there are seven synchronous machine models to represent synchronous machine dynamics with various degrees of approximations:

Type-0: classical model with constant $E'$
Type-1: 2-order model with constant $E'_q$
Type-2: 3-order model with variable $E'_q$
Type-3: 5-order model with variables $E''_q, E''_d, E'_q$
Type-4: 2-order model with constant $E''$
Type-5: 4-order model with variables $E'_q, E'_d$
Type-6: 6-order model with variables $E''_q, E''_d, E'_q, E'_d$

In this section, the per-unit mathematical equations for each type of model are given for the different purposes of stability analysis. In the per-unit system, generator angle $\delta$ is in radian. The per-unit of speed $\omega$ is based on the synchronous angular speed $2\pi f_0$ in radian/second. $f_0$ is either 50 or 60 Hz; The time $t$ is in seconds. The synchronous machine vector direction convention is given in Figure A.1.

### B.1  Type-6: 6-Order Model with Variables $E''_q, E''_d, E'_q, E'_d$

Type-6 model is a high-order synchronous machine model with variables. $E''_q, E''_d, E'_q, E'_d$ derived from Park's equations which are represented by the following equations.

$$T'_{d0}\frac{dE'_q}{dt} = E_{fd} - \left[E'_q + (x_d - x'_d)I_d + (K_G - 1)E'_q\right] \tag{A.1}$$

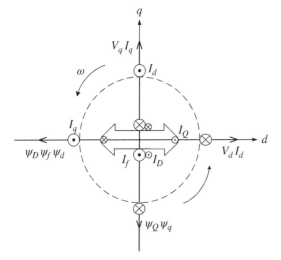

**Figure A.1**  Synchronous machine phasor direction convention diagram.

$$T''_{d0} \frac{dE''_q}{dt} = -E''_q - (x'_d - x''_d)I_d + E'_q + T''_{d0} \frac{dE'_q}{dt} \tag{A.2}$$

$$T'_{q0} \frac{dE'_d}{dt} = -E'_d + (x_q - x'_q)I_q \tag{A.3}$$

$$T''_{q0} \frac{dE''_d}{dt} = -E''_d + (x'_q - x''_q)I_q + E'_d + T''_{q0} \frac{dE'_d}{dt} \tag{A.4}$$

$$T_J \frac{d\omega}{dt} = P_T - P_e - D(\omega - 1) \tag{A.5}$$

$$\frac{d\delta}{dt} = \omega - 1 \tag{A.6}$$

where, according to Park's transformation, $\psi_d$ and $\psi_q$ are determined by

$$\begin{bmatrix} \psi_d \\ \psi_q \end{bmatrix} = \begin{bmatrix} E''_q \\ -E''_d \end{bmatrix} - \begin{bmatrix} x''_d & 0 \\ 0 & x''_q \end{bmatrix} \begin{bmatrix} I_d \\ I_q \end{bmatrix} \tag{A.7}$$

$$\begin{bmatrix} U_d \\ U_q \end{bmatrix} = \omega \begin{bmatrix} -\psi_q \\ \psi_d \end{bmatrix} - \begin{bmatrix} R_a & 0 \\ 0 & R_a \end{bmatrix} \begin{bmatrix} I_d \\ I_q \end{bmatrix} \tag{A.8}$$

Generator parameters in the above equations are listed in Table A.18.

**Table A.18** Generator parameters.

| Attributes | Annotation |
| --- | --- |
| $x_d$ | Direct-axis synchronous reactance |
| $x'_d$ | Direct-axis transient reactance |
| $x''_d$ | Direct-axis subtransient reactance |
| $x_q$ | Quadrature-axis synchronous reactance |
| $x'_q$ | Quadrature-axis transient reactance |
| $x''_q$ | Quadrature-axis subtransient reactance |
| $x_2$ | Negative sequence reactance |
| $R_a$ | Stator resistance |
| $T'_{d0}$ | Direct-axis transient open circuit time constant |
| $T''_{d0}$ | Direct-axis subtransient open circuit time constant |
| $T'_{q0}$ | Quadrature-axis transient open circuit time constant |
| $T''_{q0}$ | Quadrature-axis subtransient open circuit time constant |
| $T_J$ | Inertia time constant $T_J = 2H$ |
| $K_G$ | Saturation factor, $K_G = 1 + \frac{b}{a}E'^{n-1}_q$ |
| $a, b, n$ | Saturation parameters |
| $D$ | Damping coefficient |

Assuming $\omega = 1$, by substituting (A.7), (A.8), we build the voltage equations in the $d - q$ reference frame as follows:

$$\begin{bmatrix} U_q \\ U_d \end{bmatrix} = \begin{bmatrix} E''_q \\ E''_d \end{bmatrix} - \begin{bmatrix} R_a & x''_d \\ -x''_q & R_a \end{bmatrix} \begin{bmatrix} I_q \\ I_d \end{bmatrix} \tag{A.9}$$

Network equations are built in the $x - y$ reference frame. In the transient stability study, the generator terminal voltage and current in the $x - y$ reference frame can be transformed into the $d - q$ reference frame by

$$\begin{bmatrix} U_q \\ U_d \end{bmatrix} = \begin{bmatrix} \cos\delta & \sin\delta \\ \sin\delta & -\cos\delta \end{bmatrix} \begin{bmatrix} U_x \\ U_y \end{bmatrix} \tag{A.10}$$

$$\begin{bmatrix} I_q \\ I_d \end{bmatrix} = \begin{bmatrix} \cos\delta & \sin\delta \\ \sin\delta & -\cos\delta \end{bmatrix} \begin{bmatrix} I_x \\ I_y \end{bmatrix} \tag{A.11}$$

So, we get

$$\begin{bmatrix} \cos\delta & \sin\delta \\ \sin\delta & -\cos\delta \end{bmatrix} \begin{bmatrix} U_x \\ U_y \end{bmatrix} = \begin{bmatrix} E''_q \\ E''_d \end{bmatrix} - \begin{bmatrix} R_a & x''_d \\ -x''_q & R_a \end{bmatrix} \begin{bmatrix} \cos\delta & \sin\delta \\ \sin\delta & -\cos\delta \end{bmatrix} \begin{bmatrix} I_x \\ I_y \end{bmatrix} \tag{A.12}$$

Since

$$\begin{bmatrix} R_a & x''_d \\ -x''_q & R_a \end{bmatrix}^{-1} = \frac{1}{R_a^2 + x''_d x''_q} \begin{bmatrix} R_a & -x''_d \\ x''_q & R_a \end{bmatrix} \tag{A.13}$$

$$\begin{bmatrix} \cos\delta & \sin\delta \\ \sin\delta & -\cos\delta \end{bmatrix}^{-1} = \begin{bmatrix} \cos\delta & \sin\delta \\ \sin\delta & -\cos\delta \end{bmatrix} \tag{A.14}$$

Equation (A.12) is transformed as follows:

$$\frac{1}{R_a^2 + x''_d x''_q} \begin{bmatrix} \cos\delta & \sin\delta \\ \sin\delta & -\cos\delta \end{bmatrix} \begin{bmatrix} R_a & -x''_d \\ x''_q & R_a \end{bmatrix} \begin{bmatrix} \cos\delta & \sin\delta \\ \sin\delta & -\cos\delta \end{bmatrix} \begin{bmatrix} U_x \\ U_y \end{bmatrix} = \frac{1}{R_a^2 + x''_d x''_q}$$

$$\begin{bmatrix} \cos\delta & \sin\delta \\ \sin\delta & -\cos\delta \end{bmatrix} \begin{bmatrix} R_a & -x''_d \\ x''_q & R_a \end{bmatrix} \begin{bmatrix} E''_q \\ E''_d \end{bmatrix} - \begin{bmatrix} \cos\delta & \sin\delta \\ \sin\delta & -\cos\delta \end{bmatrix} \begin{bmatrix} \cos\delta & \sin\delta \\ \sin\delta & -\cos\delta \end{bmatrix} \begin{bmatrix} I_x \\ I_y \end{bmatrix} \tag{A.15}$$

Equation (A.15) is rewritten as

$$\begin{bmatrix} I_x \\ I_y \end{bmatrix} = \frac{1}{R_a^2 + x''_d x''_q} \begin{bmatrix} \cos\delta & \sin\delta \\ \sin\delta & -\cos\delta \end{bmatrix} \begin{bmatrix} R_a & -x''_d \\ x''_q & R_a \end{bmatrix} \begin{bmatrix} E''_q \\ E''_d \end{bmatrix}$$

$$- \frac{1}{R_a^2 + x''_d x''_q} \begin{bmatrix} \cos\delta & \sin\delta \\ \sin\delta & -\cos\delta \end{bmatrix} \begin{bmatrix} R_a & -x''_d \\ x''_q & R_a \end{bmatrix} \begin{bmatrix} \cos\delta & \sin\delta \\ \sin\delta & -\cos\delta \end{bmatrix} \begin{bmatrix} U_x \\ U_y \end{bmatrix} \tag{A.16}$$

And the generator injection current to the network in the $x - y$ reference frame can be calculated by

$$\begin{bmatrix} I_x \\ I_y \end{bmatrix} = \begin{bmatrix} g_x & b_x \\ b_y & g_y \end{bmatrix} \begin{bmatrix} E_q'' \\ E_d'' \end{bmatrix} - \begin{bmatrix} G_x & B_x \\ B_y & G_y \end{bmatrix} \begin{bmatrix} U_x \\ U_y \end{bmatrix} \tag{A.17}$$

where

$$G_x = \frac{R_a - \left(x_d'' - x_q''\right)sin\delta cos\delta}{R_a^2 + x_d'' x_q''}$$

$$B_x = \frac{x_d'' cos^2\delta + x_q'' sin^2\delta}{R_a^2 + x_d'' x_q''}$$

$$B_y = \frac{-x_d'' sin^2\delta - x_q'' cos^2\delta}{R_a^2 + x_d'' x_q''}$$

$$G_y = \frac{R_a + \left(x_d'' - x_q''\right)sin\delta cos\delta}{R_a^2 + x_d'' x_q''}$$

$$g_x = \frac{R_a cos\delta + x_q'' sin\delta}{R_a^2 + x_d'' x_q''}$$

$$b_x = \frac{R_a sin\delta - x_d'' cos\delta}{R_a^2 + x_d'' x_q''}$$

$$b_y = \frac{R_a sin\delta - x_q'' cos\delta}{R_a^2 + x_d'' x_q''}$$

$$g_y = \frac{-R_a cos\delta - x_d'' sin\delta}{R_a^2 + x_d'' x_q''}$$

## B.2   Type-5: 4-Order Model with Variables $E_q', E_d'$

For studies in which subtransients are not interesting, the 6-order model is simplified to the Type-5 model 4-order model by assuming direct-axis and quadrature-axis subtransient voltage $E_d''$ and $E_q''$ constant. The per-unit differential equations for the Type-5 model are given by

$$T_{d0}' \frac{dE_q'}{dt} = E_{fd} - \left[E_q' + \left(x_d - x_d'\right)I_d + (K_G - 1)E_q'\right] \tag{A.18}$$

$$T_{q0}' \frac{dE_d'}{dt} = -E_d' + \left(x_q - x_q'\right)I_q \tag{A.19}$$

$$T_J \frac{d\omega}{dt} = P_T - P_e - D(\omega - 1) \tag{A.20}$$

$$\frac{d\delta}{dt} = \omega - 1 \tag{A.21}$$

And the generator injection current to the network in the $x-y$ reference frame can be calculated by

$$\begin{bmatrix} I_x \\ I_y \end{bmatrix} = \begin{bmatrix} g_x & b_x \\ b_y & g_y \end{bmatrix} \begin{bmatrix} E'_q \\ E'_d \end{bmatrix} - \begin{bmatrix} G_x & B_x \\ B_y & G_y \end{bmatrix} \begin{bmatrix} U_x \\ U_y \end{bmatrix}$$  (A.22)

where

$$G_x = \frac{R_a - \left(x'_d - x'_q\right)\sin\delta\cos\delta}{R_a^2 + x'_d x'_q}$$

$$B_x = \frac{x'_d \cos^2\delta + x'_q \sin^2\delta}{R_a^2 + x'_d x'_q}$$

$$B_y = \frac{-x'_d \sin^2\delta - x'_q \cos^2\delta}{R_a^2 + x'_d x'_q}$$

$$G_y = \frac{R_a + \left(x'_d - x'_q\right)\sin\delta\cos\delta}{R_a^2 + x'_d x'_q}$$

$$g_x = \frac{R_a\cos\delta + x'_q\sin\delta}{R_a^2 + x'_d x'_q}$$

$$b_x = \frac{R_a\sin\delta - x'_d\cos\delta}{R_a^2 + x'_d x'_q}$$

$$b_y = \frac{R_a\sin\delta - x'_q\cos\delta}{R_a^2 + x'_d x'_q}$$

$$g_y = \frac{-R_a\cos\delta - x'_d\sin\delta}{R_a^2 + x'_d x'_q}$$

## B.3 Type-4: 2-Order Model with Constant $E''$

With further approximation, the Type-4 model assumes the subtransient voltage $E''$ constant to represent the physics at the generator short-circuit moment. Thus Typer-4 model is typically used for a short-circuit current analysis. The per-unit differential equations for the Type-4 model are given by the rotor swing equations.

$$T_J \frac{d\omega}{dt} = P_T - P_e - D(\omega - 1)$$  (A.23)

$$\frac{d\delta}{dt} = \omega - 1$$  (A.24)

And the generator injection current to the network in the $x-y$ reference frame is given by

$$\begin{bmatrix} I_x \\ I_y \end{bmatrix} = \begin{bmatrix} g_x \\ b_y \end{bmatrix} E'' - \begin{bmatrix} G_x & B_x \\ B_y & G_y \end{bmatrix} \begin{bmatrix} U_x \\ U_y \end{bmatrix}$$  (A.25)

where

$$G_x = \frac{R_a}{R_a^2 + x_d'' x_q''}$$

$$B_x = \frac{x_d''}{R_a^2 + x_d'' x_q''}$$

$$B_y = \frac{-x_d''}{R_a^2 + x_d'' x_q''}$$

$$G_y = \frac{R_a}{R_a^2 + x_d'' x_q''}$$

$$g_x = \frac{R_a \cos\delta + x_q'' \sin\delta}{R_a^2 + x_d'' x_q''}$$

$$b_y = \frac{R_a \sin\delta - x_q'' \cos\delta}{R_a^2 + x_d'' x_q''}$$

Since $x_d'' = x_q''$, $G_x$, $B_x$, $G_y$, $B_y$ are constant.

## B.4 Type-3: 5-Order Model with Variable $E_q''$, $E_d''$, $E_q'$

The Type-6 model is simplified to the Type-3 model by assuming $pE_d' = 0$. The per-unit differential equations for the Type-2 model are given by

$$T_{d0}' \frac{dE_q'}{dt} = E_{fd} - \left[ E_q' + (x_d - x_d') I_d + (K_G - 1) E_q' \right] \tag{A.26}$$

$$T_{d0}'' \frac{dE_q''}{dt} = -E_q'' - (x_d' - x_d'') I_d + E_q' + T_{d0}'' \frac{dE_q'}{dt} \tag{A.27}$$

$$T_{q0}'' \frac{dE_d''}{dt} = -E_d'' + (x_q' - x_q'') I_q \tag{A.28}$$

$$T_J \frac{d\omega}{dt} = P_T - P_e - D(\omega - 1) \tag{A.29}$$

$$\frac{d\delta}{dt} = \omega - 1 \tag{A.30}$$

the generator injection current to the network in the $x - y$ reference frame can be calculated by

$$\begin{bmatrix} I_x \\ I_y \end{bmatrix} = \begin{bmatrix} g_x & b_x \\ b_y & g_y \end{bmatrix} \begin{bmatrix} E_q'' \\ E_d'' \end{bmatrix} - \begin{bmatrix} G_x & B_x \\ B_y & G_y \end{bmatrix} \begin{bmatrix} U_x \\ U_y \end{bmatrix} \tag{A.31}$$

where

$$G_x = \frac{R_a}{R_a^2 + x_d'' x_q''}$$

$$B_x = \frac{x_d''}{R_a^2 + x_d''x_q''}$$

$$B_y = \frac{-x_d''}{R_a^2 + x_d''x_q''}$$

$$G_y = \frac{R_a}{R_a^2 + x_d''x_q''}$$

$$g_x = \frac{R_a\cos\delta + x_q''\sin\delta}{R_a^2 + x_d''x_q''}$$

$$b_x = \frac{R_a\sin\delta - x_d''\cos\delta}{R_a^2 + x_d''x_q''}$$

$$b_y = \frac{R_a\sin\delta - x_q''\cos\delta}{R_a^2 + x_d''x_q''}$$

$$g_y = \frac{-R_a\cos\delta - x_d''\sin\delta}{R_a^2 + x_d''x_q''}$$

$G_x$, $B_x$, $G_y$, $B_y$ are constant.

### B.5  Type-2: 3-Order Model with Variable $E_q'$

The Type-2 model is simplified from the Type-5 model by assuming $pE_q'' = 0$ and $pE_d'' = 0$. The per-unit differential equations for the Type-2 model are given by

$$T_{d0}'\frac{dE_q'}{dt} = E_{fd} - \left[E_q' + (x_d - x_d')I_d + (K_G - 1)E_q'\right] \tag{A.32}$$

$$T_J\frac{d\omega}{dt} = P_T - P_e - D(\omega - 1) \tag{A.33}$$

$$\frac{d\delta}{dt} = \omega - 1 \tag{A.34}$$

And the generator injection current to the network in the $x - y$ reference frame can be calculated by

$$\begin{bmatrix} I_x \\ I_y \end{bmatrix} = \begin{bmatrix} g_x \\ b_y \end{bmatrix} E_q' - \begin{bmatrix} G_x & B_x \\ B_y & G_y \end{bmatrix}\begin{bmatrix} U_x \\ U_y \end{bmatrix} \tag{A.35}$$

where

$$G_x = \frac{R_a - \left(x_d' - x_q'\right)\sin\delta\cos\delta}{R_a^2 + x_d'x_q'}$$

$$B_x = \frac{x_d'\cos^2\delta + x_q'\sin^2\delta}{R_a^2 + x_d'x_q'}$$

$$B_y = \frac{-x_d' \sin^2\delta - x_q' \cos^2\delta}{R_a^2 + x_d'x_q'}$$

$$G_y = \frac{R_a + \left(x_d' - x_q'\right)\sin\delta\cos\delta}{R_a^2 + x_d'x_q'}$$

$$g_x = \frac{R_a\cos\delta + x_q'\sin\delta}{R_a^2 + x_d'x_q'}$$

$$b_y = \frac{R_a\sin\delta - x_q'\cos\delta}{R_a^2 + x_d'x_q'}$$

## B.6  Type-1: 2-Order Model with Constant $E_d'$ and $E_q'$

The Type-1 model is the classic 2-order representation of a synchronous machine for large-scale and quick study. The Type-1 model is simplified by assuming $pE_d' = pE_q' = 0$. The per-unit differential equations for the Type-1 model are given by the rotor swing equations.

$$T_J\frac{d\omega}{dt} = P_T - P_e - D(\omega - 1) \tag{A.36}$$

$$\frac{d\delta}{dt} = \omega - 1 \tag{A.37}$$

## B.7  Steady-State Equations and Phasor Diagram

In power system stability analysis, the system dynamics and transients start from an equilibrium point where all the derivatives are simultaneously zero. The equilibrium point satisfies the following equations.

$$U_d + R_aI_d - x_qI_q = 0 \tag{A.38}$$

$$U_q + R_aI_q + x_dI_d = E_q \tag{A.39}$$

$$U_d + R_aI_d - x_q'I_q = E_d' \tag{A.40}$$

$$U_q + R_aI_q + x_d'I_d = E_q' \tag{A.41}$$

$$U_d + R_aI_d - x_q''I_q = E_d'' \tag{A.42}$$

$$U_q + R_aI_q + x_d''I_d = E_q'' \tag{A.43}$$

Assuming $\dot{U}_t$ and $\dot{I}_t$ are the generator terminal voltages and currents in the $x - y$ reference frame. A virtual voltage $\dot{E}_Q$ on the quadrature-axis is defined to calculate the power angle $\delta$ as

$$\dot{E}_Q = \dot{U}_t + \left(R_a + jx_q\right)\dot{I}_t \tag{A.44}$$

The corresponding phasor diagram of a synchronous machine is then given in Figure A.2.

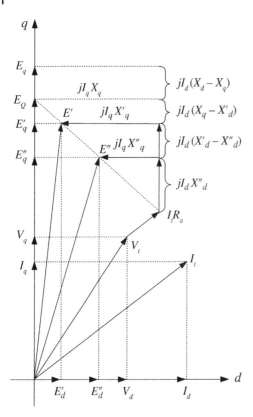

**Figure A.2** Synchronous machine steady-state phasor diagram.

## C Deep Deterministic Policy Gradient Pseudo-Code

A deep deterministic policy gradient algorithm is based on a model-free, off-policy actor-critic structure with an actor network and a critic network. To minimize correlations between samples and improve learning stability, a replay buffer is used to store and restore samples randomly. The pseudo-codes of the function implementations are listed as follows.

### C.1 Replay Buffer

```
class ReplayBuffer(object):

    def __init__(self, buffer_size, random_seed):
        self.buffer_size = buffer_size
        self.count = 0
        self.buffer = deque()

    def add_sample(self, s, a, r, s2):
        if self.count < self.buffer_size:
            self.buffer.append(s, a, r, s2)
            self.count += 1
        else:
```

```
            self.buffer.popleft()
            self.buffer.append(s, a, r, s2)


        def sample_batch(self, batch_size):
            if self.count < batch_size:
                batch = random.sample(self.buffer, self.count)
            else:
                batch = random.sample(self.buffer, batch_size)
            return batch
```

## C.2   Actor-Critic Network

```
import tensorflow as tf
import tensorlayer as tl


n_hidden_1 = 400
n_hidden_2 = 300


class Actor-Critic(object):
    def __init__(self, state_dim, action_dim, learning_rate, tau):
        self.s_dim = state_dim
        self.a_dim = action_dim
        self.learning_rate = learning_rate
        self.tau = tau


    def get_actor(self, state):
        w1 = tf.get_variable(name='w1', shape=[self.s_dim, n_hidden_1])
        b1 = tf.get_variable(name='b1', shape=[n_hidden_1])
        w2 = tf.get_variable(name='w2', shape=[n_hidden_1, n_hidden_2])
        b2 = tf.get_variable(name='b2', shape=[n_hidden_2])
        w3 = tf.get_variable(name='w3', shape=[n_hidden_2, self.a_dim])
        b3 = tf.get_variable(name='b3', shape=[self.a_dim])
        h1 = tf.nn.leaky_relu(tf.matmul(state, w1) + b1)
        h2 = tf.nn.leaky_relu(tf.matmul(h1, w2) + b2)
        outputs = tf.nn.tanh(tf.matmul(h2, w3) + b3)
        return outputs


    def get_critic(self, state, action):
        w1 = tf.get_variable(name='w1', shape=[self.s_dim, n_hidden_1])
        b1 = tf.get_variable(name='b1', shape=[n_hidden_1])
        w2 = tf.get_variable(name='w2', shape=[n_hidden_1, n_hidden_2])
        w2a = tf.get_variable(name='w2a', shape=[self.a_dim, n_hidden_2])
        b2 = tf.get_variable(name='b2', shape=[n_hidden_2])
        w3 = tf.get_variable(name='w3', shape=[n_hidden_2, 1])
        b3 = tf.get_variable(name='b3', shape=[1])
        h1 = tf.nn.leaky_relu(tf.matmul(state, w1) + b1)
        h2 = tf.nn.leaky_relu(tf.matmul(h1,w2)+tf.matmul(action, w2a)+b2)
```

```
      outputs = tf.matmul(h2, w3) + b3
      return outputs

  def learn(self, batch):
    bs = batch[:, :self.s_dim]
    ba = batch[:, self.s_dim : self.s_dim + self.a_dim]
    br = batch[:, -self.s_dim - 1:-self.s_dim]
    bs_= batch[:, -self.s_dim:]

    with tf.GradientTape() as tape:
      a_= self.get_actor(bs_)
      q_= self.get_critic(bs_, a_)
      y = br + GAMMA * q_
      q = self.get_critic(bs, ba)

      self.loss = tf.losses.mean_squared_error(y, q)
      self.optimizer = tf.train.AdamOptimizer(self.learning_rate)
      self.network_params = tf.train.MovingAverage(decay = 1-tau)

    with tf.GradientTape() as tape:
      a = self.get_actor(bs)
      q = self.get_critic([bs, a])

      self.loss = -tf.reduce_mean(q)
      self.optimizer = tf.train.AdamOptimizer(self.learning_rate)
```

## C.3  Environment

```
class powerSystem:

  def __init__(self):
    self.p_gen = None
    self.s_branch = None
    self.states = [self.p_gen, self.s_branch]

  def step(self, action):
    conn.upsertVertex(TopoND, GenP = action)
    conn.runQuery(newtonRaphsonPowerFlow(MaxIteration, Tolerance))
    conn.runQuery(calculateReward)
    Pg = conn.getVertex(TopoND, busType == 2, attributes = "GenP")
    S_BR = conn.getEdge(TopoConnect, attributes = "S_BR")
    R_BR = conn.getEdge(TopoConnect, attributes = "R_BR")
    self.p_gcn = Pg
    self.s_branch = S_BR
    reward = sum(R_BR)
    states = [self.p_gen, self.s_branch]
    return states, reward
```

# D   Security-Constrained AGC Graph Computing Pseudo-Code

## D.1   Create Flowgate Graph

To support flowgate constraint in the security-constrained automatic generation control model, the vertices FlowgateND, LineND, and edges Connected_Flowgate_Line are defined as follows:

```
CREATE VERTEX FlowgateND (PRIMARY_ID FlowgateID int, Flowgate_S double)
CREATE VERTEX LineND (PRIMARY_ID LineID int, from_bus int, to_bus int)
CREATE DIRECTED EDGE Connected_Flowgate_Line(from FlowgateND, to LineND)
```

And the flowgate graph schema is then defined as:

```
CREATE GRAPH FlowgateGraph(FlowgateND, LineND, Connected_Flowgate_Line)
```

In the flowgate graph, each flowgate and line are modeled as a vertex. The edge Connected_-Flowgate_Line connects flowgate to line, and the attribute Flowgate_S is the scheduled flowgate power flow in MW.

The flowgate information is defined in the file unitInfo.csv. And the table header of flowgateInfo.csv is shown in Table A.19.

With the new graph flowgate structure, the data file flowgateInfo.csv will be read to the flowgate graph by the following query readFlowgateInfo.

```
CREATE TABLE flowgateInfoTable (FlowgateID int, Flowgate_S double,
LineID int, from_bus int, to_bus int)

CREATE QUERY readFlowgateInfo() FOR GRAPH FlowgateGraph {
BULK INSERT flowgateInfoTable
FROM '$sys.data_root/flowgateInfo.csv'
WITH(FIRSTROW = 2,FIELDTERMINATOR = ',',ROWTERMINATOR = '\n',TABLOCK)
   SELECT flowgate FROM flowgateInfoTable WHERE unit.BusID == s.TOPOID
     ACCUM
       INSERT INTO FlowgateGraph.FlowgateND(flowgate.FlowgateID,
           flowgate. Flowgate_S)
       INSERT INTO FlowgateGraph.LineND(flowgate.LineID, flowgate.
           from_bus,flowgate.to_bus)
       INSERT INTO FlowgateGraph.Connected_Flowgate_Line(flowgate.
           FlowgateID,flowgate.LineID)
}
```

Statement BULK INSERT reads data from flowgateInfo.csv. The control parameters defined by a WITH clause WITH(FIRSTROW = 2, FIELDTERMINATOR = ',', ROWTERMINATOR = '\n', TABLOCK) tells the first row is a table header, the field delimiter is a comma, the row terminator is '\n', and the table is locked when it is read.

When the CSV file is read into the flowgateInfoTable, the data are inserted into FlowgateGraph attributes.

**Table A.19**   Flowgate information table header.

| Flowgate ID | Scheduled power (MW) | Line ID | From bus | To bus |
| --- | --- | --- | --- | --- |

### D.2   Build Simplex Graph

The graph query to build up the simplex graph for the security-constrained automatic generation control is similar to the one in 12.2.3.1 for the security-constrained economic dispatch with addition to build edges for flowgate constraints as shown as follows:

```
CREATE QUERY buildSimplexGraphforSCAGC ()
FOR GRAPH busBranchGraph and GSFGraph and FlowgateGraph and
SimplexGraph {
T1 = {GSFGraph.GenND.*}
T2 = {busBranchGraph.TopoND.*}
T3 = {FlowgateGraph.FlowgateND.*}
E2 = {busBranchGraph.TopoConnect.*}

int nUnit = T1.getCount()
int nBranch = E2.getCount()

// Build up simplex table
// Insert vertices
INSERT INTO SimplexGraph.V("V1") //Vertex for objective function
INSERT INTO SimplexGraph.V("S")  //Vertex for power balance equation
FOREACH s IN T1 DO
  INSERT INTO SimplexGraph.V("u"+str(s.GenID)) //Vertex ui
  INSERT INTO SimplexGraph.V("v"+str(s.GenID)) //Vertex vi

FOREACH i IN RANGE[1, 2*(nUnit+nBranch)] DO
  ACCUM
    INSERT INTO SimplexGraph.V("x"+str(i)) //Vertex xi

SELECT s1 FROM T1
  SELECT s2 FROM T2 WHERE s1.bus == s2.TOPOID AND s2.unitType == "Hydro"
    ACCUM
        //Vertex for water consumption equation
        INSERT INTO SimplexGraph.V("H"+str(s.GenID))

INSERT INTO SimplexGraph.V("b") //Vertex b

// Insert edges
SELECT s1 FROM T1
  ACCUM
    SELECT s2 FROM T2 WHERE s1.bus == s2.TOPOID
      //Objective function
      INSERT INTO SimplexGraph.E("V1","u"+to_str(s1.GenID),
                              2*s2.a*s2.GenP+s2.b)
      INSERT INTO SimplexGraph.E("V1","v"+to_str(s1.GenID),
                            -(2*s2.a*s2.GenP+s2.b))
      INSERT INTO SimplexGraph.E("V1","b",0)
```

```
    //Power balance constraint
    INSERT INTO SimplexGraph.E("S","u"+to_str(s1.GenID),1)
    INSERT INTO SimplexGraph.E("S","v"+to_str(s1.GenID),-1)
    INSERT INTO SimplexGraph.E("S","b",0)

    //PGmax limit constraints
    INSERT INTO SimplexGraph.E("x"+to_str(s1.GenID),
                               "u"+to_str(s1.GenID),1)
    INSERT INTO SimplexGraph.E("x"+to_str(s1.GenID),
                               "v"+to_str(s1.GenID),-1)
    INSERT INTO SimplexGraph.E("x"+to_str(s1.GenID),
                               "x"+to_str(s1.GenID),1)
    INSERT INTO SimplexGraph.E("x"+to_str(s1.GenID),
                               "b",s2.Pmax-s2.GenP)

    //PGmin limit constraints
    INSERT INTO SimplexGraph.E("x"+to_str(s1.GenID+nUnit),
                               "u"+to_str(s1.GenID),1)
    INSERT INTO SimplexGraph.E("x"+to_str(s1.GenID+nUnit),
                               "v"+to_str(s1.GenID),-1)
    INSERT INTO SimplexGraph.E("x"+to_str(s1.GenID+nUnit),
                               "x"+to_str(s1.GenID),-1)
    INSERT INTO SimplexGraph.E("x"+to_str(s1.GenID+nUnit),
                               "b",s2.Pmin-s2.GenP)

SELECT s FROM T1:s-(connected:e)->:t
  ACCUM
    //Branch flow upper limit
    INSERT INTO SimplexGraph.E("x"+to_str(2*nUnit+e.LineND),
                               "u"+to_str(e.GenND),e.hij)
    INSERT INTO SimplexGraph.E("x"+to_str(2*nUnit+e.LineND),
                               "v"+to_str(e.GenND),-e.hij)
    INSERT INTO SimplexGraph.E("x"+to_str(2*nUnit+e.LineND),
                               "x"+to_str(2*nUnit+e.LineND),1)
    //Branch flow lower limit
    INSERT INTO SimplexGraph.E("x"+to_str(2*nUnit+nBranch+e.LineND),
                               "u"+to_str(e.GenND),e.hij)
    INSERT INTO SimplexGraph.E("x"+to_str(2*nUnit+nBranch+e.LineND),
                               "v"+to_str(e.GenND),-e.hij)
    INSERT INTO SimplexGraph.E("x"+to_str(2*nUnit+nBranch+e.LineND),
                               "x"+to_str(2*nUnit+nBranch+e.LineND),-1)

POST-ACCUM
 SELECT s2 FROM E2
         WHERE e.from_bus == s2.from_bus and e.to_bus == s2.to_bus
    INSERT INTO SimplexGraph.E("x"+to_str(2*nUnit+e.LineND),
                               "b",s2.line_Q1_list-s2.P_BR)
    INSERT INTO SimplexGraph.E("x"+to_str(2*nUnit+nBranch+e.LineND),
                               "b",-s2.line_Q1_list-s2.P_BR)
```

```
//Water Consumption Equation
SELECT s1 FROM T1
  ACCUM
    SELECT s2 FROM T2 WHERE s1.bus == s2.TOPOID AND s2.unitType ==
      "Hydro"
    INSERT INTO SimplexGraph.E("H"+str(s1.GenID),
                          "u"+to_str(s1.GenID),2*s2.a*s2.GenP+s2.b)
    INSERT INTO SimplexGraph.E("H"+str(s1.GenID),
                        "v"+to_str(s1.GenID),-(2*s2.a*s2.GenP+s2.b))
    INSERT INTO SimplexGraph.E("H"+str(s.GenID),
                        "b",s1.waterConsum.Ck/s1.waterConsum.Tk
                        -(s2.a*s2.a*s2.GenP+s2.b*s2.GenP+s2.c))

//Flowgate power flow constraint
SELECT sf FROM T3:sf-(connected:ef)->:tf
 ACCUM
   SELECT s1 FROM T2:s1-(connected:el)->:tl
       WHERE s2.from_bus == tf.from_bus AND s2.to_bus == tf.to_bus
   ACCUM
     sf.@Flowgate_P = sf.@Flowgate_P + el.P_BR

  SELECT s FROM T1:s-(connected:e)->:t where e.LineND == ef.LineND
   ACCUM
     s.@SumHij = s.@SumHij + e.hij

POST-ACCUM
  INSERT INTO SimplexGraph.E("x"+to_str
        (2*nUnit+2*nBranch+sf.FlowgateID),
        "u"+to_str(e.GenND.GenID),s.@SumHij)
  INSERT INTO SimplexGraph.E("x"+to_str
        (2*nUnit+2*nBranch+sf.FlowgateID),
        "v"+to_str(e.GenND.GenID),-s.@SumHij)
  INSERT INTO SimplexGraph.E("x"+to_str
        (2*nUnit+2*nBranch+sf.FlowgateID),"b",
        sf.Flowgate_S-sf.@Flowgate_P)
}
```

In the query buildSimplexGraphforSCAGC, as same as in the query buildSimplex-Graph, all vertices are virtual. Vertex V1 specifies a vertex for the objective function. Vertices ui and vi represent the artificial nonnegative variables for generation output PGi = ui-vi. xi represents slack variables for inequality constraints. Vertices S and Hi refer to the power balance equation and the water consumption equations. Vertex-attached variables SumAccum @Flowgate_P, @ SumHij are defined to store the flowgate power flow and the total generation shift factor of a generator to every branch in the flowgate.

After vertices are created, the statement INSERT INTO SimplexGraph.E (FROM Vertex, TO Vertex, Value) inserts edges to link objective function or constraints (FROM Vertex) to variables (TO Vertex).

In the last block of the query, the edges to link flowgate power flow constraints to variables are inserted as follows:

```
//Flowgate power flow constraint
SELECT sf FROM T3:sf-(connected:ef)->:tf
ACCUM
  SELECT sl FROM T2:sl-(connected:el)->:tl
      WHERE s2.from_bus == tf.from_bus AND s2.to_bus == tf.to_bus
    ACCUM
      sf.@Flowgate_P = sf.@Flowgate_P + el.P_BR

  SELECT s FROM T1:s-(connected:e)->:t where e.LineND == ef.LineND
    ACCUM
    s.@SumHij = s.@SumHij + e.hij


POST-ACCUM
  INSERT INTO SimplexGraph.E("x"+to_str
          (2*nUnit+2*nBranch+sf.FlowgateID),
          "u"+to_str(e.GenND.GenID),s.@SumHij)
  INSERT INTO SimplexGraph.E("x"+to_str
          (2*nUnit+2*nBranch+sf.FlowgateID),
          "v"+to_str(e.GenND.GenID),-s.@SumHij)
  INSERT INTO SimplexGraph.E("x"+to_str
          (2*nUnit+2*nBranch+sf.FlowgateID),"b",
          sf.Flowgate_S-sf.@Flowgate_P)
```

Where, the graph traversal statement SELECT sf FROM T3:sf-(connected:ef)->:tf coupling with the keyword ACCUM travels all vertices flowgates in parallel.

Nested in the ACCUM, SELECT sl FROM T2:sl-(connected:el)->:tl WHERE s2.from_bus == tf.from_bus AND s2.to_bus == tf.to_bus finds the branches associated with the flowgate to calculate the total flowgate power flow by sf.@Flowgate_P = sf.@Flowgate_P + el.P_BR.

Meanwhile, SELECT s FROM T1:s-(connected:e)->:t where e.LineND == ef.LineND selects all lines in GSFGraph associated with the flowgate to calculate the total generation shift factor of a generator to every branch in the flowgate by s.@SumHij = s.@SumHij + e.hij under ACCUM in parallel.

When the actual flowgate and the total generation shift factor are calculated, the edge to link the flowgate power constraint to the generator variable vertices "u" and "v" and right-hand side vertex "b" are inserted by

```
INSERT INTO SimplexGraph.E("x"+to_str
          (2*nUnit+2*nBranch+sf.FlowgateID),
          "u"+to_str(e.GenND.GenID),s.@SumHij)
    INSERT INTO SimplexGraph.E("x"+to_str
          (2*nUnit+2*nBranch+sf.FlowgateID),
          "v"+to_str(e.GenND.GenID),-s.@SumHij)
    INSERT INTO SimplexGraph.E("x"+to_str
          (2*nUnit+2*nBranch+sf.FlowgateID),"b",
          sf.Flowgate_S-sf.@Flowgate_P)
```

When the `SimplexGraph` graph is built, solving the security-constrained automatic generation control by the simplex method and updating power flow follow the same processes in 12.2.3 till the solution is converged.

## E  Graph-Based Eigenvalue Calculation Pseudo-Code

### E.1  Graph-Based QR Decomposition

```
CREATE QUERY QR_Decomposition () FOR GRAPH SmallSignalStabilityGraph {

T0 = {SmallSignalStabilityGraph.*}
int nState = SmallSignalStabilityGraph.getVertexCount()

double normA = 0
SELECT t FROM T0:s-(connected:e)->:t WHERE (t.id == 1)
 ACCUM
   normA = normA + e.a_ij*e.a_ij
 POST-ACCUM
   normA = sqrt(normA)

SELECT s FROM T0:s-(connected:e)->:t WHERE (t.id == 1)
  ACCUM
    e.q_ij = e.a_ij/normA

FOREACH j in range(2:nState)
  ACCUM
    SELECT s FROM T0:s-(connected:e)->:t WHERE (t.id == j)
      e.q_ij = e.a_ij

    FOREACH i in range(1:j-1)
      SELECT s1 FROM T0:s1-(connected:e1)->:t1 WHERE (t1.id == j)
        ACCUM
          SELECT s2 FROM T0:s2-(connected:e2)->:t2 WHERE (t2.id == i)
            ACCUM
            e1.q_ij = e1.q_ij - e1.a_ij*e2.q_ij*e2.q_ij
  POST-ACCUM
    normA = 0
    SELECT s FROM T0:s-(connected:e)->:t WHERE (t.id == j)
    ACCUM
      normA = normA + e.a_ij*e.a_ij
    POST-ACCUM
      normA = sqrt(normA)

    SELECT s FROM T0:s-(connected:e)->:t WHERE (t.id == j)
      ACCUM
        e.q_ij = e.q_ij/normA
```

```
FOREACH i in range(1:nState)
 ACCUM
   FOREACH j in range(1:nState)
     double r_ij = 0
     ACCUM
       SELECT s1 FROM T0:s1-(connected:e1)->:t1 WHERE (t1.id == i)
         ACCUM
           SELECT s2 FROM T0:s2-(connected:e2)->:t2 WHERE (t2.id == j)
             r_ij = r_ij + e1->q_ij*e2.a_ij
         POST-ACCUM
           e1.r_ij = r_ij
}
```

In the query `QR_Decomposition`, the column-vectors of the system state matrix *A* represented by `SmallSignalStabilityGraph` are orthogonalized and normalized in order. Firstly, the Euclidean norm of the first vector is calculated using the following query statements.

```
double normA = 0
SELECT t FROM T0:s-(connected:e)->:t WHERE (t.id == 1)
 ACCUM
   normA = normA + e.a_ij*e.a_ij
 POST-ACCUM
   normA = sqrt(normA)
```

The first column-vector is selected by `SELECT t FROM T0:s-(connected:e)->:t WHERE (t.id == 1)`. Note that the clause `WHERE (t.id == 1)` ensures to select the first column-vector other than the first row-vector.

Then the elements in the first column-vector are normalized in parallel by

```
SELECT s FROM T0:s-(connected:e)->:t WHERE (t.id == 1)
 ACCUM
   e.q_ij = e.a_ij/normA
```

To orthogonalize the vector, a nested loop is developed to orthogonalize all column-vectors by looping through preceding orthogonalized vectors. The orthogonalization involves vector multiplication and subtraction. They are calculated in nodal parallel.

After *Q* is calculated, the last block of the query calculates elements of *R* by *Q′A*. In matrix multiplication, each element calculation is independent. They are calculated by nodal parallel graph computing naturally and saved as edge attributed in the small-signal stability graph.

## E.2 Graph-Based Eigenvalue Calculation

By calling *QR* decomposition iteratively, the eigenvalue calculation graph query is implemented as follows:

```
CREATE QUERY eigenvalueCalculation (int MaxIteration, double
Tolerance) FOR GRAPH SmallSignalStabilityGraph {
MaxAccum<double> @@subDiagonalRatio =-INF
```

```
T0 = {SmallSignalStabilityGraph.*}
int nState = SmallSignalStabilityGraph.getVertexCount()

WHILE (abs(@@subDiagonalRatio)>Tolerance) LIMIT MaxIteration DO
 CALL QR_Decomposition()
 FOREACH i in range(1:nState)
   ACCUM
     FOREACH j in range(1:nState)
       double a_ij = 0
       ACCUM
         SELECT s1 FROM T0:s1-(connected:e1)->:t1 WHERE (s1.id == i)
           ACCUM
             SELECT s2 FROM T0:s2-(connected:e2)->:t2 WHERE (t2.id == j)
               a_ij = a_ij + e1->r_ij*e2.q_ij
         POST-ACCUM
           e1.a_ij = a_ij

 FOREACH i in range(2:nState)
   ACCUM
     SELECT s1 FROM T0:s1-(connected:e1)->:t1
       WHERE (s1.id == i and t1.id == i-1)
     SELECT s2 FROM T0:s2-(connected:e2)->:t2
       WHERE (s2.id == i-1 and t2.id == i-1)
     SELECT s3 FROM T0:s3-(connected:e3)->:t3
       WHERE (s3.id == i and t3.id == i)
     @@subDiagonalRatio = abs(e1.a_ij)/(abs(e2.a_ij)+abs(e3.a_ij))
POST-ACCUM
   SELECT s FROM T0:s-(connected:e)->:t
     WHERE (s.id == i and t.id == i)
   ACCUM
     s.eigenvalue.a = e.a_ij
}
```

In the query `eigenvalueCalculation`, maximum iteration number `MaxIteration` and convergency tolerance `Tolerance` are passed through. Global variable `@@subDiagonalRatio` is defined to store updated subdiagonal to the diagonal ratio for the convergence check.

The `WHILE` loop statement checks the `@@subDiagonalRatio` against the `Tolerance` with the `LIMIT` clause, which counts the iterations in the `WHILE` loop against the maximum iteration number `MaxIteration`. By calling subquery `QR_Decomposition`, the matrix, $A$ is decomposed as $Q$ and $R$. And $A$ is transformed by $A = RQ$. It is a matrix multiplication implemented by a nested loop similar to the block when we calculated $R = Q'A$ in the QR decomposition query. Note that in the calculation of $A = RQ$. Row-vector of $R$ is used to multiply the column-vector of $Q$. The `WHERE` clause in the select statement `SELECT s1 FROM T0:s1-(connected:e1)->:t1 WHERE (s1.id == i)` is checking the `FROM` vector.

# F Graph-Based Transient Simulation Implementation

## F.1 Steady-State Equilibrium Point Calculation

The power system transient simulation starts with a steady-state equilibrium point at a given power flow solution. At the steady-state equilibrium point, all state variable derivatives are zero. The steady-state equilibrium condition is solved by

$$\dot{X} = f(X, U) = 0 \tag{A.45}$$

Using the extended busBranchGraph defined in 15.4.1, the pseudo-code to implement the steady-state equilibrium condition calculation is shown as follows:

```
CREATE QUERY stateInitialization() FOR GRAPH busBranchGraph {
T0 = {TopoND.*}
T1 = SELECT s FROM T0 WHERE (s.busType == 2 OR s.busType == 3)
  ACCUM
    //Generator state initialization
    s.Gen_Var.Vx  = s.Vm * cos(s.Va)
    s.Gen_Var.Vy  = s.Vm * sin(s.Va)
    s.Gen_Var.Ix  = (s.GenP * cos(s.Va) + s.GenQ * sin(s.Va)) / s.Vm
    s.Gen_Var.Iy  = (s.GenP * sin(s.Va) - s.GenQ * cos(s.Va)) / s.Vm
    s.Gen_Var.EQx = s.Gen_Var.Vx + s.Gen_Par.Ra * s.Gen_Var.Ix -
          s.Gen_Par.Xq * s.Gen_Var.Iy
    s.Gen_Var.EQy = s.Gen_Var.Vy + s.Gen_Par.Ra * s.Gen_Var.Iy +
          s.Gen_Par.Xq * s.Gen_Var.Ix
    s.Gen_Var.delta0 = atan2(s.Gen_Var.EQy, s.Gen_Var.EQx)
    s.Gen_Var.Vd = s.Gen_Var.Vx * sin(s.Gen_Var.delta0) - s.Gen_Var.
          Vy * cos(s.Gen_Var.delta0)
    s.Gen_Var.Vq = s.Gen_Var.Vx * cos(s.Gen_Var.delta0) + s.Gen_Var.
          Vy * sin(s.Gen_Var.delta0)
    s.Gen_Var.Id = s.Gen_Var.Ix * sin(s.Gen_Var.delta0) - s.Gen_Var.
          Iy * cos(s.Gen_Var.delta0)
    s.Gen_Var.Iq = s.Gen_Var.Ix * cos(s.Gen_Var.delta0) + s.Gen_Var.
          Iy * sin(s.Gen_Var.delta0)

    s.Gen_Var.omega0 = 1.0
    s.Gen_Var.Pmech0 = s.GenP
    s.Gen_Var.Efd0   = sqrt(s.Gen_Var.EQx * s.Gen_Var.EQx + s.Gen_Var.
          EQy * s.Gen_Var.EQy) + (s.Gen_Par.Xd - s.Gen_Par.Xq) *
          s.Gen_Var.Id
    s.Gen_Var.Edp0   = s.Gen_Var.Vd + s.Gen_Par.Ra * s.Gen_Var.Id -
          s.Gen_Par.Xqp * s.Gen_Var.Iq
    s.Gen_Var.Eqp0   = s.Gen_Var.Vq + s.Gen_Par.Ra * s.Gen_Var.Iq +
          s.Gen_Par.Xdp * s.Gen_Var.Id
    s.Gen_Var.Edpp0  = s.Gen_Var.Vd + s.Gen_Par.Ra * s.Gen_Var.Id -
          s.Gen_Par.Xqpp * s.Gen_Var.Iq
```

```
    s.Gen_Var.Eqpp0   = s.Gen_Var.Vq + s.Gen_Par.Ra * s.Gen_Var.Iq +
        s.Gen_Par.Xdpp * s.Gen_Var.Id

    s.Gen_Var.omega_ref = s.Gen_Var.omega0
    s.Gen_Var.Ep0   = sqrt(s.Gen_Var.Edp0 * s.Gen_Var.Edp0 + s.Gen_Var.
        Eqp0 * s.Gen_Var.Eqp0)
    s.Gen_Var.Epp0 = sqrt(s.Gen_Var.Edpp0 * s.Gen_Var.Edpp0 +
        s.Gen_Var.Eqpp0 * s.Gen_Var.Eqpp0)

    //governor state initialization
    s.Gov_Var[1].mu0 = 0
    s.Gov_Var[1].GF  = 0
    s.Gov_Var[1].GP  = s.GenP/s.Gov_Par.a23

    //self-excited AVR state initialization
    s.AVR_Var[1].VR  = 0
    s.AVR_Var[1].VA  = 0
    s.AVR_Var[1].VE  = s.Gen_Var.Efd0
    s.AVR_Var[1].Vt0 = s.Vm
    s.AVR_Var[1].Efd0= s.Gen_Var.Efd0

    // Separately excited AVR state initialization
    s.AVR_Var[2].V1  = 0
    s.AVR_Var[2].V2  = 0
    s.AVR_Var[2].V3  = 0
    s.AVR_Var[2].V4  = 0
    s.AVR_Var[2].Vt0 = s.Vm
    s.AVR_Var[2].Efd0= s.Gen_Var.Efd0

    //PSS state initialization
    s.PSS_Var[1].VA  = 0
    s.PSS_Var[1].VP1 = 0
    s.PSS_Var[1].VP2 = 0
    s.PSS_Var[1].VP3 = 0
    s.PSS_Var[1].Vs  = 0
}
```

### F.2  Generator Injection Current Calculation

The generator injection current is a function of the generator terminal voltage and generator state variables which is calculated by

$$
\begin{bmatrix} I_x \\ I_y \end{bmatrix} = \begin{bmatrix} g_x & b_x \\ b_y & g_y \end{bmatrix} \begin{bmatrix} E_q'' \\ E_d'' \end{bmatrix} - \begin{bmatrix} G_x & B_x \\ B_y & G_y \end{bmatrix} \begin{bmatrix} U_x \\ U_y \end{bmatrix}
\tag{$\Lambda$.46}
$$

where

$$G_x = \frac{R_a - \left(x_d'' - x_q''\right)\sin\delta\cos\delta}{R_a^2 + x_d'' x_q''} = \frac{R_a - \frac{\left(x_d'' - x_q''\right)}{2}\sin 2\delta}{R_a^2 + x_d'' x_q''}$$

$$B_x = \frac{x_d'' \cos^2\delta + x_q'' \sin^2\delta}{R_a^2 + x_d'' x_q''} = \frac{\left(x_d'' + x_q''\right) + \left(x_d'' - x_q''\right)\cos 2\delta}{2\left(R_a^2 + x_d'' x_q''\right)}$$

$$B_y = \frac{-x_d'' \sin^2\delta - x_q'' \cos^2\delta}{R_a^2 + x_d'' x_q''} = \frac{-\left(x_d'' + x_q''\right) + \left(x_d'' - x_q''\right)\cos 2\delta}{2\left(R_a^2 + x_d'' x_q''\right)}$$

$$G_y = \frac{R_a + \left(x_d'' - x_q''\right)\sin\delta\cos\delta}{R_a^2 + x_d'' x_q''} = \frac{R_a + \frac{\left(x_d'' - x_q''\right)}{2}\sin 2\delta}{R_a^2 + x_d'' x_q''}$$

$$g_x = \frac{R_a\cos\delta + x_q''\sin\delta}{R_a^2 + x_d'' x_q''}$$

$$b_x = \frac{R_a\sin\delta - x_d''\cos\delta}{R_a^2 + x_d'' x_q''}$$

$$b_y = \frac{R_a\sin\delta - x_q''\cos\delta}{R_a^2 + x_d'' x_q''}$$

$$g_y = \frac{-R_a\cos\delta - x_d''\sin\delta}{R_a^2 + x_d'' x_q''}$$

When generator terminal voltage is solved by the network algebraic Eq. (A.48) and generator state variables are determined by the differential Eq. (A.47),

$$\dot{X} = f(X, U) \quad X \in \mathbb{R}^n \tag{A.47}$$

$$Y = g(X, U) \quad Y \in \mathbb{R}^m \tag{A.48}$$

the injection current of each generator is calculated in parallel by the following graph query.

```
CREATE QUERY injectionCurrent() FOR GRAPH busBranchGraph {
SumAccum<double> @temp1, @temp2, @denom, @Gx, @Bx, @Gy, @By, @gx, @bx,
@gy, @by

T0 = {TopoND.*}
T1 = SELECT s FROM T0 WHERE (s.busType == 2 OR s.busType == 3)
  ACCUM
    s.@denom =s.Gen_Par.Ra *s.Gen_Par.Ra + s.Gen_Par.Xdpp *s.Gen_Par.
        Xqpp
    s.@temp1 = (s.Gen_Par.Xdpp - s.Gen_Par.Xqpp) / 2.
    s.@temp2 = (s.Gen_Par.Xdpp + s.Gen_Par.Xqpp) / 2.
```

```
    s.@Gx = (s.Gen_Par.Ra - s.@temp1 * sin(2 * Gen_Var.delta)) /
            s.@denom
    s.@Bx = (s.@temp2 + s.@temp1 * cos(2 * Gen_Var.delta)) / s.@denom
    s.@Gy = (s.Gen_Par.Ra + s.@temp1 * sin(2 * s.Gen_Var.delta)) /
            s.@denom
    s.@By = (s.@temp2 + s.@temp1 * cos(2 * Gen_Var.delta)) / s.@denom

    s.@gx = (s.Gen_Par.Ra * cos(Gen_Var.delta) - s.Gen_Par.Xqpp *
            sin(Gen_Var.delta)) / s.@denom
    s.@bx = (s.Gen_Par.Ra * sin(Gen_Var.delta) - s.Gen_Par.Xdpp *
            cos(Gen_Var.delta))  / s.@denom
    s.@gy = (-s.Gen_Par.Ra * cos(Gen_Var.delta) - s.Gen_Par.Xdpp *
            sin(Gen_Var.delta)) / s.@denom
    s.@by = (s.Gen_Par.Ra * sin(Gen_Var.delta) - s.Gen_Par.Xqpp *
            cos(Gen_Var.delta))  / s.@denom

    Gen_Var.Ix = s.@gx * s.Gen_Var.Eqpp + s.@bx * s.Gen_Var.Edpp -
            s.@Gx * s.Gen_Var.Vx - s.@Bx * s.Gen_Var.Vy
    Gen_Var.Iy = s.@by * s.Gen_Var.Eqpp + s.@gy * s.Gen_Var.Edpp -
            s.@By * s.Gen_Var.Vx - s.@Gy * s.Gen_Var.Vy
    S.GenP = s.Gen_Var.Vx * Gen_Var.Ix + s.Gen_Var.Vy * Gen_Var.Iy
    S.GenQ = s.Gen_Var.Vy * Gen_Var.Ix - s.Gen_Var.Vx * Gen_Var.Iy
}
```

### F.3  Solve Differential Equations

Differential equations are built to represent the dynamics of generators, exciters, governors, power system stabilizers, and other components. Without loss of generality, the generator 3-order model, self-excited AVR, generic governor, and type-1 power system stabilizer are modeled in the differential equations to simulate the power system transient. Various integration methods are developed to solve differential equations. To illustrate the graph-based parallel transient simulation, the Trapezoidal rule is applied.

#### F.3.1  3-Order Generator Model

The 3-order generator model is simplified by assuming direct-axis and quadrature-axis subtransient voltage $E_d''$ and $E_q''$ and direct-axis transient voltage $E_d'$ constant. The per-unit differential equations for the 3-order generator model are given by

$$T_J \frac{d\omega}{dt} = P_m - P_e - D(\omega - 1) \tag{A.49}$$

$$\frac{d\delta}{dt} = \omega - 1 \tag{A.50}$$

$$T_{d0}' \frac{dE_q'}{dt} = E_{fd} - \left[ E_q' + (x_d - x_d')I_d + (K_G - 1)E_q' \right] \tag{A.51}$$

Applying the Trapezoidal rule, (A.49)–(A.51) are converted to difference equations. Starting from (A.49), we have

$$T_J \frac{\omega_1 - \omega_0}{h} = \frac{1}{2}[(P_{m1} + P_{m0}) - (P_{e1} + P_{e0}) - D(\omega_1 + \omega_0 - 2)] \tag{A.52}$$

$$\omega_1 = \omega_0 + \frac{h}{2T_J}[(P_{m1} + P_{m0}) - (P_{e1} + P_{e0}) - D(\omega_1 + \omega_0 - 2)] \tag{A.53}$$

Equations (A.50) and (A.51) are converted to difference equations as follows:

$$\frac{1}{h}(\delta_1 - \delta_0) = \frac{1}{2}(\omega_1 + \omega_0 - 2) \tag{A.54}$$

$$\delta_1 = \delta_0 + \frac{h}{2}(\omega_1 + \omega_0 - 2) \tag{A.55}$$

$$E'_{q1} = \frac{2T'_{d0} + h}{2T'_{d0} - h}E'_{q0} + \frac{h}{2T'_{d0} - h}\left[E_{fd1} + E_{fd0} + (x_d - x'_d)(I_{d1} + I_{d0}) + \frac{b}{a}E'^{n}_{q0}\right] \tag{A.56}$$

### F.3.2 Self-Excited AVR Model

According to the self-excited AVR model transfer function diagram Figure A.3, the algebraic-differential equations for the self-excited AVR are shown as follows:

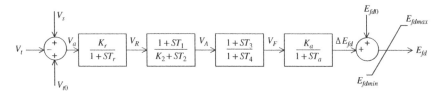

**Figure A.3** Self-excited AVR transfer function block diagram.

$$V_a = V_{to} - V_t + V_s \tag{A.57}$$

$$K_r V_a = V_R + ST_r V_R \tag{A.58}$$

$$V_A = \frac{1 + ST_1}{K_2 + ST_2}V_R \tag{A.59}$$

$$V_F = \frac{1 + ST_3}{1 + ST_4}V_A \tag{A.60}$$

$$\Delta E_{fd} = \frac{K_a}{1 + ST_a}V_F \tag{A.61}$$

$$E_{fd} = \Delta E_{fd} + E_{fd0} \tag{A.62}$$

Applying the Trapezoidal rule, (A.58) is converted as

$$V_{R1} = V_{R0} + \frac{h}{2T_r}(K_r V_{a1} - V_{R1} + K_r V_{a0} - V_{R0}) \tag{A.63}$$

$$V_{R1} = \frac{2T_r - h}{2T_r + h} V_{R0} + \frac{K_r h}{2T_r + h}(V_{a1} + V_{a0}) \tag{A.64}$$

Equation (A.59) can be rewritten as:

$$V_R + T_1 \frac{dV_R}{dt} = K_2 V_A + T_2 \frac{dV_A}{dt} \tag{A.65}$$

Substitute (A.58)–(A.65), and we get

$$T_2 \frac{dV_A}{dt} = \left(1 - \frac{T_1}{T_r}\right) V_R + \frac{T_1 K_r}{T_r} V_a - K_2 V_A \tag{A.66}$$

$$V_{A1} = V_{A0} + \frac{h}{2}\left[\frac{T_r - T_1}{T_2 T_r}(V_{R1} + V_{R0}) + \frac{T_1 K_r}{T_2 T_r}(V_{a1} + V_{a0}) - \frac{K_2}{T_2}(V_{A1} + V_{A0})\right] \tag{A.67}$$

Then, we have

$$V_{A1} = \frac{2T_2 - K_2 h}{2T_2 + K_2 h} V_{A0} + \frac{h(T_r - T_1)}{T_r(2T_2 + K_2 h)}(V_{R1} + V_{R0}) + \frac{h T_1 K_r}{T_r(2T_2 + K_2 h)}(V_{a1} + V_{a0}) \tag{A.68}$$

Similarly, Eq. (A.60) is rewritten as:

$$V_A + T_3 \frac{dV_A}{dt} = V_F + T_4 \frac{dV_F}{dt} \tag{A.69}$$

Substitute (A.66)–(A.68), and we get

$$T_4 \frac{dV_F}{dt} = \frac{T_3(T_r - T_1)}{T_2 T_r} V_R + \frac{T_1 T_3 K_r}{T_2 T_r} V_a + \frac{T_2 - T_3 K_2}{T_2 T_r} V_A - V_F \tag{A.70}$$

$$T_4(V_{F1} - V_{F0}) = \frac{h}{2}\left[\frac{T_3(T_r - T_1)}{T_2 T_r}(V_{R1} + V_{R0}) + \frac{T_1 T_3 K_r}{T_2 T_r}(V_{a1} + V_{a0})\right.$$
$$\left. + \frac{T_2 - T_3 K_2}{T_2 T_r}(V_{A1} + V_{A0}) - (V_{F1} + V_{F0})\right] \tag{A.71}$$

Then, we have

$$V_{F1} = \frac{2T_4 - h}{2T_4 + h} V_{F0} + \frac{T_3(T_r - T_1)h}{(2T_4 + h)T_2 T_r}(V_{R1} + V_{R0}) + \frac{T_1 T_3 K_r h}{(2T_4 + h)T_2 T_r}(V_{a1} + V_{a0})$$
$$+ \frac{(T_2 - T_3 K_2)h}{(2T_4 + h)T_2 T_r}(V_{A1} + V_{A0}) \tag{A.72}$$

Equation (A.59) is rewritten as:

$$\frac{d\Delta E_{fd}}{dt} = \frac{1}{T_a} K_a V_F - \frac{1}{T_a} \Delta E_{fd} \tag{A.73}$$

Applying the Trapezoidal rule, we have

$$\Delta E_{fd1} - \Delta E_{fd0} = \frac{h}{2T_a} \left[ K_a(V_{F1} + V_{F0}) - \left(\Delta E_{fd1} + \Delta E_{fd0}\right) \right] \tag{A.74}$$

$$\Delta E_{fd1} = \frac{2T_a - h}{2T_a + h} \Delta E_{fd0} + \frac{K_a h}{2T_a + h}(V_{F1} + V_{F0}) \tag{A.75}$$

### F.3.3 Generic Governor Model

According to the generic governor model transfer function diagram Figure A.4 the algebraic-differential equations for the generic governor are shown as follows:

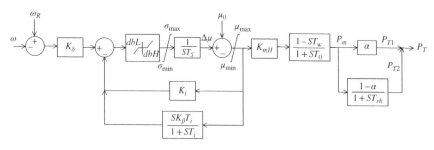

**Figure A.4** Generic speed-governor transfer function.

$$\delta = K_\delta(\omega_R - \omega) - F_1 - F_2 \tag{A.76}$$

$$\delta = T_s \frac{d\Delta\mu}{dt} \tag{A.77}$$

$$\mu = \Delta\mu + \mu_0 \tag{A.78}$$

$$F_1 = K_i\mu \tag{A.79}$$

$$F_2 + T_i \frac{dF_2}{dt} = K_\beta T_i \frac{d\mu}{dt} \tag{A.80}$$

$$P_m + T_0 \frac{dP_m}{dt} = K_{mH}\mu - K_{mH}T_W \frac{d\mu}{dt} \tag{A.81}$$

$$P_{T1} = \alpha P_m \tag{A.82}$$

$$P_{T2} + T_{rh} \frac{dP_{T2}}{dt} = P_m(1 - \alpha) \tag{A.83}$$

$$P_T = P_{T1} + P_{T2} \tag{A.84}$$

Applying the Trapezoidal rule, Eqs. (A.76)–(A.78) are converted to be

$$\delta_1 = K_\delta(\omega_R - \omega_1) - K_i\mu_0 - F_{2(0)} \tag{A.85}$$

$$\Delta\mu_1 = \Delta\mu_0 + \frac{h}{2T_s}(\delta_1 + \delta_0) \tag{A.86}$$

$$\mu_1 = \Delta\mu_1 + \mu_0 \tag{A.87}$$

Applying the Trapezoidal rule to (A.80), we have

$$\frac{F_{2(1)} + F_{2(0)}}{2} + \frac{T_i}{h}\left(F_{2(1)} - F_{2(0)}\right) = \frac{K_\beta T_i}{h}(\mu_1 - \mu_0) \tag{A.88}$$

Then, we get

$$F_{2(1)} = \frac{1}{2T_i + h}\left[(2T_i - h)F_{2(0)} + 2K_\beta T_i(\mu_1 - \mu_0)\right] \tag{A.89}$$

Applying the Trapezoidal rule to (A.81), we have

$$\frac{P_{m1} + P_{m0}}{2} + \frac{T_0}{h}(P_{m1} - P_{m0}) = \frac{K_{mH}}{2}(\mu_1 + \mu_0) - \frac{K_{mH}T_w}{h}(\mu_1 - \mu_0) \tag{A.90}$$

Then, we get

$$\left(\frac{h + 2T_0}{2h}\right)P_{m1} + \left(\frac{h - 2T_0}{2h}\right)P_{m0} = K_{mH}\left(\frac{h - 2T_w}{2h}\mu_1 + \frac{h + 2T_w}{2h}\mu_0\right) \tag{A.91}$$

Which is rewritten as

$$P_{m1} = \frac{1}{2T_0 + h}\left[(2T_0 - h)P_{m0} + K_{mH}(h - 2T_w)\mu_1 + K_{mH}(h + 2T_w)\mu_0\right] \tag{A.92}$$

Applying the Trapezoidal rule, (A.83) is converted as follows:

$$\frac{P_{T2(1)} + P_{T2(0)}}{2} + T_{rh}\frac{P_{T2(1)} - P_{T2(0)}}{h} = \frac{(1 - \alpha)}{2}(P_{m1} + P_{m0}) \tag{A.93}$$

Or

$$P_{T2(1)} = \frac{2T_{rh} - h}{2T_{rh} + h}P_{T2(0)} + \frac{(1 - \alpha)h}{2T_{rh} + h}(P_{m1} + P_{m0}) \tag{A.94}$$

### F.3.4 Power System Stabilizer Model

According to the type-1 power system stabilizer model transfer function diagram Figure A.5, the algebraic-differential equations for the power system stabilizer are shown as follows:

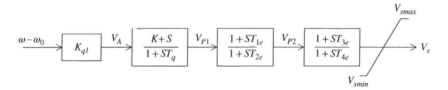

**Figure A.5** Type I power system stabilizer function block diagram.

$$V_A = K_{q1}\Delta\omega = K_{q1}(\omega - \omega_0) \tag{A.95}$$

$$V_{P1} = \frac{K + S}{1 + ST_q}V_A \tag{A.96}$$

$$V_{P2} = \frac{1 + ST_{1e}}{1 + ST_{2e}}V_{P1} \tag{A.97}$$

$$V_s = \frac{1 + ST_{3e}}{1 + ST_{4e}}V_{P2} \tag{A.98}$$

Equation (A.96) is rewritten as follows:

$$V_{P1} + T_q\frac{dV_{P1}}{dt} = KV_A + \frac{dV_A}{dt} \tag{A.99}$$

Applying the Trapezoidal rule to (A.99), we get

$$\frac{1}{2}\left(V_{P1(1)} + V_{P1(0)}\right) + \frac{T_q}{h}\left(V_{P1(1)} - V_{P1(0)}\right) = \frac{K}{2}(V_{A1} + V_{A0}) + \frac{1}{h}(V_{A1} - V_{A0}) \tag{A.100}$$

Or

$$V_{p1(1)} = \frac{1}{2T_q + h}\left[(2T_q - h)V_{P1(0)} + (Kh + 2)V_{A1} + (Kh - 2)V_{A0}\right] \tag{A.101}$$

Similarly, we rewrite Eq. (A.97) as:

$$V_{P2} + T_{2e}\frac{dV_{P2}}{dt} = V_{P1} + T_{1e}\frac{dV_{P1}}{dt} \tag{A.102}$$

Applying the Trapezoidal rule, we get

$$\frac{1}{2}\left(V_{P2(1)} + V_{P2(0)}\right) + \frac{T_{2e}}{h}\left(V_{P2(1)} - V_{P2(0)}\right) = \frac{1}{2}\left(V_{P1(1)} + V_{P1(0)}\right) + \frac{T_{1e}}{h}\left(V_{P1(1)} - V_{P1(0)}\right) \tag{A.103}$$

Or

$$V_{P2(1)} = \frac{1}{2T_{2e} + h}\left[(2T_{2e} - h)V_{P2(0)} + (2T_{1e} + h)V_{P1(1)} - (2T_{1e} - h)V_{P1(0)}\right] \tag{A.104}$$

Lastly, Eq. (A.98) is rewritten as

$$V_s + T_{4e}\frac{dV_s}{dt} = V_{P2} + T_{3e}\frac{dV_{P2}}{dt} \tag{A.105}$$

Applying the Trapezoidal rule, we get

$$\frac{1}{2}(V_{s1} + V_{s0}) + \frac{T_{4e}}{h}(V_{s1} - V_{s0}) = \frac{1}{2}\left(V_{P2(1)} + V_{P2(0)}\right) + \frac{T_{3e}}{h}\left(V_{P2(1)} - V_{P2(0)}\right) \tag{A.106}$$

Or

$$V_{s1} = \frac{1}{2T_{4e} + h}\left[(2T_{4e} - h)V_{s0} + (2T_{3e} + h)V_{P2(1)} - (2T_{3e} - h)V_{P2(0)}\right] \tag{A.107}$$

### F.3.5 Graph Query to Solve Differential Equations

Using the Trapezoidal rule, the differential equations are converted to the difference equations above which are solved in parallel by the following graph query.

```
CREAT QUERY ODESolver(h)FOR GRAPH busBranchGraph {
T0 = {TopoND.*}
T1 = SELECT s FROM T0 WHERE (s.busType == 2 OR s.busType == 3)
  ACCUM
    IF(s.Gen_Model == 2) THEN //Generator 3-order classic model
      s.Gen_Var.omega = s.Gen_Var.omega0 + h/(2.0*s.Gen_Par.Tj) *
          ((s.GOV_Var.PT - s.Gen_Var.Pe) + (s.GOV_Var.PT0 - s.Gen_Var.
          Pe0) - s.Gen_Par.D*(s.Gen_Var.omega + s.Gen_Var.omega0 - 2.0))
      s.Gen_Var.delta = s.Gen_Var.delta0 + h/2.0 * (s.Gen_Var.omega +
          s.Gen_Var.omega0 -2.0)
      s.Gen_Var.Eqp = ((2.0*s.Gen_Par.Tdp0 + h)* s.Gen_Var.Eqp0 + h*
          ((s.AVR_Var.Efd + s.AVR_Var.Efd0) + (s.Gen_Par.Xd - s.Gen_Par.
          Xdp) * (s.Gen_Var.Id1 - s.Gen_Var.Id0) + s.Gen_Par.b/s.Gen_Par.
          a *pow(s.Gen_Var.Eqp0, s.Gen_Par.n)))/(2.0*s.Gen_Par.Tdp0 - h)

    IF(s.AVR_Model == 1) THEN //Self-excited AVR model
      s.AVR_Var[1].Va = s.AVR_Var[1].Vt_ref-s.AVR_Var[1].Vt+ s.PSS_Var
          [1].Vs
      s.AVR_Var[1].VR = ((2*s.AVR_Par[1].Tr - h)*s.AVR_Var[1].VR0 +
          (s.AVR_Par[1].Kr*h*(s.AVR_Var[1].Va + s.AVR_Var[1].Va0)))/
          (2*s.AVR_Par[1].Tr + h)
      s.AVR_Var[1].VA = ((2*s.AVR_Par[1].T2 - s.AVR_Par[1].K2*h)*
          s.AVR_Var[1].VA0 + h*(s.AVR_Par[1].Tr - s.AVR_Par[1].T1)/
          s.AVR_Par[1].Tr*(s.AVR_Var[1].VR + s.AVR_Var[1].VR0) + h*s.
          AVR_Par[1].T1*s.AVR_Par[1].Kr/s.AVR_Par[1].Tr*(s.AVR_Var[1].Va
          + s.AVR_Var[1].Va0))/(2*s.AVR_Par[1].T2 + s.AVR_Par[1].K2*h)
      s.AVR_Var[1].VF = ((2*s.AVR_Par[1].T4 - h)*s.AVR_Var[1].VF0 +
          ((s.AVR_Par[1].T3*(s.AVR_Par[1].Tr - s.AVR_Par[1].T1)*h*
          (s.AVR_Var[1].VR + s.AVR_Var[1].VR0) + s.AVR_Par[1].T1*
          s.AVR_Par[1].T3*s.AVR_Par[1].Kr*h*(s.AVR_Var[1].Va +
          s.AVR_Var[1].Va0) + (s.AVR_Par[1].T2 - s.AVR_Par[1].T3*
          s.AVR_Par[1].K2)*h)*(s.AVR_Var[1].VA + s.AVR_Var[1].VA0))/
          (s.AVR_Par[1].T2*s.AVR_Par[1].Tr))/(2*s.AVR_Par[1].T4 + h)
      s.AVR_Var[1].dEfd = ((2*s.AVR_Par[1].Ta - h)*s.AVR_Var[1].dEfd0 +
          s.AVR_Par[1].Ka*h*(s.AVR_Var[1].VF + s.AVR_Var[1].VF0))/
          (2*s.AVR_Par[1].Ta + h)
      s.AVR_Var[1].Efd = s.AVR_Var[1].dEfd + s.AVR_Var[1].Efd_ref
      s.AVR_Var[1].Efd = min(max(s.AVR_Var[1].Efd, s.AVR_Par[1].Efdmin),
          s.AVR_Par[1].Efdmax)

    IF( s.GOV_Model == 1) THEN  //Generic governor model
      s.GOV_Var[1].delta = s.GOV_Par[1].Kdelt*(s.GOV.Var.omega ref -
          s.Gen.Var.omega) - s.GOV_Var[1].F1 - s.GOV_Var[1].F2
```

```
    IF (s.GOV_Var[1].delta <= s.GOV_Par[1].dbH AND s.GOV_Var[1].delta
        >= s.GOV_Par[1].dbL) THEN
    s.GOV_Var[1].delta = 0.0
    s.GOV_Var[1].dmu = s.GOV_Var[1].dmu0 + h/(2*s.GOV_Par[1].Ts)*
        (s.GOV_Var[1].delta + s.GOV_Var[1].delta0)
    s.GOV_Var[1].mu = s.GOV_Var[1].dmu + s.GOV_Var[1].mu_ref
    s.GOV_Var[1].mu = min(max(s.GOV_Var[1].mu, s.GOV_Var[1].mnMin),
        s.GOV_Var[1].muMin)
    s.GOV_Var[1].F1 = s.GOV_Var[1].mu * s.GOV_Par[1].Ki
    s.GOV_Var[1].F2 = ((2*s.GOV_Par[1].Ti - h)*s.GOV_Var[1].F20 +
        2*s.GOV_Par[1].Kbeta*s.GOV_Par[1].Ti*(s.GOV_Var[1].mu +
        s.GOV_Var[1].mu0))/(2*s.GOV_Par[1].Ti + h)
    s.GOV_Var[1].Pm = (((2*s.GOV_Par[1].T0 - h)*s.GOV_Var[1].Pm0) +
        s.GOV_Par[1].KmH*((h - 2*s.GOV_Par[1].Tw)*s.GOV_Var[1].mu1 + (h
        + 2*s.GOV_Par[1].Tw)*s.GOV_Var[1].mu0))/(2*s.GOV_Par[1].T0 + h)
        s.GOV_Var[1].PT1 = s.GOV_Var[1].Pm*s.GOV_Par[1].alpha
    s.GOV_Var[1].PT2 = ((2*s.GOV_Par[1].T0 - h)*s.GOV_Var[1].PT20 +
        (1 - s.GOV_Par[1].alpha)*(s.GOV_Var[1].Pm + s.GOV_Var[1].
        Pm0))/ (2*s.GOV_Par[1].T0 + h)
    s.GOV_Var[1].PT = s.GOV_Var[1].PT1 + s.GOV_Var[1].PT2

  IF( s.PSS_Model == 1) THEN  //Type I Power System Stabilizer model
    s.PSS_Var[1].dOmega = s.Gen_Var.omega - s.Gen_Var.omega_ref
    s.PSS_Var[1].VA = s.PSS_Par[1].Kq1 * s.PSS_Var[1].dOmega
    s.PSS_Var[1].VP1 = ((2*s.PSS_Par[1].Tq - h)*s.PSS_Var[1].VP10 +
        (s.PSS_Par[1].K*h + 2)*s.PSS_Var[1].VA + (s.PSS_Par[1].K*h -
        2)*s.PSS_Var[1].VA0)/(2*s.PSS_Par[1].Tq + h)
    s.PSS_Var[1].VP2 = ((2*s.PSS_Par[1].T2e - h)*s.PSS_Var[1].VP20 +
        (2*s.PSS_Par[1].T2e + h)*s.PSS_Var[1].VP1 + (h - 2*s.PSS_Par
        [1].T2e)*s.PSS_Var[1].VP10)/(2*s.PSS_Par[1].T2e + h)
    s.PSS_Var[1].Vs = ((2*s.PSS_Par[1].T4e - h)*s.PSS_Var[1].Vs0 +
        (2*s.PSS_Par[1].T3e + h)*s.PSS_Var[1].VP2 + (h - 2*s.PSS_Par
        [1].T3e)*s.PSS_Var[1].VP20)/(2*s.PSS_Par[1].T4e + h)
    s.PSS_Var[1].Vs = min(max(s.PSS_Var[1].Vs, s.PSS_Par[1].Vsmin),
        s.PSS_Par[1].Vsmax)
}
```

In the query ODESolver, time step h is passed through. In the query definition, the statement T1 = SELECT s FROM T0 WHERE s.busType == 2 or s.busType == 3 coupling with the keyword ACCUM selects all unit vertices s from vertex set T0 = {TopoND.*}, and updates the state variables by the difference equations in parallel.

## F.3.6  Graph-Based Transient Simulation

The query and subquery to calculate the steady-state equilibrium point, form and factorize the $Y$ matrix, update bus voltage, solve differential equations, and calculate generator injection current are developed. They are called to implement the graph-based transient simulation as shown in the following high-level pseudo-code.

```
CREAT QUERY transientSimulation(float tstart, float tend, float tstep,
string faultBus, float fstart, float fend, int MaxIteration, double
Tolerance) FOR GRAPH busBranchGraph {

MaxAccum<double> @@maxVx =-INF
MaxAccum<double> @@maxVy =-INF


T0 = {TopoND.*}
stateInitialization() FOR GRAPH busBranchGraph

FOREACH @@t IN RANGE[tstart, tend].STEP(tstep) DO
 WHILE( abs(@@maxVm) > Tolerance OR abs(@@maxVa) > Tolerance )
            LIMIT MaxIteration DO
   IF (@@t >= fstart AND @@t - fstart < tsetp) THEN  // fault happens
     T1 = SELECT s FROM T0 WHERE (s.bus_name == faultBus)
       s.G += INF
   ELSE IF (@@t <= fend && fend - @@t < tsetp) THEN  // clear the fault
     T1 = SELECT s FROM T0 WHERE (s.bus_name == faultBus)
       s.G -= INF


   newtonRaphsonPowerFlow(MaxIteration, Tolerance)
       FOR GRAPH busBranchGraph and GRAPH JacobianGraph
       and GRAPH eTreeGraph
   // Network Interface
   T1 = SELECT s FROM T0 WHERE (s.busType == 2 OR s.busType == 3)
     ACCUM
       s.Gen_Var.Vx  = s.Vm * cos(s.Va)
       s.Gen_Var.Vy  = s.Vm * sin(s.Va)
       s.Gen_Var.Ix  = (s.GenP * cos(s.Va) + s.GenQ * sin(s.Va)) / s.Vm
       s.Gen_Var.Iy  = (s.GenP * sin(s.Va) - s.GenQ * cos(s.Va)) / s.Vm
       s.Gen_Var.EQx = s.Gen_Var.Vx + s.Gen_Par.Ra * s.Gen_Var.Ix -
           s.Gen_Par.Xq * s.Gen_Var.Iy
       s.Gen_Var.EQy = s.Gen_Var.Vy + s.Gen_Par.Ra * s.Gen_Var.Iy +
           s.Gen_Par.Xq * s.Gen_Var.Ix
       double delta = atan2(s.Gen_Var.EQy, s.Gen_Var.EQx)
       s.Gen_Var.Vd = s.Gen_Var.Vx * sin(delta) - s.Gen_Var.Vy *
           cos(delta)
       s.Gen_Var.Vq = s.Gen_Var.Vx * cos(delta) + s.Gen_Var.Vy *
           sin(delta)
       s.Gen_Var.Id = s.Gen_Var.Ix * sin(delta) - s.Gen_Var.Iy *
           cos(delta)
       s.Gen_Var.Iq = s.Gen_Var.Ix * cos(delta) + s.Gen_Var.Iy *
           sin(delta)
       s.Gen_Var.Pe = S.GenP
       s.AVR_Var[1].Vt = s.Vm
```

```
  T1 = SELECT s FROM T0 WHERE (s.busType == 2 OR s.busType == 3)
    ACCUM
       @@maxVx = max(abs(s.Vm*cos(s.Va))-s.Gen_Var.Vx)
       @@maxVy = max(abs(s.Vm*sin(s.Va))-s.Gen_Var.Vy)

  ODESolver(tsetp)FOR GRAPH busBranchGraph
  updateState()FOR GRAPH busBranchGraph
}
```

In the query `transientSimulation`, simulation start time `tstart`, end time `tend`, time step `tstep`, fault bus identification `faultBus`, fault start and clear time `fstart` and `fend`, maximum iteration number `MaxIteration` and tolerance `Tolerance` are passed through. The global variables `@@maxVx` and `@@maxVy` are defined to store the maximum voltages which will be used to compare against the tolerance `Tolerance` to check the convergence.

To start the simulation, the subquery `stateInitialization` is called to initialize state variables based on the base case power flow solution. In the while loop, in each of the time steps, firstly the fault is set up based on the fault start time and clear time. Then the algebraic equations and the differential equations are solved sequentially till the generator terminal voltage is converged. The algebraic equations are solved by calling the subquery `newtonRaphsonPowerFlow`. And the differential equations are solved by calling the subquery `ODESolver`. Before the subquery `ODESolver` is called, the generator terminal voltages and injection currents in the $x - y$ reference frame are transformed into the voltages and currents in the $d - q$ reference frame.

After differential equations are solved, the state variables are updated for the next time step calculation by calling subquery `updateState`. Subquery `updateState` is implemented in nodal parallel as the following pseudo-code.

```
CREAT QUERY updateState()FOR GRAPH busBranchGraph {

T0 = {TopoND.*}
T1 = SELECT s FROM T0 WHERE (s.busType == 2 OR s.busType == 3)
  ACCUM
    IF(s.Gen_Model == 2) THEN //Generator 3-order classic model
    s.Gen_Var.omega0 = s.Gen_Var.omega
    s.Gen_Var.delta0 = s.Gen_Var.delta
    s.Gen_Var.Eqp0 = s.Gen_Var.Eqp

  IF(s.AVR_Model == 1) THEN //Self-excited AVR model
    s.AVR_Var[1].Va0 = s.AVR_Var[1].Va
    s.AVR_Var[1].VR0 = s.AVR_Var[1].VR
    s.AVR_Var[1].VA0 = s.AVR_Var[1].VA
    s.AVR_Var[1].VF0 = s.AVR_Var[1].VF
    s.AVR_Var[1].dEfd0 = s.AVR_Var[1].dEfd
    s.AVR_Var[1].Efd0 = s.AVR_Var[1].Efd

  IF(s.GOV_Model == 1) THEN  //Generic governor model
    s.GOV_Var[1].delta0 = s.GOV_Var[1].delta
    s.GOV_Var[1].dmu0 = s.GOV_Var[1].dmu
    s.GOV_Var[1].mu0 = s.GOV_Var[1].mu
```

```
      s.GOV_Var[1].F20 = s.GOV_Var[1].F2
      s.GOV_Var[1].Pm0 = s.GOV_Var[1].Pm
      s.GOV_Var[1].PT20 = s.GOV_Var[1].PT2
      s.GOV_Var[1].PT0 = s.GOV_Var[1].PT

   IF(s.PSS_Model == 1) THEN  //Type I Power System Stabilizer model
      s.PSS_Var[1].VA0 = s.PSS_Var[1].VA
      s.PSS_Var[1].VP10 = s.PSS_Var[1].VP1
      s.PSS_Var[1].VP20 = s.PSS_Var[1].VP2
      s.PSS_Var[1].Vs0 = s.PSS_Var[1].Vs
}
```

# Index

## a

Absolute stability   101, 102, 104, 123
AC line   180, 181, 189, 196, 197
AC power flow   273, 294, 295, 301
Activations   170
Actor network   151, 430, 456
Adams–Bashforth methods   99
Adams–Moulton methods   99, 100
Adaptive time step   9, 96, 99, 103–105, 117, 123, 124
Adjacent   14, 23–25, 45, 50, 51, 127, 153, 163, 201–203, 205, 248, 249, 252, 264, 279, 282, 293, 331, 367
Admittance graph   45, 51, 58, 117, 179
Admittance matrix   12, 14, 18, 20, 23, 24, 44–47, 58, 76, 79, 110, 115, 117, 179, 203, 234, 239, 242, 244, 245, 416
Aggregation   4, 12
Area control error (ACE)   7, 356
Artificial intelligence   10, 159, 199
Asymmetrical   151
Attributes   12–14, 17–25, 35, 37, 43, 46, 177, 178, 180, 182, 184, 187, 189, 190, 194–196, 198, 201, 215, 216, 218, 220, 223, 230, 236, 239, 241, 242, 245, 249, 252–254, 256, 259, 260, 275, 280–282, 290, 302, 303, 311, 312, 327, 332, 382, 383, 401–404, 432, 435, 437–447, 449, 458, 459
Auto-encoder   160, 162–170
Automatic generation control (AGC)   3, 4, 7, 8, 20, 25, 345, 355, 358, 362, 363, 436, 459, 460, 345, 355–361, 363, 459, 464

## b

Backward Euler method   96, 97, 100–102, 123
Backward-forward sweep (BFS)   28, 29
Bad data detection   6, 14, 46, 223, 224, 229
Bad data identification   224, 228, 229
Balance equations   5, 41
Base case   7, 273, 276, 282, 283, 285, 287, 288, 290, 293, 306, 341, 349, 360, 479
BCG   83, 90, 91
Bellman's equation   427
Bi-conjugate gradient   11
Binary variables   12, 267
Bisection   31–35
Blackouts   7, 273, 425
Boundary   9, 391–393, 398
Breadth-first search   34, 276
Bridge   43, 107, 186, 276, 277, 279–282
Bulk Synchronous Parallel   41–43
Busbar   180, 181, 189, 190, 196
Bus-branch model   177, 179, 186, 189, 190, 196, 198, 215, 235, 236, 245, 274, 280, 401, 446, 447

## c

Capacity rating   13, 18
Chebyshev polynomials   169
Chi-squares test   224
Cholesky decomposition   268
Cholesky factorization   80, 81, 83
CIM/E   180–182, 184, 187, 191, 196, 215, 437–439, 442–445
CIM/XML   180
Circuit breaker   17, 18, 178, 180, 190, 191, 196

*Graph Database and Graph Computing for Power System Analysis*, First Edition. Renchang Dai and Guangyi Liu.
© 2024 The Institute of Electrical and Electronics Engineers, Inc. Published 2024 by John Wiley & Sons, Inc.

# IEEE Press Series on Power and Energy Systems

**Series Editor:** Ganesh Kumar Venayagamoorthy, Clemson University, Clemson, South Carolina, USA.

The mission of the IEEE Press Series on Power and Energy Systems is to publish leading-edge books that cover a broad spectrum of current and forward-looking technologies in the fast-moving area of power and energy systems including smart grid, renewable energy systems, electric vehicles and related areas. Our target audience includes power and energy systems professionals from academia, industry and government who are interested in enhancing their knowledge and perspectives in their areas of interest.

Printed and bound by CPI Group (UK) Ltd, Croydon, CR0 4YY

16/04/2025

14658587-0004